Handbook of Experimental Pharmacology

Volume 140

Springer-Verlag Berlin Heidelberg GmbH

Proteases as Targets for Therapy

Contributors

S.S. Abdel-Meguid, R.C. Andrews, D.L. Barnard, J.D. Becherer,
E.M. Bergmann, K. Beyreuther, M.T. Brown, C.R. Caffrey,
J.C. Cheronis, C.E. Dabrowski, R. De Francesco,
Q.L. Deveraux, J. Ding, H. Fritz, E.S. Furfine, C. Haass,
M.N.G. James, H. Kawahara, D. Keppler, U. Koch, B.D. Korant,
M.H. Lambert, S.F. Lichtenthaler, K.P. Lynch, W.F. Mangel,
C.L. Masters, W.J. McGrath, J.H. McKerrow, A. Pessi,
J. Potempa, X. Qiu, M. Rabinovitch, H.S. Rasmussen,
S. Redshaw, J.C. Reed, N.A. Roberts, J.P. Salter, G.S. Salvesen,
A.H. Schmaier, W. Shao, R. Shridhar, B.F. Sloane, T. Smith,
C. Steinkühler, R. Swanstrom, R.M. Sweet, K. Tanaka,
G.J. Thomas, D.L. Toledo, J. Travis, M. Valliancourt, S. Vella,
K. von der Helm

Editors
Klaus von der Helm, Bruce D. Korant,
and John C. Cheronis

Springer

Professor Klaus von der Helm, Ph.D., M.D.
Max-von-Pettenkofer Institut
für medizinische Mikrobiologie, LM Universität
LS Virologie
Pettenkoferstr. 9a
D-80336 München
Germany
e-mail: kvdh@mvp.med.uni-muenchen.de

Bruce D. Korant, Ph.D.
DuPont Pharmaceutical Co
Research and Development
Edperimental Station, E336/31B
Wilmington, DE 19880-0336
USA
e-mail: Bruce.D.Korant@dupontpharma.com

John C. Cheronis, M.D., Ph.D.
Vice President, Drug Discovery and Development
Source Precision Medicine
2425 N. 55th Street, Suite 111
Boulder, CO 80301
USA
e-mail: jcheroni@sourcepharma.com

With 58 Figures and 16 Tables

ISBN 978-3-642-63023-1

Library of Congress Cataloging-in-Publication Data applied for
Die Deutsche Bibliothek – CIP-Einheitsaufnahme
Proteases as targets for therapy / contributors M. Abdel-Meguid . . . Ed. Klaus von der Helm . . . – Berlin;
  Heidelberg; New York; Barcelona; Hong Kong; London; Milan; Paris; Singapore; Tokyo: Springer, 2000
    (Handbook of experimental pharmacology; Vol. 140)
    ISBN 978-3-642-63023-1   ISBN 978-3-642-57092-6 (eBook)
    DOI 10.1007/978-3-642-57092-6

© Springer–Verlag Berlin Heidelberg 2000
Originally published by Springer-Verlag Berlin Heidelberg in 2000
Softcover reprint of the hardcover 1st edition 2000

The use of general descriptive names, registered names, trademarks, etc. in this publication does not imply, even in the absence of a specific statement, that such names are exempt from the relevant protective laws and regulations and therefore free for general use.

Product liability: The publishers cannot guarantee the accuracy of any information about dosage and application contained in this book. In every individual case the user must check such information by consulting the relevant literature.

Cover design: *design & production* GmbH, Heidelberg

Typesetting: Best-set Typesetter Ltd., Hong Kong
Production: ProduServ GmbH Verlagsservice, Berlin

SPIN: 10676934   27/3020 – 5 4 3 2 1 0 – Printed on acid-free paper

Foreword

Proteolytic enzymes and their natural antagonists, the protease inhibitor proteins, play a crucial role in the physiology and pathology of living organisms including humans. Remarkable advantages revealed their wide functional context.

Proteases digest food proteinase in the digestive tract and liberate polypeptide hormones, stimulating gastric and pancreatic secretion. Proteases are involved during fertilization in sperm – egg interaction, ovulation, ovum implantation and parturition. Proteases of the renin-angiotensin and kallikrein-kinin systems act synergistically to generate blood pressure regulating polypeptides. In wound healing a battery of proteases is involved in the proteolytic cascades of clotting, fibrinolysis and tissue repair. Another battery of very different proteases directs the immune defense via several routes, i.e. complement activation, antigen presentation, the generation of chemokines and chemotaxins directing phagocytes to the site of injury or infection and the generation of cell-stimulating factors such as cytokines regulating the inflammatory response of the organism. Granzymes contribute to the toxicity of lymphocytes or killer cells, caspases regulate physiological cell death and calpains intracellular signaling cascades. The energy-dependent proteasome-ubiquitin system controls highly efficiently the activity or level of intracellular proteins, including cell-cycle regulators, transcription and signal transduction factors, oncoproteins and short-lived metabolic enzymes. And this listing is far from complete.

The activity of proteases is directly controlled by potent protease inhibitors also produced by the organism, partly in several fold excess of the total amount of protease which can be liberated. Generally speaking, the diversity of existing proteases is confronted with a corresponding variety of inhibitors. Nearly every protease is faced with an antagonist limiting its proteolytic activity locally and in a timely fashion to prevent pathologies. The physiological balance between the active protease available at its target substrate(s) and inhibitor activity is regulated by various cellular mediators. They control the synthesis and location (storage in granules, secretion etc.) of the zymogen and of the inhibitor. They also control the activation of the proenzyme, which is itself triggered by a specific protease.

Major reasons for proteolysis-induced pathologies are either excessive production or liberation (e.g. from cells and microbes) of proteases or extensive consumption of protease inhibitors or both, leading to an imbalance of the physiological protease/inhibitor equilibrium. Such an acquired imbalance may be caused by traumatic or inflammatory events or infections. Whereas at the onset of such pathologies proteases are the major pathogenetically active agents, in a more advanced state of the disease often cellular inflammatory mediators also produced by proteolysis, such as cytokines or shedded soluble adhesion molecules, become the major players. In other diseases excessive local generation of proteases may be the underlying pathological event, e.g. thrombin activation leading to embolism (infarction). In still other pathologies, such as tumorigenesis or metastasis, cancer cells express and often use very efficiently various proteases for degradation of extracellular matrix components and migration through solid tissue structures, simultaneously knocking out the endogenous protective inhibitor shield of the organism that they finally kill. In a similarly elegant way bacteria and parasites often use special protease equipment to reach their goal, their own reproduction via infection of the host. Such proteases may be highly potent activators and/or inactivators of the protease zymogens or inhibitors of the host, which lacks in many cases a specific inhibitory defense system against the microbial and parasitic proteases.

This volume combines examples of diseases triggered or enhanced by cellular or microbial proteases that are of great socio-economical and medical significance due to their widespread distribution and the difficulties associated with their therapy. Protease inhibitors are promising candidates for new therapeutic approaches based on the basic pathomechanisms of these diseases. The contributing authors' detailed knowledge and profound experience in their particular research areas make this volume a most valuable tool for the identification of a new generation of therapeutics, the protease inhibitors, which might assist in controlling or even preventing disease-specific, proteolysis-induced pathomechanisms. The therapeutic success achieved so far with synthetic inhibitors of the angiotensin converting enzyme in the treatment of essential hypertension and of the HIV protease in HIV-infected patients gives hope that other approaches described in detail in this volume will also be successful in the near future.

Munich, September 1999   HANS FRITZ

Preface

"The way new ideas are going to be realized becomes clearer during the voyage" (HOMER, ODYSSEY).

Proteases are a class of enzymes that have been known about for longer than many other enzymes, and the early achieved knowledge about structure and function of proteases had inspired and eased the elicitation of many other enzymes. Interest in protease inhibitors soon grew, striking the idea of employing protease inhibitors for medical therapeutic purposes. Applications, however, were not ventured at that time. The proteases were generally understood to be metabolically and catabolically active, i.e., digested and removed aberrant proteins by cleaving a wide spectrum of substrates. Thus, inhibition of individual proteases for therapeutic purpose appeared hazardous, because of unpredictable and possibly uncontrollable consequences within the long-range chains of metabolic reactions.

The regulatory role of proteases was only slowly recognized. Among the first to envisage limited proteolytic inhibition was HANS FRITZ, who provided early and active leadership in medical applications of protease inhibitors and recognized their potential as a new class of drugs. On the basis of WERLE and FREY, he together with an initially small number of engaged pioneers (FRITZ and TSCHESCHE 1971) promoted the dedicated pursuit of protease inhibitors in the clinic, particularly those of the kininogen system.

One event that greatly changed the situation was the revelation of viral-encoded proteases. In 1977, the first viral protease was identified in a retrovirus (VON DER HELM 1977; YOSHINAKA and LUFTIG 1977). Within a short period of time, further proteases were found in other viruses (PALMENBERG et al. 1979; KORANT et al. 1980). They were shown to have – unlike 'cellular' host proteases – a very restricted range of function, limited to the viral life cycle. They were, thus, distinct from cellular metabolic enzymes. Viral proteases are processing, i.e., anabolically acting, enzymes – they mature viral protein precursors to smaller, functional proteins. By this process (not yet infectious), virion particles mature to infectious viruses. This novel insight stimulated the search for viral protease inhibitors considerably (KRAUSSLICH et al. 1989). However, as most viruses cause diseases that are self-limiting, the rising momentum was not yet sufficient for expensive (therapeutic) clinical trials that should have followed the initial encouragement of inhibitor developments of low-cost cell-culture experiments.

The sudden appearance of acquired immunodeficiency syndrome (AIDS) in the early 1980s (reviewed in GALLO and MONTAGNIER 1987) was a shock and changed this situation dramatically. After the causative agent, human immunodeficiency virus (HIV), had been found and shown to cause irreversible, fatal, destructive disease, it seemed mandatory for Western society to develop an immediate remedy. Various types of approaches – prophylactic and therapeutic – were undertaken with unprecedented efforts. Most of these activities had initially been concentrated on novel molecular and gene-technique approaches before the focus turned to the classical biochemical search for the HIV protease inhibitors. The swift and clear therapeutic success of the HIV protease inhibitor in combination with HIV reverse transcriptase inhibitors (brought about by many positive coincidences, see first part of this volume) changed the perception of employing protease inhibitors for therapeutic purpose very much (MELLORS 1996; RICHMAN 1996). What were the reasons for this dramatic progress?

First, the HIV proteases had been revealed as enzymes unique in structure and function: structurally, a symmetric homodimer, the enzyme is distinct from all other proteases and, thus, is ideally suited for symmetry-like design of inhibitor compounds; functionally this protease is limited to the processing of immature particles into infectious viruses and not involved in other reactions. Thus, the inhibition of this protease had no evident dangerous consequences for the cell's metabolism.

Second, and most essentially, AIDS – having been perceived as a fatal exemption from the typical self-limiting viral diseases – posed the strongest indication ever for antiviral therapy, even more as vaccination is still not feasible. So started a most concentrated but multidisciplinary battle of scientific research which was fortunately won – after only a decade – a few years ago. It had been the fastest development in history of an entirely de novo drug.

A momentum had started. The experience with the HIV enzyme as a distinct type of protease and target for antiviral inhibitors was convincing because of the unexpected swift success; it began to drive a development for therapeutic inhibitors to other viral proteases. Proteases of viruses causing serious, less self-limiting diseases (herpes-, hepatitis-viruses etc.) are presently under special study and, in fact, the results already achieved (as described in the second part of this volume) are very encouraging.

Consequently, the initial intention revived, namely to employ inhibitors to "cellular" (host) proteases for chemotherapeutic use against diseases mediated by action of those proteases. Although the enthusiasm about the HIV combination therapy has recently tempered – as predictable and unpredictable problems (resistance and pharmacokinetic problems, see Chaps 3 and 4) have become apparent – it seemed to us an appropriate time to put together all facts, aspects and fancies about proteases as therapeutic targets in this volume.

We have asked colleagues to describe, in the first part of this volume, all aspects of HIV protease inhibitors as therapeutic drugs (used in combination with the reverse transcriptase inhibitors). The second part has some encouraging examples of inhibitors to viruses other than HIV. In addition to the reviews on viral proteases, two chapters cover recent efforts in designing inhibitors against microbes such as bacteria and parasites. The third part of this volume deals with the question whether inhibitors against cellular proteases might be employed therapeutically.

The very first chapter reports on how the decision was born to design the first effective HIV protease inhibitor and which obstacles had to be overcome before the initial clinical trial was successfully performed. In Chap. 2, the present data of clinical results of the combination therapy are discussed together with upcoming challenges. The main problems are pharmacokinetics and the resistance that will inevitably develop during long periods of therapy (up to several years) which might be indicated (Chap. 3). The entire Chap. 3 is devoted to the discussion of this problem because it may be a principal problem for any (future) type of protease inhibitor, whether antiviral, antibacterial or antifungal. Then, in Chap. 4, aspects of how to limit and control the resistance problem in the future are discussed.

The second part of the volume covers recent encouraging work in development of other antiviral (PATICK and POTTS 1998) and antimicrobial protease inhibitors. Hepatitis C virus (HCV), for example, causes a very troublesome liver disease, many cases progressing chronically. Chapter 5 describes the beginning of a frame work for rational approaches to HCV protease inhibitors which may be useful as antiviral drugs. Some herpes viruses, such as cytomegalovirus (CMV) are responsible for fatal disease outcome. Recently, the structures of the CMV protease and other herpes viruses have been revealed, thus facilitating the design of inhibitor drug candidates; Chap. 6 outlines the state of the art. The proteases of picornaviruses were among the earliest viral proteases to be characterized. Various inhibitors have been produced since but serious efforts were lacking to apply these clinically. Nevertheless, as described in Chap. 7, hepatitis A might be a useful application for compounds with this mode of action and the rhino (common cold) viruses are still under consideration as an indication for (protease inhibitor) antiviral therapy. Chapter 8 presents the adenovirus protease. Diseases caused by the adenovirus are probably not a profitable indication for antiviral therapeutic drugs. Here, the example of the protease structure demonstrates an intriguing feature – the adenoviral protease has three active-site folds generated by the unique existence of two essential co-factors. This chapter discusses the advantage of having an inhibitory drug for different active sites and the probable benefit in preventing a general resistance.

The next two chapters summarize known proteases of some bacteria and parasites responsible for diseases that justify anti-infective drug development. The possibilities and probabilities of inhibitors against these proteases are outlined.

The third and last part of this volume presents (non-microbial) cellular proteases involved in the generation of medically serious diseases, which might be a conceivable target for therapeutic application. As pointed out in a separate overview for this part (Chap. 10), the situation for therapeutic action, here, by protease inhibitors is quite different from that of the microbial ones.

In our editorial work, we refrained from distinguishing between the (almost) synonyms: "protease" or "proteinase". Each of these words has it particular meaning but both clearly describe proteolytically active enzymes. So we left the decision to the authors.

Throughout our editorial attempts to organize, coordinate and complete this volume there were two frequent observations. First, we learned a lot more about the dynamic topic of proteolysis, which we had approached convinced we were knowledgable. For that, we are indebted to the contributing experts. Second, the numerous positive impacts of the work of Mrs Doris Walker and her colleagues at Springer Verlag were essential to maintain the quality and timeliness of the book, and their efforts deserve special thanks on behalf of all the authors.

References

Fritz H, Tschesche H (eds) (1971) Proceedings: 1. International conference on protease inhibitors. W de Gruyter, Berlin

Gallo RC, Montagnier L (1987) The chronology of AIDS research. Nature 326:435–446

Korant B, Chow N, Lively M, Powers J (1980) Proteolytic events in replication of animal viruses Ann NY Acad Sci 334:304–318

Krausslich HG, Oroszlan S, Wimmer E (eds) (1989) Viral proteinases as targets for chemotherapy. Curr Commun Mol Biol, Cold Spring Harbor Laboratory Press

Mellor JW (1996) Closing in on HIV: antiretroviral therapy back on track with HIV-1 protease inhibitors. Nat Med 2:274–276

Richmann DD (1996) HIV therapeutics. Science 272:1868–1888

Palmenberg AC, Pallansch MA, Rueckert RR (1979) Protease required for processing picornaviral coat protein resides in the viral replicate gene. J Virol 32:770–778

Patick A, Potts KE (1998) Protease inhibitors as antiviral agents. Clin Microbiol Rev 11:614–627

Von der Helm K (1977) Cleavage of Rous sarcoma virus polypeptide precursor into internal structural proteins in vitro involves viral protein p15. Proc Natl Acad Sci USA 74:911–916

Yoshinaka R, Luftig R (1977) Properties of a p70 proteolytic factor of murine leukemia viruses. Murine leukemia virus morphogenesis. Proc Natl Acad Sci USA 74:3446–3451

Munich, Germany   KLAUS VON DER HELM
Wilmington, DE, USA   BRUCE D. KORANT

List of Contributors

ABDEL-MEGUID, S.S., Department of Structural Biology, Smith Kline Beecham Pharmaceuticals, 709 Swedeland Road, Mail Code UE0447, King of Prussia, PA 19406, USA,
e-mail: Sherin_S_Abdel-Meguid@sbphrd.com

ANDREWS, R.C., Glaxo Wellcome Inc., P.O. Box 13398, Research Triangle Park, NC 27709, USA

BARNARD, D.L., Institute for Antiviral Research, Utah State University, Logan, UT 84322-5600, USA

BECHERER, J.D., Glaxo Wellcome Inc., P.O. Box 13398, Research Triangle Park, NC 27709, USA,
e-mail: jae.c.//db5617@glaxowellcome.com

BERGMANN, E.M., Department of Biochemistry and Medical Research Council of Canada, Group in Protein Structure Function, University of Alberta, Medical Sciences Building 4-25, Edmonton, Alberta T6G 2H7, CANADA,
e-mail: berg@manitou.biochem.ualberta,ca

BEYREUTHER, K., Centre for Molecular Biology Heidelberg (ZMBH), University of Heidelberg, Im Neuenheimer Feld 282, D-69120 Heidelberg, GERMANY

BROWN, M.T., Brookhaven National Laboratory, Associated Universities, Inc., Biology Department, P.O. Box 5000, Upton, NY 11973-5000, USA

CAFFREY, C.R., Department of Pathology –113B, University of California at San Francisco, 4150 Clement Street, San Francisco, CA 94121, USA

CHERONIS, J.C., Vice President, Drug Discovery and Development, Source Precision Medicine, 2425 N. 55th Street, Suite 111, Boulder, CO 80301, USA
e-mail: jcheroni@sourcepharma.com

DABROWSKI, C.E., Departments of Molecular Virology and Host Defense, Smith Kline Beecham Pharmaceuticals, 709 Swedeland Road, Mail Code UE0447, King of Prussia, PA 19406, USA

DE FRANCESCO, R., I.R.B.M., Istituto di Ricerche di Biologia Molecolare "P. Angeletti", Pomezia, I-00040 Roma, ITALIA, *e-mail:* Defrancesco@IRBM.it

DEVERAUX, Q.L., the Burnham Institute, La Jolla Cancer Research Foundation, 10901 North Torrey Pines Road, La Jolla, CA 92037, USA

DING, J., Brookhaven National Laboratory, Associated Universities, Inc., Biology Department, P.O. Box 5000, Upton, NY 11973-5000, USA

FRITZ, H., Abteilung für Klinische Chemie und Klinische Biochemie in der Chirurgischen Klinik und Poliklinik, Ludwig Maximilians Universität München, Medizinische Fakultät, Klinikum Innenstadt, Nussbaumstr. 20, D-80336 München, Germany

FURFINE, E.S., Department of Molecular Biochemistry, Glaxo Wellcome Inc., Moore Drive, Venture 298, P.O. Box 13398, Research Triangle Park, NC 27709, USA, *e-mail:* esf34284@glaxowellcome.com

HAASS, C., LM University, Adolf-Butenandt-Institute, Dep. of Metabolic Biochemistry, Schillerstraße 44, D-80336 Munich, Germany *e-mail:* CHaass@pbm.med.uni-muenchen.de

JAMES, M.N.G., Department of Biochemistry and Medical Research Council of Canada, Group in Protein Structure Function, University of Alberta, Medical Sciences Building 4-25, Edmonton, Alberta T6G 2H7, CANADA

KAWAHARA, H., Institute of Molecular and Cellular Biosciences, The University of Tokyo, 1-1-1 Yayoi, Bunkyo-ku, Tokyo 113-0032, JAPAN

KEPPLER, D., Department of Pharmacology and Barbara Ann Karmanos Cancer Institute, Wayne State University School of Medicine, 540 E. Canfield Ave., Detroit, MI 48201, USA, *e-mail:* dkeppler@med.wayne.edu

KOCH, U., I.R.B.M., Istituto di Ricerche di Biologia Molecolare "P. Angeletti", Pomezia, I-00040 Roma, ITALIA

KORANT, B.D., DuPont Merck Pharmaceutical Co., Research and
    Development, Experimental Station, E336/31B, Wilmington,
    DE 19880-0336, USA,
    e-mail: Bruce.D.Korant@dupontpharma.com

LAMBERT, M.H., Glaxo Wellcome Inc., P.O. Box 13398, Research Triangle
    Park, NC 27709, USA

LICHTENTHALER, S.F., Massachusetts General Hospital, Department of
    Molecular Biology, Wellman Bldg., 9th Floor, 50 Blossom Street, Boston,
    MA 02114, USA,
    e-mail: stefanL@frodo.mgh.harvard.edu

LYNCH, K.P., British Biotech, Inc., 201 Defense Highway, Suite 260,
    Annapolis, MD 21405, USA

MANGEL, W.F., Brookhaven National Laboratory, Associated Universities,
    Inc., Biology Department, P.O. Box 5000, Upton, NY 11973-5000,
    USA
    e-mail: Mangel@BNL.GOV

MASTERS, C.L., the Department of Pathology, The University of Melbourne,
    Parkville, Victoria 3052, AUSTRALIA

McGRATH, W.J., Brookhaven National Laboratory, Associated Universities,
    Inc., Biology Department, P.O. Box 5000, Upton, NY 11973-5000,
    USA

McKERROW, J.H., Department of Pathology –113B, University of California
    at San Francisco/VA Medical Center, 4150 Clement Street,
    San Francisco, CA 94121, USA,
    e-mail: jmck@cgl.ucsf.edu

PESSI, A., I.R.B.M., Istituto di Ricerche di Biologia Molecolare
    "P. Angeletti", Pomezia, I-00040 Roma, ITALIA

POTEMPA, J., Jagiellonian University, Institute of Molecular Biology,
    Department of Microbiology and Immunology, 31-120 Krakow, Poland,
    and University of Georgia, Department of Biochemistry and Molecular
    Biology, Life Sciences Building, Athens, GA 30602, USA,
    e-mail: potempa@arches.uga.edu

QIU, X., Department of Structural Biology, Smith Kline Beecham
    Pharmaceuticals, 709 Swedeland Road, Mail Code UE0447,
    King of Prussia, PA 19406, USA,
    e-mail: xiayang_qui-1@shphrd.com

RABINOVITCH, M., Laboratory Medicine & Pathobiology, and Cardiovascular Research, The Hospital for Sick Children, 555 University Avenue, Toronto, Ontario M5G 1X8, CANADA

RASMUSSEN, H.S., Clinical Research & Regulatory Affairs, British Biotech, Inc., 201 Defense Highway, Suite 260, Annapolis, MD 21405, USA, *e-mail:* hrasmuss@britbio.co.uk

REDSHAW, S., Medicinal Chemistry III, Roche Discovery Welwyn, Broadwater Road, Welwyn Garden City, Hertfordshire AL7 3AY, United Kingdom, *e-mail:* sally-redshaw@roche.com

REED, J.C., The Burnham Institute, La Jolla Cancer Research Foundation, 10901 North Torrey Pines Road, La Jolla, CA 92037, USA

ROBERTS, N.A., Viral Diseases, Roche Discovery Welwyn, Broadwater Road, Welwyn Garden City, Hertfordshire AL7 3AY, United Kingdom

SALTER, J.P., Department of Pathology –113B, University of California at San Francisco, 4150 Clement Street, San Francisco, CA 94121, USA

SALVESEN, G.S., The Burnham Institute, La Jolla Cancer Research Foundation, 10901 North Torrey Pines Road, La Jolla, CA 92037, USA, *e-mail:* sramey@burnham-inst.org

SCHMAIER, A.H., Department of Internal Medicine and Pathology, University of Michigan, 5301 MSRB III, 1150 W. Medical Center Drive, Ann Arbor, MI 48109-0640, USA, *e-mail:* aschmaie@umich.edu

SHAO, W., Structural Biochemistry Program, NCI-FCRDC, Frederick, MD, 21102, USA

SHRIDHAR, R., Department of Pharmacology, Wayne State University School of Medicine, 540 E. Canfield Ave., Detroit, MI 48201, USA

SLOANE, B.F., Department of Pharmacology and Barbara Ann Karmanos Cancer Institute, Wayne State University School of Medicine, 540 E. Canfield Ave., Detroit, MI 48201, USA

SMITH, T., UNC Center for AIDS Research, University of North Carolina at Chapel Hill, Chapel Hill, NC 27599-7295, USA

STEINKÜHLER, C., I.R.B.M., Istituto di Ricerche di Biologia Molecolare "P. Angeletti", Pomezia, I-00040 Roma, ITALIA

SWANSTROM, R., UNC Center For AIDS Research, Rm. 22-006, Mason Farm Road Campus Box 7295, University of North Carolina at Chapel Hill, Chapel Hill, NC 27599-7295, USA
e-mail: risunc@med.unc.edu

SWEET, R.M., Brookhaven National Laboratory, Associated Universities, Inc., Biology Department, P.O. Box 5000, Upton, NY 11973-5000, USA

TANAKA, K., The Tokyo Metropolitan Institute of Medical Science and CREST, Japan Science and Technology Corporation (JST), 3-18-22 Honkomagome, Bunkyo-ku, Tokyo 113-0021, JAPAN, *e-mail:* tanakak@rinshoken.or.jp

THOMAS, G.J., Medicinal Chemistry III, Roche Discovery Welwyn, Broadwater Road, Welwyn Garden City, Hertfordshire AL7 3AY, United Kingdom

TOLEDO, D.L., Brookhaven National Laboratory, Associated Universities, Inc., Biology Department, P.O. Box 5000, Upton, NY 11973-5000, USA

TRAVIS, J., University of Georgia, Department of Biochemistry and Molecular Biology, Life Sciences Building, Athens, GA 30602, USA

VALLIANCOURT, M., Institut de Recherches Clinique de Montreal, Bio-organic Chemistry Laboratory, 110 Avenue des Pins Quest, Montreal, Quebec, F2W 1R7, CANADA

VELLA, S., Istituto Superiore di Sanità, Laboratorio di Virologia, Reparto Retrovirus, Viale Regina Elena, 299, I-00161 Roma, ITALIA, *e-mail:* segreteria@vella.net.iss.it

VON DER HELM, K., Max-von-Pettenkofer Institut für medizinische Mikrobiologie, LS Virologie, Pettenkoferstr. 9a, D-80336 München, Germany, *e-mail:* kvdh@mvp.med.uni-muenchen.de

Contents

CHAPTER 2

**Clinical Experience with Human Immunodeficiency Virus Protease
Inhibitors: Antiretroviral Results, Questions and Future Strategies**

CHAPTER 3

**The Nature of Resistance to Human Immunodeficiency Virus Type-1
Protease Inhibitors**

CHAPTER 4

**The Next Generation of Human Immunodeficiency Virus Protease
Inhibitors: Targeting Viral Resistance**

Section II. Other Viral (Non-HIV) Protease Inhibitors

CHAPTER 5

The Proteinases Encoded by Hepatitis C Virus as Therapeutic Targets
C. STEINKÜHLER, U. KOCH, R. DE FRANCESCO, and A. PESSI.

CHAPTER 8

Adenovirus Proteinase-Antiviral Target for Triple-Combination Therapy on a Single Enzyme: Potential Inhibitor-Binding Sites

W.F. MANGEL, D.L. TOLEDO, M.T. BROWN, J. DING, R.M. SWEET,
D.L. BARNARD, and W.J. McGRATH. With 7 Figures 145

CHAPTER 10

Section III.   (Non-Viral) Proteases Involved in Diseases

CHAPTER 11

CHAPTER 12

CHAPTER 13

The Tumor Necrosis Factor-α Converting Enzyme
J.D. BECHERER, M.H. LAMBERT, and R.C. ANDREWS.

CHAPTER 14

Serine Elastases in Inflammatory and Vascular Diseases
J.C. CHERONIS and M. RABINOVITCH. With 5 Figures 259

CHAPTER 15

Inhibitors of Thrombin and Factor Xa

CHAPTER 16

Inhibitors of Papain-Like Cysteine Peptidases in Cancer

CHAPTER 17

**Caspases and Their Natural Inhibitors as Therapeutic Targets
for Regulating Apoptosis**
Q.L. DEVERAUX, J.C. REED, and G.S. SALVESEN. With 3 Figures 329

CHAPTER 18

Proteasome and Apoptosis
K. TANAKA and H. KAWAHARA. With 2 Figures 341

CHAPTER 19

**Proteolytic Processing of the Amyloid Precursor Protein of
Alzheimer's Disease**
S.F. LICHTENTHALER, C.L. MASTERS, and K. BEYREUTHER.

CHAPTER 20

**Presenilins and β-Amyloid Precursor Protein-Proteolytically Processed
Proteins Involved in the Generation of Alzheimer's Amyloid β Peptide**

Section I
Human Immunodeficiency Virus
Protease Inhibitors

CHAPTER 1

The Road to Fortovase. A History of Saquinavir, the First Human Immunodeficiency Virus Protease Inhibitor

S. Redshaw, N.A. Roberts, and G.J. Thomas

A. Background

I. Present Scale of the Acquired Immunodeficiency Syndrome Pandemic

Since the early 1980s, acquired immunodeficiency syndrome (AIDS) has evolved from a seemingly rare disease, first seen in small numbers of individuals in urban areas of the United States, into a worldwide epidemic. The syndrome is characterised by severe impairment of the immune system, resulting in infection by "opportunistic" pathogens and, ultimately, death. A recent joint report by the United Nations Programme on HIV/AIDS and the World Health Organization estimates that over 30 million people (one in every 100 sexually active adults worldwide) are living with human immunodeficiency virus (HIV) or AIDS. If the current transmission rate of around 16,000 new infections every day is not reduced, this number is predicted to exceed 40 million by the year 2000.

II. Identification of the Cause of AIDS

Until quite recently, it was widely believed that infectious diseases posed little further threat to the developed world and that the remaining medical challenges were non-infectious conditions, such as heart disease and cancer. That confidence was shattered in the early 1980s when it was discovered that AIDS was caused by an infectious agent. Although many investigators at first thought that AIDS might be caused by a new toxin or environmental chemical, the fact that the early cases occurred among homosexual men and that the main difference between people with AIDS and homosexual controls was the number and frequency of their sexual contacts suggested a sexually transmitted infectious agent. This theory gained ground when AIDS was also diagnosed in recipients of contaminated blood or blood products and in intravenous drug users who had shared syringes.

An intensive search began for the infectious agent and, in 1983, a new virus, now known as human immunodeficiency virus type 1, or HIV-1, was isolated independently by two groups (Barré-Sinoussi et al. 1983; Popovic et al. 1984). A little later, a genetically distinct virus, HIV-2, which occurs in different geographic locations, was isolated (CLAVEL et al. 1986). The identification of HIV-1 was facilitated by the prior discovery of the first human retrovirus, human T-lymphotropic virus (HTLV-1), which infects T-lymphocytes and can cause a rare and highly malignant type of cancer (POIESZ et al. 1980). Since AIDS is characterised by a severe depletion of T-lymphocytes, it seemed likely that this disease, too, might be caused by a retro-virus. This hypothesis was confirmed when reverse-transcriptase activity, a characteristic of retroviruses, was detected in a sample of lymph tissue from a patient at risk of AIDS (BARRÉ-SINOUSSI et al. 1983). The isolation of HIV allowed the development of a test that could be used to detect antibodies to the virus, and this test soon revealed that the total number of HIV infections was very much greater than number of AIDS cases so far reported.

III. Search for a Cure

When it first became established that AIDS was caused by a retrovirus, many doubted that a drug capable of directly attacking the virus would ever be found. Those doubts were at least partially dispelled when a survey of avail-able drugs at the National Cancer Institute identified several compounds capable of preventing HIV replication in vitro. One of these, 3'-azido-3'-deoxythymidine or AZT (zidovudine), was the first drug to be used for the treatment of AIDS. This compound, after conversion to the triphosphate by cellular kinases, was later shown to inhibit the viral reverse transcriptase. Although AZT has undoubtedly shown some benefit to patients (ABOULKER and SWART 1993), it quite quickly became clear that treatment is of limited efficacy, largely because of dose-limiting side-effects caused by interference with human cell metabolism (STYRT 1996) and because of the emergence of drug-resistant virus (LARDER et al. 1989). There was thus a great need for novel antiretrovirals which could be administered at sufficiently high doses, and for long enough periods, to allow recovery of patients' immune functions. Much effort has been devoted to elucidating the viral life cycle (MITSUYA and BRODER 1987) and identifying potential targets for antiviral chemotherapy: one of the most attractive of these was a virally encoded protease.

IV. Identification and Characterisation of HIV Protease

When a retrovirus enters a cell, the single-stranded viral RNA is copied to produce double-stranded DNA. The viral DNA becomes integrated into the host cell genome and is subsequently transcribed and translated by cellular enzymes to produce the viral proteins. The open reading frames for viral *gag* and *gag-pol* proteins are first translated as fusion polyproteins which are subsequently processed into mature proteins by a protease, which is itself encoded within the *gag-pol* polyprotein (Fig. 1). Molecular cloning and sequence determination of the HIV genome revealed the presence of open reading frames analogous to the *gag* and *gag-pol* open reading frames of the

k b   H I V   R N A

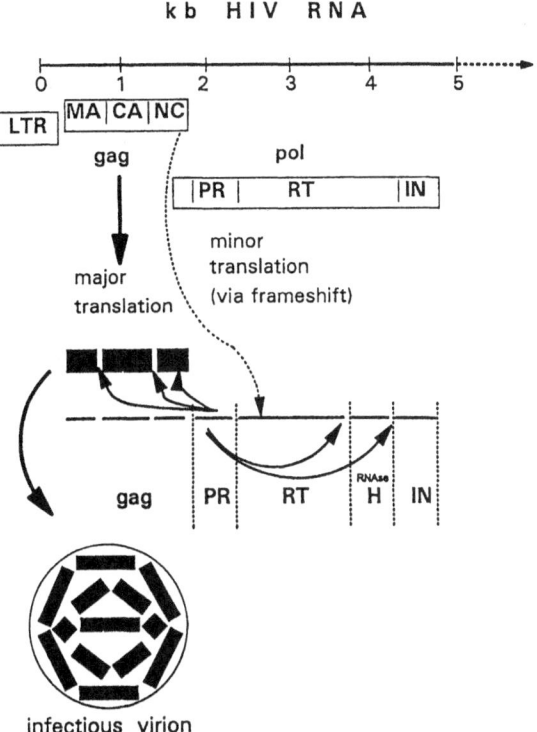

Fig. 1. The open reading frames for viral *gag* and *gag-pol* proteins

known retroviruses, and it was suggested that HIV might also encode a protease (RATNER et al. 1985).

It was realised (TOH et al. 1985) that the retroviral proteases contain a highly conserved Asp-Thr-Gly motif, and it was suggested that their catalytic mechanism might be similar to that of the cellular aspartic proteases. Apart from the conserved motif, however, there were few similarities between the viral and cellular enzymes. The cellular enzymes are relatively large proteins of more than 200 amino acids, comprising two homologous domains, each of which contains one Asp-Thr-Gly motif. The aspartic-acid residues from each domain are spatially close and interact to form the active site of the protease. It was not immediately apparent how the retroviral proteases, which are much smaller (around 100 amino acids) and contain only one Asp-Thr-Gly motif, could form a similar active site. A model of HIV protease was proposed (PEARL and TAYLOR 1987) in which the active species is a C_2-symmetric dimer, with each monomer contributing an aspartic acid to the active site. A 99-amino-acid, 11-kDa, form of protease was subsequently shown to be the minimum active domain (GRAVES et al. 1988), and further biological and crystallographic studies have confirmed that HIV protease does indeed function as a homo-dimer of 198 amino acids.

Before the catalytic mechanism had been fully clarified, the protease in HIV *pol* was shown to be essential for cleavage of the *gag* polyprotein substrate (Kramer et al. 1986). Recombinant HIV *gag-pol* was expressed in yeast cells, and processing of the *gag* polyprotein was observed. This processing was completely prevented by a frameshift mutation in the protease region of *pol*. An active-site mutation (Asp to Ala) was later shown to produce inactive protease (Mous et al. 1988; Seelmeir et al. 1988). The protease was shown shortly afterwards to be essential for viral infectivity (Kohl et al. 1988). When proviral DNA incorporating a mutant protease gene was used to transfect human colon carcinoma cells, no *gag* processing occurred and the resulting viral particles were non-infectious.

At the same time as efforts to establish the catalytic mechanism of the protease and its role in the viral life-cycle, work was underway to define the protease's substrate specificity. Even before the protease had been isolated, a study of peptides obtained from infected cells suggested Met-Met and Tyr-Pro as likely cleavage sites in the *gag* protein (Sanchez-Pescador et al. 1985). Further cleavage sites were later elucidated in the *pol* protein (Lightfoote et al. 1986; Veronese et al. 1986), and subsequent work at Roche has shown the protease to be responsible for all cleavages involved in the maturation of both the *gag-* and *pol*-gene products (Le Grice et al. 1988).

B. Roche Inhibitor Program

Although characterisation of HIV protease was far from complete in the mid 1980s, sufficient evidence was available to make the enzyme an exciting, if ambitious, target for antiviral chemotherapy. The enzyme had been provisionally classified as an aspartic protease, and some of the substrate cleavage sites had been predicted, although not yet confirmed. The protease had been shown to be necessary for some of the maturational cleavages of the viral polyproteins, but had not yet been proven to be essential for viral infectivity.

It was against this background that we began our program to design inhibitors of HIV protease in the autumn of 1986. From the outset, we were particularly intrigued by the notion that HIV protease was able to cleave substrates *N*-terminal to proline residues. Since mammalian endopeptidases are unable to carry out such cleavages, it seemed likely that inhibitors based on this motif would be selective for the viral enzyme. Such inhibitors should not, therefore, cause side effects by inhibition of human aspartic proteases.

We decided at once to verify this cleavage if possible and, if such cleavage were proven, to design our inhibitors around a Tyr(Phe)-Pro motif. Key early objectives were, obviously, to isolate the protease, to establish a suitable assay with appropriate substrates and to identify prototype inhibitors. Since the relative clinical importance of HIV-2 was unclear at the time, we felt that it was important to consider both viral proteases within our program.

I. Enzyme Assays

It would have been possible to attempt to isolate the protease from infected cells [a feat which was later achieved (LILLEHOJ et al. 1988)], but it seemed to us that recombinant-DNA technology offered the best source of adequate quantities of protein. Roche molecular biologists in Nutley, USA (GRAVES et al. 1988) and Basle, Switzerland (LE GRICE et al. 1988; MOUS et al. 1988) set out to clone, express and purify the protease and its protein substrates. These materials were used to establish an assay, to test potential inhibitors and also for detailed mechanistic studies.

As potential substrates, we prepared peptides containing Tyr(Phe)-Pro motifs based on consensus sequences around the *gag* and *pol* cleavage sites. The N- and C-termini were protected to prevent cleavage by exopeptidases produced by *Escherichia coli* when using partially purified enzyme preparations, and an N-terminal succinyl residue was included to improve solubility. Peptides with less than six residues were not processed efficiently, but we were pleased to discover that a hexapeptide, succinyl-Ser-Leu-Asn-Tyr-Pro-Ile-isobutylamide, based on the P_4-P_2' sequence in the *gag* polyprotein, was a reasonable substrate ($K_m = 1.42$ mM). Importantly, we were also able to establish that the peptide was cleaved between the tyrosine and proline residues, as we had hoped. This peptide, together with a related heptapeptide, was subsequently used for all our routine screening assays.

We wanted to establish a colorimetric assay, since this would allow us to screen potential inhibitors quickly and easily. We also recognised that an assay based on an ability to detect the N-terminal proline residue of the cleavage product would be unaffected by contaminating proteases and would have the considerable advantage of allowing us to use relatively crude enzyme preparations. These criteria were met by an assay (BROADHURST et al. 1991) based on an ability to detect the deep-blue colour which is produced on reaction of cyclic secondary amines such as proline with isatin. This reaction had been known since the end of the last century (SCHOTTEN 1891) but had not previously been adapted for the quantitative determination of proline-containing peptides.

II. Inhibitor Design

Proteases have been described as "molecular scissors" that snip large proteins into smaller pieces. Aspartic proteases achieve this by adding a water molecule to the amide bond that is to be cleaved, creating an unstable tetrahedral species (*2* in Fig. 2). This species, which is referred to as the transition state, collapses rapidly to give the cleavage products (*3* and *4* in Fig. 2).

Stable compounds that resemble the transition state, but cannot dissociate, bind tightly to the active site of the protease and so act as powerful inhibitors. Many different mimetics have been devised, each imitating some of the supposed aspects of the transition state. Inhibitors are prepared by

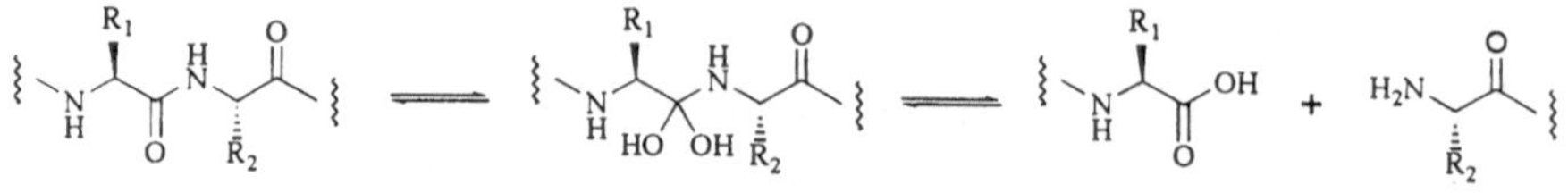

Fig. 2. Cleavage of amide bonds by aspartic proteases

Compound	Structure	IC_{50} (nM)
5	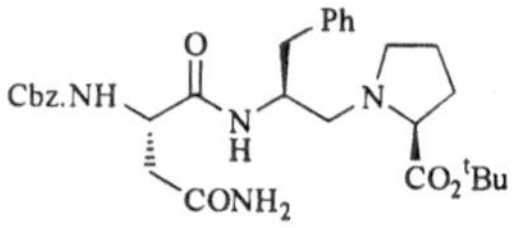	50,000
6	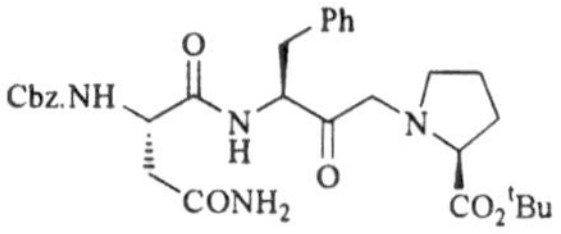	870
7 (*R*-isomer) 8 (*S*-isomer)	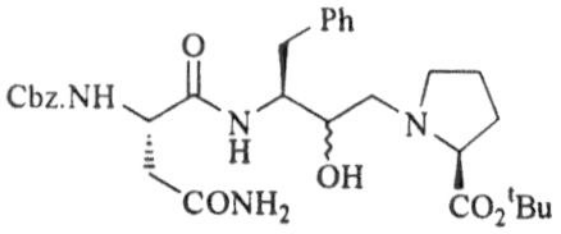	140 300

Fig. 3. Comparison of transition-state mimetics

incorporating an appropriate transition-state mimetic into a suitable peptide sequence.

Of the many possible transition-state mimetics, the reduced amide, ketomethylene derivative and hydroxyethylamine (*5–8* in Fig. 3) seemed especially suited to the scissile Tyr(Phe)-Pro motif that we had chosen for our inhibitors. In a preliminary study (ROBERTS et al. 1990), these three transition-state mimetics were incorporated into the Asn. Phe-Pro cleavage sequence of the *pol* polyprotein. The most potent inhibitors were found to be the diastereomeric hydroxyethylamines (Fig. 3). Although the more potent of the two isomers showed very encouraging activity, with 50% inhibition of HIV-1 protease at a concentration of 140 nM, we felt that for clinical evaluation a more potent compound would be needed.

At the time, no X-ray crystallographic data were available and, in the absence of structural information that might help in the design of more potent inhibitors, we set out to explore structure–activity relationships by systematic modifications. Our first task was to determine the effect of size on the activity of our inhibitors. We found that the protected dipeptide mimetics (*9* and *10* in Fig. 4) showed much-reduced activity compared with compounds *7* and *8*, while addition of residues at the N-terminus (*11* and *12*), the C-terminus (*13*

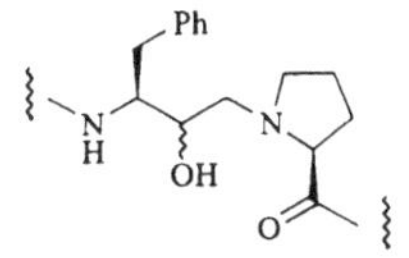

Ph-Ψ[CH(OH)CH₂N]Pro

Compound	Structure	IC_{50} (nM)
9 and 10	Cbz.Ph-Ψ[CH(OH)CH₂N]Pro.O^{t}Bu	6,500 and 30,000
7 and 8	Cbz.Asn.Ph-Ψ[CH(OH)CH₂N]Pro.O^{t}Bu	140 and 300
11 and 12	Cbz.Leu.Asn.Ph-Ψ[CH(OH)CH₂N]Pro.O^{t}Bu	600 and 1,100
13 and 14	Cbz.Asn.Ph-Ψ[CH(OH)CH₂N]Pro.Ile.NH^{i}Bu	130 and 2,400
15 and 16	Cbz.Leu.Asn.Ph-Ψ[CH(OH)CH₂N]Pro.Ile.NH^{i}Bu	750 and 10,000

Fig. 4. Effect of inhibitor size on potency

and *14*), or indeed at both ends of the molecule (*15* and *16*) gave no improvement in potency. We therefore chose the more potent hydroxyethylamine (*7* in Fig. 4), with the *R* configuration at the secondary alcohol function, for further investigation of structure–activity relationships and optimisation of activity.

At the C-terminus, medium-sized lipophilic residues appeared to be preferred, with little difference between esters and amides (*7* vs *17* in Fig. 4 and Fig. 5, respectively). The *t*-butyl amide group was chosen as the C-terminal residue for subsequent compounds on the basis of chemical and, possibly, metabolic stability. Replacement of the N-terminal benzyloxycarbonyl group by smaller non-aromatic groups, such as acetyl or tert-butoxycarbonyl, gave compounds with reduced activity, whilst introduction of bicyclic aromatic groups, such as β-naphthoyl or, especially, quinoline-2-carbonyl, led to compounds (*18* in Fig. 5) with significantly improved potency. At the P₂ and P₁ positions, conservative changes were allowed, but no significant improvements over the parent compound were identified. The most dramatic changes in potency were achieved by modifying the prolyl residue which occupies the S₁′ subsite. Ring size was found to be very important for activity – replacing the proline five-membered ring by a four-membered azetidine ring almost abolished activity, whilst incorporation of a six-membered ring improved potency approximately 12-fold. Replacement of proline by fused bicyclic imino acids led to the greatest enhancement of activity, and S,S,S-decahydroisoquinoline carboxylic acid (*19* in Fig. 5) was the best replacement for proline that we identified.

Having identified regions of the molecule in which changes substantially altered binding affinities, additional analogues that contained combinations of preferred side chains were synthesised; these compounds showed that

Compound	Structure	HIV-1 Protease IC_{50} (nM)	HIV-2 Protease IC_{50} (nM)
17	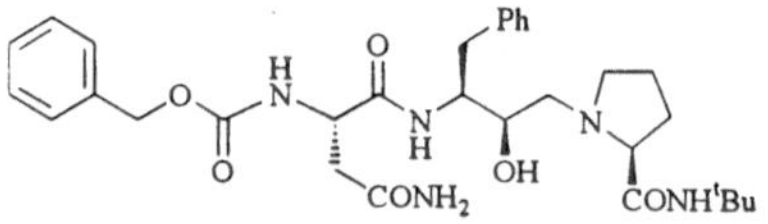	210	330
18	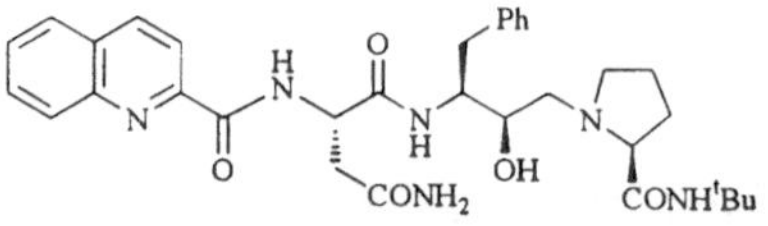	52	50
19	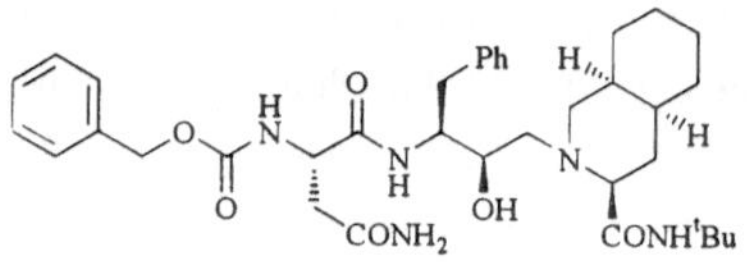	2.7	ND
20	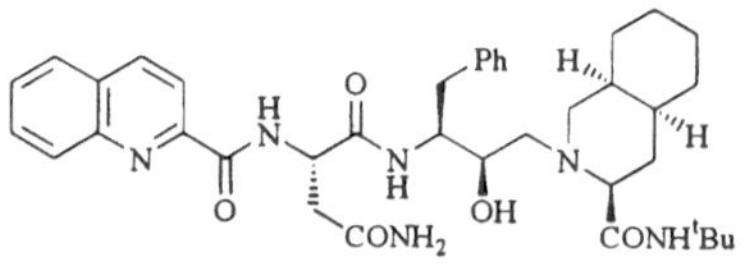	<0.4	<0.8

Fig. 5. Effect of substitutions on potency

the effects of beneficial changes were frequently more than additive. The compounds, moreover, showed the same order of potencies against HIV-2 protease, although they were somewhat more active against the HIV-1 enzyme. Within this series of compounds, the order of potencies in a preliminary antiviral assay also paralleled the enzyme-inhibitory potency very closely, probably indicating good penetration into cells. One of the most potent (antiviral EC_{50} = 2 nM) of these hydroxyethylamine derivatives was Ro 31-8959 (named saquinavir, *20* in Fig. 5).

III. Selectivity

Since selectivity for the viral protease had been an integral part of our design hypothesis, we were pleased to find that saquinavir was, indeed, extremely selective, showing less than 50% inhibition of the human aspartic proteases, renin, pepsin, gastricsin, or cathepsins D and E even at a concentration of $10 \mu M$ (ROBERTS et al. 1990). When tested against representatives from the three other mechanistic classes of mammalian proteases (serine, cysteine and metallo), the compound similarly showed less than 50% inhibition at a concentration of $10 \mu M$. These results indicated that, as intended, saquinavir would be unlikely to have effects on human proteases, which we hoped would limit side effects in patients.

IV. Antiviral Activity

The antiviral activities of our protease inhibitors were initially determined through a collaboration with St. Mary's Hospital Medical School, London, and later in a high-containment laboratory suitable for HIV work, which was commissioned at Roche in the UK. In addition to assays carried out in these two centres, saquinavir was also included in a Medical Research Council multi-centre, blinded testing program (HOLMES et al. 1991). Antiviral EC_{50} values (viral growth inhibitory concentrations) reported by all of these laboratories were typically in the range 1–10 nM against HIV-1, with very similar potencies against HIV-2 (HOLMES et al. 1991) and simian immunodeficiency virus (MARTIN et al. 1991). The tests also demonstrated that saquinavir was effective against both laboratory strains of HIV and clinical isolates, including AZT-resistant strains (JOHNSON et al. 1992; GALPIN et al. 1994), and moreover was active in both lymphoblastoid and macrophage-derived cell lines, as well as in primary cells.

Unlike reverse-transcriptase inhibitors, which are ineffective if added more than 2–3 h post-infection, saquinavir was fully active following delayed addition to infected cells (CRAIG et al. 1991a; GALPIN et al. 1994), and also in assays using chronically infected cells (CRAIG et al. 1991a,b). Both of these observations confirm its late point of action in the infectious cycle. More directly, in one-step acute infection of MT-4 cells, saquinavir was found to have no effect on synthesis of cDNA, integration into cellular DNA or on transcription, although production of progeny virus was prevented (JACOBSEN et al. 1992). Electron microscopy of virions produced in chronically infected cells treated with saquinavir confirmed that the particles had failed to mature (CRAIG et al. 1991a), and the immature morphology was also found to be associated with a lack of infectivity (ROBERTS et al. 1992). The consistently high efficacy and breadth of activity shown by saquinavir have not yet been surpassed by any other protease inhibitor.

V. Combination Studies

A combination of antiviral drugs with different mechanisms of action should enhance suppression of viral replication (HALL and DUNCAN 1988) and potentially allow a reduction in dose of the individual drugs whilst maintaining efficacy. Two- or three-way combinations of saquinavir with other antiretrovirals (e.g. reverse-transcriptase inhibitors) produced effects that were additive to synergistic (JOHNSON et al. 1992; CRAIG et al. 1993a; CONNELL et al. 1994; CRAIG et al. 1994; TAYLOR et al. 1995), irrespective of the mathematical model used to analyse the data. These results suggest likely benefits for drug combinations, with possibilities for improved efficacy as well as control over drug resistance resulting from reduced viral replication.

VI. Resistance

The efficacy of any anti-infective agent may be severely compromised by the development of resistance. This is a particular problem in antiretroviral therapy because of the very high rate of retroviral replication and also because mutations are not corrected by the reverse transcriptase, which is intrinsically error prone. Resistance to nucleoside analogue reverse-transcriptase inhibitors was well documented and it was therefore important to elicit whether protease inhibitors could also select for resistant mutants. In separate experiments, HIV-1 strain$_{GB8}$ was serially passaged in CEM cells in the presence of increasing concentrations of saquinavir, AZT and a non-nucleoside reverse-transcriptase inhibitor, the TIBO compound (R82150). In each case, it proved possible to select virus with reduced sensitivity to the test compound, although this occurred to a lesser extent and at a later time point with saquinavir than with either of the reverse-transcriptase inhibitors (CRAIG et al. 1993b).

To elucidate the molecular basis for reduced sensitivity to saquinavir, protease from mutant virus was cloned and sequenced. The first mutation occurred at position 48, where a glycine residue was replaced by valine (G48V). In some instances, a second mutation, L90M, followed the G48V substitution. The double mutant was associated with substantially increased loss of sensitivity compared with the single substitution (JACOBSEN et al. 1995). In a study of the kinetic properties of mutant proteases, the relative processing activities of G48V, L90M, and the double mutant G48V/L90M have been estimated to be around 10, 7, and 3% that of wild-type HIV-1 protease, respectively (ERMOLIEFF et al. 1997), hopefully indicating that the mutant virus might have impaired growth characteristics compared with wild-type virus (see also Chaps. 2 and 3, this volume).

VII. Safety and Pharmacokinetics

We had thus shown that saquinavir showed extremely good activity in a range of antiviral tests, as well as a low propensity to cause resistance. We had already shown that, although there was some slight variation in sensitivity in different cell types, there was a difference of at least three orders of magnitude between the antiviral and cytotoxic effects of saquinavir. Saquinavir also had no effect in mutagenicity or genotoxicity assays.

The in vivo toxicity of saquinavir also proved to be minimal. In several animal species, only very slight effects in general pharmacology were seen on oral administration of high doses of saquinavir. Oral toxicity and toxicokinetic studies of up to 12-months duration showed excellent tolerability even at high plasma exposure levels. Saquinavir is not teratogenic, and no reproductive or developmental defects were seen in statutory segment I, II or III reproductive-toxicity studies. These results all supported our original belief that a selective inhibitor of the viral protease would show little toxicity and be well suited to long-term therapy.

Since our ultimate goal had always been an orally active compound, pharmacokinetic evaluation of selected compounds had formed an integral part of our screening cascade. These preliminary studies indicated that it should be possible to achieve clinically relevant concentrations of saquinavir following oral administration.

C. Early Clinical Studies

In 1991, it was time to begin clinical trials with saquinavir. Although we had shown extremely potent antiviral activity in a wide range of in vitro test systems as well as an extremely good safety profile, the lack of an animal model meant that we now had to go forward into the clinic with no true efficacy data. We therefore needed to extrapolate from in vitro antiviral concentrations to the dose needed to achieve a clinical effect. Representative EC_{50} and EC_{90} values of 2 nM and 16 nM, respectively, were chosen for the extrapolation to a clinical dose, and at the same time a twofold increase in concentration was allowed to compensate for plasma binding of the drug in vivo.

I. Absorption and Metabolism

The mean absolute bioavailability of saquinavir in its initial hard-gel formulation, Invirase, was found to be governed by its limited absorption and, more especially, by extensive first-pass metabolism (WILLIAMS et al. 1992). The absorption level was found to be increased approximately sixfold when administered to volunteers after food, leading to the stipulation that the drug be taken within 2 h of a meal. It has been found that metabolism of saquinavir (and of other HIV protease inhibitors) is mediated by the cytochrome $P450_{3A4}$ isoenzyme (FARRAR et al. 1994; FITZSIMMONS and COLLINS 1997). Therapeutic or recreational drugs which induce or block cytochrome $P450_{3A4}$ would thus be expected to reduce or increase levels of saquinavir correspondingly. As might be expected, co-administration with other protease inhibitors, such as ritonavir (MERRY et al. 1997) or nelfinavir (MERRY et al. 1998), leads to enhanced levels of saquinavir. The inhibition by HIV protease inhibitors on different cytochrome P450 isoforms has been studied in vitro using human-liver microsomes (EAGLING et al. 1997). Saquinavir was found to be 100-fold less potent than ritonavir and 10-fold less potent than indinavir as an inhibitor of $P450_{3A4}$. In line with these observations, ritonavir shows the most drug interactions and saquinavir the least.

II. Efficacy

Because progression of disease is relatively slow in AIDS patients, assessment of efficacy in early clinical trials of saquinavir was based on surrogate markers of disease, namely viral load (as determined by quantitative amplification of plasma viral RNA), and CD4+ T-cell counts (as a general marker of immune status). Invirase was initially assessed as monotherapy at doses ranging from 25 mg to 600 mg three times daily (tid) in randomised double-blind studies in Europe. These studies showed a positive dose–response relationship (Kitchen et al. 1995; Noble and Faulds 1996; Vella et al. 1996) and supported the 600-mg (t.i.d., thrice daily) regimen. Subsequently, a study at Stanford University (Schapiro et al. 1996) showed increased efficacy at higher doses. Decreases in viral load of 1.1 and 1.3 $\log_{10}$ units as well as elevated CD4+ T-cell counts were maintained to 24 weeks at doses of 3.6 g/day and 7.2 g/day, respectively.

In vitro studies supported the concept of using saquinavir in combination with other antiretrovirals, and this has been confirmed in clinical practice. In previously untreated patients with advanced disease, both the peak reduction in viral load (measured by means of polymerase chain reaction) and the median increase in CD4+ T-cell counts over 16 weeks were greater in patients receiving saquinavir in combination with AZT than in patients receiving monotherapy (Noble and Faulds 1996; Vella et al. 1996). A second study compared the effect of saquinavir + zalcitabine (2′,3′-dideoxycytidine, ddC) with that of zalcitabine + AZT and a combination of all three drugs in advanced, heavily AZT-pretreated patients. The median increase in CD4+ T-cell counts over 24 weeks or 48 weeks, the mean normalised area under the plot of CD4+ T-cell counts over time and the reduction in viral load were all greater with triple therapy than with either of the double therapies (Noble and Faulds 1996). Combination treatment with saquinavir plus zalcitabine has been shown to reduce the risk of progression to AIDS by 49% and to reduce deaths by 68% compared with zalcitabine monotherapy (Haubrich et al. 1998).

III. Tolerability

Invirase was found to be very well tolerated in the clinic, resulting in exceptionally good compliance. Side effects were uncommon; gastrointestinal effects (mostly diarrhoea) occurred most frequently and were seen in 3.8% of patients (Noble 1996). Other adverse events (headache, paraesthesia, asthenia, skin rash or musculoskeletal pain) were reported at 1% or lower incidence. All adverse effects were fully reversible.

D. Approval and Beyond

The studies described above formed the basis for the approval of the first inhibitor of HIV protease as an antiretroviral for human use. Invirase was approved on 6 December 1995 in the United States, and approvals in the European Union and other countries followed rapidly.

I. Incidence of Resistance in Clinical Use

Since we had shown that selection of saquinavir-resistant mutant virus could be induced in vitro, it was important to look for evidence of emerging resistance in the clinic. Virus isolated from patients during monotherapy with saquinavir did show reduced susceptibility, ranging from threefold to, in one case, 100-fold after approximately 1 year of therapy (IVES et al. 1997). In this trial, the G48V resistance mutation was not observed but, in keeping with in vitro data, isolates from five of eight subjects showed a L90M/I mutation. This substitution had only a modest effect on sensitivity: three of the isolates showed a less than eightfold reduction in susceptibility. Other mutations occurred at the naturally polymorphic positions 36 and 71, and at position 84. A similar study in patients undergoing long-term therapy with saquinavir found both G48V and L90M mutations (JACOBSEN et al. 1996). It has become clear that the key mutations that result from saquinavir therapy differ from those seen with other protease inhibitors (MELLORS et al. 1995), and this should lead to a low frequency of cross-resistance between saquinavir and the other agents. Indeed, long-term treatment with saquinavir does not, in most cases, induce a significant decrease in sensitivity to saquinavir itself or to other protease inhibitors (ROBERTS et al. 1998) and, because of this, saquinavir has been suggested as a good first-choice protease inhibitor for combination therapy (BOUCHER 1996). Continued viral replication is critical to the emergence of resistance, and the major treatment goal must thus be to reduce viral load to as low a level as possible. Prolonged drug failure may be associated with the emergence of more extensive cross-resistance, and the British HIV Association has recommended (GAZZARD et al. 1997) that if control over viral load cannot be maintained by a particular combination therapy, then at least two new agents should be added or substituted. Switching therapy at the first sign of virological failure should help to ensure maximum benefit from the next regimen.

II. Immune Function and Opportunistic Infections

Among the clinical benefits seen following the introduction of protease inhibitors is an improvement in immune function. It is known that humoral immune response to HIV infection plays an important role in disease progression, and recent studies using saquinavir alone or in combination with AZT suggest that these treatments improve neutralising-antibody activity against autologous virus (SARMATI et al. 1997). Changes in immune status could also account for the resolution of opportunistic infections, which has been seen in patients treated with saquinavir. Complete resolution of oral candidiasis, which had previously been refractory to treatment with antifungal agents, has been reported (ZINGMAN 1996), while combination antiretroviral therapy that includes a protease inhibitor has led to a complete clinical response in patients with chronic microsporidiosis or cryptosporidiosis infection (CARR et al. 1998). Combinations that include saquinavir have also led to regression of tumours in cases of AIDS-related Kaposi's sarcoma (PARRA et al. 1998).

Recently, it has been shown that an increase in CD4$^+$ T-cell counts can occur even in the face of "detectable" virus (KAUFMAN et al. 1998) indicating that measurable viral load is not necessarily an indicator of treatment failure. The finding that highly active antiretroviral therapy (HAART) may have a prolonged effect on CD4$^+$ T-cell counts, even without suppression of viraemia, challenges current understanding of the mechanisms of immune damage caused by HIV and may open the way to new therapeutic approaches.

III. Fortovase – a New Formulation

Encouraged by the increased antiviral effect at higher doses of Invirase (SCHAPIRO et al. 1996) and by its excellent tolerability, we have developed a second formulation of saquinavir, which improves the exposure to this intrinsically very potent compound. At the approved dose of 1200 mg t.i.d., the new soft gelatin formulation, Fortovase, gives exposure levels eightfold higher than those achieved with Invirase (600 mg t.i.d.).

A head-to-head study of the triple combinations AZT+lamivudine (3TC) + Crixivan versus AZT + 3TC + Fortovase (BORLEFFS 1998) showed both regimens to be equivalent in terms of reduction of viral load, but the CD4$^+$ T-cell count increased more rapidly in the Fortovase arm. Importantly, increased exposure to saquinavir due to Fortovase does not alter the signature of key mutations found with Invirase, and accumulation of accessory mutations associated with cross-resistance remains rare (CRAIG et al. 1998).

Fortovase was approved for marketing in the USA on 17 November 1997, and applications for approval have been submitted in other countries. Recently, a panel of experts comprising AIDS specialists from the US Department of Health and Human Services' Office of HIV/AIDS Policy, the National Institutes of Health and the Centers for Disease Control, as well leading AIDS practitioners and treatment activists, has announced the decision to designate Fortovase a "preferred" therapy in revised federal-government AIDS-treatment guidelines.

E. Outlook

Although protease inhibitors have been on the market for only 4 years, they have already had a considerable impact on the treatment of HIV infections. Studies such as NV15355 (SLATER et al. 1998), SUN (SENSION et al. 1998) so-named because it was conducted in Florida and California, CHEESE (a Comparative trial in HIV-infected patients Evaluating Efficacy and Safety of saquinavir-Enhanced oral formulation and indinavir given as part of a triple therapy) (BORLEFFS et al. 1998) and Study of Protease Inhibitor Combination in Europe (OPRAVIL et al. 1998) have shown that highly active antiretroviral therapy comprising Fortovase together with two reverse-transcriptase inhibitors can reduce plasma viral load below the level of quantification (400 copies/ml, Amplicor assay) in the majority (80%, 90%, 100% and 70%, respectively) of study participants. Such reductions in viral load have been shown to be associated with significant clinical benefit in terms of delayed disease progression and reduced mortality rates (EGGER et al. 1997; PALLELA et al. 1998). Thus, whilst it cannot be claimed that HAART can cure HIV disease, it is clear that such treatment can transform a progressive and ultimately fatal disease into a manageable chronic condition, and can offer the prospect of enhanced survival and a relatively normal life.

Recent studies (CHUN et al. 1997; FINZI et al. 1997; WONG et al. 1997) have shown that latently infected CD4$^+$ T-cells carrying integrated proviral DNA can persist and are capable of producing infectious virions upon activation in vitro, even when viral replication is undetectable. Whether the conditions used to reactivate virus from latently infected cells in vitro are representative of the situation in vivo is not known, but these studies do show that caution must be exercised before HAART is withdrawn from patients who have no evidence of ongoing viral replication. It may be that a "cure" can only be achieved if latent reservoirs of replication-competent provirus are also eliminated. Although HAART has recently been shown to bring about an increase in CD4$^+$ T-cell counts (KAUFMANN et al. 1998), the effectiveness of these new cells remains to be demonstrated, and the current goal of HIV therapy must remain the maximal suppression of viral load for the longest possible duration.

References

Aboulker JP, Swart AM (1993) Preliminary analysis of the Concorde trial. Lancet 341:889–890

Barré-Sinoussi F, Chermann JC, Rey F, Nugeyre MT, Chamaret S, Gruest J, Axler-Blin C, Vézinet-Brun F, Rouzioux C, Rozenbaum W, Montagnier L (1983) Isolation of a T-lymphotropic retrovirus from a patient at risk for acquired immune deficiency syndrome (AIDS). Science 220:868–871

Borleffs JC on behalf of the CHEESE Study Team (1998) First comparative study of saquinavir soft-gel capsules vs indinavir as part of triple therapy regimen (CHEESE) (abstract). Proceedings of the 5th Conference on Retroviruses and Opportunistic Infections, Chicago 1–5 Feb 1998, no.387b

Boucher C (1996) Rational approaches to resistance: using saquinavir. AIDS 10[Suppl 1]:S15–S19

Broadhurst AV, Roberts NA, Ritchie AJ, Handa BK, Kay C (1991) Assay of HIV-1 proteinase: a colorimetric method using small peptide substrates. Anal Biochem 193: 280–286

Carr A, Marriott D, Field A, Vasak E, Cooper DA (1998) Treatment of HIV-1-associated microsporidiosis and cryptosporidiosis with combination antiretroviral therapy. Lancet 351:256–261

Chun T-W, Stuyver L, Mizell SB, Ehler LA, Mican JAM, Baseler M, Lloyd AL, Nowak MA, Fauci AS (1997) Presence of an inducible HIV-1 latent reservoir during highly active antiretroviral therapy. Proc Natl Acad Sci USA 94:13193–13197

Clavel F, Guyader M, Guetard D, Salle M, Montagnier L, Alizon M (1986) Molecular cloning and polymorphism of the human immune deficiency virus type 2. Nature 324:691–695

Connell EV, Hsu M-C, Richman DD (1994) Combinative interactions of a human immunodeficiency virus (HIV) tat antagonist with HIV reverse transcriptase inhibitors and an HIV proteinase inhibitor. Antimicrob Agents Chemother 38: 348–352

Craig JC, Duncan IB, Hockley D, Grief C, Roberts NA, Mills JS (1991a) Antiviral properties of Ro 31-8959, an inhibitor of human immunodeficiency virus (HIV) proteinase. Antiviral Res 16:295–305

Craig JC, Grief C, Mills JS, Hockley D, Duncan IB, Roberts NA (1991b) Effects of a specific inhibitor of HIV proteinase (Ro 31-8959) on virus maturation in a chronically infected promonocytic cell line (U1). Antiviral Chem Chemother 2:181–186

Craig JC, Duncan IB, Whittaker LN, Roberts NA (1993a) Antiviral synergy between inhibitors of HIV proteinase and reverse transcriptase. Antiviral Chem Chemother 4:161–166

Craig JC, Whittaker L, Duncan IB, Roberts NA (1993b) In vitro resistance to an inhibitor of HIV proteinase (Ro 31-8959) relative to inhibitors of reverse transcriptase (AZT and TIBO). Antiviral Chem Chemother 4:335–339

Craig JC, Whittaker LN, Duncan IB, Roberts NA (1994) In vitro anti-HIV and cytotoxicological evaluation of the triple combination: AZT and ddC with HIV proteinase inhibitor saquinavir (Ro 31-8959). Antiviral Chem Chemother 5:380–386

Craig C, O'Sullivan E, Cammack N (1998) Increased exposure to the HIV protease inhibitor saquinavir (SQV) does not alter the nature of key resistance mutations (abstract). Proceedings of the 5th Conference on Retroviruses and Opportunistic Infections, Chicago 1–5 Feb 1998, no. 398

Eagling VA, Back DJ, Barry MG (1997) Differential inhibition of cytochrome P450 isoforms by the protease inhibitors ritonavir, saquinavir and indinavir. Br J Pharmacol 44:190–194

Egger M, Hirschel B, Francioli P, Sudre P, Wirz M, Flepp M, Rickenbach M, Malinverni R, Vernazza P, Battegay M (1997) Impact of new antiretroviral combination therapies in HIV infected patients in Switzerland: prospective multicentre study. Swiss HIV Cohort Study. BMJ 315:1194–1199

Ermolieff J, Lin X, Tang J (1997) Kinetic properties of saquinavir-resistant mutants of human immunodeficiency type 1 protease and their implications in drug resistance in vivo. Biochemistry 36:12364–12370

Farrar G, Mitchell AM, Hooper H, Stewart F, Malcolm SL (1994) Prediction of potential drug interactions of saquinavir (Ro 31-8959) from in vitro data. Br J Clin Pharmacol 38:162P

Finzi D, Hermankova M, Pierson T, Carruth LM, Buck C, Chaisson RE, Quinn TC, Chadwick K, Margolick J, Brookmeyer R, Gallant J, Markowitz M, Ho DD, Richman DD, Siliciano RF (1997) Identification of a reservoir for HIV-1 in patients on highly active antiretroviral therapy. Science 278:1295–1300

Fitzsimmons ME, Collins JM (1997) Selective biotransformation of the human immunodeficiency virus protease inhibitor saquinavir by human small-intestinal cytochrome P450 3A4. Potential contribution to high first-pass metabolism. Drug Metab Dispos 25:256–266

Galpin S, Roberts NA, O'Connor T, Jeffries DJ, Kinchington D (1994) Antiviral properties of the HIV-1 proteinase inhibitor Ro 31-8959. Antiviral Chem Chemother 5:43–45

Gazzard BG, Moyle GJ, Weber J, Johnson M, Bingham JS, Brettle R, Churchill D, Fisher M, Griffin G, Jefferies D, King E, Gormer R, Lee C, Pozniak A, Smith JR, Tudor-Williams G, Williams I (1997) British HIV Association guidelines for antiretroviral therapy of HIV seropositive individuals. Lancet 349:1086–1092

Graves MC, Lim JJ, Heimer EP, Kramer R (1988) An 11-KD form of human immunodeficiency virus protease expressed in Escherichia coli is sufficient for enzymatic activity. Proc Natl Acad Sci USA 85:2449–2453

Hall MJ, Duncan IB (1988) Antiviral drug and interferon combinations. In: Field HJ (ed) Antiviral agents: the development and assessment of antiviral chemotherapy. CRC, Boca Raton, 2:29–84

Haubrich R, Lalezari J, Follansbee SE, Gill M J, Hirsch M, Richman D, Mildvan D, Burger HU, Beattie D, Donatacci L, Salgo MP (1998) Improved survival and reduced clinical progression in HIV-infected patients with advanced disease treated with saquinavir plus zalcitabine. Antiviral Ther 3:33–42

Holmes IN, Mahmood N, Karpas A, Petrik J, Kinchington D, O'Connor T, Jeffries DJ, Desmyter J, De Clercq E, Pauwels R, Hay A (1991) Screening of compounds for activity against HIV: a collaborative study. Antiviral Chem Chemother 2:287–293

Ives KJ, Jacobsen H, Galpin SA, Garaev MM, Dorrell L, Mous J, Bragman K, Weber JN (1997) Emergence of resistant variants of HIV in vivo during monotherapy with the protease inhibitor saquinavir. J Antimicrob Chemother 39:771–779

Jacobsen H, Ahlborn-Laake L, Gugel R, Mous J (1992) Progression of early steps of human immunodeficiency virus type 1 replication in the presence of an inhibitor of viral protease. J Virol 66:5087–5091

Jacobsen H, Yasargil K, Winslow DL, Craig JC, Krohn A, Duncan IB, Mous J (1995) Characterisation of human immunodeficiency virus type 1 mutants with decreased sensitivity to proteinase inhibitor Ro 31-8959. Virology 206:527–534

Jacobsen H, Hanggi M, Ott M, Duncan IB, Owen S, Andreoni M, Vella S, Mous Jan (1996) In vivo resistance to a human immunodeficiency virus type 1 proteinase inhibitor: mutations, kinetics, and frequencies. J Infect Dis 173:1379–1387

Johnson VA, Merrill DP, Chou T-C, Hirsch MS (1992) Human immunodeficiency virus type-1 (HIV-1) inhibitory interactions between protease inhibitor Ro 31-8959 and zidovudine, 2,3-dideoxycytidine or recombinant interferon αA against zidovudine-sensitive or -resistant HIV-1 in vitro. J Infect Dis 166:1143–1146

Kaufmann D, Pantaleo G, Sudre P, Telenti A (1998) CD4+ T-cell count in HIV-1-infected individuals remaining viraemic with highly active antiretoviral therapy (HAART). Swiss HIV Cohort Study. Lancet 351:723–724

Kitchen VS, Skinner C, Ariyoshi K, Lane EA, Duncan IB, Burckhardt J, Burger HU, Bragman K, Pinching AJ, Weber JN (1995) Safety and activity of saquinavir in HIV infection. Lancet 345:952–955

Kohl NE, Emini EA, Schlief WA, Davis LJ, Heimbach JC, Dixon RA, Scolnick EM, Sigal IS (1988) Active human immunodeficiency virus protease is required for viral infectivity. Proc Natl Acad Sci USA 85:4686–4690

Kramer RA, Schaber MD, Skalka AM, Ganguly K, Wong-Staal F, Reddy EP (1986) HTLV-III gag- protein is processed in yeast cells by the virus pol-protease. Science 231:1580–1584

Larder BA, Darby G, Richman DD (1989) HIV with reduced sensitivity to zidovudine (AZT) isolated during prolonged therapy. Science 243:1731–1734

Le Grice SFJ, Mills J, Mous J (1988) Active site mutagenesis of the AIDS virus protease and its alleviation by trans complementation. EMBO J 7:2547–2553

Lightfoote MM, Coligan JE, Folks TM, Fauci AS, Martin MA, Venkatesan S (1986) Structural characterization of reverse transcriptase and endonuclease polypeptides of the acquired immunodeficiency syndrome retrovirus. J Virol 60:771–775

Lillehoj EP, Salazar FHR, Mervis RJ, Raum MG, Chan HW, Ahmad N, Venkatesan S (1988) Purification and structural characterization of the putative gag-pol protease of human immunodeficiency virus. J Virol 62:3053–3058

Martin JA, Mobberley MA, Redshaw S, Burke A, Tyms AS, Ryder TA (1991) The inhibitory activity of a peptide derivative against the growth of simian immunodeficiency virus in C8166 cells. Biochem Biophys Res Comm 176:180

Mellors JW, Larder BA, Schinazi RF (1995) Mutations in HIV-1 reverse transcriptase and protease associated with drug resistance. Int Antiviral News 3:8–13

Merry C, Barry MG, Mulcahy F, Ryan M, Heavey J, Tija JF, Gibbons SE, Breckenridge AM, Back DJ (1997) Saquinavir pharmacokinetics alone and in combination with ritonavir in HIV-infected patients. AIDS (London) 11:F29–F33

Merry C, Barry MG, Mulcahy FM, Back DJ (1998) Saquinavir pharmacokinetics alone and in combination with nelfinavir in HIV infected patients (abstract). Proceedings of the 5th Conference on Retroviruses and Opportunistic Infections, Chicago 1–5 Feb 1998, no. 352

Mitsuya H, Broder S (1987) Strategies for antiviral therapy in AIDS. Nature 325: 773–778

Mous J, Heimer EP, Le Grice SFJ (1988) Processing protease and reverse transcriptase from human immunodeficiency virus type 1 polyprotein in Escherichia coli. J Virol 62:1433–1436

Noble S, Faulds D (1996) Saquinavir: a review of its pharmacology and clinical potential in the management of HIV infection. Drugs 52:93–112

Opravil M on behalf of the SPICE Study Team (1998) Study of protease inhibitor combination in Europe (SPICE); saquinavir soft gelatin capsule (SQV-SGC) and nelfinavir in HIV infected individuals (abstract). Proceedings of the 5th Conference on Retroviruses and Opportunistic Infections, Chicago 1–5 Feb 1998, no. 394b

Palella FJ Jr, Delaney KM, Moorman AC, Loveless MO, Fuhrer J, Satten GA, Aschman DJ, Holmberg SD (1998) Declining morbidity and mortality among patients with advanced human immunodeficiency virus infection. HIV Outpatient Study Investigators. N Engl J Med 338:853–860

Parra R, Leal M, Delgado J, Macías J, Rubio A, Gómez F, Soriano V, Sanchez-Quijano A, Pineda JA, Lissen E (1998) Regression of invasive AIDS-related Kaposi's sarcoma following antiretroviral therapy. Clin Infect Dis 26:218–219

Pearl LH, Taylor WR (1987) A structural model for the retroviral proteases. Nature 329:351–354

Popovic M, Sarngadharan MG, Read E, Gallo RC (1984) Detection and continuous production of cytopathic retroviruses (HTLV-III) from patients with AIDS and pre-AIDS. Science 224:497–500

Poiesz BJ, Ruscetti FW, Gazdar AF, Bunn PA, Minna JD, Gallo RC (1980) Detection and isolation of type C retrovirus particles from fresh and cultured lymphocytes of a patient with cutaneous T-cell lymphoma. Proc Nat Acad Sci USA 77: 7415–7419

Ratner L, Haseltine W, Patarca R, Livak KJ, Starcich B, Josephs SF, Doran ER, Rafalski JA, Whitehorn EA, Baumeister K, Ivanoff L, Petteway SR[JR], Pearson ML, Lautenberger JA, Papas TS, Ghrayeb J, Chang NT, Gallo RC, Wong-Staal F (1985) Complete nucleotide sequence of the AIDS virus, HTLV-III. Nature 313:277–284

Roberts NA, Martin JA, Kinchington D, Broadhurst AV, Craig JC, Duncan IB, Galpin SA, Handa BK, Kay J, Kröhn A, Lambert RW, Merrett JM, Mills JS, Parkes KEB, Redshaw S, Ritchie AJ, Taylor DL, Thomas GJ, Machin PJ (1990) Rational design of peptide-based HIV proteinase inhibitors. Science 248:358–361

Roberts NA, Craig JC, Duncan IB (1992) HIV proteinase inhibitors. Biochem Soc Trans 20:513–516

Roberts NA, Craig JC, Sheldon J (1998) Resistance and cross-resistance with saquinavir and other HIV protease inhibitors: theory and practice. AIDS 12: 453–460

Sanchez-Pescador R, Power MD, Barr PJ, Steimer KS, Stempien MM, Brown-Shimer SL, Gee WW, Renard A, Randolph A, Levy JA, Dina D, Luciw PA (1985) Nucleotide sequence and expression of an AIDS-associated retrovirus (ARV-2). Science 227:484–492

Sarmati L, Nicastri E, el-Sawaf G, Ventura L, Salanitro A, Ercoli L, Vella S, Andreoni M (1997) J Med Virol 53:313–318

Schapiro JM, Winters MA, Stewart F, Efron B, Norris J, Kozal MJ, Merigan TC (1996) The effect of high-dose saquinavir on viral load and CD4[+] T-cell counts in HIV-infected patients. Ann Intern Med 124:1039–1050

Schotten C (1891) Ueber isatinblau, einen aus der verbindung von isatin und piperidin enstehenden farbstoff. Ber Deutsch Chem Gesellschaft 24:1366–1373

Sension M, Farthing C, Palmer Pattison T, Pilson R, Siemon-Hryczyk P (1998) Fortovase (Saquinavir soft gel capsule, SQV-SGC) in combination with AZT and 3TC in antiretroviral-naive HIV-1 infected patients (abstract). Proceedings of the 5th Conference on Retroviruses and Opportunistic Infections, Chicago, 1–5 Feb 1998, no. 369

Slater L on behalf of the NV15355 study group (1998) Activity of a new formulation of saquinavir in combination with two nucleosides in treatment naive patients (abstract). Proceedings of the 5th Conference on Retroviruses and Opportunistic Infections, Chicago 1–5 Feb 1998, no. 368

Styrt BA, Piazza-Hepp TD, Chikami GK (1996) Clinical toxicity of antiretroviral nucleoside analogs. Antiviral Res 31:121–135

Taylor DL, Brennan TM, Bridges CG, Kang MS, Tyms AS (1995) Synergistic inhibition of human immunodeficiency virus type 1 in vitro by 6-O-butanoyl-castanospermine (MDL 28574), in combination with inhibitors of the virus encoded reverse transcriptase and proteinase. Antiviral Chem Chemother 6:143–152

Toh H, Ono M, Saigo K, Miyata T (1985) Retroviral protease-like sequence in the yeast transposon. Nature 315:691–692

Vella S, Galluzzo C, Giannini G, Pirillo MF, Duncan I, Jacobsen H, Andreoni M, Sarmati L, Ercoli L (1996) Saquinavir/zidovudine combination in patients with advanced HIV infection and no prior antiretroviral therapy: CD_4^+ lymphocyte/plasma RNA changes, and emergence of strains with reduced phenotypic sensitivity. Antiviral Res 29:91–93

Veronese F di M, Copeland TD, DeVico AL, Rahman R, Oroszlan S, Gallo RC, Sarngadharan MG (1986) Characterization of highly immunogenic p66/p51 as the reverse transcriptase of HTLV-III/LAV. Science 231:1289–1291

Williams PEO, Muihead GJ, Madigan MJ, Mitchell AM, Shaw T (1992) Disposition and bioavailability of the HIV proteinase inhibitor, Ro 31-8959, after single doses in healthy volunteers. Br J Clin Pharmacol 34:155P–156P

Wong JK, Hezareh M, Gunthard HF, Havlir DV, Ignacio CC, Spina CA, Richman DD (1997) Recovery of replication-competent HIV despite prolonged suppression of plasma viremia. Science 278:1291–1295

Zingman BS (1996) Resolution of refractory AIDS-related mucosal candidiasis after initiation of didanosine plus saquinavir. N Engl J Med 334:1674–1675

CHAPTER 2

Clinical Experience with Human Immunodeficiency Virus Protease Inhibitors: Antiretroviral Results, Questions and Future Strategies

S. VELLA

A. Introduction

In the last 3 years, the availability of new drugs, in particular the human immunodeficiency virus (HIV) protease inhibitors (PIs) and new combinations to combat HIV infection, has translated into progressive clinical benefit for patients (HAMMER et al. 1997; PALELLA et al. 1998). We clearly entered an era of therapeutic success, with significantly reduced rates of opportunistic infections, hospitalization and mortality.

Recent reports described the reduction in mortality and the health improvement of HIV-infected patients seen in industrialized countries coincidental with the use and the introduction of PIs and more potent combinations. Exemplary for these publications, data by PALELLA et al. (1998), collected over 42 months from 1255 outpatients in eight U.S cities, show a dramatic reduction in morbidity and mortality among patients who had a disease marker of $CD4^+$ T-cell counts under $100 \mu l^{-1}$. The mortality declined from 29.4 per person-years in 1995 to 8.8 per person-years in the second quarter of 1997. The reductions were seen regardless of gender, race, age and risk factors for HIV transmission. In a "failure-rate" model the reductions in death and disease were clearly linked to the increasing use of PIs in antiviral combination therapy. In this analysis, increases of steps in the intensity of antiviral therapy (no therapy, monotherapy, combination without PI and combination with PI) were associated with stepwise reduction of morbidity and mortality. Combination with PI conferred the most benefit. The dramatic decrease in death rates is summarized in Fig. 1.

Comparable results had been reported by LALEZARI et al. (1996), BOULTON (1997), CHAISSONS et al. (1997), HOGG et al. (1997), MOCROFT et al. (1997), MOUTON et al. (1997), TORRES et al. (1997) and in the MMWR (Morbidity and Mortality Weekly Report, 1997). At least in countries that can afford the cost of the new antiretroviral therapies (CARPENTER et al. 1998), the perspective has changed from a view of HIV disease as inevitably fatal to a view of it as a disease that is potentially manageable for several decades (HAMMER and YENI 1998).

Following the discovery that the high replication rate of HIV is the leading pathogenetic force that drives the progression of HIV disease and the characterization of mechanisms of HIV resistance to antiretroviral drugs, new

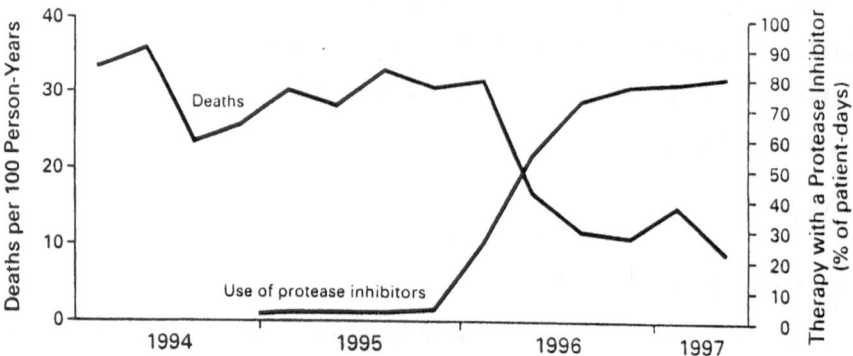

Fig. 1. Mortality and frequency of use of combination antiretroviral therapy, including a protease inhibitor among HIV-infected patients with fewer than 100 CD4⁺ T-cell counts per cubic millimeter. According toCalender Quarter, from January 1994 through June 1997 (from Palella et al. 1998)

principles of antiretroviral therapy were defined. We believe today that HIV infection is almost always harmful and that true long-term survival free of significant immune dysfunction is extremely rare. We also know that therapy-induced inhibition of HIV replication predicts clinical benefit and that combination therapy that suppresses HIV replication to undetectable levels (possibly below 50copies RNA/ml) can delay or prevent the emergence of drug-resistant variants.

B. Activity on Immunological and Virological Markers, and Clinical Efficacy

PIs had been evaluated initially as monotherapy but quickly appeared in combination regimens. Clinical trials have shown that disease markers such as CD4⁺ T-cell counts and viral-load responses (Saag et al. 1996; Mellors et al 1997), although evident in monotherapy, are sustained only when PIs are combined with other antiretroviral drugs (Vanhove et al. 1996). This effect is more evident in antiretroviral-naive patients with relatively preserved CD4 counts, where it can be expected that starting treatment with a triple combination of a PI and two RTIs (reverse-transcriptase inhibitors) will produce significant changes in CD4 counts (average increases between 100cells/mm³ and 150cells/mm³) and viral load (decline below the threshold of HIV-RNA detection) in 60–80% of treated patients (Hammer et al. 1997; Gulick et al. 1998). Both responses appear to be sustained for at least 2years of treatment, a result previously not achieved with any other antiretroviral regimens. Although in more-advanced and previously treated patients CD4 and RNA responses are less striking, the strategy of adding one PI to current nucleoside therapy has proven clinically effective in such patients in delaying disease progression and increasing survival. However, according to the more recent therapeutic guidelines (CARPENTER et al. 1998; U.S. DEPARTMENT OF HEALTH AND HUMAN SERVICES AND THE HENRY J. KAISER FAMILY FOUNDATION 1998), adding a new antiretroviral drug to current regimens is today considered a suboptimal strategy, while starting from the beginning with completely suppressive regimens should be considered the treatment of choice (PRINS et al. 1998).

Indeed, at the beginning of the PI era (LAM et al. 1994; SAHAI et al. 1996; VANHOVE et al. 1996) significant, although short-term, increases in progression-free time and survival have been observed by adding the PI ritonavir to current antiretroviral treatments in patients below 100 CD4 cells/mm³ using a triple combination of the PI indinavir, and the RTIs zidovudine and lamivudine (compared with zidovudine and lamivudine only) in patients with CD4⁺ T-cell counts below 200/mm³; (for names and structures of PIs see Fig. 1 of Chap. 3 of this volume) The delay in progression was also generally associated with positive changes in quality-of-life scores. These findings reflect the high antiviral activity of PI compared with NRTIs (non-nucleotide reverse-transcriptase inhibitors), which have generally proved less effective in advanced HIV disease (HAMMER et al. 1997; GULICK et al. 1998). They also indicate that even minor changes in viral load and CD4 have a significant, although not durable, impact on clinical outcome. The better CD4 and viral RNA responses observed in patients with less severe immune deficiencies suggested that the possible clinical benefits of starting treatment with triple combination regimens can be significantly greater and more durable (DEEKS 1997).

The articles included in this chapter will address some of the issues raised by the dramatic change in HIV disease management, particularly the problem of resistance to PI and the design of clinical trials in the era of maximally suppressive treatment in view of the changing face of opportunistic infections, the revised concept of HIV eradication and, finally, the current and future strategies of antiretroviral research. Recent results of clinical research on new agents and new combinations are also discussed in Chap. 3 of this volume.

C. Clinical Implications of Resistance to PIs

The resistance profile of all drugs being used in a specific combination regimen may have important consequences for the long-term outcome of the overall therapeutic strategy because of the limitations in the sequential use of antiretrovirals if a common resistance profile is present (RICHMAN 1993; DEEKS et al. 1998). As for other classes of antiretrovirals, reduced sensitivity has been reported for all available PIs (discussed in Chaps. 2 and 3). Overall, the patterns of mutations for PI appear to be more complex than those observed for RTIs, with a high natural polymorphism, a higher number of sites involved and higher variability in the temporal patterns and in the combinations of mutations leading to "phenotypic" resistance.

Some mutations reduce inhibitor-enzyme binding; other mutations have a "compensatory" effect by improving the "fitness" of the virus in addition to causing disadvantageous changes in the functionality of the protease enzyme. Compensatory mutations may include those that codify for new changes in the protease enzyme, mutations that drive the increased production of the "less-fit" enzyme, and even mutations that modify the protein cleavage sites (see also Chap. 2).

To cope with the problem of PI resistance a final, hypothetical and still unproved issue is the possibility of increasing antiviral efficacy by using protease/protease-combination regimens based on mutually counteracting, drug-induced mutations that could convert the unavoidable selection of mutant viruses into an at least partially favorable phenomenon. This possibility is currently addressed by the development of compounds designed to act on other PI-resistant viruses.

The dynamics of HIV-1 replication in vivo strongly suggest that aggressive antiretroviral therapy be started very early to minimize the negative consequences of HIV replication. Indeed, the best way to accomplish maximal suppression of virus replication and to minimize the risk of drug-resistance is to use potent combinations in individuals with no prior history of antiretroviral use (PRINS et al. 1998). However, very early intervention is unfortunately not possible for the majority of our patients. Careful therapeutic intervention should therefore be designed with the aim of keeping HIV at undetectable or minimal levels indefinitely with no negative consequences in terms of disease progression, transforming HIV disease into a chronic disease with minimal negative impact on the duration and quality of life of the infected persons. Rationale criteria must then be adopted to select drug combinations with the best expectancy of long-term efficacy, but also considering the possibility of preserving subsequent therapeutic options if the initial choice fails to achieve its desired results (DEEKS et al. 1998). Because any choice will impact subsequent options, whenever possible therapy should be initiated bearing in mind a pre-defined salvage antiretroviral regimen. Unfortunately, new resistance mutations continue to be discovered, and many of the promising new drugs seems to be better for first-line use and cannot, therefore, address the issue of the many patients that are failing aggressive antiretroviral regimens.

As far as resistance is concerned, a number of rules should be followed in planning therapeutic strategies: (1) combinations or sequential use of drugs which share clear cross-resistance should be avoided, (2) potent antiretroviral drugs to which HIV readily develops high-level resistance should not be used in regimens that are expected to yield incomplete suppression of viral replication and (3) decisions to alter antiretroviral therapy need to be made carefully, because the number of effective drugs available is still very limited. In fact, an increase in HIV-RNA levels in persons receiving fully suppressive antiretroviral therapy can be due to a number of different factors, one of which is represented by the lack of full adherence to a particular drug combination.

In fact, as with other antiretrovirals, but particularly with PIs, the development of resistance is strictly linked to compliance: both dose reductions and missing doses increase the risk of rapid development of resistance by reducing the pressure on viral replication. As a consequence, full dosage should always be maintained and tolerability should be carefully checked to avoid loss of efficacy dependent on low adherence to the prescribed regimen.

D. Place of PIs in Current Treatment Strategies

After the recently published reports on the declining HIV/acquired immune deficiency syndrome(AIDS) morbidity and mortality, the overall perspective has finally changed from viewing HIV as an inevitably fatal disease to one that is potentially manageable. All treatment decisions should be based on the understanding of HIV virology and pathogenesis: the ongoing HIV replication leads to immune-system damage and progression to AIDS, and HIV is the driving force of the pathogenesis of the diseases.

An important impact on treatment strategies has been the revisiting of the concept of HIV eradication. This hypothesis (Ho et al. 1995; PERELSON et al. 1996) was based on the assumption that cellular reservoirs of HIV have a short half-life and that, if antiretroviral therapy could completely stop HIV replication and new rounds of infection, therapy could possibly be discontinued after all infected cells had died (after approximately 3 years). Recent studies showed that the virus indeed persists in memory-resting cells in the face of undetectable plasma levels of HIV RNA (CHUN et al. 1997; WONG et al. 1997). The longevity of these cells is actually unknown, but may range from months to years. In practice, HIV eradication is unlikely with current regimens, and patients should therefore be prepared for a long-term commitment to antiretroviral therapy. Investigations on the possibility of eliminating these long-lived cells are underway.

We currently individualize treatment decisions according to the level of risk indicated by plasma HIV RNA and CD4+ T-cell counts. Although some expert opinions suggest the need to start treatment for any patient that has detectable RNA, others consider the many problems that we are facing today with actual regimens (FLEXNER 1998).

Available regimens are indeed complex; they have short-term and long-term toxicities that can impact quality of life and they may induce cross-resistant viruses to emerge. In the face of these problems, the potential benefits and the possible risks of very early initiation must be weighed carefully.

Current guidelines suggest (CARPENTER et al. 1998; U.S. DEPARTMENT OF HEALTH AND HUMAN SERVICES AND THE HENRY J. KAISER FAMILY FOUNDATION 1998) that the balancing of these two factors should lead to the decision to treat any symptomatic infection, regardless of plasma HIV RNA or CD4+ T-cell counts, and all asymptomatic patients with a definitive risk of progression (those with more than 10,000 copies of HIV RNA in their plasma or a CD4+ T-cell count below 500/mm^3). For persons at low risk of progression, therapy may be deferred, particularly if the patient is not committed to complex regimens.

I. How to Start Antiretroviral Therapy and When to Make the Decision to Start Treatment

We know today that therapy-induced inhibition of HIV replication predicts clinical benefit and that combination therapy that suppress HIV replication to undetectable levels (below cut-off of the most sensitive assays) can delay or even prevent the emergence of drug-resistant viral variants. Today, we have 12 antiretroviral drugs that are licensed in Western countries, and more are coming out in clinical trials. They all belong to one of the three classes: NRTIs, non-nucleoside reverse transcriptase inhibitors (NNRTIs) and PIs).

The most widely used regimen is a combination of one PI with two NRTIs; this combination has the advantage of being the one with which we have the most experience and clearly represents the first choice. It can be applicable to all viral levels, but regimens are complex (adherence may be a challenge), and the emergence of drug-resistant strains may limit the effectiveness of future treatment with other PIs.

Another concrete possibility is the use of an NNRTI combined with two NRTIs. The possible advantages of this combination include the deferral of PI use and the lower daily pill burden, but potential disadvantages include the chance of not having a second chance with NNRTIs in case of failure.

Other possibilities include the use of a combination of 2PI with or without a NRTI. This combination exploits the pharmacokinetic interactions of PI and may have very high potency. Disadvantages may include the potential for broad PI resistance, high pill count and long-term toxicities.

Finally, a combination of all three classes of antiretrovirals may possess a very high potency and attack HIV on multiple targets. However, multi-class resistance may emerge in case of failure and compromise all future options. Furthermore, adherence may be a true challenge for the patient.

In conclusion, in the choice of the initial regimen, a number of important issues remain to be clarified: is a PI-containing regime always preferable? Can we use NNRTIs as a first line of treatment? Should we adjust the potency of the starting regimen according to the disease stage? Are regimens that attack the virus at a single target, better than regimens that attack different targets? These questions can only be answered by long-term strategic clinical trials, such as the European INITIO trial, or the AIDS Clinical Trial Group (ACTG) 384 and ACTG 388 trials.

E. Future Directions

Because no trial information is available to clearly indicate the optimal timing of the therapeutic intervention, treatment decisions should be individualized according to the level of risk indicated by plasma HIV-RNA levels and CD4+ T-cell counts. However, for persons at low risk of progression, therapy initiation can be deferred; a relatively conservative approach emerged because the potential to control HIV-1 replication in the long term is hampered by the limitations of current regimens, including the problems of adherence, incomplete response rates, cross-resistance and long-term toxicities.

For the future, important issues should be addressed by viro-pathogenic research. In particular:

- Refining the knowledge of viral turnover (in plasma and cell/tissue reservoirs) and of cell turnover, particularly of latently infected cell reservoirs, and better defining the replication competence of residual virus in persons who are maximally suppressed
- Elucidating the relative roles of viral and host determinants of outcome
- Understanding and modulating the pathogenic events following primary infection
- Defining the mechanisms of immune reconstitution and better defining the role of HIV-specific immunity

The most important issue in clinical practice is addressing adherence. Because adherence is directly related to the convenience of antiretroviral regimens, the development of lower-frequency schedules, which may exploit the pharmacokinetic interactions between some of the available or future drugs, are eagerly awaited.

Another possible strategy for increasing long-term adherence to antiretroviral therapy would be to start with an aggressive regimen (induction) and follow up with a less-intense regimen (maintenance). Recently published initial trial data from the US ACTG 343 trial (HAVLIR et al. 1998; REIJERS et al. 1998) and the French Trilege trial (PIALOUX et al. 1998) do not support this strategy. However, additional studies aimed at testing this hypothesis should be undertaken, since the negative results of these two trials could be due to their design.

As these controlled clinical trials are advancing, so too is research for the development of new compounds. The development of new compounds includes the NRTIs abacavir, cis-5-fluoro-1-[2-(hydroxymethyl)-1,3-oxathiolan-5-yl]cytosine (2′-deoxy-3′-thia-5-fluorocytosine) (FTC), adefovir and 9-R-(2-phosphonomethoxypropyl)adenine (PMPA), the NNRTI MCK-443 and the PIs amprenavir, tipranavir, ABT-378 and BMS-232632 (see also Chap. 3)

These new drugs may offer several advantages over existing antiretroviral agents. It is hoped several of these compounds will be available in the near future and will retain the profiles presently seen in limited clinical work. Examples of the compounds are: FTC, which is from the same nucleoside series as FTC but is believed to be more potent; PMPA, being investigated in a topical gel formation for prevention of HIV-1 transmission; MCK-442, which appears to be synergistic with NRTIs, NNRTIs and PIs; the PI tipranavir, which is being investigated in patients who are both naive and resistant to presently available PI therapy, and ABT-378 which may represent a step forward in overcoming HIV resistance because of the very high plasma levels that can be obtained with this compound in combination with low doses of ritonavir.

While some research is focused on compound development, still other clinical-research objectives should include:

- Exploring the potential of the new monitoring tools such as ultrasensitive HIV-1-RNA assays and resistance testing (HIRSCH et al. 1998)
- Understanding how to better manage drug failure and resistance, and defining the proper sequence of drugs and combinations
- Testing new strategic approaches, including induction maintenance and intensification
- Defining the potential role of immunomodulatory therapy in the era of highly active antiretroviral therapy (ANGEL et al. 1997; AUTRAN et al. 1997; KELLEHER et al. 1997)
- Defining the consequences of potential long-term toxicities
- Developing a better understanding of the determinants of adherence

Research is continuing, but it appears that increasingly effective agents and strategies will become available to clinicians and thus continue the benefit of the first really effective therapy against HIV. The end results should be improvements in care that should translate into better health and improved quality of life for people living with HIV infection.

References

Angel JB, Parato K, Kumar A, et al. (1997) Rapid improvement in cell mediated immune function with initiation of ritonavir plus saquinavir in HIV immune deficiency (abstract 33). 4th Conference on Retroviruses and Opportunistic Infections, Washington

Autran B, Mathez D, Carcelain G, et al. (1997) Dynamics of the CD4T helper cell reconstitution after combined antiretroviral therapies (abstract 34). 4th Conference on Retroviruses and Opportunistic Infections, Washington

Boulton A (1997) HIV trial stopped early after good results. BMJ 314:699

Carpenter CCJ, Fischl MA, Hammer SM, et al. (1998) Antiretroviral therapy for HIV infection in 1998: updated recommendations of the International AIDS Society-USA panel. JAMA 280:78–86

Chaissons MA, Berensen L, Li W, Schwartz S, Mojica B, Hamburg M (1997) Declining AIDS mortality in New York city (abstract 133). 4th Conference on Retroviruses and Opportunistic Infections, Washington

Chun T-W, Carruth L, Finzi D, Shen X, DiGiuseppe J, Taylor H, et al. (1997) Quantification of latent tissue reservoirs and total body viral load in HIV-1 infection. Nature 387:183–188

Cox SR, Ferry JJ, Batts DH, et al. (1997) Delavirdine and marketed protease inhibitors: pharmacokinetic interaction studies in healthy volunteers (abstract 372). 4th Conference on Retroviruses and Opportunistic Infections, Washington ••

Deeks SG (1997) HIV-1 protease inhibitors: a review for clinicians. JAMA 277:145–149

Deeks SG, Grant RM, Beatty G, Horton C, Detmer J, Eastman S (1998) Activity of ritonavir plus saquinavir-containing regimen in patients with virologic evidence of indinavir or ritonavir failure. AIDS 12:F97–F102

Flexner C (1998) HIV-protease inhibitors. N Engl J Med 388:1281–1292

Gulick RM, Mellors JW, Havlir D, et al. (1998) Simultaneous vs sequential initiation of therapy with indinavir, zidovudine, and lamivudine for HIV-1 infection: 100-week follow-up. JAMA 280:35–41

Hammer S, Yeni P (1998) Antiretroviral therapy: where are we? AIDS 12[Suppl A]:S181–S188

Hammer SM, Squires KE, Hughes MD, et al. (1997) A controlled trial of two nucleoside analogues plus indinavir in persons with human immunodeficiency virus infection and CD4$^+$ T-cell counts of 200/μL or less. N Engl J Med 337:725–733

Havlir DV, Marschner IC, Hirsch MS, et al. (1998) Maintenance antiretroviral therapies in HIV-infected subjects with undetectable plasma HIV RNA after triple drug therapy. N Engl J Med 339:1261–1268

Hirsch MS, Conway B, D'Aquila RT, et al. (1998) Antiretroviral drug resistance testing in HIV infection of adults: implications for clinical management. JAMA 279:1984–1991

Hogg RS, O'Shaughnessy MV, Gaterie N, et al. (1997) Decline in death from AIDS due to new antiretrovirals. Lancet 349:1294–1299

Kelleher AD, Zaunders J, Sewell W, et al. (1997) Increased proliferative and cytokine responses following ritonavir therapy; relative contribution of lymphocyte subsets (abstract 536). 4th Conference on Retroviruses and Opportunistic Infections, Washington

Lalezari J, Haubrich R, Burger HU, et al. (1996) Improved survival and decreased progression of HIV in patients treated with saquinavir plus zalcitabine (abstract LB.B 6033). XI International Conference on AIDS, Vancouver

Lam PY, Jadhav PK, Eyermann CJ, et al. (1994) Rational design of potent, bio-available, non peptide cyclic ureas as HIV protease inhibitors. Science 263:380–384

Mocroft AJ, Lundgren JD, dármino-Montforte A, Ledergerber B,Barton SE, Vella S, Katlama C, Gerstoft J, Pedersen C, Phillips AN (1997) Survival of AIDS patients according to type of AIDS-defining event. The AIDS in Europe Study Group. Int J Epidemiol 26:400–407

Mouton Y, Catiert F, Dellaminica P, et al. (1997) Dramatic cut in AIDS defining events and hospitalisation for patients under protease inhibitors (PI) and tritherapie (TTT) in 9 AIDS reference centers (ARC) and 7.391 patients (abstract 208). 4th Conference on Retroviruses and Opportunistic Infections, Washington

Palella FJ, Delaney KM, Moorman AC, et al. (1998) Declining morbidity and mortality among patients with advanced human immunodeficiency virus infection. N Engl J Med 338:853–860

Perelson AS, Essunger P, Cao Y, et al. (1997) Decay characteristics of HIV-1-infected compartments during combination therapy. Nature 387:188–191

Pialoux G, Raffi F, Brun-Vezinet F, et al. (1998) A randomized trial of three maintenance regimens given after three months of induction therapy with zidovudine, lamivudine and indinavir in previously untreated HIV-1 infected patients. N Engl J Med 339:1269–1277

Prins J, Jurriaans S, Roos M, de Wolf F, Miedema F, Lange J (1998) An attempt at maximally suppressive anti-HIV therapy (abstract 385). 5th Conference on Retroviruses and Opportunistic Infections, Chicago

Reijers MME, Weverling GJ, Ten Kate RW, Frissen PMJ, de Wolf F, Scmuitemaker M, Lange JMA (1998) ADAM study: induction-maintenance therapy in HIV-1 infection: early results. 12th World AIDS Conference, Geneva

Richman DD (1993) Resistance of clinical isolates of human immunodeficiency virus to antiretroviral agents. Antimicrob Agents Chemother 37:1207–1213

Sahai J (1996) Risks and synergies from drug interactions. AIDS 10[Suppl 1]: S21–25

Torres RA, Barr M (1997) Impact of combination therapy for HIV infection on patient census. N Engl J Med 336:1531–1532

U.S. Department of Health and Human Services and the Henry J. Kaiser Family Foun-
    dation (1998) Guidelines for the use of antiretroviral agents in HIV-infected adults
    and adolescents. MMWR Morb Mortal Wkly Rep 47:43–82
Vanhove GF, Schapiro JM, Winters MA, et al. (1996) Patient compliance and drug
    failure in protease inhibitor monotherapy. JAMA 276:1955–1956
Wong J, Hezareh M, Gunthard HF, et al. (1997) Recovery of replication-competent
    HIV despite prolonged suppression of plasma viremia. Science 278:1295–1300

CHAPTER 3

The Nature of Resistance to Human Immunodeficiency Virus Type-1 Protease Inhibitors

M. Valliancourt, W. Shao, T. Smith, and R. Swanstrom

A. Introduction

The availability of highly potent human immunodeficiency virus (HIV)-1-protease inhibitors has revolutionized both our ability to treat people infected with HIV-1 and our view of how to use antiretroviral agents. Some clinical benefit was obtained with the less potent nucleoside analogs, but resistance almost always appeared and clinical progression resumed. With the availability of potent protease-inhibitor therapies, it has become possible to suppress virus replication to undetectable levels of viral RNA in blood plasma (virus load). However, strong therapy that does not achieve suppression leads to resistance, even to the potent protease inhibitors. Thus, the goal of therapy is to combine drugs so that the total potency is sufficient to suppress detectable virus.

The selection for resistance occurs because of residual virus replication in the presence of the inhibitor. Variants, either pre-existing or appearing because of residual replication, that replicate more efficiently than the wild-type virus in the presence of the drug increase in their percentage of the population. The available evidence indicates that this process of mutation and selection is repeated sequentially until the virus load returns to its pre-therapy level. In this review, we will examine the consequences of failing to suppress virus replication, which leads to the development of resistance. The discussion will cover the genetic markers of resistance seen after selection in vitro or in vivo, the biological and biochemical consequences of resistance and the clinical implications of therapy failure after treatment with a potent protease inhibitor.

B. Selection for Resistance: in Vitro and in Vivo Comparison

The identification and classification of the viral protease as being in the family of aspartic proteinases led to the design of specific inhibitors related to this class of enzymes. A key feature of the inhibitor-design strategy of renin inhibitors proved useful: i.e. the addition of a hydroxyl group at the scissile bond, mimicking the tetrahedral transition-state intermediate. Even though many different early inhibitors were potent, the hydroxyl group presented in the form of hydroxyethylene, hydroxyethylamine, or hydroxyethylamino

	Saquinavir	Ritonavir	Indinavir	Nelfinavir	Amprenavir
L10	X		X	X	X
K20		X	X		X
D30				X	
M36	X	X		X	
M46	X	X	X		X
G48	X				
I50					X
I54		X	X		
L63	X		X	X	
A71	X	X	X	X	
V82		X	X		X
I84	X	X	X		X
L90	X	X	X		

Fig. 1. Summation of residues that are frequently mutated after therapy failure with human immunodeficiency virus-1 protease inhibitors. Data for sequence changes after treatment with ritonavir are taken from MOLLA et al. (1996). Data for sequence changes after treatment with indinavir are taken from CONDRA et al. (1996). Data for sequence changes after treatment with saquinavir are included from JACOBSEN et al. (1996), SCHAPIRO et al. (1996) and Swanstrom (unpublished data). Data for sequence changes after treatment with nelfinavir are taken from MARKOWITZ et al. (1998). Data for sequence changes after treatment with amprenavir are taken from DE PASQUALE et al. (1998)

sulfonamide was pursued and resulted in inhibitors that are now approved or in clinical evaluation.

While tight binding is enhanced through direct interaction of the hydroxyl group with the aspartic acids, specificity of the inhibitor is achieved by specific interactions at each of the binding pockets. The structure of ritonavir (Norvir, Abbott) evolved from earlier C_2-symmetric inhibitors, while saquinavir (Invirase/Fortovase, Roche), indinavir (Crixivan, Merck) and nelfinavir (Viracept, Agouron) share other common features (for names and structures see Chap. 4, Fig. 1). Indinavir and nelfinavir both have a proline-like structure similar to saquinavir at the P1′ position and a *tert*-butyl group at the P2′ position. The *R* stereochemistry of the hydroxyl is maintained for all three. The hydroxyl of ritonavir is of the *S* configuration flanked by two phenyl groups at the P1–P1′ position. All of the compounds have a phenyl group or a phenylthio group (nelfinavir) at the P1 position. According to the described structural similarities between these drugs, one might predict that cross-resistance could develop.

Different approaches have been used to evaluate resistance associated with anti-HIV drugs. First, identification of mutations arising after selection in vitro has been used by DNA-sequence analysis of the HIV protease gene *pro*. Once identified, viruses carrying relevant mutations are tested to determine changes in IC_{50}/IC_{95} (50% and 95% viral growth inhibitory concentrations, respectively) and compared to the parental strain. An increase in the IC_{50} demonstrates reduced sensitivity (resistance). A second approach is to clone the *pro* gene and produce recombinant protease carrying the relevant mutations to be tested. Changes in the inhibition constant (K_i) reflect the effect of such mutations on the sensitivity of the enzyme to inhibition. The K_i changes and IC_{50}/IC_{95} changes should, in theory, directly correlate. However, such a direct correlation is not always observed (KLABE et al. 1998).

The early availability of C_2-symmetric inhibitors (ERICKSON et al. 1990) led to their use in initial selections of resistance. OTTO et al. (1993) reported first that V82A-mutant viruses were six- to eightfold less sensitive to P9941, the inhibitor used to select the mutant. They also reported that recombinant V82A protease is less sensitive to this compound. Other groups selected resistant viruses using A-77003 and reported virus stocks with 10- to 30-fold resistance (HO et al. 1994; KAPLAN et al. 1994; TISDALE et al. 1995). Molecular clones of viruses with mutations at positions 8, 32 and 82 were produced and showed IC_{50} increases. KAPLAN et al. (1994) also measured K_i changes in the range of 7- to 50-fold for active-site mutants V32 and V82. An I84V mutation was also reported for a non-symmetrical inhibitor, suggesting a role for this active site residue in resistance (EL-FARRASH et al. 1994).

Saquinavir, the first approved protease inhibitor, has also been shown to select for resistant viruses. CRAIG et al. (1993) and DIANZANI et al. (1993) reported reduced sensitivity of viral stocks after selection with saquinavir in the range of 10- to 30-fold. Subsequent studies identified a virus pool with the G48V, I54V and L90M mutations as being responsible for the resistance to saquinavir. This pool of virus was shown to be 50-fold resistant to saquinavir (EBERLE et al. 1995). Other mutations were identified in various virus pools, and the addition of M36I and L63P led to 30-fold resistance to saquinavir (JACOBSEN et al. 1995). Much attention has been given to the G48V and L90M mutations in later studies. These single mutations introduced into viral clones resulted in 8- to 3-fold resistance, respectively, while the double mutant generated 20-fold resistance (JACOBSEN et al. 1995; MASCHERA et al. 1995; TISDALE et al. 1995). Patients failing the initial hard-capsule formulation of saquinavir frequently have sensitive viruses, probably due to poor drug bioavailability, but some patients had viruses that were resistant up to 160-fold (JACOBSEN et al. 1996; IVES et al. 1997; WINTERS et al. 1998).

Initial selection with indinavir resulted in the identification of four mutations in a virus pool that led to an increase in resistance of 15-fold (TISDALE et al. 1995). A molecular clone with the four mutations proved to be sixfold resistant (TISDALE et al. 1995). Patients who fail indinavir therapy show a wide range of mutations with resistance greater than 30-fold (CONDRA et al. 1995; CONDRA et al. 1996). In contrast, mutated molecular clones with up to five inserted mutations were less resistant, in the range of four- to eightfold. These mutants included the active-site mutations V82T/I84V. These studies also emphasise the fact that there is a strong correlation between the number of mutations and increasing resistance.

Virus pools selected with ritonavir showed 30-fold resistance with the M46I, V82F and I84V mutations being the predominant changes (MARKOWITZ et al. 1995). When V82F and I84V were introduced in molecular clones, five and tenfold resistance was seen, respectively (Markowitz et al. 1995). When viruses with mutations seen in patients failing ritonavir therapy were tested, resistance from six to several hundredfold was observed, with mutations at position 82 usually appearing first (Molla et al. 1996; Schmit et al. 1996). Like indinavir, those studies showed increasing resistance with increasing numbers of mutations.

In vitro selection with the approved protease inhibitor, nelfinavir, resulted in the initial appearance of a D30N mutation. This mutant was shown to be sevenfold less sensitive than the parental strain. It is interesting to note that further selection, up to 30-fold resistance, reversed the initial mutation concomitant with the appearance of M46 and I84 mutations (Patick et al. 1996). However, patients failing nelfinavir therapy were shown to be resistant between 5- and 80-fold, with D30N and non-active-site mutations (Markowitz et al. 1998).

Figure 1 summarises the mutations of the protease that appear in vivo. It also contains data for in vitro selection with the most recently approved amprenavir (currently in clinical trials), which selects for mutations at positions I47, I50, M46, V82 and I84 (Partaledis et al. 1995; De Pasquale et al. 1998). Data for high-level drug resistance to nelfinavir and amprenavir are limited due to the smaller number of patients who have failed therapy, while data for saquinavir are limited due to the poor bioavailability of the initial formulation.

As noted above, a typical feature of the HIV-1-protease inhibitors is the presence of large hydrophobic side chains at the positions equivalent to the P1 and P1′ amino acids. These large side chains fit into the large S1 and S1′ subsites in the nearly symmetric protease. However, these symmetrical interactions create a situation where a single mutation in the protease impacts on two sites of inhibitor-enzyme contact. In our own work, we have explored the nature of resistance mutations using similar inhibitors that differ in their P1 and P1′ side chains. The inhibitor SKF-108922 contains Phe/Ala at these positions (Lambert et al. 1993), while the similar inhibitor SKF-108842 contains Phe/Phe (they also differ in one of their end blocking groups). We used each of these inhibitors in a culture-selection scheme to determine if they would select for different resistance mutations. Both of these inhibitors selected first for changes at position 82, in both cases with a change from valine to threonine. Thus, there was no difference in the initial marker of resistance in these two cases. The presence of a large hydrophobic side chain at position P1 appears to provide a dominant interaction between the enzyme and the inhibitor and does not require an equivalent interaction with the P1′ amino acid. Also, the small side chain in P1′ does not provide a counter-selection when the enzyme mutates to reduce interactions with a large side chain in P1.

There is no evidence for pre-existing resistance among drug-naive patients. In Fig. 2, we summarise data from two studies that examined protease-sequence variability (Kozal et al. 1996; Lech et al. 1996). Mutations associated with resistance appear infrequently in this list, and when they do

Fig. 2. Protease sequence variability. Data from KOZAL et al. (1996) and LECH et al. (1996) were combined to determine the variability within the protease gene of protease-inhibitor-naïve patients. The *bold letters* indicate the sites associated with resistance to protease inhibitors. The *capital letters* indicate amino-acid substitutions that appear in 21–49% of the sequences. The *capital italic letters* indicate amino-acid substitutions that appear in 6–20% of the sequences. The *lowercase letters* indicate amino-acid substitutions that appear at a frequency of 5% or less in the sequences. The *underlined* substitutions are amino-acid changes that are associated with resistance

occur in vivo it is probably in the absence of other resistant mutations. Presumably, drug potency is sufficient, on average, to inhibit this background of pre-existing mutations.

In summary, the in vitro data suggest that all protease inhibitors can select for resistance. Resistance will be most easily achieved when viral growth is partially inhibited, allowing protease mutations to appear faster than when the virus is totally inhibited. Optimal therapy will completely block viral replication, suppressing the onset of resistance.

C. Biochemical Basis for Resistance

Enzymatic assays provide a tool for the direct analysis of substitutions associated with resistance. Recombinant protease produced in bacteria is the most widely used source of enzyme. Most studies have looked at collections of mutant proteases by targeting previously identified amino acids from either in vitro or in vivo selections. Assessment of the inhibitory activity of inhibitors toward mutated enzymes clearly indicates the effect of these mutations on enzyme sensitivity. One of the major concerns has been to correlate K_i increases with IC_{50}/IC_{95} increases. While K_i increases with mutant proteases are usually clear, the increases in IC_{50}/IC_{95} can be less evident. The differences between biological and biochemical data of mutant proteases may be explained as follows: (1) in enzymatic assays, the enzyme concentration is limited, while in virus, there may be an excess of enzyme concentration (Konvalinka et al. 1995; Rose et al. 1995), (2) in enzymatic assays, there is essentially a constant amount of substrate, since less than 10–15% of the substrate is consumed; this is done to keep the enzyme activity linear. In virus, the concentration of substrate becomes limiting, since substrate should be cleaved almost entirely. Finally, (3) in enzymatic assays, the data are collected for a limited period of time; in virus, the enzyme may be active over an extended period of time. Thus, these three differences may contribute to the reported K_i and IC_{50}/IC_{95} discrepancies (Klabe et al. 1998).

Biochemical analysis of the mutations responsible for resistance complements structural information of mutant proteases. This gives a clearer picture of the nature of the resistance. Mutations like V82A lead to changes in the enzyme backbone. The shorter side chain of alanine, which would lead to less favorable interactions with the inhibitors is displaced to fill, in part, the void left by the longer side chain (Baldwin et al. 1995). More drastic changes like V82F should significantly disrupt the binding pocket, while V82T modifies the hydrophobic environment of the active site (Chen et al. 1995). Similarly, I84V creates a larger unfilled space that less tightly binds the inhibitor (Chen et al. 1995). Other mutations, described later as compensatory mutations, at positions M46 and L63, affect the flap conformation or slightly modify the backbone of the enzyme (Chen et al. 1995). Those mutations enhance enzyme activity in the presence of other mutations. Usually, K_i increases correlate with viral resistance. However, the order of magnitude of the resistance differs. In an attempt to correlate kinetic changes (K_is) versus antiviral activity (IC_{50}/IC_{95}), a new approach was proposed by GULNIK et al. (1995). They introduced the vitality value, which compares the kinetic data of mutated enzymes with the wild-type (wt) enzymes [$(K_i\, k\mathrm{cat}/Km_{\mathrm{mutant}})/(K_i\, k\mathrm{cat}/Km_{\mathrm{wt}})$, where $k$cat is the catalysis rate constant and Km is the Michaelis-Menten constant]. They reported a vitality value increase of twofold for saquinavir (I84V mutant). A more drastic increase of 38-fold (R8Q) in the vitality value was reported with A-77003.

Valine at position 82 and isoleucine at position 84 are the most common active-site mutations shown to affect drug sensitivity. I84V is associated with K_i increases of five- to tenfold for ritonavir, indinavir and saquinavir (GULNIK et al. 1995; PARTALEDIS et al. 1995; VACCA et al. 1996), although some variation in the magnitude of increase has been reported (WILSON et al. 1997). V82 mutants A and F have been shown to decrease sensitivity to ritonavir and indinavir (GULNIK et al. 1995, WILSON et al. 1997). The I50V mutation in the flap affects the K_i of saquinavir and indinavir on the order of 10- and 20-fold, respectively (PARTALEDIS et al. 1995). The effects of L90M and G48V on saquinavir have also been studied. K_i increases from 3- to 20-fold are associated with L90M, 13- to 200-fold for G48V and 400- to 1000-fold for the double mutant (MASCHERA et al. 1996; ERMOLIEFF 1997; WILSON et al. 1997; VAILLANCOURT and SWANSTROM, unpublished data). MASCHERA et al. (1996) reported that the basis for resistance was a higher rate of inhibitor dissociation.

D. Different Classes of Resistance Mutations in the Protease

The first mutations arising are frequently common among patients undergoing therapy with a given drug. These mutations are clustered within a few amino acids, usually V82 and I84 for ritonavir/indinavir, L90 for saquinavir and D30 for nelfinavir. Other mutations arising later further decrease drug sensitivity but appear after the first mutations and may be referred to as secondary mutations. A third class of mutations can be referred to as compensatory mutations. These mutations have no effect on drug sensitivity but increase enzyme efficiency to compensate for the deleterious effects of the primary and secondary mutations on enzyme activity (SCHOCK et al. 1996).

The hard-capsule formulation of saquinavir (Invirase) had weak pharmacokinetic properties, thus leading to low drug exposure in patients. Mutations at position 90 (L90M) were frequently seen as the only mutation for patients with this formulation (JACOBSEN et al. 1996). Mutations at position 82 are initially seen in patients failing ritonavir therapy (MOLLA et al. 1996; EASTMAN et al. 1998). Finally, patients failing nelfinavir usually have a D30N mutation (MARKOWITZ et al. 1998). In patients, differences among secondary or compensatory mutations are due in part to differences in viral load and the drug levels achieved for a given compound. We have observed that in vitro selections with saquinavir can also lead to the appearance of an I84V mutation as the primary mutation (Smith and Swanstrom, unpublished observation) or the G48V mutation (Tisdale et al. 1995), which emphasises the fact that higher selective pressure may change the temporal pattern of mutations seen with saquinavir. In the case of nelfinavir, the D30N mutation seen in patients represents the primary mutation and is likely to represent an intermediate level of resistance since selection in vitro can proceed to higher levels of resistance with the loss of the D30N mutation (Patick et al. 1996). In the case of L90M and D30N, these mutations generated the best viral replication properties for the selection applied. As such, these mutations represent the initial stopping point and are not followed by extensive mutational pathways. The evolution of resistance to indinavir appears to follow a greater variety of starting paths (Condra et al. 1995; Condra et al. 1996). It is not clear whether this is due to differences in the starting protease sequences that restrict the range of useful mutations or due to random events that select from among a series of mutations that can each contribute a moderate level of resistance. However, V82 and I84 mutations are seen as the more common paths for primary mutations leading to indinavir resistance (Condra et al. 1996).

It is generally accepted that active-site mutations at positions 82 and 84 are deleterious to enzyme activity (Vacca et al. 1994; Gulnik et al. 1995; Schock et al. 1996; Wilson et al. 1997), as is a mutation at position 48 (Maschera et al. 1996; Ermolieff et al. 1997; Wilson et al. 1997), providing the selective pressure for compensatory mutations. Mutations compromising viral-replication capacity have also been demonstrated for protease sequences that have undergone selection in vivo (Zennou et al. 1998) and in vitro (Markowitz et al. 1995; Croteau et al. 1997). Numerous mutations outside of the active site are associated with resistance, and it is tempting to attribute them to compensatory effects. They still remain largely unexplored, and in only a few cases has a compensatory effect been documented. A substitution at position 10 has been shown to generate a clear phenotype in enhancing viral replication with several resistance-associated mutations (Rose et al. 1996). M46I and L63P were also shown to confer an improvement in catalytic efficiency (Schock et al. 1996).

E. Cleavage-Site Mutations

Cleavage-site mutations can compensate for the reduced enzymatic activity of mutant enzymes. In theory, these mutated sites could be more specific for the mutated enzyme compared to the wild type. However, this appears not to be the case. Mutations have been found within the NC-p1-p6 Gag cleavage sites. Extensive selection with the compound BILA 2185 BS led to the following mutations: in the NC-p1 cleavage site, RQAN-FLG to RRVN-FLG; in the p1-p6 cleavage site, PGNF-LQS to PGNF-FQS (Doyon et al. 1996). These mutations enhance cleavage not only by the mutant enzyme but also by the wild-type enzyme, suggesting that they are "better" sites than the natural ones. The mutations also improved viral-replication kinetics. The reason why the wild-type virus does not evolve these "better" cleavage sites remains unknown. The effect is perhaps to keep in sync the order of cleavage of the Gag precursor necessary for the assembly/maturation process. ZHANG et al. (1997) have detected similar mutations in patients failing indinavir therapy and have shown that the presence of cleavage-site mutations enhances the replication of resistant virus in vitro.

F. Cross-Resistance

The clearest and most relevant demonstration of cross-resistance is seen when therapy-naive patients are treated sequentially with two different protease inhibitors. Under these circumstances, the assessment of potency and duration of effect of the second inhibitor compared to its effect in drug-naive patients can be analysed. Such studies are only beginning, but the initial impression is that sequential therapy with protease inhibitors after therapeutic failure will be challenging.

Cross-resistance was shown to be possible for most of the inhibitors tested in vitro, even under low-level selection (TISDALE et al. 1995). We have shown that selection with high levels of drug leads to very high levels of resistance and increasing cross-resistance (SMITH and SWANSTROM, unpublished observations). The potential for cross-resistance in virus has been demonstrated for patients who have failed indinavir (CONDRA et al. 1995), ritonavir (MOLLA et al. 1996) and saquinavir (WINTERS et al. 1998). In a small study, patients failing nelfinavir therapy had viruses that remained sensitive to the other protease inhibitors (MARKOWITZ et al. 1998). A useful marker for resistance or cross-resistance may be phenotypic or genotypic analysis of a patient's isolate prior to a change in therapy which may predict drug failure in patients (DEEKS 1998; HARRIGAN et al. 1998).

G. Concepts for Salvage Therapy

Salvage therapy is defined as the need to alter the therapeutic regimen after virologic failure (return of detectable virus load). Virologic failure is starting to be understood as a more complex phenomenon and must be viewed in at least three contexts. First, the initial appearance of virus in the face of drug may occur with the virus displaying little or no resistance. It may be that in some patients detection of rebounding virus while it is at low levels will permit successful intensification of therapy rather than abandonment of the current drugs. At this point, the successful use of this approach is anecdotal. It is not known how often patients need to have their virus load monitored to assure that they can be caught in the initial act of virus rebound. However, the concept of intensification with early failure may prove to be clinically important. Second, a more extreme case of virologic failure in the absence of apparent resistance has recently been reported (Havlir et al. 1998), in which virus load rebounds initially but in the absence of resistance mutations. The best available rationalisation for this seeming paradox is that in the presence of strong therapy and a drop in virus load, there is a dramatic increase in the available target cells. The residual virus replication in the presence of drug is greatly enhanced by the increase in target cells, and virus rebound occurs. At the heart of this phenomenon lie important issues of virus–host interactions that are poorly understood. However, as in the first case, it may be that in the subset of patients that initially fail therapy, here with apparent complete failure as evidenced by the return of virus load, therapy intensification rather than therapy change may be a viable strategy. The third case of virologic failure is where the virus that reappears carries resistance mutations. Time may also be an issue here, since there is some evidence that the number of resistance mutations increases during treatment after the reappearance of virus (Molla et al. 1996; Eastman et al. 1998). Thus, salvage strategies even in this case may, on average, be more successful if started earlier rather than later.

Initial attempts have been made to use genotypic or phenotypic data about the reappearing virus to predict subsequent therapy success or failure (Deeks et al. 1998; Harrigan et al. 1998). These results show some promise in being able to predict that a certain regimen will fail based on pre-existing resistance to one or more of the drugs to be used. Prospective studies are being planned that will allow this information to be used to select new therapies that may be more efficacious than those designed based solely on treatment history.

Salvage therapy for virus that does not carry resistance mutations represents the simpler case of intensification on top of the pre-existing therapy. Under these circumstances, the full range of antiviral agents can be used. However, options are significantly limited when true therapy failure has occurred.

Two approaches to salvage therapy are available. The first is intensification to overcome resistance/cross-resistance. In this approach, less potent therapy is used initially, with the hope that it will be effective in suppressing virus replication. In those cases where there is therapy failure, intensification is used to suppress the now partially resistant virus. Part of this concept is the notion that the rebounding virus is not fully resistant to the drugs being used, and if drug levels can be raised, clinical benefit can still be realised. This reasoning has led to the use of two protease inhibitors in patients who have failed treatment with one protease inhibitor. Of special note here is that ritonavir has the ability to increase significantly the blood level of other protease inhibitors by inhibiting their metabolism through the p450 cytochrome-oxidase system (Kempf et al. 1997). The effect is that pairs of protease inhibitors that include ritonavir are able to attain significantly higher drug levels. Obviously, this strategy is strengthened if other drugs in the regimen, i.e. the reverse-transcriptase inhibitors, can be changed with benefit.

The second approach is sequential use of inhibitors with different resistance patterns. This is, of course, the ideal strategy for salvage therapy, but it has thus far proven elusive for the protease inhibitors. As seen in Fig. 1, there is significant overlap in resistance mutations associated with each of the available inhibitors. This fact has been borne out with the observation that a change in therapy from low-dose saquinavir to the more potent indinavir resulted in a poor response to indinavir [Executive Summary of Interim Analysis of Acquired Immune Deficiency Syndrome (AIDS) Clinical Trial Group (ACTG) 333 1993], even in a subset of patients who had no detectable resistance mutations associated with saquinavir resistance (DULIOST et al. 1997). The one potential combination in this strategy is the initial use of nelfinavir followed by other protease inhibitors. This possibility exists because of nelfinavir's propensity to initially select for a D30N mutation (MARKOWITZ et al. 1998). This mutation does not confer cross-resistance to other protease inhibitors (PATICK et al. 1996) and, thus, if this were the only mutant form of the protease present, the other protease inhibitors would still be effective. However, it is clear that nelfinavir can select for other mutations (PATICK et al. 1996) and, as in the case of saquinavir cited above, these mutations may be in the background of the predominant D30N population. If this is the case, then the selection of cross-resistance by nelfinavir would limit the utility of sequential protease-inhibitor use. The potential for this strategy with these inhibitors needs to be assessed critically in a clinical trial supported by extensive sequence analysis.

In summary, it is usually observed that when one or several mutations (primary and compensatory) are selected, it is concomitant with moderate cross-resistance levels. When cross-resistance has not been demonstrated with a drug, the possibility of low drug exposure must be considered. When incomplete cross-resistance leads to residual potency with other inhibitors, care must be taken to use the residual potency of these inhibitors in the context of multi-drug therapy to enhance the chance of obtaining a state of complete suppression of virus replication (see also Chap. 4 "The two strategies to reduce viral resistance to protease inhibitors").

H. Summary

The potent HIV-1-protease inhibitors have provided important therapy options for people infected with HIV-1. However, therapy failure has required the development of a deeper understanding of why therapy succeeds and of the biological and clinical consequences of therapy failure. This deeper understanding is leading to improved initial therapies, as defined by higher frequencies of success in suppressing virus load, and evolving strategies for salvage therapies for patients who have virologic rebound. Salvage therapies based on increased potency or altered resistance patterns can be envisioned but have not yet been proven. Because of its susceptibility to potent inhibitor design, the HIV-1 protease has become one of the most important proteins for intensive study. Understanding the HIV-1 protease and the chemical basis of inhibitor action, developing new concepts and strategies for inhibitor design and exploring the biochemical and biological nature of resistance will remain important goals for the foreseeable future.

Acknowledgements. We acknowledge the support of this research by NIH RO1-AI32892 grant.

References

Baldwin ET, Bhat TN, Liu B, Pattabiraman N, Erickson JW (1995) Structural basis of drug resistance for the V82A mutant of HIV-1 proteinase. Nat Struct Biol 2:244–249

Cameron DW, Heath-Chiozzi M, Danner S, Cohen C, Kravcik S, Maurath C, Sun E, Henry D, Rode R, Potthoff A, Leonard J (1998) Randomized placebo-controlled trial of ritonavir in advanced HIV-1 disease. The Advanced HIV Disease Ritonavir Study Group. Lancet 351:543–549

Chen Z, Li Y, Schock HB, Hall D, Chen E, Kuo LC (1995) Three-dimensional structure of a mutant HIV-1 protease displaying cross-resistance to all protease inhibitors in clinical trials. J Biol Chem 270:21433–21436

Condra JH, Schleif WA, Blahy OM, Gabryelski LJ, Graham DJ, Quintero JC, Rhodes A, Robbins HL, Roth E, Shivaprakash M, Titus D, Yang T, Teppler H, Squires KE, Deutsch PJ, Emini EA (1995) In vivo emergence of HIV-1 variants resistant to multiple protease inhibitors. Nature 374:569–571

Condra JH, Holder DJ, Schleif WA, Blahy OM, Danovich RM, Gabryelski LJ, Graham DJ, Laird D, Quintero JC, Rhodes A, Robbins HL, Roth E, Shivaprakash M, Yang T, Chodakewitz JA, Deutsch PJ, Leavitt RY, Massari FE, Mellors JW, Squires KE, Steigbigel RT, Teppler H, Emini EA (1996) Genetic correlates of in vivo viral resistance to indinavir, a human immunodeficiency virus type 1 protease inhibitor. J Virol 70:8270–8276

Craig JC, Whittaker L, Duncan IB, Roberts NA (1993) In vitro resistance to an inhibitor of HIV proteinase (Ro 31–8959) relative to inhibitors of reverse transcriptase (AZT and TIBO). Antiviral Chem Chemother 4:335–339

Croteau G, Doyon L, Thibeault D, McKercher G, Pilote L, Lamarre D (1997) Impaired fitness of human immunodeficiency virus type 1 variants with high-level resistance to protease inhibitors. J Virol 71:1089–1096

Danner SA, Carr A, Leonard JM, Lehman LM, Gudiol F, Gonzales J, Raventos A, Rubio R, Bouza E, Pintado V, Aguado AG, Garcia de Lomas J, Delgado R, Borleffs, JCC, Hsu A, Valdes JM, Boucher CAB, Cooper DA (1995) A short-term study of the safety, pharmacokinetics, and efficacy of ritonavir, an inhibitor of HIV-1 protease. European-Australian Collaborative Ritonavir Study Group. N Engl J Med 333:1528–1533

De Pasquale MP, Murphy R, Kuritzkes D, Martinez-Picado J, Sommadossi J-P, Gulick R, Smeaton L, DeGruttola V, Caliendo A, Sutton L, Savara AV, D'Aquila RT (1998) Mutations selected in HIV plasma RNA during 141W94 therapy (abstract 71). 2nd International Workshop on HIV Drug Resistance and Treatment Strategies, Lake Maggiore, Italy. June 24–27

Deeks SG, Parkin N., Pertopoulos CJ, Grant RM, Volberding PA, Whitcomb J. Tian H, Wrin T, Limoli K, Drews B, Warmerdam M, Hellmann NS (1998) Correlation of baseline phenotypic drug susceptibility with 16 week virologic response in a pilot combination therapy study in HIV-infected patients who failed indinavir therapy (abstract 53). 2nd International Workshop on HIV Drug Resistance and Treatment Strategies, Lake Maggiore, Italy. June 24–27

Dianzani F, Antonelli G, Turriziani O, Riva E, Dong G, Bellarosa D (1993) In vitro selection of human immunodeficiency virus type 1 resistant to Ro 31–8959 proteinase inhibitor. Antiviral Chem Chemother 4:329–333

Doyon L, Croteau G, Thibeault D, Poulin F, Pilote L, Lamarre D (1996) Second locus involved in human immunodeficiency virus type 1 resistance to protease inhibitors. J Virol 70:3763–3769

Duliost A, Paulos S, Guillemot L, Boue G, Galanaud P, Clavel F (1997) Selection of saquinavir-resistant mutants by indinavir following a switch from saquinavir. International Workshop on HIV Drug Resistance, Treatment Strategies and Eradication, St. Petersburg

Eastman PS, Mittler J, Kelso R, Gee C, Boyer E, Kolberg J, Urdea M, Leonard JM, Norbeck DW, Hongmei M, Markowitz M (1998) Genotypic changes in human immunodeficiency virus type 1 associated with loss of suppression of plasma viral RNA levels in subjects treated with Ritonavir (Norvir) monotherapy. J Virol 72:5154–5164

Eberle J, Bechowsky B, Rose D, Hauser U, Von Der Helm K, Gurtler L, Nitschko H (1995) Resistance of HIV type 1 to proteinase inhibitor Ro 31–8959. AIDS Res Hum Retroviruses 11:671–676

El-Farrash MA, Kuroda MJ, Kitazaki T, Masuda T, Kato K, Hatanaka M, Harada S (1994) Generation and characterization of a human immunodeficiency virus type 1 (HIV-1) mutant resistant to an HIV-1 protease inhibitor. J Virol 68:233–239

Erickson J, Neidhart DJ, VanDrie J, Kempf DJ, Wang XC, Norbeck DW, Plattner JJ, Rittenhouse JW, Turon M, Wideburg N, Kohlbrenner WE, Simmer R, Hlefrich R, Paul DA, Knigge M (1990) Design, activity, and 2.8A crystal structure of a C2 symmetric inhibitor complexed to HIV-1 protease. Science 249:527–533

Ermolieff J, Lin X, Tang J (1997) Kinetic properties of saquinavir-resistant mutants of human immunodeficiency virus type 1 protease and their implications in drug resistance. Biochemistry 36:12364–12370

Executive Summary of Interim Analysis of ACTG 333 (1997) The anti-viral effect of switching from the hard capsule saquinavir (SQVhc) to the soft gelatin capsule of saquinavir (SQVsgc) vs. switching to indinavir (IDV) after one year of saquinavir use

Gulick R, Mellors J, Havlir, D, Eron J, Gonzalez C, McMahon D, Jonas L, Meibohm A, et al. (1998) Simultaneous vs. sequential initiation of therapy with indinavir, zidovudine, and lamivudine for HIV-1 infection: 100 week follow-up. JAMA 280:35–41

Gulnik SV, Suvorov LI, Liu B, Yu B, Anderson B, Mitsuya H, Erickson JW (1995) Kinetic characterization and cross-resistance patterns of HIV-1 protease mutants selected under drug pressure. Biochemistry 34:9282–9287

Hammer SM, Squires K, Hughes M, Grimes J, Demeter L, Courier J, Eron JJ, Feinberg J, Balfour H, Deyton L, Chodakewitz J, Fischl M, Nguyen B-Y, Spreen W, Pedneault L, Kuritzkes D (1997) A randomized, placebo-controlled trial of indinavir in combination with two nucleoside analogs in human immunodeficiency virus infected persons with CD4[+] T-cell counts less than or equal to 200 per cubic millimeter. N Engl J Med 337:725–733

Harrigan P, Montaner J, Hogg R, et al. (1998) Baseline resistance profile predicts response to ritonavir/saquinavir therapy in community setting (abstract 55). 2nd International Workshop on HIV Drug Resistance and Treatment Strategies, Lake Maggiore, Italy. June 24–27

Havlir DD, Petropoulos CJ, Hellmann NS, Whitcomb JM, Richman DD, and the ACTG 343 Team (1998) Evolution of drug resistance associated with loss of viral suppression in patients treated with indinavir, lamivudine and zidovudine (abstract 74). 2nd International Workshop on HIV Drug Resistance and Treatment Strategies, Lake Maggiore, Italy. June 24–27

Ho DD, Toyoshima T, Mo H, Kempf DJ, Norbeck D, Chen CM, Wideburg NE, Burt SK, Erickson JW, Singh MK (1994) Characterization of human immunodeficiency virus type 1 variants with increased resistance to a C2-symmetric protease inhibitor. J Virol 68:2016–2020

Ives KJ, Jacobsen H, Galpin SA, Garaev MM, Dorrell L, Mous J, Bragman K, Weber JN (1997) Emergence of resistant variants of HIV in vivo during monotherapy with the proteinase inhibitor saquinavir. J Antimicrob Chemother 39:771–779

Jacobsen H, Yasargil K, Winslow DW, Craig C, Krohn A, Duncan IB, Mous J (1995) Characterization of human immunodeficiency virus type 1 mutants with decreased sensitivity to proteinase inhibitor Ro 31–8959. Virology 206:527–534

Jacobsen H, Hanggi M, Ott M, Duncan IB, Owen S, Andreoni M, Vella S, Mous J (1996) In vivo resistance to a human immunodeficiency virus type 1 proteinase inhibitor: mutations, kinetics, and frequencies. J Infect Dis 173:1379–1387

Kaplan AH, Zack JA, Knigge M, Paul DA, Kempf DJ, Norbeck DW, Swanstrom R (1993) Partial inhibition of the human immunodeficiency virus type 1 protease results in aberrant virus assembly and the formation of noninfectious particles. J Virol 67:4050–4055

Kaplan AH, Michael SF, Wehbie RS, Knigge MF, Paul DA, Everitt L, Kempf DJ, Norbeck DW, Erickson JW, Swanstrom R (1994) Selection of multiple human immunodeficiency virus type 1 variants that encode viral proteases with decreased sensitivity to an inhibitor of the viral protease. Proc Natl Acad of Sci USA 91:5597–5601

Kempf DJ, Marsh KC, Kumar G, Rodrigues RD, Denissen JF, McDonald E, Kikulka MJ, Hsu A, Granneman GR, Baroldi PA, Sun E, Pizzuti D, Plattner JJ, Norbeck DW, Leonard JM (1997) Pharmacokinetic enhancement of inhibitors of the human immunodeficiency virus protease by coadministration with ritonavir. Antimicrob Agents Chemother 41:654–660

Klabe R, Bacheler L, Ala P, Erickson-Viitanen S, Meek J (1998) Resistance to HIV protease inhibitors: a comparison of enzyme inhibition and antiviral potency. Biochemistry 37:8735–8742

Konvalinka J, Litterst MA, Welker R, Kottler H, Rippmann F, Heuser AM, Krausslich HG (1995) An active-site mutation in the human immunodeficiency virus type 1 proteinase (PR) causes reduced PR activity and loss of PR-mediated cytotoxicity without apparent effect on virus maturation and infectivity. J Virol 69:7180–7186

Kozal MJ, Shah N, Shen N, Yang R, Fucini R, Merigan TC, Richman DD, Morris D, Hubbell E, Chee M, Gingeras TR (1996) Extensive polymorphisms observed in HIV-1 clade B protease gene using high-density oligonucleotide arrays. Nat Med 2:753–759

Lambert DM, Bartus H, Fernandez C, Bratby-Anders C, Leary JJ, Dreyer GB, Metcalf BW, Petteway SR Jr (1993) Synergistic drug interactions of an HIV-1 protease inhibitor with AZT in different in vitro models of HIV-1 infection. Antiviral Res 21:327–342

Lech WJ, Wang G, Yang YL, Chee Y, Dorman K, McCrae D, Lazzeroni LC, Erickson JW, Sinsheimer JS, Kaplan AH (1996) In vivo sequence diversity of the protease of human immunodeficiency virus type 1: presence of protease inhibitor-resistant variants in untreated subjects. J Virol 70:2038–2043

Markowitz M, Mo H, Kempf DJ, Norbeck DW, Bhat TN, Erickson JW, Ho DD (1995) Selection and analysis of human immunodeficiency virus type 1 variants with increased resistance to ABT-538, a novel protease inhibitor. J Virol 69:701–706

Markowitz M, Conant M, Hurley A, Schluger R, Duran M, Peterkin J, Chapman S, et al. (1998) A preliminary evaluation of nelfinavir mesylate, an inhibitor of Human Immunodeficiency Virus (HIV)-1 protease, to treat HIV infection. J Infect Dis 177:1533–1540

Maschera B, Furfine E, Blair ED (1995) Analysis of resistance to human immunodeficiency virus type 1 protease inhibitors by using matched bacterial expression and proviral infection vectors. J Virol 69:5431–5436

Maschera B, Darby G, Palu G, Wright LL, Tisdale M, Myers R, Blair ED, Furfine ES
(1996) Human immunodeficiency virus. Mutations in the viral protease that confer
resistance to saquinavir increase the dissociation rate constant of the protease-
saquinavir complex. J Biol Chem 271:33231–33235

Molla A, Korneyeva M, Gao Q, Vasavanonda S, Schipper PJ, Mo H-M, Markowitz M,
Chernyavskiy T, Niu P, Lyons N, Hsu A, Granneman R, Ho DD, Boucher CAB,
Leonard JM, Norbeck DW, Kempf DJ (1996) Ordered accumulation of mutations
in HIV protease confers resistance to ritonavir. Nat Med 2:760–766

Molla A, Vasavanonda S, Denissen J, Kumar G, Grabowski B, Sham H, Norbeck D,
Kohlbrenner W, Plattner J, Kempf D, Leonard J (1997) Effect of human serum
proteins on the antiretroviral activity of ritonavir and ABT-378, potent inhibitors
of HIV protease (abstract 104). 4th Conference on Retroviruses and Opportunistic Infections, Washington

Otto MJ, Garber S, Winslow DW, Reid CD, Aldrich P, Jadhav PK, Patterson CE, Hodge
CN, Cheng Y-SE (1993) In vitro isolation and identification of human immunodeficiency virus (HIV) variants with reduced sensitivity to C-2 symmetrical
inhibitors of HIV type 1 protease. Proc Natl Acad Sci USA 90:7543–7547

Partaledis JA, Yamaguchi K, Tisdale M, Blair EE, Falcione C, Maschera B, Myers RE,
Pazhanisamy S, Futer O, Cullinan AB, Stuver CM, Byrn RA, Livingston DJ (1995)
In vitro selection and characterization of human immunodeficiency virus type 1
(HIV-1) isolates with reduced sensitivity to hydroxyethylamino sulfonamide
inhibitors of HIV-1 aspartyl protease. J Virol 69:5228–5235

Patick AK, Mo H, Markowitz M, Appelt K, Wu B, Musick L, Kalish V, Kaldor S, Reich
S, Ho D, Webber S (1996) Antiviral and resistance studies of AG1343, an orally
bioavailable inhibitor of human immunodeficiency virus protease. Antimicrob
Agents Chemother 40:292–297

Rose JR, Babe LM, Craik CS (1995) Defining the level of human immunodeficiency
virus type 1 (HIV-1) protease activity required for HIV-1 particle maturation and
infectivity. J Virol 69:2751–2758

Rose RE, Gong Y-F, Greytok JA, Bechtold CM, Terry BJ, Robinson BS, Alam M,
Colonno RJ, Lin P-F (1996) Human Immunodeficiency virus type 1 viral back-
ground plays a major role in development of resistance to protease inhibitors. Proc
Natl Acad Sci USA 93:1648–1653

Schapiro JM, Winters MA, Stewart F, Efron B, Norris J, Kozal MJ, Merigan TC (1996)
The effect of high-dose saquinavir on viral load and CD4$^+$ T-cell counts in HIV-
infected patients. Ann Intern Med 124:1039–1050

Schmit JC, Ruiz L, Clotet B, Raventos A, Tor J, Leonard J, Desmyter J, De CE,
Vandamme AM (1996) Resistance-related mutations in the HIV-1 protease gene
of patients treated for 1 year with the protease inhibitor ritonavir (ABT-538).
AIDS 10:995–999

Schock HB, Garsky VM, Kuo LC (1996) Mutational anatomy of an HIV-1
protease variant conferring cross-resistance to protease inhibitors in clinical
trials. Compensatory modulations of binding and activity. J Biol Chem 271:
31957–31963

Tisdale M, Myers RE, Maschera B, Parry NR, Oliver NM, Blair ED (1995) Cross-
resistance analysis of human immunodeficiency virus type 1 variants individually
selected for resistance to five different protease inhibitors. Antimicrob Agents
Chemother 39:1704–1710

Vacca JP, Dorsey BD, Schleif WA, Levin RB, McDaniel SL, Darke PL, Zugay J,
Quintero JC, Blahy OM, Roth E, Sardana VV, Schlabach AJ, Graham PI, Condra
JH, Gotlib L, Holloway MK, Lin J, Chen I-W, Vastag K, Ostovic D, Anderson PS,
Emini EA, Huff JR (1994) L-735,524: an orally bioavailable human immun-
odeficiency virus type 1 protease inhibitor. Proc Natl Acad Sci USA 91:4096–4100

Wilson SI, Lowri HP, Mills JS, Gulnik SV, Erickson JW, Dunn BM, Kay J (1997) Escape
mutants of HIV-1 proteinase: enzymatic efficiency and susceptibility to inhibition.
Biochim Biophys Acta 1339:113–125

Winters MA, Schapito JM, Lawrence J, Merigan TC (1998) Human immunodeficiency
    virus type 1 protease genotypes and in vitro protease inhibitor susceptibilities of
    isolates from individuals who were switched to other protease inhibitors after
    long-term saquinavir treatment. J Virol 72:5303–5306
Zennou V, Mammano F, Paulous S, Mathez D and Clavel F (1998) Loss of viral fitness
    associated with multiple gag and gag-pol processing defects in human immun-
    odeficiency virus type 1 variants selected for resistance to protease inhibitors in
    vivo. J Virol 72:3300–3306
Zhang Y-M, Imamichi H, Imamichi T, Lane HC, Falloon J, Vasudevachari MB, Salzman
    NP (1997) Drug resistance during indinavir therapy is caused by mutations in the
    protease gene and in its gag substrate cleavage sites. J Virol 71:6662–6670

The Next Generation of Human Immunodeficiency Virus Protease Inhibitors: Targeting Viral Resistance

E.S. FURFINE

A. Human Immunodeficiency Virus Protease Inhibitors: Advancements in the Treatment of Human Immunodeficiency Virus Disease

I. Current Status of Human Immunodeficiency Virus Protease Inhibitors

Numerous chemotherapeutic agents for the treatment of human immunodeficiency virus (HIV) disease have been approved as drugs in the past 10 years (reviewed by MOLLA and KOHLBRENNER 1998). In all cases, viral resistance to these compounds has begun to develop. The two viral enzymes most commonly targeted by chemotherapeutic agents are the reverse transcriptase (the viral DNA polymerase) and the protease, which hydrolyzes viral polyprotein-translation products into their mature and active protein components. Both enzymes are essential for viral replication. Protease inhibitors (PIs), nucleoside reverse-transcriptase inhibitors, and non-nucleoside reverse-transcriptase inhibitors compose the primary classes of HIV-chemotherapeutic agents for these viral enzymes. The discovery of clinically effective HIV-PIs is a classic example of rational/structure-based drug design (reviewed by VACCA and CONDRA 1998 and WLODAWER and VONDRASEK 1998). These compounds were created using strategies of mechanistic enzymology and medicinal chemistry to design potent inhibitors that mimic enzymatic-reaction transition states and convert these entities into drug-like molecules.

As a class, PIs are the most potent anti-retroviral agents (reviewed by HOETELMANS et al. 1998; VELLA and PALMISANO 1997). The declining mortality and morbiditiy of HIV disease has been significantly impacted by the use of HIV-PIs (PALELLA et al. 1998). The Food and Drug Administration-approved inhibitors are Norvir (ritonavir from Abbott), Fortavase (saquinavir from Roche), Crixivan (indinavir from Merck), and Viracept (nelfinavir from Agouron) (Fig. 1). Amprenavir (141W94 discovered by Vertex and developed by GlaxoWellcome) is in the late stages of phase-III clinical trials. This review will discuss short-term and long-term strategies to combat resistance to PIs, and the future developmental needs of HIV-PIs.

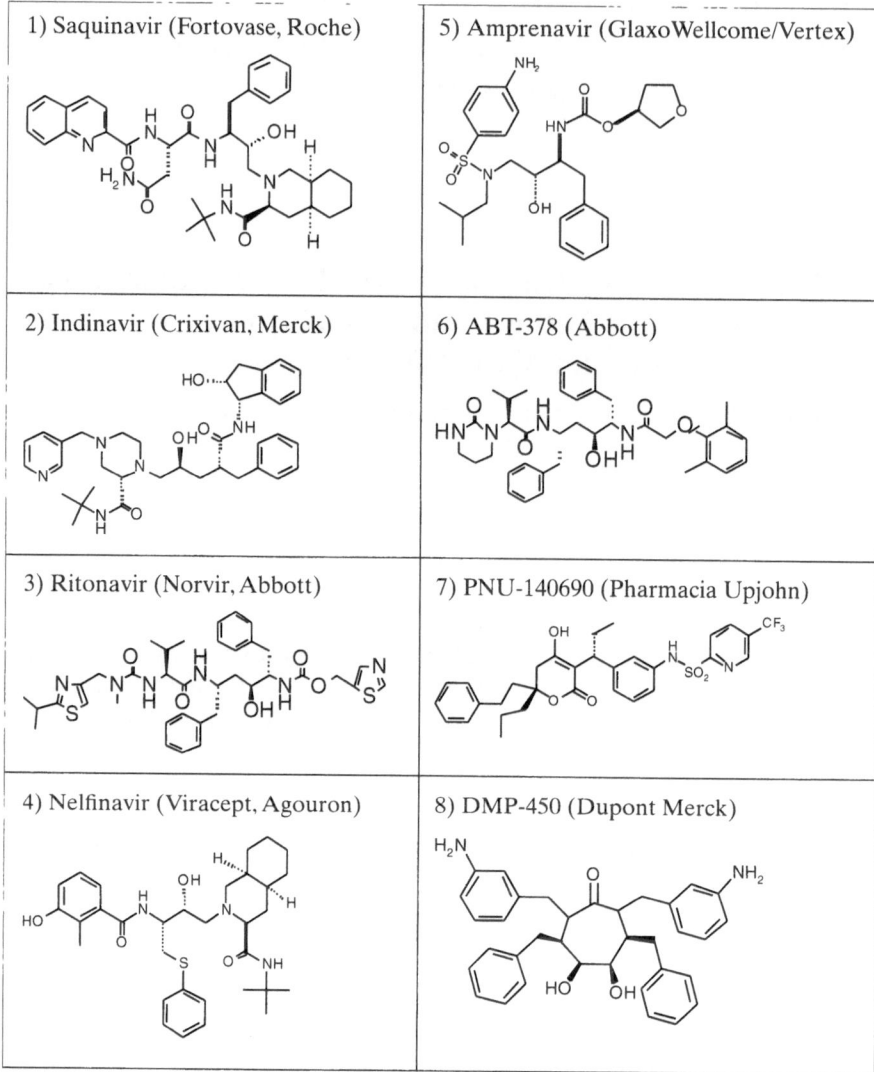

Fig. 1. The Food and Drug Administration-approved inhibitors, as well as many of the inhibitors under current clinical evaluation. The list is not intended to be exhaustive. The numbering system is utilized in Table 1

II. Two Strategies to Reduce Viral Resistance to PIs

Mechanisms for overcoming viral resistance to PIs can be catagorized into two strategies. The first (short-term) strategy utilizes current (and future) agents to minimize "wild-type" (WT) viral replication. The second (long-term) strategy is to design agents that target resistant viruses that emerge during therapy.

The rationale behind the first strategy is based on the Darwinian-selection principle. Theoretically, the development of resistance requires viral replication (as outlined in three reviews: RICHMAN 1997; MOYLE 1998; ROBERTS et al. 1998). In this model, the probability of developing resistance is described by a "gaussian" curve with increasing drug exposure. At low drug concentrations, selection pressure for resistance is absent because virus growth is not significantly inhibited. At intermediate drug concentrations, the occurrence of resistance is maximal because DNA replication occurs at a rate sufficient to generate mutations that provide drug resistance (growth advantage) for the mutant virus. At extremely high drug concentrations viral replication is inhibited, thus slowing the generation of mutant populations from which to select resistant virus. Therefore, at high concentrations of drug, resistance develops much more slowly than at low or intermediate concentrations of drug. However, resistance will likely develop eventually, even in extremely successful viral-load-inhibition schemes. Once resistant virus is established, the currently available PIs are unlikely to serve as effective therapeutic agents (CONDRA 1998; discussed in detail in Sect. C.I). Therefore, the second strategy – to understand the molecular mechanism of resistance and directly target appropriate resistant strains of HIV with new chemical entities – appears needed for effective long-term treatment of HIV disease. However, because our understanding of the mechanism of viral resistance is limited (discussed in Sect. C.II), this strategy is unlikely to yield effective agents in the very near future.

B. Strategy 1: Combination Therapy. Maximal Reduction of Viral Load to Retard Development of Resistance

I. Theory and Background

Viral load (levels of plasma viral RNA) is the accepted surrogate marker for disease progression during anti-retroviral therapy (reviewed by CARPENTER et al. 1998 and GOLDSCHMIDT et al. 1998). Alternatively, the CD4T cell level is used as a surrogate marker for disease progression but is probably a better prognosticator of immune reconstitution (reviewed by CARPENTER et al. 1998 and GOLDSCHMIDT et al. 1998). Reducing viral load delays the development of resistance, increases the durability of therapy (life expectancy), and is clinically practical with the existing therapeutic agents. For example, ritonavir monotherapy reduces viral load, with a concomitant increase in life-expectancy (LEONARD 1996; VELLA and PALMISANO 1997). The durability of ritonavir therapy correlates inversely with the viral-load level at the nadir (lowest level of virus in plasma observed during therapy; KEMPF et al. 1997). That is, patients who achieved the lowest viral load during therapy maintained viral suppression the longest. Finally, patients on ritonavir, indinavir, saquinavir, or nelfinavir develop mutations in the viral-protease gene that confer resistance (reduced sensitivity) to that inhibitor and sometimes to multiple PIs, resulting in increased viral load (Jacobsen 1995, Condra et al 1995, Markowitz et al. 1998, and Molla et al. 1996, Eastman et al.1998). Resistance development can be delayed by combination therapy. Numerous examples of PIs reducing viral load to undetectable levels, particularly when used in combination with one or more other antiviral agents have been reported (reviewed by Kakuda et al. 1998). For example, the combination of 3'-azido-3'-deoxythymidine (AZT), 3TC, and indinavir reduced viral load to undetectable levels (<500copies/ml plasma) for more than 2years in over 50% of the patients in the trial (Conway et al. 1998). These results support the strategy of maximally reducing viral load by the use of multiple anti-retroviral agents to minimize the development of resistance and thus prolong the life of HIV patients.

In addition to minimizing the viral load, combination therapy may allow targeting of more diverse tissue types, and these agents together may have decreased toxicity due to the lower required doses than with the individual agents (Manion et al. 1998). Additional reasons for employing combination therapy are discussed in more detail in Sect. B.IV.

Given the success of combination therapy, the current standard for HIV is simultaneous treatment with multiple drugs (Fauci et al. 1998; Goldschmidt et al. 1998). Typically, the combination treatment strategy, often referred to as highly active anti-retroviral therapy (HAART), simultaneously utilizes one or more PIs with one or more nucleoside reverse-transcriptase inhibitors. However, some HAARTs do not utilize PIs. Nonetheless, PIs are widely used as the cornerstone of HAART because these compounds are the most potent class of HIV chemotherapeutic agents for the treatment of HIV disease (reviewed by Hoetelmans et al. 1998; Vella and Palmisano 1997).

II. Limitations

While the advances in therapy attributed to PIs are substantial, up to 60% of patients utilizing HAART may ultimately fail therapy (Conway et al. 1998). Many factors contribute to this therapeutic failure rate. Foremost is the importance of a patient adherence to the drug regimen (Molla and Japour 1997; Flexner 1998; Goldschmidt et al. 1998; Moyle 1998). Decreases in adherence are positively correlated with increases in development of resistance and viral load, resulting in the loss of durability of therapy.

Physiological and psychological factors make adherence to protease-inhibitor regimens difficult. For example, adverse events (physiological factors), such as gastrointestinal disturbances, decrease patient adherence are associated with all approved inhibitors (reviewed by Hoetelmans et al. 1997 and Flexner 1998). Furthermore, PI therapy is sometimes associated with various metabolic disturbances, such as abnormal fat distribution, hyperlipidemia, and glucose intolerance (reviewed by Carr et al. 1998; Flexner 1998). Other side effects specific to a given PI include paresthesias from ritonavir therapy and nephrolithiasis from indinavir therapy (reviewed by HOETELMANS et al. 1997 and FLEXNER 1998). Another physiological factor that may effect adherence is that all approved PIs inhibit cytochrome P4503A4. This inhibition results, to some extent, in drug interaction problems with all the approved agents, ritonavir being the most potent inhibitor of P4503A4 (reviewed by HOETELMANS et al. 1997 and KAKUDA 1998; VON MOLTKE et al. 1998). A psychological factor affecting patient adherence is the complexity of the dosing regimen. Therapy with PIs typically requires a large pill burden and a complex time- and food-dependence on dosing (HOETELMANS et al. 1997; MOYLE 1998; KAKUDA 1998). For example, saquinavir, indinavir and nelfinavir regimens are t.i.d. (thrice daily) with different food effects. These regimens become particularly complicated when coupled with other antivirals, especially when multiple PIs are used. The possible exception to this rule is the use of ritonavir with other PIs. Ritonavir enhances the pharmacokinetics (exposure) of other PIs, thus reducing the required dose of the other inhibitors (reviewed by MOYLE 1998 and KEMPF et al. 1998). Interestingly, the ritonavir-dependent inhibition of P4503A4, which causes drug-interaction problems, enables it to enhance its own and other PIs' plasma levels by inhibiting their metabolism.

Another challenge to HAART is that some HIV reservoirs turn over very slowly, perhaps due to the slow turnover of the tissue (CHUN et al. 1997; PERELSON et al. 1997; reviewed by COEN 1998). Because these reservoirs turn over slowly and are a source of infectious virus, patients must be maintained on a viral-load-suppression regimen for extremely long periods of time, even though plasma virus is undetectable. Furthermore, some tissues (such as brain) are not optimally exposed to drug. This type of "drug-excluded viral sanctuary" may be a breeding ground for resistant virus due to sub-optimal exposure to the drugs (CHUN et al. 1997; PERELSON et al. 1997; reviewed by COEN 1998).

III. Improvement of Strategy 1: Exploiting Currently Available Inhibitors

1. Approaches to Improving Patient Adherence

Because patient adherence to PI regimens appears to play such a large part in the durability of existing PI-containing HAART, improving patient adherence should improve the success rate of these therapies. In fact, the adherence issue is being addressed with approved inhibitors and those undergoing clinical evaluation. Improvements in patient compliance accompany reductions in the number of daily doses patients must take and the pill burden (reviewed by MOYLE 1998). Consequently, both Merck and Agouron have clinical trials comparing the current approved t.i.d. regimen (for indinavir and nelfinavir, respectively) with a b.i.d. (twice daily) dosing regimen (JOHNSON et al.1998; NGUYEN et al. 1998). These studies demonstrated similar levels of drug exposure and similar reduction in viral load with the b.i.d. or the t.i.d. regimen.

Abbott has tried to reduce the daily dosing of ABT-378 (an investigational PI) by combining it with ritonavir. Because low doses of ritonavir (50–100 mg, minimally therapeutic) inhibit P4503A4, increased exposure of ABT-378 (dosed 200–600 mg) is observed, such that dosing once per day may be possible (LAL et al. 1998). Because ABT-378 is tenfold more potent than ritonavir (in vitro), the pill-burden and the frequency of dosing may be reduced by this strategy. Furthermore, clinical studies combining ritonavir with saquinavir or nelfinavir demonstrated reductions in the doses per day and the size of the dose required for significant viral-load suppression (CAMERON et al. 1998; GALLANT et al. 1998). Amprenavir, which is being developed by Glaxo Wellcome and Vertex, may have advantages in patient compliance, because its b.i.d. regimen and its efficacy has been demonstrated with patients taking the drug without regard to food intake (ADKINS and FAULDS 1998, GOODGAME et al. 1998).

2. Reducing Resistance Development by Treatment with Multiple PIs

Treatment of patients with two or more PIs simultaneously may lower the rate of resistance development and increase the durability of antiviral therapy. There are several reasons to expect that regimens with dual (or multiple) PIs may offer advantages. First, most antiviral agents are dosed to maximize viral-load reduction until dose-limiting toxicity or adverse effects occur (strategy 1). If different PIs have different adverse effects (or different mechanisms of causing the same adverse effect), then two inhibitors might be combined to increase inhibition of the target without increasing any one adverse event. Second, as mentioned earlier, ritonavir inhibits P4503A4, thus increasing exposure to other PIs. Finally, multiple-PI therapy may reduce the rate of resistance development by inhibiting a broader population of "low-resistance" viruses. Many of the approved PIs acquire mutations through unique "preferred" pathways, leading to significant clinical resistance and viral-load rebound. For example, ritonavir selects for mutations in an ordered fashion, with mutations at V82 typically appearing first (MOLLA et al. 1996). Similarly, indinavir selects for V82 mutations, but may not have a clear preference for the V82 pathway (CONDRA et al. 1995; CONDRA et al. 1996). Nelfinavir typically selects for D30N first (MARKOWITZ 1998), saquinavir selects for L90M (ROBERTS et al. 1998), and amprenavir therapy selects for I50V (DE PASQUALE et al. 1998). Viruses with single mutations typically have low-level resistance to the "selecting" inhibitor and often retain or increase their sensitivity to the other inhibitors not used in the selection process (MASCHERA et al. 1995; TISDALE et al. 1995; MARKLAND et al. 1998; MOLLA et al. 1996). Furthermore, compounds from Pharmacia Upjohn (UP140690; TARPLEY 1998) or Parke Davis (PD178390; DOMAGALA et al. 1998) have resistant profiles distinct from any of the currently approved inhibitors and are in early clinical evaluation or late preclinical evaluation, respectively. Many viruses with mutations that yield a single amino acid substitution exist in the WT population (patients that are PI naive) at a low level

(YAMAGUCHI and BYRN 1995; LECH et al. 1996; TUCKER et al. 1998). However, if they did not exist in the native population, one could easily imagine their generation in minimal time given the rapid rate of viral replication and the low fidelity of the polymerase (MOLLA and KOHLBRENNER 1998; ROBERTS 1998). Together, this suggests that multiple-PI therapy would not only inhibit WT virus, but may also result in increased inhibition of the "single-mutant" diversity (low-level resistant virus) present in the population. Better inhibition of the virus population with low-level PI-resistance should increase the durability of therapy.

Dual-PI therapy is presently under clinical investigation. In addition to the trials with ritonavir mentioned earlier, amprenavir has been coupled with indinavir, nelfinavir, and saquinavir (ERON et al. 1998). In all cases, potent viral suppression was demonstrated. The main drawback to most combinations of multiple-PI therapy with currently approved agents is the challenge of patient adherence. Alone, these agents have the largest pill burdens and dosing restrictions of any class of antiviral agents, and the combination therapy can only be more challenging.

IV. The Next Generation of Inhibitors: the Benefits of Increasing Potency

There are at least three areas in which the currently approved compounds (and compounds in late-stage clinical evaluation) could be improved using available technologies. The goal would be to improve treatment strategy 1 (minimizing WT viral replication). The first improvement is to reduce dose-limiting adverse effects. The second improvement is to reduce inhibition of P4503A4, thus reducing drug-interaction effects (reviewed by HOETELMANS et al. 1997 and KAKUDA 1998; VON MOLTKE et al. 1998). The third improvement is to reduce the challenge of the dosing regimen by reducing the pill burdens, food effects, and/or dosing frequency (t.i.d.).

One approach to attaining these goals is to increase the potency of PIs against WT virus while maintaining the pharmacokinetic properties of compounds, such as amprenavir or ritonavir, that have b.i.d. regimens (ADKINS and FAULDS 1998; KEMPF et al. 1998). First, increasing potency against a specific target protein (HIV protease in this case) with chemical strategies that utilize structural information specific to HIV protease will likely result in compounds with a higher degree of selectivity. Improvements in selectivity would allow increases in the concentration of drug above its antiviral median effective concentration value while having fewer adverse events and drug-interaction limitations. Second, increasing inhibitory potency should allow dosing with smaller amounts of compound to reach similar levels of efficacy. Hopefully, these reduced compound requirements will result in lower pill burden (smaller and fewer pills). While the ultimate goal of reducing the pill burden is to increase patient compliance in general for combination therapy, PIs may benefit from this more than other anti-retroviral agents. As described earlier, the rationale for multiple PIs in combination is well-established (Sect. B.III.2). However, the pill burden and dosing regimen make these therapies difficult to adhere to. There would be a great advantage to dosing multiple PIs with a level of convenience similar to that of nucleoside analogues, such as the retrovir/lamivudine combination (AZT/3TC; combivir). The next generation of PIs must have a dosing regimen that is easy to adhere to in combination with any other anti-retroviral agents. Inhibitors with increased potency have the potential to yield such characteristics.

C. Strategy 2: Designing Drugs to Inhibit PI-Resistant Viruses

I. Viral Resistance to PIs

Even given HAART approaches that maximize potency and patient adherence, resistance to PIs (and other classes of anti-retroviral agents) will develop eventually. HIV strains that are highly resistant to PIs are present in the patient population (CONDRA 1995; MOLLA 1996; Shafer 1998) and are being transmitted through the population (COHEN and FAUCI 1998; HECHT et al 1998). Furthermore, patients who fail specific PI therapy as a result of PI-resistant virus have typically not done well when other PIs were employed in salvage-therapy regimes (reviewed by CONDRA 1998). That is, failure of therapy on one PI typically limits the utility of all approved PIs, thus eliminating perhaps the most potent arm of combination therapy. Furthermore, if 60% of patients on PI-containing HAART will ultimately fail therapy (CONWAY et al. 1998) and the majority of patients are utilizing PI-containing HAART, then a large population of patients who can not use the currently available PI-therapy will emerge. Ultimately, it will be necessary to treat PI-resistant HIV with a new class of inhibitors (PIs or agents based on other targets). However, these new inhibitors are not rapidly forthcoming, and the understanding of the molecular mechanism of PI resistance is limited. The remainder of this review will focus on the state of our knowledge of the molecular mechanism of resistance to HIV PIs, and potential strategies to target these resistant viruses.

II. PI-Resistant Virus: What's the Real Target?

At least 23 of the 100 residues of HIV are "mutatable" in response to selective pressure by HIV PIs (VACCA and CONDRA 1997). These mutations were observed in in vitro and in vivo studies. As discussed in Sect. B.III.2, multiple mutations are acquired stepwise, starting with single mutations. However, no protease-coding region having all 23 mutations has been observed, although protease genes that encode protein with as many as nine altered residues have been identified from clinical samples (CONDRA et al. 1995). Some of these residues are altered more commonly than others; nonetheless, the potential combinations available to the virus appear endless. With so many possibilities for resistance, the relevant question is "What is the target virus for HIV with high-level resistance to PIs?" First, let's define high-level resistance. High-level resistance to PIs usually requires multiple (three or more) missense mutations in the protease gene. High-level resistance is described by a significant (usually over ten-fold) increase in median inhibitory concentration in a cell-based assay in vitro and a rebound in viral load in vivo (CONDRA et al. 1995; MOLLA et al. 1996; CONDRA 1998). Resistance to ritonavir and indinavir results in lost sensitivity to numerous other clinically relevant PIs (CONDRA et al. 1995; TISDALE et al. 1995; MOLLA et al. 1996; SHAFER et al. 1998). However, not all HIV that is highly resistant to a given inhibitor results in resistance to all PIs. For example, some HIV strains resistant to indinavir (in vitro) experience increased sensitivity to saquinavir and retain sensitivity to amprenavir (TISDALE et al. 1995). Also, mutations conferring resistance to saquinavir retain sensitivity to amprenaivr and vice versa (MARKLAND et al. 1998). Given all the possibilities for acquiring resistance to PIs, it is difficult to select any one virus as a target for new inhibitors. Perhaps a better understanding of the interactions between inhibitors and various "mutant" proteases will uncover some common themes employed by these resistant viruses and thus allow design of a compound that inhibits most PI-resistant viruses.

III. The Role of Mutations

1. Mutations in the Protease Gene

Numerous biochemical and structural studies of substituted ("mutant") proteases have begun to elucidate the molecular role of these substitutions (reviewed by CARROL and KUO 1998 and RIDKY and LEIS 1995). Many amino-acid substitutions result in a protease that binds inhibitors with reduced affinity. Examples of this mode of resistance are: (1) the G48V and L90M substitutions both reduce affinity for saquinavir (MASCHERA et al. 1995; MASCHERA et al. 1996a; ERMOLIEFF 1997), (2) the I50V substitution reduces affinity for amprenavir (PAZHANISAMY et al.1996), (3) the I84V substitution reduces affinity for indinavir, ritonavir, saquinavir, and cyclic urea inhibitors such as DMP323 (CHEN et al. 1995; GULNIK et al. 1995; SCHOCK et al. 1996; NILROTH et al. 1997), and (4) the V82T substitution reduces affinity for indinavir, ritonavir, and saquinavir (GULNICK et al. 1995; SCHOCK et al. 1996). Some substitutions decrease the affinity for an inhibitor when present in combination with another mutation but, alone, do not change the enzyme binding. For example, M46I and I47V alone do not change the protease affinity for amprenavir but, when coupled with I50V, they decrease inhibitor affinity by 80% compared with the I50V substitution (PAZHANISAMY et al. 1996). Mechanisms by which these substitutions reduce inhibitor affinity for protease are discussed in section C.IV.

Typically, substitutions that decrease affinity for an inhibitor decrease the catalytic efficiency of the enzyme (on at least some substrates). G48V, I84V, V82T, and I50V all reduce the protease catalytic efficiency (k_{cat}/K_m) to as low as 1/50th that of WT enzyme (dependent on substrate and reaction conditions, GULNIK et al. 1995; MASCHERA et al. 1996a; PAZHANISAMY et al. 1996; SCHOCK et al. 1996; WILSON et al. 1997). This loss of catalytic efficiency reduces the viral fitness and thus negates some of the selective advantage gained through reductions in protease affinity for inhibitor. PI-resistance mutations that lower protease catalytic efficiency also inhibit processing of viral proteins and reduce viral-growth rates (fitness) in vitro (CROTEAU et al. 1997; MASCHERA et al. 1996b; ZENNOU et al. 1998). Furthermore, a minimal amount of enzyme catalytic efficiency (2–25%) is required to keep viral replication viable (ROSE et al. 1995). Mathematical modeling suggests that the protease must maintain an average of at least 61% catalytic efficiency on all substrates (assuming eight successive cleavage sites) to remain viable (RASNICK 1997). Gulnik et al. (1995) were the first to describe a measure of the "vitality" of the enzyme $[(K_i k_{cat}/K_m)_{mut}/(K_i k_{cat}/K_m)_{wt}]$, which normalized the reduction in affinity for the loss in catalytic efficiency. In addition, ERMOLEIFF et al. (1997) demonstrated that mutant proteases are more catalytically efficient than WT proteases in the presence of drug. However, vitality values may be too simple an explanation. First, vitality values depended upon the substrate used for the test (MASCHERA et al. 1996a; PAZHANISAMY et al. 1996; SCHOCK et al. 1996). Second, reduction of binding affinity, even when normalized as a vitality factor, does not always predict the rank order of viral resistance in vitro (KLABE et al. 1998). Nonetheless, to determine the resistance role of a given mutation, it is useful to measure the kinetic properties of the substituted proteases, because changes in these kinetic parameters are often more readily observed than changes in mutant-virus susceptibility in vitro (MASCHERA et al. 1995; TISDALE et al. 1995; MOLLA 1996).

While some mutations reduce the catalytic efficiency of the protease (thus impairing viral fitness), other mutations partially restore catalytic efficiency to hampered enzymes. M46I and L63P have little effect on the affinity of the enzyme for indinavir but improve the catalytic efficiency of the I84V/V82T mutant that has reduced affinity for indinavir and reduced catalytic efficiency (Schock et al. 1996). Similarly, M46I enhances the catalytic efficiency of the catalytically deficient I50V-enzyme (PAZHANISAMY et al. 1996), and A71T enhances the catalytic efficiency of G48V/L90M enzyme (ROSE et al. 1996). Because M46I enhances the catalytic efficiencies of two different proteases (I50V and V82T/I84V), this effect may be a general property of the M46I mutation (SCHOCK et al. 1996). Alternatively, there might be some specificity to the effects of M46I, because both deficient proteases have mutations in the P1/P2'- and P1'/P2-binding sub-site (Fig. 3). That is, I50V and I84V interact with analogous regions of the inhibitor, but I50 resides in the "flap" or "lid" region of the sub-site and I84 resides in the "pan" region of the sub-site.

2. Mutations Outside of the Protease Gene

Compensatory mutations that restore catalytic efficiency to HIV protease and thus restore viral fitness are not limited to the protease gene. Mutations also occur in the genes of the protease substrates. Substitutions in the processing sites for the Gag precursor p1/p6 and/or NC (p7/p1) result in increased substrate efficiency of mutant protease (DOYON et al. 1996). Furthermore, these mutations, in combination with PI-resistant protease mutations, improve viral processing of protein precursors and increase viral-growth rates (fitness) in vitro (DOYON et al. 1996). Because all processing sites are not mutated in a compensatory response, the two sites that are mutated are probably partially rate determining for viral growth and viability. Interestingly, DOYON et al. point out that these two sites have the lowest substrate efficiency (based on in vitro assays; DARKE et al. 1988; TOZSER et al. 1991; WONDRAK et al. 1993). Further evidence that cleavage at these sites is a rate-determining step in viral growth derives from compensatory mutations in the protease gene (M46I; SCHOCK et al. 1996) that increase HIV-protease processing of p1/p6 substrate compared with other processing sites (the p1/p7 site was not examined). Isolates of PI-resistant virus from patients on indinavir therapy also contain compensatory mutations in the p1/p6 and p1/p7 processing sites (ZHANG et al. 1997), demonstrating the relevance of this pathway in vivo.

3. Viral Fitness

As discussed previously (Sect. C.III.1, C.III.2.), many mutations that confer resistance to PIs yield virus that is less fit than WT (the virus grows more slowly in the absence of inhibitor in vitro). The primary reason for this loss in fitness is the reduction in catalytic efficiency of the protease. As noted above, some mutations compensate for the loss of fitness by enhancing the catalytic efficiency for those protease-catalyzed processing steps that are rate limiting for viral growth. Even though PI-resistant virus strains are less fit than WT in the absence of inhibitor, RAYNER et al. (1997) demonstrated that mutant virus strains that are resistant to PIs grow faster than WT in the presence of inhibitor. Furthermore, ERMOLEIFF et al. (1997) demonstrated that the catalytic efficiencies of mutant proteases are higher than those of WT proteases in the presence of inhibitor.

Even though these PI-resistant virus strains grow more slowly than WT in vitro, it has not been established that these mutant viruses are less virulent in vivo. Because the rate of viral growth in vivo (viral load) is the best surrogate marker for disease progression (as discussed earlier), it is likely that the less fit PI-resistant virus would be less virulent. In support of this idea, the partial rebound of viral load of some patients on PI therapy (usually correlated with the appearance of PI-resistant virus) does not result in a significant loss of CD4 cells (DEEKS et al. 1998). These data suggest that mutant virus strains are less virulent to CD4 cells. While patients with viral load rebound would be predicted to progress to acquired immune deficiency syndrome eventually, this data suggests that the process may be retarded.

IV. The Mechanism of Reduction of PI Binding Affinity to Resistant Protease

1. Structural Evaluation

Residues in HIV protease that confer resistance to PIs can be divided into two general structural classes: (1) residues in the active site that make direct interactions with the inhibitor and (2) residues distal to the active site that do not interact directly with the inhibitor. Figure 2 is a schematic representation of the protease that defines three domains that form the "pan" of an active site and two "flap" domains that form a "lid" over the "pan". Figure 3 is a schematic representation of the inhibitor-binding sub-sites that the active-site mutations reside in, using amprenavir as the model inhibitor. Table 1 catalogs many of the residues that confer resistance to PIs, their structural interactions with inhibitors, and the effects of mutations on catalytic efficiency. As suggested by SCHOCK et al. (1996) and CARROL and KUO (1998), most of the active-site substitutions cause reductions in inhibitor binding affinity and a decreased catalytic efficiency. Mutations distal to the active site tend to have minimal effects on inhibitor binding (an exception being L90M) but instead partially compensate for the lost catalytic efficiency of the active-site substitutions.

Several mechanisms describe the reduction of enzyme-inhibitor affinity caused by active-site substitutions. First, removal of van der Waals contacts or hydrophobic interactions directly reduce affinities. I84V removes a methyl group from the (P1/P1′) pocket, thus creating an unfilled cavity in the active site and losing van der Waals' contacts with indinavir (CHEN et al. 1995). Similarly, V82A reduces van der Waals' contacts in the P1′ pocket (BALDWIN et al. 1995) but, interestingly, the flexibility of the enzyme permits P1-pocket "repacking" to maintain contacts. V82F reduces van der Waals' contacts with cyclic urea analogues only when combined with I84V (ALA et al. 1997). Alone, V82F slightly increases interactions. In another example, I50V (in the P2/P2′ pocket) reduced hydrophobic interactions with the phenyl sulfonamide ring of amprenavir (PAZHANISAMY et al. 1996). These substitutions all reduce van der Waals' contacts by creating unfilled space in the active site.

Another mechanism whereby active-site substitutions reduce affinity is by creating repulsive (or unfavorable) interactions. The V82I and V32I mutations are modeled to lead to repulsive (steric overlap) van der Waals contacts (KAPLAN et al. 1994). In a second example, V82T places a hydrophilic residue (hydroxyl) in a hydrophobic environment of the P1 pocket (CHEN et al. 1995), which results in an unfavorable interaction.

Active-site substitutions can also alter the structure outside of the active site by altering "flap" H-bonding and changing flap rigidity (HONG et al. 1997;

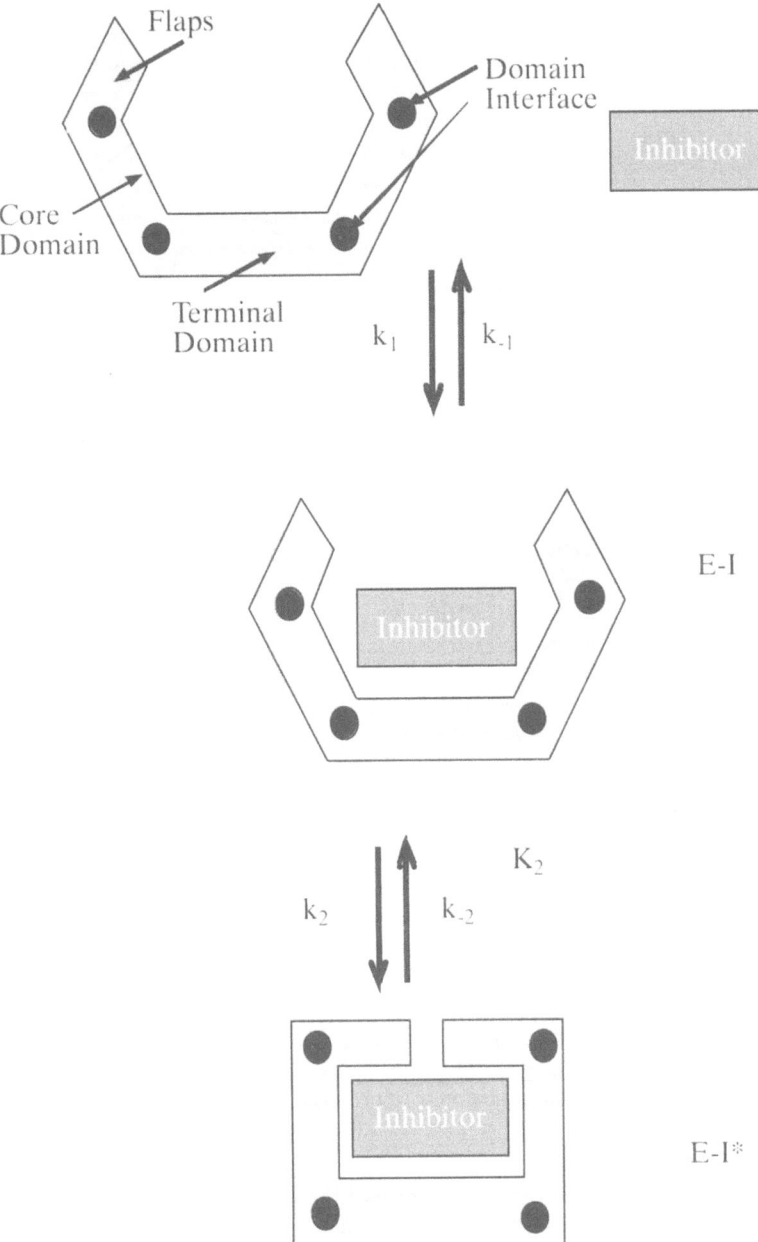

Fig. 2. A model for the two-step binding of human immunodeficiency virus protease and its inhibitors. The dark circles on the enzyme represent the interface between rigid domains of the protein (ROSE et al. 1998). The regions separated by the interfaces are the rigid domains. The conformational change that occurs with the K_2 equilibrium includes the "flaps" closing as well as other rigid-body movements. It is this equilibrium that is primarily affected by substitutions that reduce protease affinity for inhibitors. This change in $K2$ equilibrium is primarily observed as an increased dissociation constant for the protease-inhibitor complex (MASCHERA et al. 1996a)

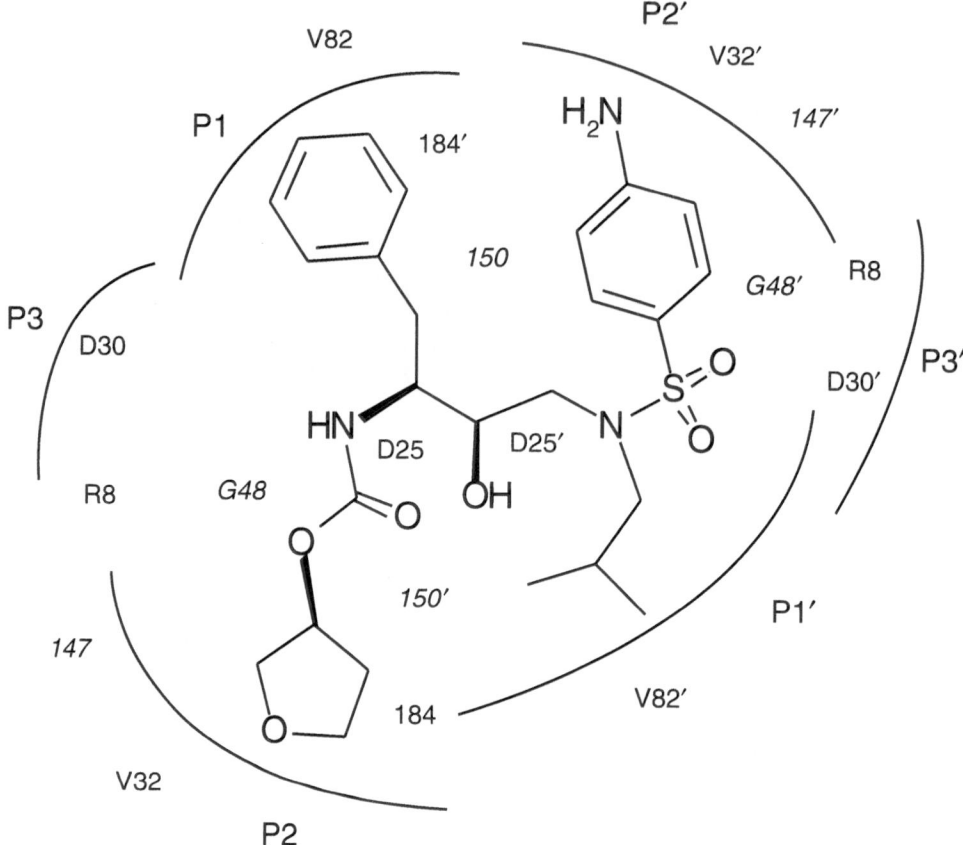

Fig. 3. Schematic representation of the inhibitor-binding sub-sites and the position of the resistance mutations that make direct interactions with the inhibitor in those sub-sites. The model shown is based on amprenavir binding, but other inhibitors make analogous interactions. Amprenavir does not make interactions in the P3 sub-sites. The active site is analogous to a "pan" with a "lid" ("flap" residues); see Fig. 2. The residues that are in the "lid" ("flap") of a sub-site are designated by *italic text*, whereas the residues in the "pan" of a sub-site are designated by *normal text*. The catalytic aspartates (D25 and D25') coordinate (H-bond) the central hydroxyl of the inhibitor and are not residues that mutate in response to protease inhibitors

Hoog et al. 1998). G48 substitutions create flaps that are less mobile and that have a reduction in overall van der Waals' contacts with inhibitor (Hong et al. 1997). Even though these substitutions actually increase van der Waals' contact of the flap itself, the overall interactions and affinity are decreased.

The mechanism by which substitutions outside of the active site operate is typically more difficult to understand, because they often do not grossly affect the structure of the enzyme, nor do they typically affect the affinity of the enzyme for inhibitor. Early studies suggested that L63 and M46 substitu-

Table 1. Most of the mutations are summarized by VACCA and CONDRA (1997) as being selected for by one or more of the approved inhibitors (Fig. 1)

Residue	Structural class	Domain interface	Affinity	Inhibitor studied (Fig. 1)	Catalytic efficiency
R8Q	Active site, P3 (5, 6)	Core:Term (15)	Decreases (5–7)	(1–3)	Decreases (7)
L10F,R	Distal	Core:Flap (1, 2)	Minimal effect (15)	1, 2, 5	Minimal effect (15)
K20	Distal	Core:Term (1, 2)			
L24	Distal				
D30N	Active Site P3 (1)		Decreases	4	
V32	Active Site, P2	Core:Flap (1, 2)	Decreases (7)	(1–3)	Minimal effect (7)
M36	Distal				
N37	Distal (1, 2)				
M46I	Distal (13, 15)		Minimal effect (8, 15)	(1–3, 5)	Increases (8)
I47V	Active site, P2 (1, 2, 15)		Decreases (15)	5	Minimal effect (15)
G48V	Active Site (1) P2		Decreases (10)	(1–3)	Decreases (10, 11)
I50	Active Site P1'/P2 (15)		Decreases (15)	5	Decreases (15)
D60	Distal (1, 2)				
L63P	Distal	Core:Flap (1, 2, 13)	Minimal effect (7)	2	Increases (7)
I64	Distal (1, 2)				
A71T	Distal (1, 2)	Core:Term (1, 2)	Minimal effect (12)	1	Increases (12)
G73	Distal (1, 2)				
V77	Distal	Core:Term (1, 2)			
V82A,F,T,V	Active Site, P1 (1, 2, 5, 13, 14)		Decreases (7–9, 16, 17)	(1–5, 8)	Decreases (7, 8, 9)
I84V	Active Site, P1'/P2 (1, 2, 13)		Decreases (7–9, 16, 17)	(1–5, 8)	Decreases (7–9)
N88S,D	Distal (4)	Core:Term (1, 2)	Decreases (10)	(1–3)	
L90M	Distal				Minimal effect (9–11)

The table is meant to serve as list of examples, not an exhaustive study. Some results may be contradictory. Blanks indicate that I am not aware of a detailed analysis. P1, P2, and P3 refer to the active-site sub-sites at which the residue makes contact with the inhibitor (see Fig. 3) either via H-bonds or van der Waals' contact. Under "domain", the terms *Core, Term* (for terminal), or *Flap* refer to the domain that a residue resides in (also see Fig. 2 or reference 2 below). A *colon* between these terms refers to the domain interface that this reside participates in. For example, *Core:Term* defines the residue as part of the Core domain and interacting at the interface of the Core and Terminal domains. Decreases in binding affinity may be observed for some but not all inhibitors. Increases in catalytic efficiency are defined as increasing the catalytic efficiency of a protease with diminished efficiency on at least some substrates but not necessarily improving wild-type protease.

References (parentheses in table) are: (1) COVELL et al. 1998z; (2) ROSE et al. 1998; (3) HONG et al. 1997; (4) SMIDT et al. 1997z; (5) KAPLAN et al. 1994; (6) HO et al. 1994; (7) GULNIK et al. 1995; (8) SCHOCK et al. 1996; (9) NILROTH et al. 1997; (10) MASCHERA et al. 1996a; (11) WILSON et al. 1997; (12) ROSE et al. 1996; (13) CHEN et al. 1995; (14) BALDWIN et al. 1995; (15) PAZHANISAMY et al. 1996; (16) PATICK et al. 1996z; (17) ALA et al. 1997.

tions result in small perturbations of the flaps (Chen et al. 1995) and/or alter the open/closed equilibrium of the flaps in the free enzyme (M46I; Collins et al. 1995). Recently, Rose et al. (1998) have proposed an intriguing hypothesis that identifies a common theme among many of the substitutions that occur outside of the active site. The model is based on the hypothesis that inhibitor binding occurs in a two- or multi-step process (Furfine et al. 1992; Maschera et al. 1996a). Rose et al. (1998) propose that there are five "rigid" domains of the protease. The domains include the two flaps, the two core domains, and the "central" domain composed of the N- and C-termini (see Fig. 2 for a schematic representation). These domains have interfaces with each other (represented by *black circles* in the Fig. 1 scheme). Many of the resistance substitutions outside of the active site are at the interface of these rigid domains. These domains move/rotate in order to open and close the active site (more movement than simple flaps opening or closing). It is suggested that these domain-interface substitutions result in alterations of the open/closed equilibrium. It is also possible that perturbing the interfaces of the rigid domains would alter the overall active-site architecture by slightly perturbing the juxtaposition of the rigid domains. In some (many?) cases, this altered active site might increase catalytic efficiency without affecting inhibitor binding. For example, in my opinion, some shifts in rigid domains might extend the active site, allowing the enzyme to recognize the more extended binding substituents of substrates (larger molecules capable of more interactions). Alterations in the juxtaposition of rigid domains may also explain how the L90M substitution reduces inhibitor binding.

2. Kinetic Evaluation

Reduction in binding affinity [increased inhibition constant (K_i) values] is the result of changes in the values of association-rate constants and dissociation-rate constants. In the case of protease binding to saquinavir, the affinities of L90M-, G48V-, and G48V/L90M-protease binding were reduced to 1/20, 1/160, and 1/1000 of the respective values for WT (Maschera et al. 1996a). These reductions in affinity were primarily due to increases in the value of the dissociation rate constant of 14-, 90-, and 390-fold, respectively, for the protease-saquinavir complex. The values of the association-rate constants were similar to those for the WT protease. Maschera et al. (1996a) proposed a model for the data assuming a two-step binding mechanism (Fig. 2). The data suggested that $k_2 >>> k_{-1}$. It was proposed that the equilibrium constant K_2 was large, favoring EI*, the activated enzyme–inhibitor complex (Fig. 2), primarily because this is the enzyme conformation observed in X-ray structures of all E–I complexes. This model indicates that the observed dissociation-rate constant (k_{off}) equals k_{-1}/K_2. Therefore, any substitution that decreased K_2 would result in an increased k_{off} and thus a decreased affinity for inhibitor. Substitutions that altered van der Waals' interactions of the final tight complex (EI*) would likely affect the K_2 equilibrium more than the formation of the loose or "collision" (EI) complex (described by k_1 and k_{-1}) where the interactions are lower in quantity and specificity. Therefore, most of the substitutions affecting inhibitor binding would not likely affect the association-rate constant (k_1), as was observed (MASCHERA et al. 1996a). It was originally suggested that the transition from EI to EI* consisted of flap opening and closing; however, the studies of ROSE et al. (1998) indicate that this conformational change might be a more global rotation/movement of rigid domains. The scheme in Fig. 1 is intended to model that interpretation. This model (substitutions that reduce inhibitor affinity primarily affect the K_2 equilibrium) may describe a general mechanism by which substitutions reduce HIV-protease affinity for inhibitors.

V. Chemical Strategies to Inhibit Resistant HIV Protease

There are three strategies adopted to target WT HIV protease and proteases with high-level resistance to current PIs. The first strategy is to make larger inhibitors that: (1) create many H-bond interactions and protein-backbone interactions and (2) create more interactions over a larger area of the active site (JADHAV et al. 1997). Backbone interactions are thought to remain more constant than interactions with amino-acid side chains and are likely to be maintained in "substituted" proteases. If the inhibitor makes interactions over a larger area, losing a few interactions will perturb the binding to a lesser extent. The strategy has been successful in producing inhibitors potent against WT and highly resistant (five mutations) virus. However, to make these extended interactions, these compounds have significantly increased molecular weights that may limit their "drug-like" properties (LIPINSKI et al. 1997). The second strategy is to start with a molecule with sufficiently different chemical properties to which resistant viruses will be sensitive. This strategy is exemplified by PNU-140690 and PD178390, which are effective against clinical isolates having resistance to ritonavir (CHONG and PAGANO 1997; DOMAGALA et al. 1998). Because PNU-140690 is not as potent as ritonavir against WT virus, it remains to be demonstrated that PNU-140690 is effective in a clinical environment. Furthermore, its profile vs other PI-resistant strains has not been investigated. The third strategy used to inhibit resistant virus is to inhibit association of the protease monomers to form the active dimer. HIV protease requires a dimeric form for enzyme activity (JORDAN et al. 1992). The interface between the monomers is likely less altered by resistance-substitutions than the active site. Therefore, this interface is likely to be similar in both WT and substituted (PI-resistant) proteases, and inhibitors of the association should work equally well against WT and substituted proteases. Some inhibitors of the dimerization have been identified (ZUTSHI et al. 1997). While these three strategies are showing some promise, they are considerably less developed than current strategies presented earlier (Sect. B). Furthermore, as discussed earlier (Sect. C.II), it remains unclear which HIV mutant should be the target virus for studies of resistance to PIs.

D. Suggestions for Future Therapeutic Strategies

What are the most effective ways to utilize currently available anti-retroviral agents? Combination therapy has most commonly mixed agents with multiple mechanisms of action, such as NRTIs and PIs. Patients failing these therapeutic regimens have cross-resistance to most available agents and are left with few options for continued therapy (SHAFER et al. 1998). Previously (Sect. B.III.2), we discussed the use of multi-PI combinations and the advantages this may provide for the development of resistance. Perhaps using multiple inhibitors that inhibit a single therapeutic target will increase the durability of therapy compared with combinations of multiple targeted therapeutics. This requires that (1) the virus suffers greater fitness loss by adapting to multiple agents that target one activity than by adapting to multiple single agents that target multiple activities, or (2) acquisition of resistance to multiple agents targeting the same enzyme occurs more slowly than resistance to multiple agents targeted to multiple enzymes. For example, HIV can gain resistance to two agents that target separate enzymes through homologous recombination (YUSA et al. 1997). This is a separate pathway through which to acquire resistance to multiple agents and may increase the rate of resistance acquisition. However, it is possible that this pathway may also be available for acquisition of resistance to multiple agents targeting one enzyme. Even if there is no difference in durability between these two therapeutic strategies, there is a second advantage to using multiple agents targeting the same enzyme. If resistance develops to one class of inhibitors, the second class will still be available for therapy. It has been suggested that patients utilize PI-sparing regimens so that they may use them after failing other treatments. It is just as reasonable to first use multiple-PI therapy and then move to multiple-NRTI therapy if PIs fail. Current clinical trials of anti-retroviral agents will hopefully define the best strategies in the next few years.

Acknowledgements. I thank Drs. David Porter, Andrew Spaltenstein, Webb Andrews, and Steve Short for enlightening discussions and comments during the preparation of this manuscript.

References

Adkins JC, Faulds D (1998) Amprenavir. Drugs 55:837–842
Ala PJ, Hutson EE, Klabe RM, McCabe DD, Duke JL, Rizzo CJ, Korant BD, DeLoskey RJ, Lam PYS, Hodge CN, Chang C-H (1997) Molecular basis of HIV-1 protease drug resistance: structural analysis of mutant proteases complexed with cyclic urea inhibitors. Biochemistry 36:1573–1580
Baldwin ET, Bhat TN, Liu B, Pattabiraman N, Erickson JW (1995) Structural basis of drug resistance for the V82A mutant of HIV-1 proteinase. Nat Struct Biol 2:244–249
Cameron DW, Japour A, Mellors J, Farthing C, Cohen C, Markowitz M, Poretz D, Follansbee S, Ho D, McMahon D, Berg J, Nieves J, Xu Y, Rode R, Salgo M, Leonard

J, Sun E (1998) Antiretroviral safety and durability of ritonavir (RTV)-saquinavir (SQV) in protease inhibitor-naïve patients in year two of follow-up (abstract 388). 5th Conference on Retroviruses and Opportunistic Infections, Chicago, IL February 1–5

Carpenter C, Feinberg M, Aubry W, Averitt D (NIH panel) (1998) Report of the NIH panel to define principles of therapy of HIV infection. Ann Intern Med 128: 1057–1078

Carr A, Samaras K, Chisholm DJ, Cooper DA (1998) Pathogenesis of HIV-1-protease inhibitor-associated peripheral lipodystrophy, hyperlipidaemia, and insulin resistance. Lancet 351:1881–1883

Carrol SS, Kuo LC (1998) Viral resistance at the enzyme level. Int Antiviral News 6:103–107

Chen Z, Li Y, Schock H, Hall D, Chen E, Kuo LC (1995) Three-dimensional structure of a mutant HIV-1 protease displaying cross-resistance to all protease inhibitors in clinical trials. J Biol Chem 270:21433–21436

Chong K-T, Pagano PJ (1997) In vitro combination of PNU-140690, a human immunodeficiency virus type 1 protease inhibitor, with ritonavir against ritonavir-sensitive and -resistant clinical isolates. Antimicrob Agents Chemother 41:2367–2373

Chun T-W, Stuyver L, Mizell SB, Ehler LA, Mican JA, Baseler M, Lloyd AL, Nowak MA, and Fauci A (1997) Presence of an inducible HIV-1 reservoir during highly active retroviral therapy. Proc Natl Acad Sci USA 94:13193–13197

Coen DM (1998) The persistence of HIV in memory T cells. Trends Microbiol 6:129–130

Cohen OJ, Fauci A (1998) Transmission of mutidrug-resistant human immunodeficiency virus–the wake-up call. N Engl J Med 339:341–343

Collins JR, Burt SK, Erickson JW (1995) Flap opening in HIV-1 protease simulated by 'activated' molecular dynamics. Nat Struct Biol 2:334–338

Condra JH (1998) Resistance to HIV protease inhibitors. Haemophilia 4:610–615

Condra JH, Schleif WA, Blahy OM, Gabryelski LJ, Graham DJ, Quintero JC, Rhodes A, Robbins HL, Roth E, Shivaprakash M, Titus D, Yang T, Teppler H, Squires KE, Deutsch PJ, Emini EA (1995) In vivo emergence of HIV-1 variants resistant to multiple protease inhibitors. Nature 374:569–571

Condra JH, Holder DJ, Schleif WA, Blahy OM, Danovich RM, Gabryleski LJ, Graham DJ, Laird D, Quintero JC, Rhodes A, Robbins HL, Roth E, Shivaprakash M, Yang T, Chodakewitz JA, Deutsch PJ, Leavitt RY, Massari FE, Mellors JW, Squires KE, Steigbigel RT, Teppler H, Emini EA (1996) Genetic corrleates of in vivo resistance to indinavir, a human immunodeficiency virus type 1 protease inhibitor. J Virol 70:8270–8276

Conway B, Routy J-P, Sekaly R-P (1998) Combination therapy for HIV: towards long term control of disease progression. Exp Opin Invest Drugs 7:941–961

Covell DG, Jernigan RL, Walquist A (1998) J. Molec Struct (Theochem) 423, 93–100. Structural Analysis of inhibitor binding to HIV-1 Protease: identification of a common binding motif

Croteau G, Doyon L, Thibealt D, McKercher G, Piloto L, La Marre D (1997) ••. J Virol 71:1089–1096

Darke PL, Nutt RF, Brady SF, Garsky VM, Ciccarone TM, Leu CH, Lumma PK, Freidlinger RM, Verber DF, Sigal IS (1988) HIV-1 protease specificity of peptide cleavage is sufficient for processing of Gag and Pol polyproteins. Biochem Biophys Res Comm 156:297–303

De Pasquale MP, Murphy R, Gulick R, Smeaton L, Sommadossi J-P, Degruttola V, Caliendo A, Kuritzkes D, Sutton L, Savara A, D'Aquila R (1998) Mutations selected in HIV plasma RNA during 141W94 therapy (abstract 406A). 5th Conference on Retroviruses and Opportunistic Infections, Chicago, IL, 1–5 February

Deeks S, Beatty G, Cohen PT, Grant R, Volberding P (1998) Viral load and CD4⁺ T-cell counts changes in patients failing potent protease inhibitor therapy (abstract

419). 5th Conference on Retroviruses and Opportunistic Infections, Chicago, IL, 1–5 February

Domagala JM, Boyer F, Ellsworth E, Gajda C, Hagen S, Hamilton H, Markoski L, Prasad V, Steinbaugh B, Tait B, Lunney E, Broadfuehrer J, Gracheck S, Hupe D, Iyer K, Pavlovsky S, Saunders J, Sharmeen L, Tummino P (1998) PD178390: a novel potent non peptide protease inhibitor of the 5,6-dihydro-4-hydroxy-2-pyrone class (abstract 638). 5th Conference on Retroviruses and Opportunistic Infections, Chicago, IL, 1–5 February

Doyon L, Croteau G, Thibeault D, Poulin F, Pilote L, Lamarre D (1996) Second locus involved in human immunodeficiency virus type 1 resistance to protease inhibitors J Virol 70:3763–3769

Eastman PS, Mittler J, Kelso R, Gee C, Boyer E, Kolberg J, Urdea M, Leonard JM, Norbeck DW, Mo H, Markowitz M (1997) Genotypic changes in human immunodeficiency virus type 1 associated with loss of suppression of plasma viral RNA level in subjects treated with ritonavir (Norvir) monotherapy. J Virol 72:5154–5164

Ermolieff J, Lin X, Tang J (1997) Kinetic properties of saquinavir-resistant mutants of human immunodeficiency virus type 1 protease and their implications in drug resistance in vivo. Biochemistry 36:12364–12370

Eron J, Haubrich R, Richman D, Lang W, Tisdale M, Meyers R, Pagano G, Rogers M (1998) Preliminary assessment of 141W94 in combination with other protease inhibitors (abstract 6). 5th Conference on Retroviruses and Opportunistic Infections, Chicago, IL, 1–5 February

Fauci AS, Bartlett JG, Goosby EP, Smith MD (NIH Panel) (1998) Guidelines for the use of antiretroviral agents in HIV-infected adults and adolescents. Ann Intern Med 128:1079–1100

Flexner C (1998) HIV-protease inhibitors. N Engl J Med 338:1281–1292

Furfine ES, D'Souza ED, Ingold KJ, Leban JJ, Spector T, Porter DJT (1992) Two-step binding mechanism for HIV protease inhibitors. Biochemistry 31:7886–7891

Gallant JE, Heath-Chiozzi M, Raines C, Anderson R, Katz T, Fields C, Flexner C (1998) Safety and efficacy of nelfinavir-ritonavir combination therapy (abstract 394A). 5th Conference on Retroviruses and Opportunistic Infections, Chicago, IL, 1–5 February

Goldschmidt RH, Balano KB, Legg JJ (1998) Individualized strategies in the era of combination antiretroviral therapy. J Am Board Fam Pract 11:158–164

Goodgame J, Stein A, Pottage J, Jablonowski H, Vafidis I, Hanson C (1998) Amprenavir/3TC/ZDV is superior to 3TC/ZDV in HIV-1 infected antiretroviral therapy-naive subjects (abstract 191 LB5). 38th Interscience Conference on Antimicrobial Agents and Chemotherapy24–27 September 1998, San Diego

Gulnik SV, Suvorov LI, Liu B, Yu B, Anderson B, Mitsuya H, Erickson JW (1995) Kinetic characterization and cross-resistance patterns of HIV-1 protease mutants selected under drug pressure. Biochemistry 34:9282–9287

Hecht FM, Grant RM., Petropoulos CJ, Dillon B, Chesney MA, Tian H, Hellmann NS, Bandrapalli NI, Digilio L, Branson B, Kahn JO (1998) Sexual transmission of an HIV-1 variant resistant to multiple reverse-transcriptase and protease inhibitors. N Engl J Med 339:307–311

Ho DD, Neumann, AU, Perelson AS, Chen W, Leonard JM, Markowitz M (1995) Rapid turnover of plasma virions and CD4 lymphocytes in HIV-1 infection. Nature 373:123–126

Hoetelmans RMW, Meenhirst PL, Mulder JW, Burger DM, Koks CHW, Beijnen JH (1997) Clinical pharmacology of HIV protease inhibitors: focus on saquinavir, indinavir, and ritonavir. Pharm World Sci 19:159–175

Hong L, Zhang X-J, Foundling S, Hartsuck JA, Tang J (1997) Structure of a G48H mutant of HIV-1 protease explains how glycine-48 replacements produce mutants resistant to inhibitor drugs. FEBS Lett 420:11–16

Hoog SS, Abdel-Meguid SS (1998) Towards the understanding of retroviral proteases, implications for the design of human immunodeficiency virus type 1 and 2 protease inhibitors. J Mol Struct 423:59–65

Jacobsen H, Hanggi M, Ott M, et al. (1996) In vivo resitance to a human immunodeficiency virus type 1 proteinase inhibitor: mutations, kinetics and frequencies. J Infect Dis 173:1379–1387

Jadhav PK, Ala P, Woerner FJ, Chang C-H, Garber SS, Anton ED, Bacheler LT (1997) Cyclic urea amides: HIV-1 protease inhibitors with low nanomolar potency against both wild-type and protease inhibitor resistant mutants of HIV. J Med Chem 40:181–191

Johnson M, Peterson A, Winslade J, Clendeninn N (1998) Comparison of BID and TID dosing of viracept (nelfinavir, NFV) in combination with stauvudine (d4T) and lamivudine (3TC) (abstract 373). 5th Conference on Retroviruses and Opportunistic Infections, Chicago, IL, 1–5, February

Jordan SP, Zugay J, Darke PL, Kuo LC (1992) Activity and dimerization of human immunodeficiency virus protease as a function of solvent composition and enzyme concentration. J Biol Chem 267:20028–20032

Kakuda TN, Strubble KA, Piscitelli SC (1998) Protease inhibitors for the treatment of human immunodeficiency virus infection. Am J Health Syst Pharm 55:233–253

Kaplan AH, Michael SF, Wehbie RS, Knigge MF, Paul DA, Everitt L, Kempf DJ, Norbeck DW, Erickson JW, Swanstrom R (1994) Selection of multiple human immunodeficiency virus type 1 variants that encode viral proteases with decreases sensitivity to an inhibitor of the viral protease. Proc Natl Acad Sci USA 91:5597–5601

Kempf DJ, Rode RA, Xu Y, Sun E, Heath-Chiozzi ME, Valdes J, Japour AJ, Danner S, Boucher C, Molla A, Leonard JM (1997) The duration of viral suppression during protease inhibitor therapy for HIV-1 infection is predicted by the plasma HIV-1 RNA at the nadir. AIDS 12:F9–F14

Kempf DJ, Sham HL, Marsh KC, Flentge CA, Betebenner D, Green BE, McDonald E, Vasavononda S, Saldivar A, Wideburg NE, Kati WM, Ruiz L, Zhao C, Finno LM, Patterson J, Molla A, Plattner JJ, Norbeck DW (1998) Discovery of ritonavir, a potent inhibitor of HIV protease with high oral bioavailability and clinical efficacy. J Med Chem 41:602–617

Klabe RM, Bacheler LT, Ala PJ, Ericksonviitanen S, Meek JL (1997) Resistance to HIV protease inhibitors – a comparison of enzyme inhibition and antiviral potency. Biochemistry 37:8735–8742

Lal R, Hsu A, Granneman GR, El-Shourbagy T, Johnson M, Lam W, Manning L, Japour A, Sun E (1998) Multiple dose safety, tolerability and pharmacokinetics of ABT-378 in combination with ritonavir (abstract 647). 5th Conference on Retroviruses and Opportunistic Infections, Chicago, IL, 1–5 February

Lech WJ, Wang G, Yang YL, Chee Y, Dorman K, McCrae D, Lazzeroni LC, Erickson JW, Sinsheimer JS, Kaplan AH (1996) In vivo sequence diversity of the protease of human immunodeficiency virus type 1: presence of protease inhibitor-resistant variants in untreated subjects. J Virol 70:2038–2043

Leonard J (1996) Prolongation of life and prevention of AIDS in advanced immunodeficiency with ritonavir. 3rd Conference on Retroviruses and Opportunistic Infections, Washington, DC, 28 January – 1 February

Lipinski CA, Lombardo F, Dominy BW, Feeny PJ (1997) Experimental and computational approaches to estimate solubility and permeability in drug discovery and development settings. Adv Drug Delivery Rev 23:3–25

Manion DJ, Hirsch MS (1997) Combination chemotherapy for human immunodeficiency virus-1. Am J Med 102:76–80

Markland W, Zuchowski L, Black J, Rao BG, Parsons JD, Pazhanisamy S, Griffith JP, Tisdale M, Tung R (1998) Kinetic and Structural Analysis of HIV-1 protease mutations: amprenavir resistance, cross-resistance and resensitization (abstract). International Conference on AIDS, Geneva, Switzerland

Markowitz M, Conant M, Hurley A, Schluger R, Duran M, Peterkin J, Chapman S, Patick A, Hendricks A, Yuen GJ, Hoslins W, Clendeninn N, Ho DD (1998) A preliminary evaluation of nelfinavir mesylate, an inhibitro of human immunodeficiency virus (HIV)-1 protease, to treat HIV infection. J Infect Dis 177:1533–1540

Maschera B, Furfine E, Blair ED (1995) Analysis of human immunodeficiency virus type 1 protease inhibitors by using matched bacterial expression and proviral infection vectors. J Virol 69:5431–5436

Maschera B, Darby G, Palu G, Wright LL, Tisdale M, Meyers R, Blair ED, Furfine ES (1996a) Human immunodeficiency virus: mutations in the viral protease that confer resistance to saquinavir increase the dissociation rate constant of the protease-saquinavir complex. J Biol Chem 271:33231–33235

Maschera B, Tisdale M, Darby G, Meyers R, Palu' G, Blair ED (1996b) In vitro growth characteristics of HIV-1 variants with reduced sensitivity to saquinavir explain the appearance of the L90M escape mutant in vivo. 5th International Workshop on HIV Drug Resistance, 3–6 July, Whistler, Canada. Antiviral Ther 1[Suppl 1]:53

Molla A, Japour A (1997) HIV protease inhibitors. Curr Opin Infect Dis 10:491–495

Molla A, Kohlbrenner WE (1997) Resistance to antiretroviral drug therapy. Ann Rep Med Chem 32:131–140

Molla A, Korneyeva, M, Gao Q, Vasavanonda S, Schipper PJ, Mo H-M, Markowitz M, Chernyavskiy T, Niu P, Lyons, N, Hsu A, Granneman GR, Ho DD, Boucher CAB, Leonard JM, Norbeck DW, Kempf DJ (1996) Ordered accumulation of mutations in HIV protease confers resistance to ritonavir. Nat Med 2:760–766

Moyle G (1998) The role of combinations of HIV protease inhibitors in the management of persons with HIV infection. Exp Opin Invest Drugs 7:413–426

Nguyen B-Y, Haas DW, Ramirez-Ronda C, Thompson MA, Gallant J, Currier J, Paar D, White C, Collier A, Mehrotra D, Chung M, Harvey C, Chodakewitz J (1998) Thirty-two week follow-up of indinavir sulfate (IDV) administered Q8 hours (H) versus Q12 H in combinations with zidovudine (ZVD) and lamivudine (3TC) (abstract 374). 5th Conference on Retroviruses and Opportunistic Infections, Chicago, IL, 1–5 February

Nilroth U, Vrang L, Markgren P-O, Hulten' J, Hallberg A, Danielson UH (1997) Human immunodeficiency virus type 1 proteinase resistance to symmetric cuclic urea inhibitor analogs. Antimicrob Agents Chemother 41:2383–2388

Palella FJ, Delaney KM, Moorman AC, Loveless MO, Fuhrer J, Statten GA, Aschman DJ, Holmberg SD (1998) Declining morbidity and mortality among patents with advanced human immunodeficiency virus infection. N Engl J Med 338:853–860

Patick AK, Mo H, Markowitz M, Appelt K, Wu B, Musick L, Kalish V, Kaldor S, Reich S, Ho D, Webber S (1996) Antimicrob. Agents Chemotherapy 40, 292–297. Antiviral and resistance studies of AG1343, an orally bioavailable inhibitor of human immunodeficiency virus protease

Pazhanisamy S, Stuver CM, Cullinan AB, Margolin N, Rao BG, Livingston DJ (1996) Kinetic characterization of human immunodeficiency virus type-1 protease-resistant variants. J Biol Chem 271:17979–17985

Perelson AS, Essunger P, Cao Y, Vesanen M, Hurley A, Saksela K, Markowitz, M, Ho DD (1997) Decay characteristics of HIV-1-infected compartments during combination therapy. Nature 387:188–191

Rasnick D (1997) Kinetics analysis of consecutive HIV proteolytic cleavages of the Gag-Pol polyprotein. J Biol Chem 272:6348–6353

Rayner MM, Cordova B, Jackson DA (1997) Populations dynamics studies of wild-type and drug-resistant mutant HIV in mixed infections. Virology 236:85–94

Richman DD (1997) The impact of human immunodeficiency virus drug resistance on treatment efficacy. Antiviral Ther 2[Suppl 2]:11–15

Ridky T, Leis J (1995) Development of drug resistance to HIV-1 protease inhibitors. J Biol Chem 270:29621–29623

Roberts NA, Craig C, Sheldon J (1998) Resistance and cross-resistance with saquinavir and other HIV protease inhibitors: theory and practice. AIDS 12:453–460

Rose JR, Babe LM, Craik CS (1995) Defining the level of human immunodeficiency type 1 (HIV-1) protease activity required for HIV-1 particle maturation and infectivity. J Virol 69:2751–2758

Rose RE, Gong Y-F, Greytok JA, Bechtold CM, Terry BJ, Robinson BS, Alam M, Colonno RJ, Lin P-F (1996) Human immunodeficiency virus type 1 viral background plays a major role in development of resistance to protease inhibitors. Proc Natl Acad Sci USA 93 1648–1653

Rose RB, Craik CS, Stroud RM (1998) Domain flexibility in retroviral proteases: structural implications for drug resistant mutations. Biochemistry 37:2607–2621

Schock H, Garsky VM, Kuo LC (1996) Mutational anatomy of an HIV-1 protease variant conferring cross-resistance to protease inhibitors in clinical trials. J Biol Chem 271 31957–31963

Smidt ML, Potts KE, Tucker SP, Blystone L, Stiebel TR, Stallings WC, MacDonald JJ, Pillay D, Richman DD, Bryant ML (1997) Antimicrob. Agents Chemotherapy 41, 515–522. A mutation in human immunodeficiency virus type 1 protease at position 88, located outside of the active site, convers resistance to SC-55389A

Shafer RW, Winters MA, Palmer S, Merigan TC (1998) Multiple concurrent reverse transcriptase and protease mutations and multidrug resistance of HIV-1 isolates from heavily treated patients. Ann Intern Med 128:906–911

Tarpley WG (1998) •• (abstract L6). In: Proceedings of the 5th Conference on Retroviruses and Opportunistic Infections, 1–5 February, Chicago, IL.

Tisdale M, Myers RE, Maschera B, Parry NR, Oliver NM, Blair ED (1995) Cross-resistance analysis of human immunodeficiency virus type 1 variants individually selected for resistance to five different protease inhibitors. Antimicrob Agents Chemother 39:1704–1710

Tozser J, Blaha I, Copeland TD, Wondrak EM, Oroszlan S (1991) Comparison of HIV-1 and HIV-2 proteinases using oligopeptide substrates representing cleavage sites in Gag and Gag-pol polyproteins. FEBS Lett 281:77–80

Tucker SP, Stiebel TR Jr, Potts KE, Smidt ML, Bryant M (1998) Estimate of the frequency of human immunodeficiency virus type 1 protease inhibitor resistance within unselected virus populations in vitro. Antimicrob Agents Chemother 42:478–480

Vacca JP, Condra JH (1997) Clinically effective HIV protease inhibitors. Drug Discovery Today 2:261–272

Vella S, Palmisano L (1997) Update on protease inhibitors. Antiviral Ther 2[Suppl 2]:29–37

Von Moltke LL, Greenblatt DJ, Grassi JM, Granda BW, Duan SX, Fogelman SM, Daily JP, Harmatz JS, Shader RI (1998) Protease inhibitors as inhibitors of cytochromes P450: high risk associated with ritonavir. J Clin Pharmacol 38:106–111

Wilson SI, Phylip LH, Mills JS, Gulnik SV, Erickson JW, Dunn BM, Kay J (1997) Escape mutants of HIV-1 proteinase: enzymic efficiency and susceptibility to inhibition. Biochim Biophys Acta 1339:113–125

Wlodawer A, Vondrasek J (1998) Inhibitors of HIV-1 protease: a major success of structure-assisted drug design. Annu Rev Biophys Biomol Struct 27:249–284

Wondrak EM, Louis JM, de Rocquigny H, Chermann JC, Roques BP (1993) The Gag precursor contains a specific HIV-1 cleavage site between the NC (p7) and p1 proteins. FEBS Lett 333:21–24

Yamaguchi K, Byrn RA (1995) Clinical isolates of HIV-1 contain few pre-existing proteinase inhibitor conferring mutations. Biochem Biophys Acta 1253:136–140

Yusa k, Kavlick MF, Kosalaraska P, Mitsuya H (1997) HIV-1 acquires resistance to two classes of antiviral drugs through homologous recombination. Antiviral Res 36:179–189

Zennou V, Mammano F, Paulous S, Mathez D, Clavel F (1998) Loss of viral fitness associated with multiple Gag and Gag-Pol processing defects in human immunodeficiency virus type 1 variants selected for resistnce to protease inhibitors in vivo. J Virol 72:3300–3306

Zhang Y-M, Imamichi H, Imamichi T, Lane HC, Falloon J, Vasudevachari MB, Salzman NP (1997) Drug resistance during indinavir therapy is caused by mutations in the protease gene and in its Gag substrate cleavage sites. J Virol 71:6662–6670

Zutshi R, Franciskovich J, Shultz M, Schweitzer B, Bishop P, Wilson M, Chmielewski J (1997) Targeting the dimerization interface of HIV-1 protease: inhibition with cross-linked interfacial peptides. J Am Chem Soc 119:4841–4845

Section II
Other Viral (Non-HIV)
Protease Inhibitors

CHAPTER 5

The Proteinases Encoded by Hepatitis C Virus as Therapeutic Targets

C. Steinkühler, U. Koch, R. De Francesco, and A. Pessi

A. Introduction

The hepatitis-C virus (HCV), first identified in 1989 (Choo et al. 1989), is the leading causative agent of blood-borne and community-acquired non-A, non-B viral hepatitis (Houghton 1996). According to the World Health Organization, more than 170 million people worldwide may be infected with HCV. The main route of transmission of HCV is parenteral; in the past, transfusion of blood and blood products were an important source of HCV transmission. The incidence of transfusion-associated infections has greatly diminished as a consequence of the development of reliable diagnostic assays and an effective screening of blood or blood products. Currently, the use of intravenous drugs and sexual transmission account for most HCV exposure (Alter 1997). Up to 30% of patients have no obvious risk factors for infections, and these cases of hepatitis C are termed sporadic.

Typically, infection with HCV occurs without overt clinical symptoms, causing jaundice only in a minority of cases (Hoofnagle 1997). However, at least 85% of patients who become infected with HCV develop chronic hepatitis. It is estimated that 20–30% of patients with chronic hepatitis eventually develop liver cirrhosis, but the process is insidious and may take 3–20 years (Hoofnagle 1997). A large number of these patients will develop complications of end-stage liver disease, such as liver failure, portal hypertension and hepatocellular carcinoma (Alberti and Realdi 1991).

Current therapy for infection with HCV involves treatment with interferon-α, alone or in combination with the nucleoside analogue ribavirin (reviewed in Lindsay 1997). The efficacy of interferon-α therapy is, however, rather low, with response rates of about 50%. More than half of these responders will relapse after cessation of the treatment, resulting in less than 20% sustained biochemical and virological response (Davis et al. 1989; Marcellin et al. 1991; Carithers and Emerson 1997). While ribavirin monotherapy revealed no consistent effect on HCV viremia relative to placebo (Di Bisceglie et al. 1995; Dusheiko et al. 1996), the efficacy of interferon-α appears to be enhanced in a combination therapy with ribavirin (reviewed in Reichard 1997). Recent phase-III clinical trials involving patients with hepatitis C who had relapsed following interferon-α therapy showed that combination therapy resulted in a significant increase in the number of individuals showing eradication of detectable virus compared with a second course of interferon-α alone. Combination therapy also appears to be more effective than interferon monotherapy for the treatment of naive hepatitis-C patients.

A wide array of adverse side effects of interferon therapy have been described (DUSHEIKO 1997). The most common are flu-like symptoms (fever, fatigue, headaches, nausea, arthralgias and myalgias), but also anorexia, alopecia, thrombocytopenia, leukopenia and neuropsychiatric disorders have been reported. Another important side effect associated with the addition of ribavirin to interferon-α monotherapy is reversible hemolysis.

Considering that no vaccine is available to prevent HCV infection, there is an urgent need for a safe and efficacious treatment of the disease by a novel anti-viral drug. Virally encoded enzymes that are essential for replication are the choice targets for an anti-HCV therapy (BARTENSCHLAGER 1997). The search for such targets has been severely hampered by the lack of a reliable system to grow HCV in cultured cells and by the lack of an animal model other than the chimpanzee. In spite of these limitations, the molecular cloning of the viral genome combined with the powerful tools of recombinant-DNA technology has led to the identification of several viral enzymatic functions that are believed to be essential for viral replication (Neddermann et al. 1997) (Fig. 1).

The viral genome of HCV is a 9.6-kb single-stranded positive-sense RNA molecule containing a single open-reading frame (ORF) that encodes a polyprotein of 3010–3033 amino acids (MAJOR and FEINSTONE 1997). HCV has a similar genomic organization to the pesti- and flaviviruses, and has now been classified as a separate genus of the Flaviviridae family (MILLER and PURCELL 1990; FRANCKI et al. 1991). The HCV polyprotein undergoes proteolytic processing in the cytoplasm or in the endoplasmic reticulum (ER) of the infected cell to give rise to at least ten mature proteins (reviewed in LOHMAN et al. 1996). The polypeptides arising from proteolytic processing of the amino-terminal quarter of the polyprotein are thought to be the structural proteins. The structural proteins include the core protein (p21/p19) and two envelope glycoproteins, E1 (gp31–35) and E2 (gp70). Processing of the structural proteins is effected by cellular signal peptidases associated with the lumen of the ER. Host signal peptidases are also responsible for the biogenesis of p7, a 7-kDa protein of unknown function generated by the post-translational limited processing of an E2-p7 precursor (LOHMAN et al. 1996). The remainder of the polyprotein contains the so-called nonstructural (NS) proteins believed to be part of the viral replication apparatus (NEDDERMANN et al. 1997). The NS region is proteolytically processed by two virally-encoded enzymes. The NS2-NS3 junction is cleaved by a zinc-dependent proteinase associated with NS2 (p23) and the N-terminus of NS3, the so-called NS2/3 proteinase (GRAKOUI et al. 1993a; HIJIKATA et al. 1993). The C-terminal remainder of the HCV polyprotein is further processed to give rise to mature NS3 (p70), NS4A (p8), NS4B (p27), NS5A (p58) and NS5B (p68) proteins via a serine proteinase contained within the NS3 protein (BARTENSCHLAGER et al. 1993; ECKART et al. 1993;

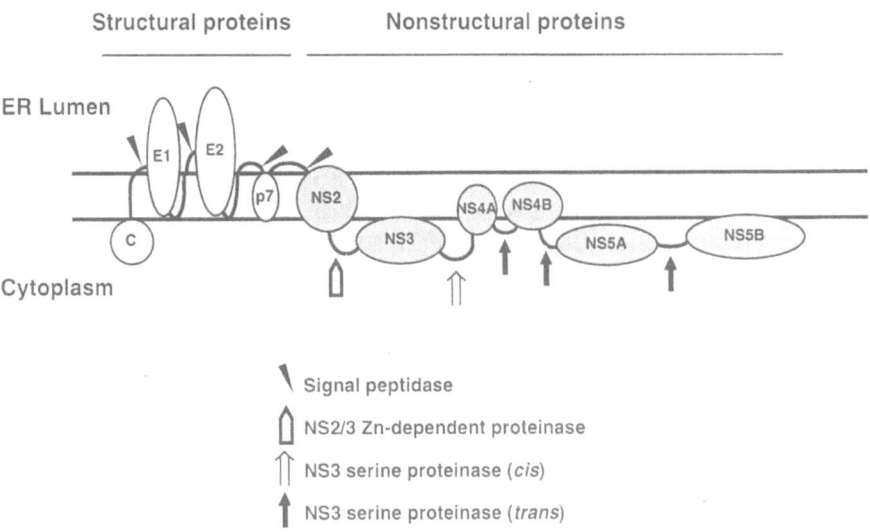

Fig. 1. Schematic representation of the polyprotein proteolytic processing and hypothetical membrane topology of the hepatitis-C-virus-encoded proteins. The proteins are arranged in order (*left* to *right*) of their appearance in the polyprotein. The nonstructural (NS) proteins are shown in *gray*. *ER*, endoplasmic reticulum

GRAKOUI et al. 1993b; HIJIKATA et al. 1993; TOMEI et al. 1993; D'SOUZA et al. 1994; MANABE et al. 1994). The serine-proteinase domain is contained within the N-terminal 180 amino acids of NS3, residues 1027–1206 of the HCV polyprotein (BARTENSCHLAGER et al. 1994; LIN et al. 1994; TANJI et al. 1994; FAILLA et al. 1995; HAHM et al. 1995; HAN et al. 1995). The C-terminal portion of the protein contains an RNA helicase and an RNA-stimulated ATPase (Neddermann et al. 1997).

Although the NS3 proteinase has proteolytic activity of its own, interaction with a second viral protein, NS4A, is essential for efficient processing of all the NS3-dependent polyprotein cleavage sites (Bartenschlager et al. 1994; Failla et al. 1994; Lin et al. 1994; Tanji et al. 1995). NS3 and NS4A have been shown to form a stable complex in cells expressing the HCV polyprotein (Bartenschlager et al. 1995; Failla et al. 1995; Lin et al. 1995; Satoh et al. 1995). A 14-amino acid, hydrophobic region of NS4A (polyprotein residues 1678–1691), has been identified as necessary and sufficient for the activation of the NS3 proteinase (Lin et al. 1995; Butkiewicz et al. 1996; Koch et al. 1996; Shimizu et al. 1996; Tomei et al. 1996); synthetic peptides encompassing this region of NS4A can efficiently activate the NS3 proteinase in vitro via the formation of a 1:1 complex (Lin et al. 1995; Shimizu et al. 1996; Steinkühler et al. 1996; Tomei et al. 1996; Bianchi et al. 1997).

Specific inhibition of the proteolytic activities of the virally encoded NS2/NS3 and NS3 proteinases is presently regarded as a promising strategy through which to interfere with HCV replication (Bartenschlager 1997). The aim of this review is to summarize the progress made in the characterization of the two HCV proteinases for the purpose of developing potent and selective proteinase inhibitors as efficient anti-HCV drugs.

B. The NS3 Proteinase

I. Structure of the NS3 Proteinase

1. The NS3 Proteinase is a Chymotrypsin-Like Serine Proteinase

The three-dimensional structure of the NS3 proteinase domain (Love et al. 1996) and that of the NS3 proteinase domain complexed with an NS4A-derived peptide (Kim et al. 1996; Yan et al. 1998) were recently solved by X-ray crystallography. The structure of the complex is considered to represent the enzymatically active form of the HCV serine proteinase and, therefore, it will be discussed first.

The NS3 N-terminal domain complexed with NS4A-derived peptides adopts a typical chymotrypsin-like fold (Lesk and Fordham 1996) consisting of two nearly symmetrical β-barrel domains with the residues of the catalytic triad, His-1083, Asp-1107 and Ser-1165, located in a crevice at the interface between the two domains (Fig. 2A). The C-terminal domain contains a canonical six-stranded β-barrel. The N-terminal β-barrel instead contains eight β-strands: the two additional strands are contributed by the N-terminus of the proteinase (A_0) and by the NS4A peptide, respectively. The NS4A cofactor forms a β-strand that intercalates in an antiparallel fashion between the two β-strands A_0 and A_1 in a β-sheet within the N-terminal domain of the enzyme (Fig. 2B). The 22 N-terminal residues of NS3, which were implicated by dele-

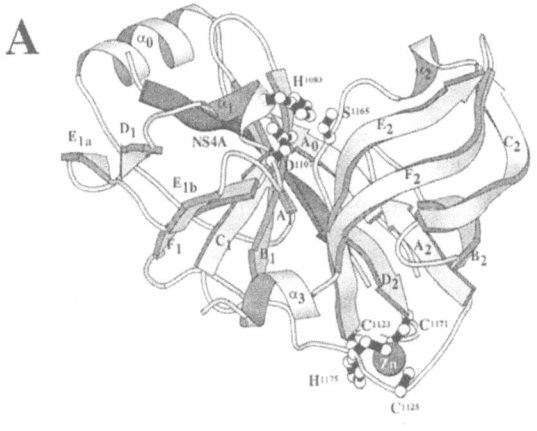

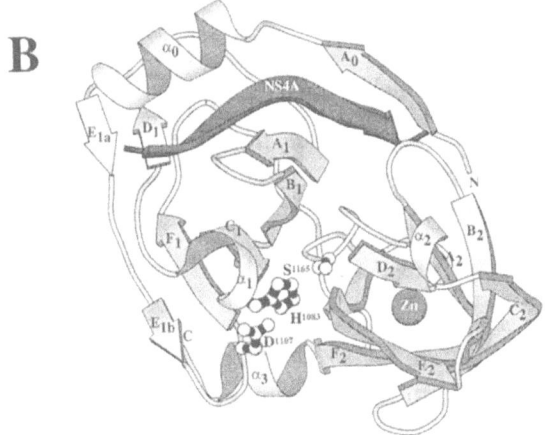

Fig. 2A,B. Schematic representations of the three-dimensional structure of the hepatitis-C-virus NS3 proteinase domain complexed with an NS4A-derived peptide. **A, B** Two different orientations. The images were generated using the MOLSCRIPT software. The coordinates used to generate the molecular models have been deposited in the Protein Database (PDB; file name: 1JXP)

tion mapping in the stabilization of the NS3-NS4A complex, encompass strand A_0 and helix α_0. In good agreement with the deletion mapping results, these two structural elements form a sort of molecular clamp that locks the NS4A cofactor onto the NS3 proteinase domain. It is worth pointing out that strand A_0 and helix α_0 have no counterparts in other structurally characterized serine proteinases: this is in line with the earlier finding that deletion of the corresponding region of NS3 resulted in an enzyme that retains the basal NS4A-independent proteolytic activity (FAILLA et al. 1995; SATOH et al. 1995; KOCH et al. 1996).

In the crystal structure without NS4A, the C-terminal β-barrel of the NS3 proteinase domain adopts essentially the same fold observed in the complex. In contrast, the N-terminal barrel of the protein assumes a significantly different conformation: the N-terminal 30 amino acids of NS3 extend away from the protein core and interact with hydrophobic surface patches of neighboring molecules in the crystallographic asymmetric unit (LOVE et al. 1996). In solution, this N-terminal region is likely to be disordered when not engaged in the interaction with NS4A.

Together with the rearrangement of the N-terminal part of the molecule, a subtle but crucial conformational change at the proteinase active site takes place upon cofactor binding. Two residues of the NS3 catalytic triad, namely His-1083 and Asp-1107, are located within the NS3 N-terminal β-barrel. In the NS3 proteinase crystal structure, the carboxyl group of Asp-1107 points away from His-1083 and forms a hydrogen bond with the guanidinium moiety of Arg-1181, a residue that is strictly conserved in all HCV genotypes. Conversely, in the crystal structure of the NS3-NS4A complex, the side chain of Asp-1107 is hydrogen bonded to the His-1083 imidazole. In this way, the imidazole ring of His-1083 is polarized and can act as a general base catalyst to extract a proton from the enzyme nucleophile, the hydroxyl group of Ser-1165. It has thus been suggested that positioning of Asp-1107 as a member of the catalytic triad is induced by the presence of the NS4A cofactor (LOVE 1998). This rearrangement would result in the proper alignment of the enzyme active-site residues in order to make efficient catalysis possible.

2. A Zinc-Binding Site in the NS3 Serine-Proteinase Domain

Comparative analysis of the polyprotein sequence of the different HCV genotypes and of the related viruses GB-A, -B and -C led to the identification of three strictly conserved Cys residues and one His residue within the NS3 proteinase domain. In a homology model of the HCV NS3 proteinase domain, the conserved residues clustered in space, suggesting that they could serve as ligands of a metal-binding site (DE FRANCESCO et al. 1996; GORBALENYA and SNIJDER 1996). This prediction was confirmed experimentally by the finding that the purified NS3 proteinase domain contains stoichiometric amounts of zinc (DE FRANCESCO et al. 1996; STEMPNIAK et al. 1997). The assignment of the metal ligands was ultimately confirmed by X-ray crystallography; the zinc ion was shown to be tetrahedrally coordinated by Cys-1123, Cys-1125 and Cys-1171 and, through a bridging water molecule, by His-1175 (KIM et al. 1996; LOVE et al. 1996; YAN et al. 1998). The metal-binding site is located opposite the active site and is believed to play a structural role (Figs. 2A,B). Interestingly, picornavirus-2A proteinases (see chap. 6 of this vol.) contain two sequence motifs, Cys-X-Cys and Cys-X-His, that are located in a region which is topologically equivalent to the one containing the metal-binding site of HCV NS3 proteinase (DE FRANCESCO et al. 1996; GORBALENYA and SNIJDER 1996). The 2A proteinase of rhinovirus has also been experimentally demonstrated to contain a tightly bound zinc ion that is required for the formation of an active enzyme and is an essential component of the native structure (Voss et al. 1995). Picornavirus-2A proteinases adopt a fold similar to that of chymotrypsin-like serine proteinases and contain a triad of residues that are spatially equivalent to the catalytic Asp-His-Ser. However, the catalytic Ser is replaced by a Cys in these viral enzymes (GORBALENYA and SNIJDER 1996). The structural conservation of a metal-binding site between HCV NS3 and picornavirus 2A is striking from an evolutionary point of view, since the zinc-binding site appears to be more conserved than the catalytic-triad residues.

3. Substrate Specificity of NS3 Serine Proteinase

The location of the sites cleaved by the NS3 proteinase within the HCV polyprotein was obtained by N-terminal sequencing of the mature NS4A, NS4B, NS5A and NS5B proteins (GRAKOUI et al. 1993b; PIZZI et al. 1994). Based on a comparative analysis of the sequences flanking the cleaved peptide bonds (Table 1), it has been possible to derive the following consensus sequence for the NS3-dependent cleavage site (P_6 to P_1'): Asp/GluXaa$_4$Cys/Thr-Ser/Ala. It is interesting to note that cleavage occurs after a Cys residue in all *trans* cleavage sites, whereas the intramolecular site between NS3 and NS4A is unique in this respect, having a Thr residue in the P_1 position. Remarkably, a Thr in the P_1 position has been shown to be suboptimal when incorporated into peptide substrates (URBANI et al 1997), suggesting that factors different from cleavage efficiency must have acted in the selection of a Thr residue in the P_1 position of the NS3/NS4A junction. Other conserved features of the different cleavage sites are a negatively charged residue in the P_6 positions and a Ser or Ala residue in the P_1' positions. In addition, the amino acid present in the P_4' position of all sites, although not strictly conserved, always possesses a bulky, hydrophobic side chain. Alanine-scanning experiments performed on peptide substrates derived from either the NS4A/NS4B (URBANI et al. 1997) or from the NS5A/NS5B (ZHANG et al. 1997) cleavage sites confirmed that the P_1 and, to a lesser extent, the P_6 and P_4' residues contribute to substrate recognition by the NS3 proteinase.

Table 1. Sequence of the NS3-dependent cleavage sites within the viral polyprotein (hepatitis-C-virus genotype 1b)

Site	Sequence	Mechanism
NS3-NS4A	**DLEVVT-STWV**	*cis*
NS4A-NS4B	**DEMEEC-ASHL**	*trans*
NS4B-NS5A	**DCSTPC-SGSW**	*trans*
NS5A-NS5B	**EDVVCC-SMSY**	*trans*

The preference displayed by the NS3 serine proteinase for cleavage after a Cys residue can be rationalized on the basis of the structure of the enzyme S_1 specificity pocket. The S_1 specificity pocket of the NS3 proteinase is small and lipophilic, lined by the hydrophobic side chains of Leu-1161, Phe-1180 and Ala-1183. The shape of the relatively small and lipophilic Cys side chain is complementary to the pocket. In addition, the sulfhydryl group can interact in a rather specific way with the aromatic ring of Phe-1180. The positively polarized S–H hydrogen could interact favorably with the π-cloud of the aromatic ring. Alternatively, the sulfur lone pairs could form hydrogen-bond-type interactions with the aromatic C–H. Similar interactions have been observed in a number of protein and small-molecule crystal structures (BURLEY et al. 1986; LEVITT et al. 1988). Consistent with the lack of apparent specificity over the P_2-P_5 region, the structure of the NS3 proteinase reveals a substrate-binding cleft that is rather shallow and solvent exposed (Fig. 3, s. appendix, page 398/399). All major loops connecting the β-strands that in other serine proteinases form the S_2, S_3 and S_4 subsites are very short or absent in NS3 (LOVE et al. 1996; KIM et al. 1996; YAN et al. 1998). In the case of other chymotrypsin-like proteinases of known structure, β-strand E_2 forms β-sheet-like hydrogen bonds with the substrate over P_2-P_3. In the NS3 serine proteinase, this β-strand is 2–3 residues longer than usually observed in the homologous enzymes (LOVE et al. 1996). It has been suggested that peptide main-chain interactions between the NS3 proteinase strand E_2 and the substrates continue for three more residues, until P_6 (Fig. 4, s. appendix, page 398/399), and that the continuous backbone interactions might compensate for the apparent lack of well-defined S_2-S_5 subsites. The two positively charged side-chains of Arg-1187 and Lys-1191 have been suggested to engage in specific interactions with the conserved negatively charged residue in the P_6 position of the substrates (Fig. 3; LOVE et al. 1996). In addition, on the P' site of NS3, an extended lipophilic surface is formed by NS4A and the NS4A-binding region (Fig. 3). This surface may be responsible for the preference for large hydrophobic amino acids in the substrate P4' position (LOVE et al 1988). In summary, substrates appear to be stabilized in the enzyme active site via interactions that extend over several peptide bonds.

II. Inhibitors of the NS3 Proteinase

1. Noncompetitive Inhibitors

To date, the efforts of pharmaceutical companies to discover novel drugs for hepatitis-C infection by screening libraries of compounds for activity against the NS3 serine proteinase have made little progress. The only small-molecule inhibitors disclosed so far that were found by screening include a phenanthrenequinone from a fermentation broth; this compound is active in a cell-free translation assay via an undefined mechanism (CHU et al. 1996), nitrophenols (SUDO et al. 1997b), thiazolidines (SUDO et al. 1997a) and halogenated benzanilides (KAKIUCHI et al. 1998); all these compounds are rather weak, noncompetitive inhibitors of the enzyme. Nitrophenols and thiazo-lidines are also poorly selective with respect to proteinases unrelated to NS3 (SUDO et al. 1997a, b). Further studies will be necessary to evaluate the significance of these results for the development of an antiviral drug.

2. Active-Site-Directed Inhibitors

Due to the absence of well-defined P$_2$, P$_3$ or P$_4$ binding pockets and to the requirement for large peptide substrates spanning at least P6-P4', the design of active-site inhibitors based on the knowledge of the enzyme structure appears exceedingly difficult in the case of NS3 proteinase (STEINKÜHLER et al. 1996). Despite these difficulties, considerable progress has been made in understanding the interactions between the enzyme, the NS4A cofactor and the substrate, and reports of potent peptide inhibitors are beginning to appear.

a) Substrate Analogues

LANDRO et al. (1997) described non-cleavable sub-micromolar decapeptide inhibitors based on the P$_6$-P$_4'$ sequence of the NS5A-NS55B cleavage site (Table 1). Substitution of secondary amino acids [i.e., Pro, Tic (tetrahydroiso-quinoline-3-carboxylic acid) or Pip (pipecolinic acid)] for the P$_1'$ Ser in the context of a substrate decapeptide, yielded peptides that were no longer cleaved, but were potent inhibitors of the enzyme. The decapeptide Glu-Asp-Val-Val-Abu-Val-Leu-Cys-Tic-Nle-Ser-Tyr (Abu, 2-aminobutyric acid; Nle, norleucine) was thus shown to be a competitive inhibitor of the NS3-NS4A proteinase complex with an inhibition constant (K_i) of 340 nM. The occupation of the S$_1'$ pocket by a bulky residue such as Tic in the most potent inhibitor reported by LANDRO and colleagues is somewhat surprising, since the natural substrate consensus for this position is either Ser or Ala (Table 1). However, modelling studies (LANDRO et al. 1997) suggested that S$_1'$ is not fully occupied by the P$_1'$-Ser of a docked substrate; thus, it has the potential to accommodate larger residues. Modeling also suggests that the hydrophobic part of the Lys-1162 side chain can engage in lipophilic interactions with hydrophobic amino acids in P$_1'$ and P$_2'$, such as Tic and Nle. One of the most interesting aspects of the work reported by LANDRO and colleagues is that the affinity of the decapeptide inhibitor for the enzyme showed an 80-fold drop in the absence of the NS4A cofactor. Deletion of up to three prime residues of this decapeptide (P$_4'$-P$_2'$) resulted in a 40-fold decrease in affinity for the NS3-NS4A complex, but only in a 4-fold decrease of the affinity for the free NS3 proteinase. In contrast, the progressive deletion of amino acids from the P side of the molecule led to roughly the same loss of affinity for both com-plexed and uncomplexed NS3 proteinase. Overall, this study has suggested that, while the interaction of the enzyme with the P side of the substrate con-tributes most of the inhibitor-binding energy, the formation of auxiliary P'-binding subsites is promoted by the presence of the NS4A cofactor.

b) Product Analogues

Two groups of researchers (INGALLINELLA et al. 1998; LLINÀS-BRUNET et al. 1998a,b) have recently exploited P-region binding to develop potent peptide inhibitors of the enzyme. Their work stemmed from the observation that NS3 undergoes competitive inhibition by the N-terminal cleavage products of synthetic peptides corresponding to its cleavage sites (LLINÀS-BRUNET et al. 1998a; STEINKÜHLER et al. 1998). A detailed study of all the natural cleavage sites (STEINKÜHLER et al. 1998) showed that, remarkably, the NS3 proteinase displays a higher affinity for the cleavage products than it does for the corresponding substrates; under the conditions used for the activity assay, the observed K_i values for the hexapeptide products arising from the NS4A-NS4B, NS4B-NS5A and NS5A-NSA5B sites were 0.6 μM, 180 μM and 1.4 μM, respectively, while the Michaelis-Menten constant values for the corresponding substrates were 10 μM, >1 mM and 3.8 μM, respectively. The structure–activity relationship (SAR) of the product inhibitors was studied in depth (INGALLINELLA et al. 1998; LLINÀS-BRUNET et al. 1998a, b). Optimal binding was found to require a dual anchor: a "P_1 anchor", including the C-terminal α-carboxylate, and an "acid anchor" at the N-terminus of the molecule. The P_1 residue contributed to the binding energy through both its side chain and its free carboxylate. Not surprisingly, the side chain providing optimal occupancy of the S_1 pocket of the enzyme was Cys; substitution of the P1 Cys with Val, Ala, Ser, Ile, Leu, Nle or Gly led to a potency decrease that ranged between 20- and 60-fold. Nva (norvaline), *allo*-Ile (*allo*-isoleucine) and Abu were found to be somewhat better substitutes for Cys, but still displayed a drop in potency of five- to tenfold. A crucial role was played by the C-terminal α-carboxylate; deletion of this group with maintenance of the side chain, substitution with a primary amide or reduction to alcohol invariably caused a greater than tenfold reduction in inhibitor potency. Interestingly, methyl and benzyl amide derivatives of the product inhibitor showed a less marked decrease in potency of three to ninefold (LINÀS-BRUNET et al. 1998b).

These studies concluded that the carboxylic acid functionality at the C-terminus contributes considerably to potency and imparts great specificity to product-derived peptide inhibitors of the HCV NS3 serine proteinase. By analogy to product-binding observed crystallographically for other serine proteinases (JAMES et al. 1980; STEITZ and SHULMAN 1982; CHOI et al. 1991; MARTIN et al. 1992, NIENABER et al. 1993; TONG et al. 1993; NIENABER et al. 1996), it has been proposed that one of the α-carboxylate oxygen atoms could bind in the enzyme oxyanion hole, while the other could be engaged in hydrogen bonding with the imidazole moiety of the active-site His (Fig. 4). Moreover, inspection of the structure of the active site of the NS3 proteinase, site-directed mutagenesis experiments and the pH-dependence of the observed product inhibition have suggested that the side chain of Lys-1162 may be selectively contributing to the stabilization of the bound product inhibitor, possibly by establishing an ionic interaction with the P_1 α-carboxylate negative charge (STEINKÜHLER et al. 1998).

The second anchor of the product inhibitors resided in the P_6-P_5 acidic pair, whose simultaneous deletion also yielded a greater than 100-fold decrease in activity. Being electrostatic in nature, this interaction was found to be permissive regarding the exact nature of the negatively charged moiety; achiral diacids with a four or five-membered carbon chain are good substitutes for Asp in P_6 or Glu in P_6-deleted pentapeptides. Inspection of the surface of NS3 shows that there is a set of basic amino acids without nearby neutralizing residues (Fig. 3) spatially close to the two basic amino acids (Arg-1187 and Lys-1191) that were proposed to be responsible for the preference for acidic amino acids in the P_6 substrate position (LOVE et al. 1996). The presence of this cluster of positive charges should further strengthen a favorable electrostatic environment. High side-chain flexibility and solvation of charged amino acids usually make ionic interactions on a protein surface rather undirectional; this would explain why binding does not depend on the exact nature of the charged residues in P_5 and P_6. The importance of this distal anchor is best emphasized by the finding that both the P1-deleted pentapeptides and the C-terminally amidated hexapeptides were still competitive inhibitors of NS3, albeit with much-reduced potency.

The study of the SAR in the P_2–P_3 portion of the inhibitors was pursued by the synthesis of a series of combinatorial peptide libraries with a large set of residues including D- and non-coded amino acids. P_2 showed a preference for either negatively charged or hydrophobic residues, while polar and positively charged residues were not accepted; the best P_2 residue was Cha (β-cyclohexylalanine). P_4 showed a strong preference for hydrophobic amino acids, both aliphatic and aromatic, the best residue being Dif (3,3-diphenylalanine), followed by Leu; again, positively charged and polar amino acids were detrimental to binding. In both positions the chirality had to be L-, since no D-amino acid was found active in either position. The best combination for P_2/P_4 (Cha/Dif) gave a potent competitive inhibitor of the NS3 proteinase, with a K_i of 40 nM. The results for position P_3 were very clear-cut: only two residues yielded a potency comparable with Glu in the P_3 position: Val and Ile.

The preference for hydrophobic amino acids such as Leu and Cha in P_2 is not immediately obvious. The side chain of Ala-1182 in S_2 is solvent exposed in the crystal structure and will be covered by the P_2 side chain (KOCH, unpublished observation). The space available to the P_2 side chain is limited due to a cleft formed by His-1183 and Ala-1182, suggesting that β-branched hydrophobic amino acids such as Val and Ile are less active for steric reasons. Another contribution could be due to an indirect electrostatic effect, since the His-1083-Asp-1107 ion pair of the catalytic triad is covered by the Leu and Cha side chains, thus becoming less solvent exposed. According to calculations this should increase the pK_a of His-1083, thus stabilizing the protonated form of the imidazole and, indirectly, the α-carboxylate of product inhibitors in the oxyanion hole. The P_3 amino acid of the product inhibitor is likely to form hydrogen bonds to Ala-1183 of the β-strand E_2. The S_3 region accessible to the P_3 side chain is formed by a number of hydrophobic side chains of amino acids Leu-1161, Val-1158, Ala-1183 and Cys-1185 (KOCH, unpublished observation). Rather than an isolated hydrophobic pocket, the S_3 region seems to be an extension of the P_1 specificity pocket (Fig. 3). Thus the preference for hydrophobic amino acids in P_3 seems to be due to the lipophilic nature of the S_3 region. Also, the Glu side-chain can form some hydrophobic contacts, while the acid group can interact favourably with two basic amino acids: Lys-1162 and Arg-1187. In S_4, the solvent-exposed side-chain of Val-1184 forms a small hydrophobic patch (Fig. 3) that should favor hydrophobic amino acids in this position (Koch, unpublished observation). Intramolecular, hydrophobic-collapse-like interactions between the hydrophobic P_2 and P_4 side chains can minimize the solvent-accessible lipophilic surface area and thus stabilize the extended conformation, which binds to NS3.

Another unexpected feature of the inhibitor SAR was that the parent Glu residue in the P5 position could be substituted by a hydrophobic amino acid (tyrosine, 4-nitrophenylalanine) or by a D-amino acid; D-Glu was threefold more active than L-Glu, and D-Gla (γ-carboxyglutamic acid) was 30-fold more active. Ac-Asp-(D-)Gla-Leu-Ile-Cha-Cys-OH is the most potent inhibitor of NS3 reported to date, with a median inhibitory concentration (IC_{50}) of 1.5 nM. Modeling suggests that the substitution of L-Glu by D-Glu in P_5 changes the backbone conformation, thus positioning the D-Glu side chain roughly in the place of the P_6 Asp side chain, while the P_6 Asp takes the place of the original L-Glu side chain.

Overall, electrostatics seems to play a dominant role in NS3–inhibitor binding. A remarkable feature of NS3 is the accumulation of basic amino acids along the P site of the substrate-binding region, especially around the active site and in the S_5-S_6 region and, indeed, analysis of the electrostatic properties of product inhibitors and the NS3 surface in the S_1-S_6 region reveal a high degree of electrostatic complementarity (KOCH, unpublished observation). Accordingly, an increase in ionic strength lowers considerably the potency of the inhibitors, depending on the number of ionizable residues in the molecule (INGALLINELLA et al. 1998).

The study by LLINÀS-BRUNET et al. (1998a), using as a parent compound the sequence Asp-Asp-Ile-Val-Pro-Cys-OH, is in excellent agreement with that of INGALLINELLA et al. (1998). First, the optimal length for the inhibitor is six residues, and acetylation of the N-terminus is beneficial. Second, alanine-scanning experiments confirm the importance of valine in P_3. Third, the P_5 Asp can be substituted by either a hydrophobic residue like t-butylglycine or Val, or by a D-amino acid, D-Asp and D-Val giving seven and fourfold higher potency, respectively; interestingly, the beneficial effect of a D-amino acid is not present in the context of a pentapeptide. Finally, the P_6 Asp cannot be deleted (fivefold potency loss), but the Ac-Asp moiety can be substituted by the achiral 3-carboxypropanoyl moiety.

c) Serine-Trap Inhibitors

The presence of a Cys residue in the P_1-residue position constitutes a major obstacle for the medicinal chemist. In fact, a thiol-containing side chain complicates the preparation of classical mechanism-based serine-proteinase inhibitors ("serine traps") because of the mutual incompatibility of the thiol nucleophile with the serine-targeting electrophile. Despite using a large panel of amino-acid replacements for the P_1 Cys, no satisfactory substitute has yet been identified (URBANI et al. 1997; ZHANG et al. 1997; LANDRO et al. 1997; LLINÀS-BRUNET et al. 1998a, b). The first mechanism-based inhibitor described (LANDRO et al. 1997), a hexapeptide aldehyde based on the NS5A/5B cleavage site with Val in P1 (Glu-AspVal-Val-Abu-Val-CHO), showed a disappointingly modest potency ($IC_{50} = 50\,\mu M$). More recently, LLINÀS-BRUNET et al. (1998b) studied the effects that introducing an electrophilic ("activated") carbonyl had on the potency of two hexapeptide inhibitors based on the sequences Asp-Asp-Ile-Val-Pro-Nva and Asp-(D-)Asp-Ile-Val-Pro-Nva. Aldehyde derivatives bound the NS3 enzyme 15-fold better than the corresponding acid analogues. Fluorine-containing carbonyl derivatives such as trifluoromethylketone (TFMK) derivatives and pentafluoroethylketone (PFEK) derivatives displayed only modest potencies, whereas α-ketoamide derivatives were found to be 25- to 75-fold stronger inhibitors of the NS3 proteinase than the corresponding carboxylic acids. It is worth pointing out that, in contrast to what was observed for other serine proteases, peptide sequences containing "activated" carbonyls were only marginally more active than the corresponding carboxylic acids.

Despite the impressive speed at which new data are generated, it is difficult to define the criteria for successful design of NS3-targeted drugs. The increasing understanding of the way in which this enzyme works reinforces pessimism about the feasibility of using a small-molecular-weight (MW) molecule to block an enzyme that uses electrostatics as the main driving force for substrate binding and features a unique specificity for a Cys in the P1 pocket. However, it has been shown that low-nanomolar inhibitors of NS3 can indeed be developed by extensive cumulative optimization of weak initial leads. Further improvement by applying the power of combinatorial non-peptide chemistry to this problem could thus be envisaged. Moreover, the most potent of these inhibitors should be amenable to co-crystallization or nuclear magnetic resonance (NMR) experiments, yielding valuable information for structure-based "rational" inhibitor design.

C. The NS2/3 Proteinase

The mature N-terminus of NS3 is generated by a viral proteinase encoded within the NS2 and NS3 regions of the HCV polyprotein. Mutagenesis experiments showed that the NS3 serine-proteinase domain, but not its catalytic activity were required for processing at this junction (GRAKOUI et al. 1993a;

HIJIKATA et al. 1993). In fact, mutagenesis of the residues of the catalytic triad of the NS3 serine proteinase, which abolishes all cleavages downstream of NS3, has no effect on hydrolysis at the NS2-NS3 site. Conversely, it was not possible to replace the NS3 moiety in the NS2-NS3 proteinase by other sequences from HCV without abolishing activity. A deletion-mutagenesis analysis performed in order to map the minimum domain required for efficient processing at the NS2-NS3 site indicated an N-terminal boundary between residues 898 and 923 within NS2, whereas a sharp drop in cleavage efficiency was observed upon C-terminal truncations beyond residue 1207 within NS3 (GRAKOUI et al. 1993a; HIJIKATA et al. 1993; REED et al. 1995; SANTOLINI et al. 1995). Radiosequencing of the mature cleavage products showed that processing at the NS2-NS3 junction occurs between Leu-1026 and Ala-1027 (polyprotein numbering) within the sequence Gly-Trp-Arg-Leu-Leu/Ala-Pro-Ile (GRAKOUI et al. 1993). Cleavage at the NS2-NS3 site is remarkably resistant to single amino-acid substitutions, the only dramatically inhibiting mutations being those likely to cause conformational alterations, simultaneous deletion of the residues flanking the cleavage site or their concomitant substitution with alanine residues (REED et al. 1995). Membranes were shown to enhance cleavage efficiency at the NS2-NS3 junction in a viral-strain-specific way (SANTOLINI et al. 1995), and membrane-activation could be mimicked by detergent micelles (PIERONI et al. 1997).

Thiol-reactive agents such as iodacetamide or N-ethylmaleimide or N-tosyl-l-phenylalanine chloromethyl ketone were found to inhibit the enzyme (PIERONI et al. 1997). Also, metal chelators such as ethylene diamine tetracetic acid or phenanthroline abolished cleavage, and subsequent addition of zinc or cadmium was able to restore activity (HIJIKATA et al 1993; PIERONI et al. 1997). Based on these findings, HIJIKATA and co-workers (1993) proposed that NS2-NS3 might be a zinc-dependent metalloproteinase. An extensive mutagenesis of all conserved His, Cys and Glu residues within NS2-NS3 was performed, aimed at identifying possible metal ligands. His-952 and Cys-993, which are localized within NS2, were recognized as essential for proteolytic activity. Although these residues were originally proposed to participate in the coordination of the zinc ion, the finding of a structural zinc binding site in the NS3 proteinase domain (DE FRANCESCO et al. 1996; KIM et al. 1996; LOVE et al. 1996; YAN et al. 1998) suggests that zinc could, in fact, be required for the NS2-NS3 proteinase activity because it stabilizes the fold of NS3. His-1175, Cys-1123, Cys-1125 and Cys-1171 were identified as the zinc ligands in NS3. Interestingly, mutation of these Cys residues into alanine abolished both the NS3 serine-proteinase activity and led to an impairment of the processing at the NS2-NS3 site (HIJIKATA et al. 1993). In contrast to these findings, REED et al. (1995) reported that a truncated precursor protein spanning residues 827 through 1137 of the HCV polyprotein still possesses some NS2-NS3 cleavage activity even though the truncation eliminated the zinc ligating Cys-1171 and His-1175 in NS3. Interestingly, addition of the proteinase domain of NS3 in trans to this precursor resulted in an improved processing efficiency.

Wu and co-workers (1998) have recently discussed the possible mechanisms by which the cleavage between NS2 and NS3 might occur. They conclude that the existing data are compatible with either a Cys proteinase or a zinc-dependent metalloproteinase being responsible for the processing event. Thus, mutagenesis data are consistent with the notion that the cleavage could be catalyzed by a Cys proteinase having His-952 and Cys-993 as a catalytic diad. In addition, it was noticed that mutation of Glu-972 also impaired NS2/3 proteinase activity, leading to the suggestion that this residue might serve as a putative third component of a catalytic triad. In this case, zinc would serve the purpose of conferring structural stability and would not be directly involved in the mechanism of catalysis. In line with this interpretation, GORBALENYA and SNIJDER (1996) have classified NS2/3 as a Cys proteinase.

The alternative view invokes a zinc-dependent metalloproteinase as a catalyst in the scission of the NS2-NS3 junction. It has been noticed that the zinc-binding site of NS3 is located in remarkable proximity to the scissile peptide bond in the NS2-NS3 junction (Wu et al. 1998). Zinc-dependent hydrolases usually contain the catalytic metal ion bound to three nitrogen and oxygen ligands provided by the protein, and to an activated water molecule (VALLEE and AULD 1990a, b). The zinc-binding site in NS3 differs considerably from this consensus, containing a metal ion coordinated to three Cys and one His ligand (His-1175), which participates in the coordination via a bridging water molecule. However, both X-ray crystallography (LOVE et al. 1996) and NMR studies (URBANI et al. 1998) have shown that this His residue is endowed with a considerable flexibility, compatible with a switch of the metal-binding site between an "open" and a "closed" conformation. It could be speculated that the "open" conformation with the His moving apart, leaving a water-bound zinc ion located in the proximity of the NS2-NS3 junction, may play a role in hydrolysis. In their recent review, Wu and co-workers (1998) have suggested that the role of His-1175 may indeed be reminiscent of the "Velcro" mechanism by which certain metalloproteinases are activated. According to the "Velcro" hypothesis, a prometalloproteinase exists in a latent state characterized by a tetradentate chelation of the catalytic metal. Activation occurs upon removal of one ligand and its replacement by a water molecule. In the case of the purified NS3 proteinase domain, this conformational switch is triggered by pH changes (URBANI et al. 1998). Other factors might influence the process in the context of the NS2-NS3 precursor. Intriguingly, it was found that in the "open" conformation the affinity of the NS3 proteinase domain for its cofactor, NS4A, was impaired. NS4A binding involves the N-terminus of NS3, where the NS2-NS3 cleavage site is located, thus suggesting that conformational changes occur in this region concomitantly with the removal of His-1175 from the coordination sphere of zinc.

Investigation of the mechanistic details of the NS2-NS3 processing is presently hampered by the unavailability of enzymatically active, purified proteins. REED et al. (1995) and WILKINSON (1997) have reported that cleavage at the NS2-NS3 junction may occur in *trans* between an NS2-NS3 precursor carrying a mutation in the cleavage site (which would act as an enzyme in the *trans* cleavage reaction) and another molecule containing a mutation which abolishes enzymatic activity (acting as a substrate in the intermolecular reaction). Bimolecular cleavage was only observed if the substrate polypeptide contributed at least one functional domain to the formation of an active proteinase. Precursors that contain a NS2-NS3 site that is cleavable in *trans* are capable of supplying either the N-terminal domain of NS3 or a functional NS2 region to form the NS2/3 proteinase. Although this behavior is suggestive of dimerization of precursor molecules occurring prior to cleavage, dimer formation could not be detected in immunoprecipitation experiments, indicating that relatively weak interactions are operative in the binding reaction of the two precursor molecules. It thus appears that the NS2-NS3 cleavage site is physiologically processed in an intramolecular reaction that can be forced to proceed in a bimolecular fashion in mutant precursors.

Autocleavage of the NS2-NS3 site could be inhibited in *trans* by the addition of polypeptides capable of participating in bimolecular cleavage reactions, i.e. those precursors having either a functional NS2 portion or an intact NS3 serine-proteinase region (REED et al. 1995). This finding indicates that low-MW inhibitors may also interfere with this processing event. Further work will be required to determine whether interference with the NS2/3 proteinase activity is a viable strategy for the development of anti-HCV therapeutics.

References

Alberti A, Realdi G (1991) Parentally acquired non-A, non-B (type C) hepatitis. In: McIntyre N, Benhamou J-P, Bircher J, Rizzetto M, Rodes J (eds) Oxford textbook of clinical hepatology. Oxford University, Oxford

Alter MJ (1997) Epidemiology of hepatitis C. Hepatology 26[Suppl 1]:62S–65S

Bartenschlager R, Ahlborn-Laake L, Mous J, Jacobsen H (1993) Nonstructural protein 3 of the hepatitis C virus encodes a serine-type proteinase required for cleavage at the NS3/4 and NS4/5 junctions. J Virol 67:3835–3844

Bartenschlager R, Ahlborn Laake L, Mous J, Jacobsen H (1994) Kinetic and structural analyses of hepatitis C virus polyprotein processing. J Virol 68:5045–5055

Bartenschlager R, Lohman V, Wilkinson T, Koch JO (1995) Complex formation between the NS3 serine-type proteinase of the hepatitis C virus and NS4A and its importance for polyprotein maturation. J Virol 69:7519–7528

Bartenschlager R (1997) Candidate targets for hepatitis C virus-specific antiviral therapy. Intervirology 40:378–393

Bianchi E, Urbani A, Biasiol G, Brunetti M, Pessi A, De Francesco R, Steinkühler C (1997) Complex formation between the hepatitis C virus serine proteinase and a synthetic NS4A cofactor peptide. Biochemistry 36:7890–7897

Burley SK, Petsko GA (1986) Amino-aromatic interactions in proteins. FEBS Lett 203:139–143

Butkiewicz NJ, Wendel M, Zhang R, Jubin R, Pichardo J, Smith EB, Hart AM, Ingram R, Durkin J, Mui PW, Murray MG, Ramanathan L, Dasmahapatra B (1996) Enhancement of hepatitis C virus NS3 proteinase activity by association with NS4A-specific synthetic peptides: identification of sequence and critical residues of NS4A for the cofactor activity. Virology 225:328–338

Carithers RL Jr, Emerson SS (1997) Therapy of hepatitis C: meta-analysis of interferon alfa-2b trials. Hepatology 26[Suppl 1]:83S–88S

Choi HK, Tong L, Minor W, Dumas P, Boege U, Rossmann MG, Wengler G (1991) Structure of Sindbis virus core protein reveals a chymotrypsin-like serine proteinase and the organization of the virion. Nature 345:37–43

Choo QL, Kuo G, Weiner AJ, Overby LR, Bradley DW, Houghton M (1989) Isolation of a cDNA clone derived from a blood-borne non-A, non-B viral hepatitis genome. Science 244:359–362

Chu M, Mierzwa R, Truumees I, King A, Patel M, Berrie R, Hart A, Butkiewicz N, Das-Mahapatra B, Chan TM, Puar MS (1996) Structure of Sch 68631: a new hepatitis C virus proteinase inhibitor from Streptomyces sp. Tetrahedron Lett 37:7229–7232

D'Souza EDA, O'Sullivan E, Amphlett EM, Rowlands DJ, Sangar DV, Clarke BE (1994) Analysis of NS3-mediated processing of the hepatitis C virus non-structural region in vitro. J Gen Virol 75:3469–3476

Davis GL, Balart LA, Schiff ER, Lindsay K, Bodenheimer HC Jr, Perrillo RP, Carey W, Jacobson IM, Payne J, Dienstag JL, Van Thiel DH, Tamburro C, Lefkowitch J, Alberth J, Meschievitz C, Orrego TJ, Gibas A (1989) Treatment of chronic hepatitis C with recombinant interferon-alfa. A multicenter randomized, controlled trial. Hepatitis Interventional Therapy Group. N Engl J Med 321:1501–1506

De Francesco R, Urbani A, Nardi MC, Tomei L, Steinkühler C, Tramontano A (1996) A zinc binding site in viral serine proteinases. Biochemistry 35:13282–13287

Di Bisceglie AM, Conjeevaram HS, Fried MW, Sallie R, Park Y, Yurdaydin C, Swain M, Kleiner DE, Mahaney K, Hoofnagle JH (1995) Ribavirin as therapy for chronic hepatitis C. A randomized, double-blind, placebo-controlled trial. Ann Intern Med 123:897–903

Dusheiko G, Main J, Thomas H, Reichard O, Lee C, Dhillon A, Rassam S, Fryden A, Reesink H, Bassendine M, Norkrans G, Cuypers T, Lelie N, Telfer P, Watson J, Weegink C, Sillikens P, Weiland O (1996) Ribavirin treatment for patients with chronic hepatitis C: results of a placebo-controlled study. J Hepatol 5:591–598

Dusheiko G (1997) Side effects of alpha interferon in chronic hepatitis C. Hepatology 26[Suppl 1]:112S–121S

Eckart MR, Selby M, Masiarz F, Lee C, Berger K, Crawford K, Kuo G, Houghton M, Choo QL (1993) The hepatitis C virus encodes a serine proteinase involved in processing of the putative nonstructural proteins from the viral polyprotein precursor. Biochem Biophys Res Commun 192:399–406

Failla C, Tomei L, De Francesco R (1994) Both NS3 and NS4A are required for proteolytic processing of hepatitis C virus nonstructural proteins. J Virol 68:3753–3760

Failla C, Tomei L, De Francesco R (1995) An amino-terminal domain of the hepatitis C virus NS3 proteinase is essential for interaction with NS4A. J Virol 69:1769–1777

Francki RIB, Fauquet CM, Knudson DL, Brown F (1991) Classification and nomenclature of viruses: fifth report of the International Committe on Taxonomy of Viruses. Arch Virol Suppl 2:223–233

Gorbalenya A, Snijder E (1996) Viral cystein proteinases. Perspect Drug Discovery Design PD3 6:64–86

Grakoui A, McCourt DW, Wychowski C, Feinstone SM, Rice CM (1993a) A second hepatitis C virus-encoded proteinase. Proc Natl Acad Sci USA 90:10583–10587

Grakoui A, McCourt DW, Wychowski C, Feinstone SM, Rice C (1993b) Characterization of the hepatitis C virus-encoded serine proteinase: determination of proteinase-dependent polyprotein cleavage sites. J Virol 67:2832–2843

Hahm B, Han DS, Back SH, Song OK, Cho MJ, Kim CJ, Shimotohno K, Jang SK (1995) NS3–4A of hepatitis C virus is a chymotrypsin-like proteinase. J Virol 69:2534–2539

Han DS, Hahm B, Rho HM, Jang SK (1995) Identification of the serine proteinase domain in NS3 of the hepatitis C virus. J Gen Virol 76:985–993

Hijikata M, Mizushima H, Akagi T, Mori S, Kakiuchi N, Kato N, Tanaka T, Kimura K, Shimotohno K (1993) Two distinct proteinase activities required for the processing of a putative nonstructural precursor protein of hepatitis C virus. J Virol 67:4665–4675

Hoofnagle JH (1997) Hepatitis C: the clinical spectrum of disease. Hepatology 26[Suppl 1]:15S–20S

Houghton M (1996) Hepatitis C viruses. In: Fields BN, Knipe DM, Howley PM (eds) Fields' virology, 3rd edn. Lippincott-Raven, Philadelphia, pp 1035–1058

Ingalinella P, Altamura S, Bianchi E, Taliani M, Ingenito R, Cortese R, De Francesco R, Steinkühler C, Pessi A (1998) Potent peptide inhibitors of human hepatitis C virus NS3 proteinase are obtained by optimising the clevage products. Biochemistry 37:8906–8914

James MN, Sielecki AR, Brayer GD, Delbaere LT, Bauer CA (1980) Structures of product and inhibitor complexes of *Streptomyces griseus* protease A at 1.8 Å resolution. A model for serine protease catalysis. J Mol Biol 144:43–88

Kakiuchi N, Komoda Y, Komoda K, Takeshita N, Okada S, Tani T, Shimotono K (1998) Non-peptide inhibitors of HCV serine proteinase. FEBS Lett 421:217–220

Kim JL, Morgenstern KA, Lin C, Fox T, Dwyer MD, Landro JA, Chambers SP, Markland W, Lepre CA, O'Malley ET, Harbeson SL, Rice CM, Murcko MA, Caron PR, Thomson JA (1996) Crystal structure of the hepatitis virus NS3 proteinase domain complexed with a synthetic NS4A cofactor peptide. Cell 87:343–355

Koch JO, Lohmann V, Herian U, Bartenschlager R (1996) In vitro studies on the activation of the hepatitis C virus NS3 proteinase by the NS4A cofactor. Virology 221:54–66

Landro JA, Raybuck S A, Luong YPC, O'Malley ET, Harbeson SL, Morgenstern KA, Rao G, Livingston DJ (1997) Mechanistic role of an NS4A peptide cofactor with the truncated NS3 proteinase of hepatitis C virus: elucidation of the NS4 A stimulatory effect via kinetic analysis and inhibitor mapping. Biochemistry 36: 9340–9348

Lesk AM, Fordham WD (1996) Conservation and variability in the structure of serine proteinases of the chymotrypsin family. J Mol Biol 258:501–537

Levitt M, Perutz MF (1988) Aromatic rings act as hydrogen bond acceptors. J Mol Biol 201:751–754

Lin C, Pragai BM, Grakoui A Xu J, Rice CM (1994) Hepatitis C virus NS3 serine proteinase: *trans*-cleavage requirements and processing kinetics. J Virol 68:8147–8157

Lin C, Thomson JA, Rice CM (1995) A central region in the hepatitis C virus NS4A protein allows formation of an active NS3-NS4A serine proteinase complex in vivo and in vitro. J Virol 69:4373–4380

Lindsay KL (1997) Therapy of hepatitis C: overview. Hepatology 26[Suppl 1]:71S–77S

Llinas-Brunet M, Bailey M, Fazal G, Goulet S, Halmos T, Laplante S, Maurice R, Poirier M, Poupart MA, Thibeault D, Wernic D, Lamarre D (1998a) Peptide-based inhibitors of the hepatitis C virus serine protease. Bioorg Med Chem Lett 8:1713–1718

Llinas-Brunet M, Bailey M, Deziel R, Fazal G, Gorys V, Goulet S, Halmos T, Maurice R, Poirier M, Poupart MA, Rancourt J, Thibeault D, Wernic D, Lamarre D (1998b) Studies on the C-terminal of hexapeptide inhibitors of the hepatitis C virus serine protease. Bioorg Med Chem Lett 8:2719–2724

Lohman V, Koch JO, Bartenschlager R (1996) Processing pathways of the hepatitis C virus proteins. J Hepatol 24[Suppl 2]:11–19

Love RA, Parge HE, Wickersham JA, Hostomsky Z, Habuka N, Moomaw EW, Adachi T, Homstomska Z (1996) The crystal structure of hepatitis C virus NS3 proteinase reveals a trypsin-like fold and a structural zinc binding site. Cell 87:331–342

Love RA, Parge HE, Wickersham JA, Hostomsky Z, Habuka N, Moomaw EW, Adachi T, Margosiak S, Dagostino E, Homstomska Z (1998) The conformation of hepatitis C virus NS3 proteinase with and without NS4A: a structural basis for the activation of the enzyme by its cofactor. Clin Diagn Virol 10:151–156

Major ME, Feinstone SM (1997) The molecular virology of hepatitis C. Hepatology 25:1527–1538

Manabe S, Fuke I, Tanishita O, Kaji C, Gomi Y, Yoshida S, Mori C, Takamizawa A, Yosida I, Okayama H (1994) Production of nonstructural proteins of hepatitis C virus requires a putative viral proteinase encoded by NS3. Virology 198:636–644

Marcellin P, Boyer N, Giostra E, Degott C, Courouce AM, Degos F, Coppere H, Cales P, Couzigou P, Benhamou JP (1991) Recombinant human alpha-interferon in patients with chronic non-A, non-B hepatitis: a multicenter randomized controlled trial from France. Hepatology 13:393–397

Martin PD, Robertson W, Turk D, Huber R, Bode W, Edwards BF (1992) The structure of residues 7–16 of the A alpha-chain of human fibrinogen bound to bovine thrombin at 2.3-Å resolution. J Biol Chem 267:7911–7920

Miller RH, Purcell RH (1990) Hepatitis C virus shares amino acid sequence similarity with pestiviruses and flaviviruses as well as members of two plant virus supergroups. Proc Natl Acad Sci USA 87:2057–2061

Neddermann P, Tomei L, Steinkühler C, Gallinari P, Tramontano A, De Francesco R (1997) The nonstructural proteins of the hepatitis C virus: structure and functions. Biol Chem 378:469–476

Nienaber VL, Breddam K, Birktoft JJ (1993) A glutamic acid specific serine protease utilizes a novel histidine triad in substrate binding. Biochemistry 32:11469–11475

Nienaber VL, Mersinger LJ, Kettner CA (1996) Structure-based understanding of ligand affinity using human thrombin as a model system. Biochemistry 35: 9690–9699

Pieroni L, Santolini E, Fipaldini C, Pacini L, Migliaccio G, la Monica N (1997) In vitro study of the NS2–3 protease of hepatitis C virus. J Virol 71:6373–6380

Pizzi E, Tramontano A, Tomei L, La Monica N, Failla C, Sardana M, Wood T, De Francesco R (1994) Molecular model of the specificity pocket of the hepatitis C virus proteinase: implications for substrate recognition. Proc Natl Acad Sci USA 91:888–892

Reed KE, Grakoui A, Rice CM (1995) Hepatitis C virus-encoded NS2–3 proteinase: cleavage-site mutagenesis and requirements for bimolecular cleavage. J Virol 69:4127–4136

Reichard O, Schvarcz R, Weiland O (1997) Therapy of hepatitis C: alpha interferon and ribavirin. Hepatology 26[Suppl 1]:108S–111S

Santolini E, Pacini L, Fipaldini C, Migliaccio G, la Monica N (1995) The NS2 protein of hepatitis C virus is a transmembrane polypeptide. J Virol 69:7461–7471

Satoh S, Tanji Y, Hijikata M, Kimura K, Shimotohno K (1995) The N-terminal region of hepatitis C virus nonstructural protein 3 (NS3) is essential for stable complex formation with NS4A. J Virol 69:4255–4260

Shimizu Y, Yamaji K, Masuho Y, Yokota T, Inoue H, Sudo K, Satoh S, Shimotohno K (1996) Identification of the sequence of NS4A required for enhanced cleavage of the NS5A/5B site by hepatitis C virus NS3 proteinase. J Virol 70:127–132

Steinkühler C, Urbani A, Tomei L, Biasiol G, Sardana M, Bianchi E, Pessi A., De Francesco R (1996) Activity of purified hepatitis C virus proteinase NS3 on peptide substrates. J Virol 70:6694–6700

Steinkühler C, Biasiol G, Brunetti M, Urbani U, Koch U, Cortese R, Pessi A, De Francesco R (1998) Product inhibition of the hepatitis C virus NS3 proteinase. Biochemistry 37:8899–8905

Steitz TA, Shulman RG (1982) Crystallographic and NMR studies of the serine proteases Annu Rev Biophys Bioeng 11:419–444

Stempniak M, Hostomska Z, Nodes BR, Hostomsky ZO (1997) The NS3 proteinase domain of hepatitis C virus is a zinc-containing enzyme. J Virol 71:2881–2886

Sudo K, Matsumoto Y, Matsushima M, Fujiwara M, Konno K, Shimotohno K, Shigeta S, Yokota T (1997a) Novel hepatitis C virus proteinase inhibitors: thiazolidine derivatives. Biochem Biophys Res Commun 238:643–647

Sudo K, Matsumoto Y, Matsushima M, Konno K, Shimotohno K, Shigeta S, Yokota T (1997b) Novel hepatitis C virus proteinase inhibitors: 2,4,6-trihydroxy, 3-nitrobenzamide derivatives. Antiviral Chem Chemother 8:541–544

Tanji Y, Hijikata M, Hirowatari Y, Shimotohno K (1994) Identification of the domain required for trans-cleavage activity of hepatitis C viral serine proteinase. Gene 145:215–219

Tanji Y, Hijikata M, Satoh S, Kaneko T, Shimotohno K (1995) Hepatitis C virus-encoded nonstructural protein NS4A has versatile functions in viral protein processing. J Virol 69:1575–1581

Tomei L, Failla C, Santolini E, De Francesco R, la Monica N (1993) NS3 is a serine proteinase required for processing of hepatitis C virus polyprotein. J Virol 67:4017–4026

Tomei L, Failla C, Vitale RL, Bianchi E, De Francesco R (1996) A central hydrophobic domain of the hepatitis C virus NS4A protein is necessary and sufficient for the activation of the NS3 proteinase. J Gen Virol 77:1065–1070

Tong L, Wengler G, Rossmann MG (1993) Refined structure of Sindbis virus core protein and comparison with other chymotrypsin-like serine proteinase structures. J Mol Biol 230:228–247

Urbani A, Bazzo R, Nardi MC, Cicero D, de Francesco R, Steinkühler C, Barbato G (1998) The metal binding site of the hepatitis C virus NS3 proteinase. A spectroscopic investigation. J Biol Chem 273:18760–18769

Urbani A, Bianchi E, Narjes F, Tramontano A, De Francesco R, Steinkühler C, Pessi A (1997) Substrate specificity of the hepatitis C virus serine proteinase NS3. J Biol Chem 272:9204–9209

Vallee B, Auld DS (1990a) Active site zinc ligands and activated H20 of zinc enzymes. Proc Natl Acad Sci USA 87:220–224

Vallee B, Auld DS (1990b) Zinc coordination, function and structure of zinc enzymes and other proteins. Biochemistry 29:5647–5659

Voss T, Meyer R, Sommergruber WO (1995) Spectroscopic characterization of rhinoviral proteinase 2A: Zn is essential for structural integrity. Protein Sci 4:2526–2531

Wilkinson CS (1997) Hepatitis C virus NS2-3 proteinase. Biochem Soc Trans 25:S611

Wu Z, Yao N, Le HV, Weber PC (1998) Mechanism of autoproteolysis at the NS2-NS3 junction of the hepatitis C virus polyprotein. Trends Biochem Sci 23:92–94

Yan Y, Li Y, Munshi S, Sardana V, Cole J, Sardana M, Steinkühler C, Tomei L, De Francesco R, Kuo L, Chen Z (1998) Complex of NS3 proteinase and NS4A peptide of BK strain hepatitis C virus: a 2.2-Å resolution structure in a hexagonal crystal form. Protein Sci 7:837–847

Zhang R, Durkin J, Windsor WT, McNemar C, Ramanathan L, Le HV (1997) Probing the substrate specificity of hepatitis C virus NS3 serine proteinase by using synthetic peptides. J Virol 71:6208–6213

CHAPTER 6

The Human Herpes-Virus Proteases

C.E. Dabrowski, X. Qiu, and S.S. Abdel-Meguid

A. Introduction

Viruses within the herpes-virus family are identified based primarily on morphology, as all herpes viruses contain a linear double-stranded DNA genome within an icosadeltahedral capsid, surrounded by a tegument and enclosed within a viral envelope. The tegument is a region between the capsid and envelope that has no distinct features in electron micrographs (ROIZMAN and FURLONG 1974). These viruses have both a lytic phase of their life cycle, resulting in the generation of infectious virus that may cause disease in the susceptible host, and a latent phase characterized by limited gene expression, during which no infectious virus can be isolated. Latent virus can be reactivated in response to a variety of stimuli, which may again result in disease manifestations (ROIZMAN 1993).

The herpes viruses are divided into three subfamilies (α, β and γ) based on biologic features and the structure and homology of the viral genomes. The α-herpes viruses include herpes simplex virus types 1 and 2 (HSV-1, HSV-2) and varicella zoster virus (VZV). These neurotropic viruses typically infect and replicate in mucosal epithelium, then travel by retrograde transport to the sensory ganglia, where latency is established (ROIZMAN 1993). The β-herpes viruses include human cytomegalovirus (CMV) and human herpes viruses 6 and 7 (HHV-6, 7). While CMV can be isolated from various tissues within the human host, HHV-6 and HHV-7 preferentially replicate in T cells (LOPEZ 1993; GRIFFITHS and EMERY 1997). The γ-herpes viruses include Epstein-Barr virus (EBV) and the newly identified human herpes virus type 8 (HHV-8), also known as Kaposi's sarcoma-associated herpes virus (KSHV; MOORE et al. 1996). The prototype of this family, EBV, infects and replicates in epithelial cells, with latency established following infection of B-lymphocytes (LIEBOWITZ and KIEFF 1993).

Current disease therapies primarily target the α- and β-herpes viruses. HSV-1 and HSV-2 are the primary causes of oro-labial and genital herpes-virus infections, respectively, while VZV infection is manifested as chickenpox in children following primary infection, and shingles in adults (WHITLEY and ROIZMAN 1997; GERSHON and SILVERSTEIN 1997). CMV is a serious cause for concern in the immunocompromised populations, including acquired-immunodeficiency-syndrome patients and transplant recipients, and causes congenital disease in infants (Griffiths and Emery 1997). HHV-6 is the causative agent of roseola in infants, with other disease associations under investigation (Lopez 1993). HHV-7 may also cause roseola in a minority of patients (Braun et al. 1997). EBV is the causative agent of infectious mononucleosis, and HHV-8 has recently been linked to the development of Kaposi's sarcoma (Beaulieu and Sullivan 1997; Moore and Chang 1997).

Antiviral therapies for several human herpes-virus infections are available, all of which ultimately target the viral DNA polymerase. Acyclovir and its prodrug, valacyclovir, as well as penciclovir and its prodrug, famciclovir, have been shown to be safe and effective against the α-herpes virus, HSV-1, HSV-2 and VZV, infections. However, these therapies are only partly efficacious, in that disease parameters such as lesion severity, time to healing and pain could be improved (Griffiths 1995). Approved antivirals for CMV include ganciclovir, foscarnet and cidofovir. Cross-resistance to two of the three or, less frequently, all three compounds has been documented in CMV-infected, immunocompromised patients (Eizuru 1998). Improved safety and bioavailability with an improvement in efficacy are needed in future therapies for CMV infection.

New targets for antiviral therapy are under investigation, with the expectation of little or no cross-resistance to current therapies and the potential of increased efficacy against the viral infection. With the success of the protease inhibitors against HIV infection, efforts were directed toward inhibition of the herpes-virus-encoded proteases.

B. Background

The herpes-virus protease and substrate were first identified in 1991 by Liu and Roizman, with the characterization of the UL26 and UL26.5 open reading frames in HSV-1 (Liu and Roizman 1991a,b). The UL26 and UL26.5 genes are transcribed independently from separate promotors, with UL26 encoding a 635-amino-acid protein and UL26.5 encoding a 329-amino-acid protein translated in-frame with the C-terminal half of UL26 (Fig. 1). UL26 encodes an 80-kDa protein (Pra) which cleaves itself at two sites, the M (maturation) site (A610/S611) and the R (release) site (A247/S248). The final products from protease cleavage include the protease, a 25-kDa protein (VP24, Prn, No) encoded between the UL26 ATG and the R site, as well as a 45-kDa minor scaffolding protein (VP21, ICP35b, Nb) encoded between the R- and M sites and a 25-amino-acid C-terminal domain (Deckman et al. 1992; DiIanni et al. 1993a; Person et al. 1993). The primary UL26.5 gene product (Fig. 1) is a 40-kDa protein (ICP35 c,d) which, when cleaved, results in generation of the major scaffolding protein (VP22a, ICP35 e,f) and the 25-amino-acid C-terminal domain (Liu and Roizman 1991b; Deckman et al. 1992).

Homologs of these proteins have been identified in all herpes viruses sequenced to date (Tigue et al. 1996; Steffy et al. 1995; see Holwerda for

Herpesvirus Life-cycle

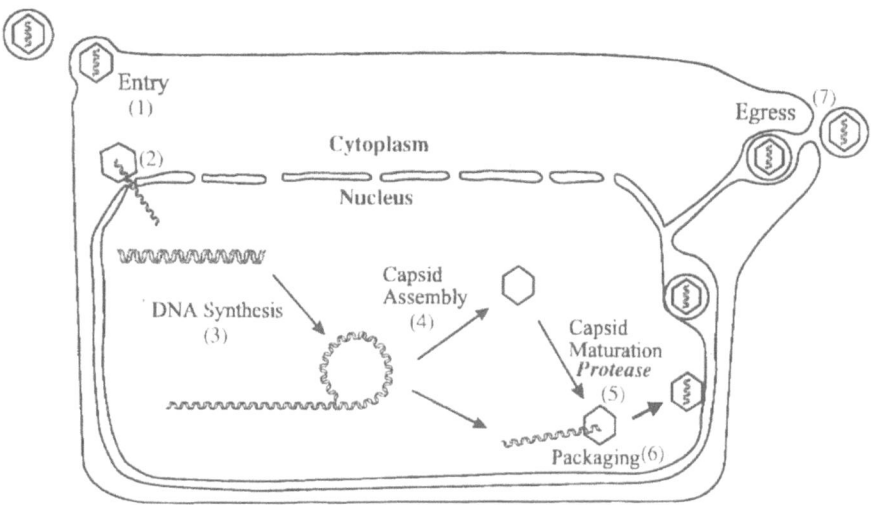

Fig. 1. Herpesvirus life cycle. The virus enters the following attachment and fusion of the viral envelope with the cell membrane (*1*). The capsid is transported to the nuclear membrane where the viral DNA is extruded into the nucleus (*2*). The viral immediate-early and early genes are expressed, followed by DNA synthesis resulting in replication of the viral genome (*3*). Follwoing expression of the late viral proteins, the capsid is assembled in the nuclues (*4*) and matures, a step dependent on the viral protease (*5*). The newly synthesized viral genome is packaged (*6*) and the viral particles egres from the host cell (*7*). Adapted from ROIZMAN and BATTERSON (1985)

review). Indeed, much of the early understanding of the relationship between these proteins came from studies of CMV, which encodes a 708-amino-acid polyprotein, the product of the UL80a open reading frame (WELCH et al. 1991a,b; BAUM et al. 1993). Following cleavage at the R site (A256/S257) and the M site (A643/S644), products comparable with those identified in HSV-1 are generated, including the 30-kDa protease catalytic domain encoded within the N-terminus. The UL80a gene product also contains a third cleavage site within the protease catalytic domain between amino acids A143 and A144, termed the I (inactivation) site, which results in a two-chain, enzymatically active protein. The I site appears to be specific for the CMV, as other herpes viruses lack a homologous cleavage site (BAUM et al. 1993; WEINHEIMER et al. 1993; WELCH et al. 1993; BURCK et al. 1994). Cleavage at the CMV-encoded UL80.5 M site results in the generation of the viral assembly protein, analogous to the HSV-1 scaffolding protein (WELCH et al. 1991b; BAUM et al. 1993).

Studies using HSV-1 viral mutants have demonstrated the essentiality of the protease for viral replication. The protease and scaffolding (assembly) proteins form the inner structure of the viral capsid during the generation of infectious virions. A temperature-sensitive mutant in HSV-1 strain 17syn+, ts1201, had previously been shown to be defective in viral DNA encapsidation and processing of the protease-related proteins at non-permissive temperatures (PRESTON et al. 1983; PRESTON et al. 1992). The mutations in ts1201 were later mapped to nucleotide changes in UL26 that resulted in two amino acid substitutions, Y30F and A48V, within the N-terminal domain shown to be necessary and sufficient for protease activity (LIU and ROIZMAN 1993; GAO et al. 1994; WEINHEIMER et al. 1993). The essential nature of the protease was confirmed following the generation of a null UL26 mutant in HSV-1, m100, which was similarly defective in viral-DNA encapsidation and protease processing (GAO et al. 1994).

C. Three-Dimensional Structures

The first report of the crystal structure of a herpes virus protease was that of the CMV-protease catalytic domain, with four different groups publishing simultaneously (CHEN et al. 1996; QIU et al. 1996; SHIEH et al. 1996; TONG et al. 1996). This was followed by reports of the crystal structures of the protease catalytic domains from the α-herpes viruses, VZV, HSV-1 and HSV-2 (QIU et al. 1997; HOOG et al. 1997), with the HSV-2-protease crystal structure determined in both the liganded and unliganded forms. These structures revealed a three-dimensional fold that has not been observed in any other protein, and a novel catalytic triad in which the third member of the triad is a histidine instead of an aspartic acid (QIU et al. 1996).

I. Overall Fold

Unlike the structures of classical serine proteases, which have two distinct β-barrel domains, the herpes-virus proteases are single-domain proteins (Fig. 2). These proteases can be described as having a seven-stranded, orthogonally-packed, β-barrel core surrounded by eight helices and connecting loops (Fig. 2). The core β-barrel is mostly antiparallel with the exception of strands B2 and B6, which are parallel to each other (Fig. 3).

The overall fold of the four herpes-virus proteases with known three-dimensional structures is very similar (QIU et al. 1997; HOOG et al. 1997). As expected, the structures of the three α-herpes-virus (HSV-1, HSV-2 and VZV) proteases are nearly identical. Superposition of the 197 pairs of equivalent α-carbon atoms of VZV and HSV-2 proteases gives a root-mean-squared (rms) deviation of 0.9 Å. Despite their limited sequence identity (Fig. 4), the overall fold of the α- and β-herpes-virus proteases is very similar. The core β-barrel of the VZV protease superimposes well with that of the CMV protease (Fig. 5), having only 0.7 Å rms difference between the 52 pairs of α-carbon atoms comprising the two β-barrels. Superposition of the 142 equivalent α-carbon atoms from the two structures gives a rms deviation of 1.3 Å, with the most

Gene	Protein	kDa	Structure

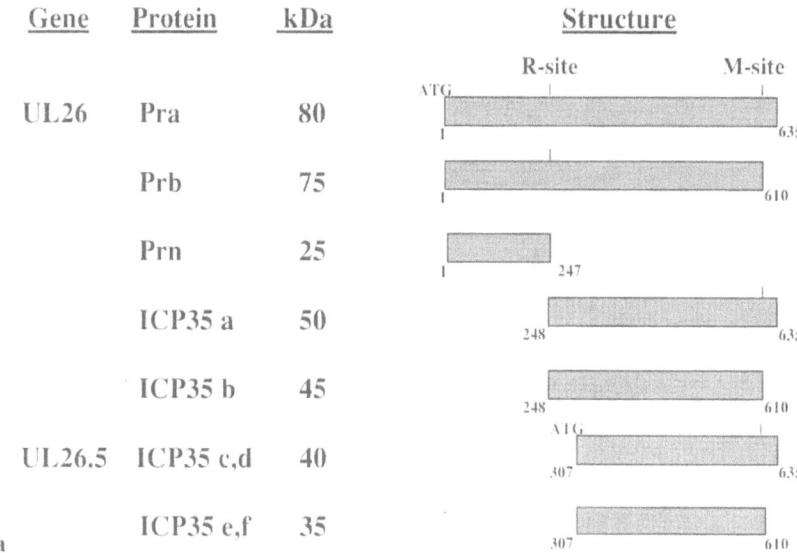

Fig. 2. a The herpes simplex virus-1 UL26 and UL26.5 open reading frames encoding the protease (*A*) and substrate (*B*), respectively. Also shown are their cleavage products. **b** Herpesvirus preotease and substrate open reading frames. Protease (*top*): Shown is the general format of the herpesvirus protease opend reading frams comprising an N-terminal catalytic domain, a minor scaffolding protein domain, and a C-terminal peptide. The amino acid resudies flaking the *R* (release) site and the *M* (maturation) site are shown for HSV and HCMV. The amino acid residues flanking the I (previously inactivation, no internal) site ar shown for HCMV. *S*, position of the active site serine. Substrate (*bottom*): The general format of the predominant substrate of the protease, the scaffolding protein, is shown. The protein is translated in-frame with the C-terminalhalf of the protease and therefore also contains an M-site, with identical flanking amino acid residues

significant conformational differences between the two structures being in the helical and loop regions. These differences, particularly those of surface loops, are proposed to influence substrate recognition of the herpes virus proteases (HOOG et al. 1997).

II. Dimer Interface

Activity of the herpes-virus proteases is greatly enhanced in the presence of co-solvents, such as sucrose, glycerol or sodium citrate (DARKE et al. 1996; MARGOSIAK et al. 1996; SCHMIDT and DARKE 1997). This has been attributed to increased dimerization, as shown by analytical ultracentrifugation studies (COLE 1996) suggesting that the herpes-virus proteases are active as homodimers. A dimer interface related by a twofold crystallographic axis was first identified in the structure of CMV protease (QIU et al. 1996; Fig. 6). The most distinct structural element in the CMV-dimer interface is helix A6, with the

Herpesvirus Protease and Substrate
Open Reading Frames

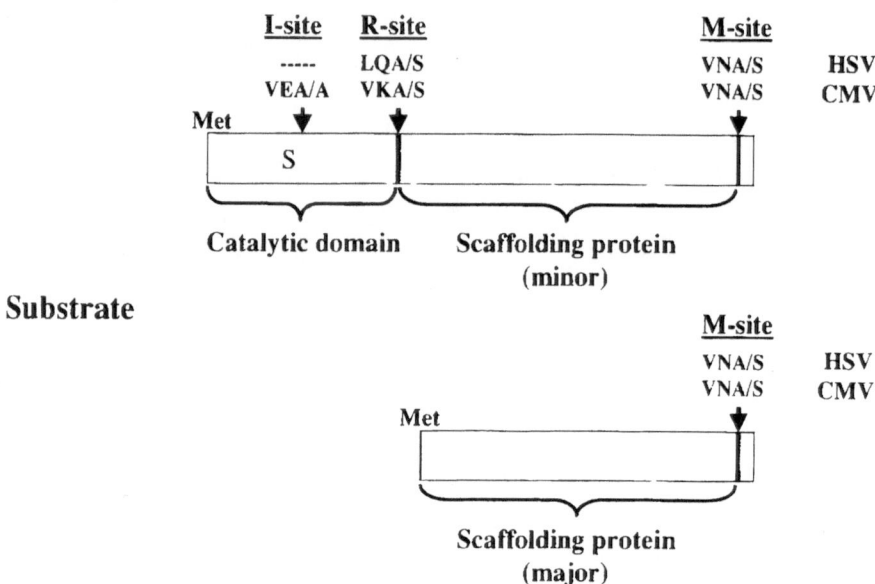

Fig. 2. *Continued*

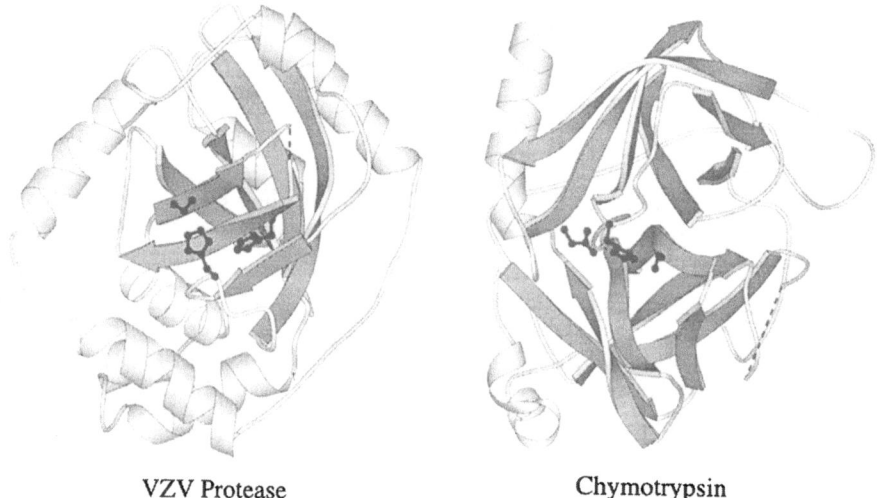

VZV Protease   Chymotrypsin

Fig. 3. Ribbon diagrams comparing the three-dimensional folds of varicella zoster virus (VZV) protease and chymotrypsin. Shown are helices and loops in *lighter shades*, strands in *darker shades* and catalytic residues in *black*

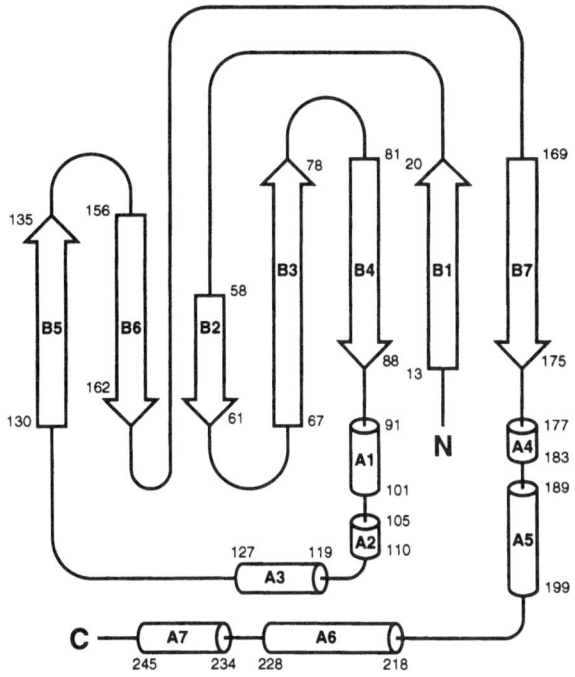

Fig. 4. Topology diagram of the cytomegalovirus protease monomer with helices represented as *cylinders* and strands as *arrows*. Note that strands B5 and B7 interact to form a closed β-barrel

two A6 helices almost parallel and helix A6 of one monomer interacting with helices A1, A2, A3 and A6 of the other monomer (QIU et al. 1996). Similar dimer interfaces have been reported for the α-herpes virus proteases (HOOG et al. 1997; QIU et al.1997; Fig. 6). There are structural differences between the dimer interfaces of the α-herpes-virus proteases and CMV protease, the most notable being the relative orientation of helix A6 of each monomer. Unlike CMV protease, in which the two A6 helices are almost parallel (QIU et al. 1996), in α-herpes-virus protease they show an approximately 30° twist in the α-herpes virus-protease structures (Fig. 6). Furthermore, the amino-acid residues at the dimer interface are not highly conserved, although the relevance of this observation to the structure-function relationship of these proteases is not yet clear.

III. Catalytic Triad

The most interesting discovery resulting from the determination of the crystal structure of CMV protease was the identification of a novel catalytic triad (QIU et al. 1996; SHIEH et al. 1996; TONG et al. 1996; CHEN et al. 1996). Unlike

```
              B1                        AA                   B2
     MTMDEQQSQA VAPVYVGGFL ARYDQSPDEA ELLLPRDVVE HWLHAQGQGQ PSLSVALPLN    60
                MSKVWVGGFL CVYGEEPSEE CLALPRDTVQ KEL---GSGN ----IPLPLN
                METVLVAGFL CVYDDNDIND NFYLPRRTIQ EEINS-GNG- ----LNIPLN
            MVQ APSVYVCGFV ERPDAPPKDA CLHLDPLTVK SQLPLK---- ----KPLPLT
              M AQGLYVGGFV DVVSCPKLEQ ELYLDPDQVT DYLPVT---- ----EPLPIT
    MAADAPG DRMEEPLPDR AVPIYVAGFL ALYDSGDSGE -LALDPDTVR AALPPD---- ----NPLPIN
    MASAEMR ERLEAPLPDR AVPIYVAGFL ALYDSGDPGE -LALDPDTVR AALPPE---- ----NPLPIN
       MAAEADEEN CEALYVAGYL ALY-SKDEGE -LNITPEIVR SALPPT---- ----SKIPIN


        B3        B4          A1           A2              A3
INHDDTAVVG HVAAMQSVRD GLFCLGCVTS PRFLEIVRRA SEKS-ELVSRG PVSPLQPDKV VEFLSGSYAG 130
INHNEKATIG MVRGLFDLEH GLFCVAQIQS QTFMDIIRNI AGKSKLITAGS VIEPLPPDPE IECLSSSFPG
INHNENAVIG TVSSLSVYST VCF-VARVQS KEFLTIIKKI AAKSKLITNTE EK-TLPPDPE IECLNSIFPG
VEHLPDAPVG SVFGLYQSRA GLFSAASITS GDFLSLLDSI YHDC-DIAQSQ RL-PLPREPK VEALHAWLPS
IEHLPETEVG WTLGLFQVSH GIFCTGAITS PAFLELASRL ADTS-HVARAP VK-NLPKEPL LEILHTWLPG
VDHRAGCEVG RVLAVVDDPR GPFFVGLIAC VQLERVLETA ASAAIFERRGP --PLSREERL LYLITNYLPS
VDHRARCEVG RVLAVVNDPR GPFFVGLIAC VQLERVLETA ASAAIFERRGP --ALSREERL LYLITNYLPS
IDHRKDCVVG EVIAIIEDIR GPFFLGIVRC PQLHAVLFEA AHSNFFGNRDS V--LSPLERA LYLVTNYLPS
 *


B5        B6          B7     A4         A5
LSLSSRRCDD VEAATSLSGS ETTPFKHVAL CSVGRRRGTL AVYGRDPEWV TQRFPDLTAA DRDGLRAQWQ 200
LSLSSK---- VLQDENLDGK --PFFHHVSV CGVGRRPGTI AIFGREISWI LDRFSCISES EKRQVLEGVN
LSLSNR---- VGGNERD--- --PFFKHVSI CGVGRRPGTI AIFGRNLNWI LDRFSSITEA EKEKILSTDQ
LSLASLHPD- -IPQTTADGG KLSFFDHVSI CALGRRRGTT AVYGTDLAWV LKHFSDLEPS IAAQIENDAN
LSLSSIHPRE -LSQTPSG-- --PVFQHVSL CALGRRRGTV AVYGHDAEWV VSRFSSVSKS ERAHILQHVS
VSLATKRLGG EAHPDR---- --TLFAHVAL CAIGRRLGTI VTYDYTGLDAA IAPFRHLSPA SREGARRLAA
VSLSTKRRGD EVPPDR---- --TLFAHVAL CAIGRRLGTI VTYDTSLDAA IAPFRHLDPA TREGVRREAA
VSLSSKRLSP NEIPDG---- --NFFTHVAL CVVGRRVGTV VNYDCTPESS IEPFRVLSME SKARLLSLVK
 *                       *


          A6              A7
RCGSTAVDAS GDP-FRSDSYG LLGNSVDALY IRERLPKLRY DKQLVGVTER ESYVKA    CMV   256
VYSQGFDENL FS----ADLYD LLADSLDTSY IRKRFPKLQL DKQLCGLS-K CTYIKA    HHV-6
SCVQFFAEEQ FK----VDLYD LLADSLDTSY IKVRFPKLQS DKQLSGIS-K STYIKA    HHV-7
AAKRESGCPE DHP---LPLTK LIAKAIDAGF LRNRVETLRQ DRGVANIPA- ESYLKA    EBV
SCRLEDLSTP NFV---SPLET LMAKAIDASF IRDRLDLLKT DRGVASILS- PVYLKA    HHV-8
EAELALSGRT WAPGVEALTHT LLSTAVNNMM LRDRWSLVAE RRRQAGIAGH -TYLQA    HSV-1
EAELALAGRT WAPGVEALTHT LLSTAVNNMM LRDRWSLVAE RRRQAGIAGH -TYLQA    HSV-2
DYAGLN--KV WKVSEDKLAKV LLSTAVNNML LRDRWDVVAK RRREAGIMGH -VYLQA    VZV
```

Fig. 5. Sequence alignments of known human herpes virus protease. Helices and strands are *underlined* and *labeled*; catalytic triad residues are marked with *asterisks*

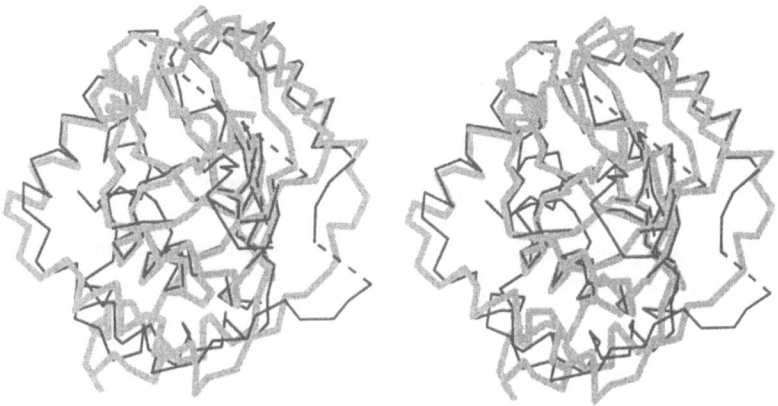

Fig. 6. Stereoview of the superposition of the varicella zoster virus (*light thick lines*) and cytomegalovirus (*dark thin lines*) proteases

all previously known serine proteases, which have a catalytic triad comprised of a serine, a histidine and an aspartic acid (PERONA and CRAIK 1995), the catalytic triad of the herpes-virus proteases consists of a serine and two histidines. Biochemical and mutational analysis of both α- and β-herpes-virus proteases (LIU and ROIZMAN 1992; HALL et al. 1995; DIIANNI et al. 1994) had identified Ser132 (in this work we will use CMV-protease numbering, as shown in Fig. 4, to describe all residues) and His63 as members of the catalytic triad, but was unable to detect the identity of the third member. The crystal structure of the CMV protease, however, suggested His157 as the third catalytic triad residue, instead of an aspartic or glutamic acid (QIU et al. 1996). As expected, residues of the catalytic triad are absolutely conserved amongst all herpes-virus proteases. The presence of a histidine instead of an aspartic acid as the third member of the catalytic triad suggests decreased catalytic efficiency of the herpes-virus proteases relative to classical serine proteases, owing to the weaker electronegativity of the histidine compared with the aspartic acid. Indeed, this has been found to be true. For example, the catalytic efficiency of CMV protease is about 10^4 times lower than that of digestive serine proteases (BABE and CRAIK 1997).

Although the herpes-virus proteases are active as homodimers, each monomer has a well-defined active site containing all the residues necessary for catalytic activity, with the two active sites on opposite sides of the dimer (Fig. 6). Thus, it is not yet clear why dimerization is necessary for activity. One possibility is that dimerization influences enzymatic activity indirectly by stabilization of the conformation of helix A6. This helix has been proposed (QIU et al. 1997) to be involved in the formation of the active-site cavity, namely the S' subsite. In the absence of dimer formation, helix A6, the core of the dimer interface, could move toward the active site to block substrate access, thus rendering the enzyme either inactive or much less active.

The active site of the herpes-virus proteases is very shallow, with the catalytic residues mostly exposed to solvent (Fig. 6). This shallowness is not unreasonable considering that P1-P1' (Ala-Ser) are small residues. However, a shallow active-site cavity for an enzyme, particularly for a serine protease, is uncommon. In trypsin and chymotrypsin, for example, the active-site residues are found in a deep groove between two domains. This suggests that the C-terminal domain of the herpes-virus protease catalytic domains may play an unrecognized role in defining the active-site cavity. The C-terminus of one monomer is proximal to the active site of the second monomer, suggesting that the C-terminal domains of the protease may shield the active sites from bulk solvent and help define the active-site cavities.

IV. The Oxyanion Hole

In addition to a catalytic triad, another important element of serine-protease catalysis is the existence of an oxyanion hole. Functional groups comprising the oxyanion hole stabilize the transition-state intermediate by forming hydrogen bonds to the negatively charged oxygen atom of the substrate. In CMV protease, Arg165 and Arg166 are involved in stabilization of the oxyanion intermediate (Liang et al. 1998). Overlays of the catalytic triad of any of the herpes-virus-protease structures with that of trypsin result in superposition of Arg165 backbone atom with that of Gly193 of trypsin. The latter is known to stabilize the oxyanion intermediate through a hydrogen bond with its backbone NH. Ser195 of trypsin is also known to stabilize the enzyme active-site oxyanion intermediate through a hydrogen bond with its backbone NH. The equivalent residue in the herpes proteases is absent; instead, a water molecule (Fig. 7) held by the side chain of Arg166 in the viral proteases was proposed to form a hydrogen bond with the oxyanion (Qiu et al. 1996, 1997; Hoog et al. 1997). The role of Arg165 and Arg166 in catalysis is further supported by the fact that both residues are absolutely conserved amongst all herpes proteases (Fig. 4) and that activity of the CMV R165A mutant protease is reduced to 30% that of the of wild type while the R166A mutant is about four orders of magnitude less active than the wild type (Liang et al. 1998).

V. Implications for the Catalytic Mechanism

There are two common models for the mechanism of classical serine proteases (Perona and Craik 1995). One is the "two-proton transfer model", in which the aspartic acid accepts the second proton to become uncharged in the transition state. In such a model, it would be quite difficult for a histidine to play the role of the aspartic acid. In the second model, supported by recent data (Perona and Craik 1995), the most important role for the aspartic acid seems to be the ground-state stabilization of the required tautomer and rotamer of the catalytic proximal histidine. This appears to be a role that is played by His157. A His–His interaction may have fewer rotameric orientations than that of His–aspartic acid, which might be relevant to the stability of the triad in such an exposed catalytic cavity. In either mechanistic model, His63 would acquire a proton in the transition state and thus become positively charged (Fig. 8). Unlike an aspartic acid, His157 will not be able to compensate for this developing positive charge but could further delocalize it. However, it is reasonable to assume that having a second histidine instead of an aspartic acid in the triad would result in decreased catalytic efficiency, which is supported by the fact that all herpes-virus proteases are rather slow enzymes (Hall and Darke 1995; Darke et al. 1996; Margosiak et al. 1996).

D. Ligands

A major goal of studies on herpes-virus proteases is the identification of inhibitors that can be used as drugs against herpes viruses. A number of pharmaceutical companies have developed research programs aimed at the identification of such inhibitors for the treatment of HSV-1, HSV-2 and CMV infections. Many of the inhibitors identified to date were either derived from

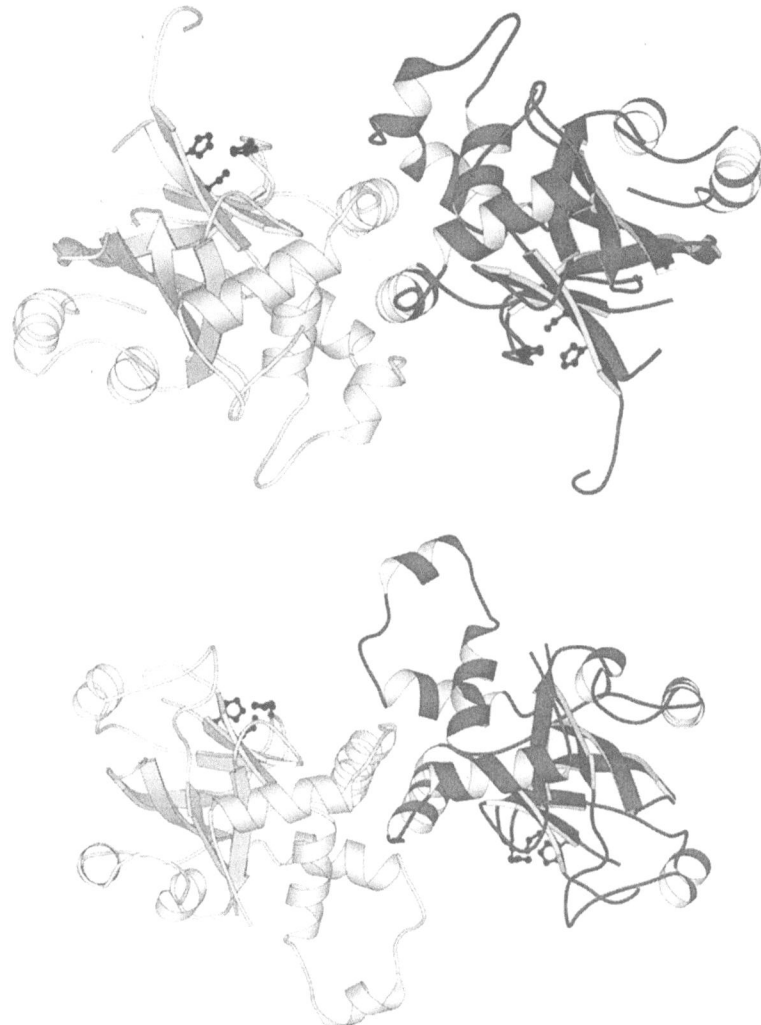

Fig. 7. The cytomegalovirus (*top*) and varicella zoster virus (*bottom*) protease dimers viewed parallel to the crystallographic twofold axis. Catalytic residues are depicted in *black*; helix A6 is at the core of the dimer interface

a substrate, or designed based on molecules known to be classical inhibitors of serine proteases, which act by covalently and reversibly binding to the active-site serine hydroxyl.

I. Substrates

CMV protease differs in catalytic activity at the R and M sites, with the turnover rate of the M-site (GVVNA ↓ SCRLA) cleavage an order of mag-

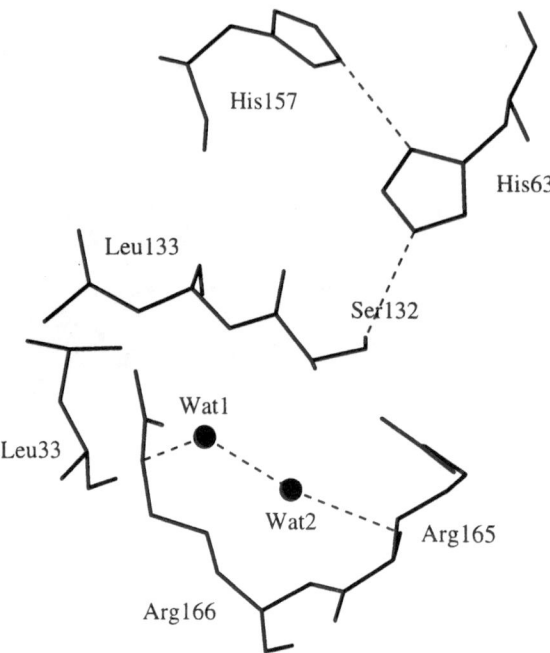

Fig. 8. Residues in the active site of herpes simplex virus-2 protease. Key hydrogen bonds are shown as *dashed lines*. The oxyanion hole is predicted to be between the backbone nitrogen of Arg165 and the oxygen of Wat2

nitude faster than that of the R site (SYVKA ↓ SVSPE) while having similar Michaelis-Menten constant (K_m) values (Burck et al. 1994). Unlike CMV protease, HSV-1 protease does not have a preference for cleavage of the M over the R site, with cleavage about ten times slower than CMV protease (DiIanni et al. 1993b; Darke et al. 1994; Hall and Darke 1995).

Evidence exists that substrate recognition/cleavage by the herpes proteases depends on more than a few residues around the cleavage site. Despite sharing the same core M-site sequence (VNA ↓ S), HSV-1 protease will not cleave at the CMV M site; however, CMV protease will cleave at the corresponding HSV-1 site (Welch et al. 1995). The smallest peptide mimic of the CMV M site that is cleaved by that protease is P4-P4′ (Sardana et al. 1994), whereas 13 residues from P5-P8′ are required for cleavage by HSV-1 protease (DiIanni et al. 1993b). This is surprising, given the high sequence homology of residues lining the active-site cavity of the two enzymes, and suggests that HSV-1 protease has a more extended substrate-binding pocket, with differences in substrate specificity between the two enzymes resulting from differences in loop conformations around the active-site cavity. These loops show low sequence homology and are of differing lengths.

II. Assays

Fluorescence-based assays have been developed for the screening of potential herpes-virus protease inhibitors in a high-throughput format. In the case of CMV protease, substrates have been designed based on the M-site sequence, covering approximately P5-P5' (GVVNA ↓ SSRLA; PINKO et al. 1995; DARKE et al. 1996; BONNEAU et al. 1997; FLYNN et al. 1997; LIANG et al. 1998). The substrates typically carry an internally quenched fluorescence system, with a 4-dimethylaminoazobenzene-4'-sulfonyl (DABSYL) group at the N-terminus and an N-acetyl-N'-(5-sulfo-1-naphthyl)ethylenediamine or dansyl-II group at the C-terminus (PINKO et al. 1995; LIANG et al. 1998). Upon substrate cleavage by the protease, the quenching effect of DABSYL is removed, and the increase in fluorescence signal at about 490nm corresponds to the amount of consumed substrate. Typically, the peptidolytic assays are performed in the presence of co-solvent (20–30% glycerol or 30% sucrose) with 20–60nM protease and substrate at micromolar concentrations, resulting in observed K_m and catalysis rate constant values of approximately $100\mu M$ and $20\,min^{-1}$, respectively.

For HSV-1 and HSV-2 proteases, no high-throughput fluorescence assays have been described. Instead, high-performance liquid chromatography (HPLC)-based assays have been used, in which the product is analyzed by HPLC reverse-phase column chromatography (DARKE et al. 1994). However, as the HSV-1 protease is approximately ten times slower than CMV protease, high (about $1\mu M$) enzyme concentrations are necessary for screening (DiIANNI et al. 1993b; DARKE et al. 1994; HALL and DARKE 1995). Fluorescence-based assays for HHV-6 protease and HHV-8 have also been reported (TIGUE et al. 1996; UNAL et al. 1997).

III. Inhibitors

Commercially available inhibitors of proteases tested for activity against the HSV and CMV proteases demonstrated little to no inhibition by compounds typically active against cysteine, aspartyl or metalloproteases. Typical serine-protease inhibitors showed weak activity in vitro at high concentrations against the HSV and CMV proteases (LIU and ROIZMAN 1992; DiIanni et al. 1993b; BAUM et al. 1993; BURCK et al. 1994). More recently, work has focused on the rational design of inhibitors initially based on peptide or peptidomimetic inhibitors, followed by a number of reports of non-peptidic inhibitors and inhibitors from natural products.

1. Peptide and Peptidomimetic Inhibitors

Peptide inhibitors were the first to be reported. Work in this area primarily focused on defining the minimal element in the substrate that could act as a competitive inhibitor, i.e., the smallest peptide that binds but is not processed. Using peptides encompassing the sequence of the natural M-site substrate of

CMV protease, LaFemina et al. (1996) identified VVNA (P4-P1 of the substrate) as such a minimal element, with an inhibition constant (K_i) of 1.36 mM against CMV protease. Further, substitution of the P1′ serine of an M-site P6-P5′ peptide by an alanine improved the K_i by about threefold over the unaltered peptide, which resulted in their most potent peptide inhibitor having a K_i of 72 μM.

Holskin et al. (1995) reported the first peptidomimetic inhibitor designed for CMV protease. Starting with the M site of CMV protease, they also prepared a reduced peptide-bond inhibitor (RGVVNAψ[CH₂NH]SSRLA-OH) having an inhibition constant of over 500 μM against CMV protease. This peptide, spanning from P6 to P5′, differs from the amino-acid sequence of the M site at two positions, namely P6 (substituting an arginine for alanine to increase solubility) and P2′ (substituting a serine for cysteine to prevent disulfide-bond formation). A number of CMV-protease inhibitors containing an activated carbonyl moiety (Fig. 9), such as fluoromethyl ketones and α-ketoamides, have also been reported (Bonneau et al. 1997).

2. Non-Peptidic Inhibitors

Several non-peptidic classes of herpes-protease inhibitors have been described, some with reported antiviral activity. Jarvest et al. (1996) were the first to report the inhibition of a herpes protease by benzoxazinones, which are a class of heterocyclic molecules (Fig. 9) initially identified as general mechanism-based inhibitors of serine proteases (Teshima et al. 1982), and which inhibit by acylation of the active-site serine through their carbonyl group (Radhakrishnan et al. 1987). A number of benzoxazinones were identified (Fig. 9) that specifically inhibited HSV-1 protease with micromolar potency and were shown to have a wide range of half-lives (1–171 h) at pH 7.5 in aqueous solutions. The most stable compound was also one of the most potent, with a median inhibitory concentration (IC_{50}) of 5 μM.

Jarvest et al. (1997) also reported the design and synthesis of a number of thienoxazinones (Fig. 9) and showed that they are potent, selective, mechanism-based inhibitors of the herpes proteases, with good aqueous stability. These compounds were found to be submicromolar inhibitors of HSV-1 and HSV-2 proteases and moderate inhibitors of CMV protease.

Flynn et al. (1997) identified a class of sulfhydryl-modifying benzimidazolylmethyl sulfoxide (Fig. 9) inhibitors, one of which inhibited CMV protease with IC_{50} of 2 μM. This compound exhibited selectivity against mammalian serine proteases and was antiviral in a CMV antiviral assay. Site-directed-mutagenesis studies suggested oxidative modification of surface-accessible CMV protease Cys138 (and possibly Cys161) by this class of inhibitors.

Targeted screening of compounds that can acylate the active-site serine of the herpes proteases identified the spirocyclopropyl oxazolones (Fig. 9) as submicromolar inhibitors of HSV-2 and CMV proteases (Pinto et al. 1996). These compounds were shown to be better inhibitors of herpes-virus proteases than

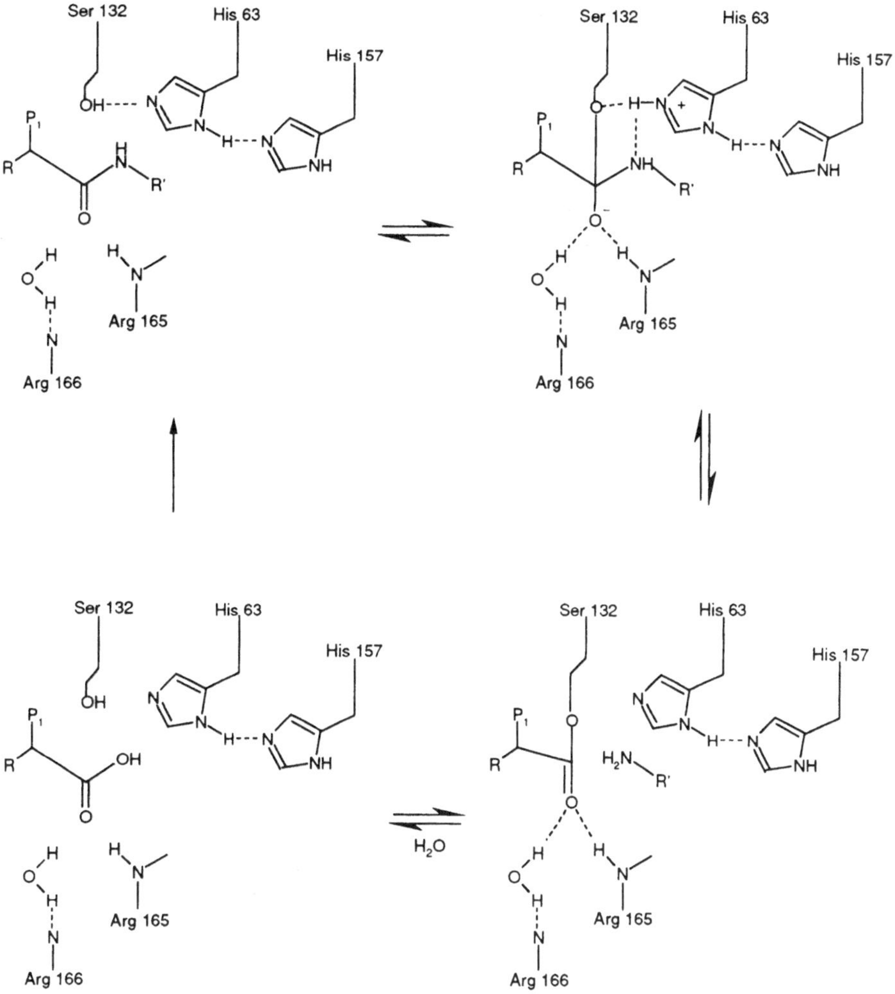

Fig. 9. A schematic representation of the cytomegalovirus protease catalytic mechanism

other enzymes of the chymotrypsin superfamily. To enhance the stability of these compounds, the imidazolones (Fig. 9) were prepared and found to be selective for CMV protease, with little inhibition of HSV-2 protease, elastase, trypsin and chymotrypsin (PINTO et al. 1996).

3. Natural-Product Inhibitors

Three natural-product inhibitors of CMV protease have been identified. A fungal metabolite (Fig. 9) was found to inhibit the enzyme with an IC_{50} of $9.8\,\mu g/ml$ (CHU et al. 1996). A second inhibitor, bripiodionen (Fig. 9), was isolated from *Streptomyces* and shown to have an IC$_{50}$ of 30 μM against CMV protease (SHU et al. 1997). Furthermore, a cycloartanol sulfate (Fig. 9) from the green alga *Tuemoya spp.* was identified as a 4- to 7-μM inhibitor of both VZV and CMV proteases (PATIL et al. 1997).

E. Cell-Based Activity of Protease Inhibitors

Novel compounds from different chemical classes have been identified as potent inhibitors of HSV and CMV proteases in in vitro mechanism-based assays, as discussed above. A key hurdle for such compounds is the ability to inhibit protease processing in the milieu of the infected cell, with such inhi-bition clearly differentiated from the effects of compound cytotoxicity. The first report of cell-based inhibition against a herpes-virus-encoded protease described the activity of benzoxazinone analogs in cells infected with a recom-binant CMV expressing the β-galactosidase reporter gene (ABOOD et al. 1997). A series of 6-substituted benzoxazinones, shown to inhibit the CMV protease at low or submicromolar concentrations in in vitro peptide-based assays, were tested for antiviral activity by measuring the resultant β-galactosidase activity following incubation of recombinant-CMV-infected cells in the presence of compound. These compounds demonstrated apparent antiviral activity against the recombinant CMV, with 50% effective concentrations (EC$_{50}$) ranging from 8.5 μM to 63 μM, with the most potent compounds limited by cytotoxicity, as demonstrated in uninfected (mock) cultures.

Recently, a series of thieno[2,3-d]oxazinones was identified capable of inhibiting the HSV-2 protease at submicromolar levels in in vitro peptide-based assays (DABROWSKI et al. 1998). Compounds in this series demonstrated mechanism-based inhibition of protease processing in infected cells, as deter-mined by analysis of protease-related bands by pulse-chase assay following infection of cells with HSV-2. The thieno[2,3-d]oxazinones were also assessed for antiviral activity by exposure of infected cells to increasing concentrations of compound for a time approximately equivalent to one viral life-cycle of the virus (20 h) and subsequent quantitation of the number of infectious virions produced. One compound demonstrated antiviral activity, with an EC$_{50}$ of 0.75 μM. The antiviral activity of this compound was well separated from its cytotoxic effects. These results indicate that potent inhibitors of viral proteases (as determined by in vitro, peptidolytically based assays) can be identified that mechanistically inhibit viral-protease processing in the infected cell, and that such inhibition results in antiviral activity.

F. Perspective

The herpes proteases have been classified as atypical serine proteases that play an essential role in the viral life cycle. As such, these enzymes have been identified as potential antiviral targets. Comparison of the crystal structures of

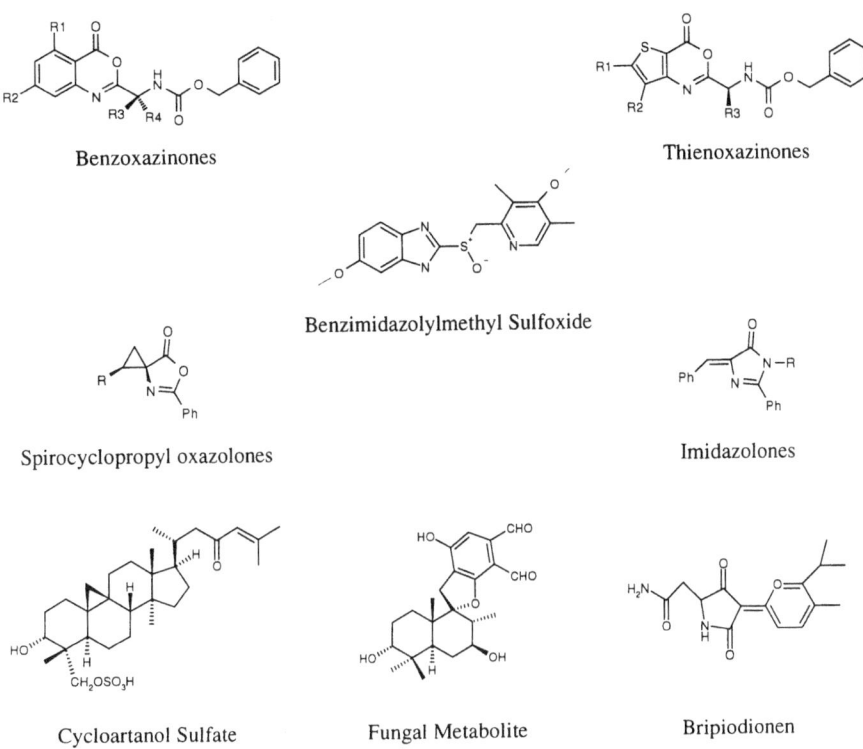

Benzoxazinones

Thienoxazinones

Benzimidazolylmethyl Sulfoxide

Spirocyclopropyl oxazolones

Imidazolones

Cycloartanol Sulfate

Fungal Metabolite

Bripiodionen

Fig. 10. Some of the known inhibitors of human herpes virus proteases

four herpes-virus proteases has revealed very similar structures, including the presence of a novel catalytic triad and fold. These proteases are active as dimers, with each monomer containing a full complement of the active-site residues. Potent inhibitors of the HSV and CMV proteases have been identified in in vitro peptidolytic assays, some of which have been identified as mechanism-based inhibitors in the context of the infected cell. The first evidence of antiviral activity against HSV and CMV by these proteases has now been reported, suggesting that the herpes-virus proteases may indeed prove to be viable targets for therapeutic intervention.

References

Abood NA, Schretzman LA, Flynn DL, Houseman KA, Wittwer AJ, Dilworth VM, Hippenmeyer PJ, Holwerda BC (1997) Inhibition of human cytomegalovirus protease by benzoxazinones and evidence of antiviral activity in cell culture. Bioorg Med Chem Lett 7:2105–2108

Babe LM, Craik CS (1997) Viral proteases: evolution of diverse structural motifs to optimize function. Cell 91:427–430

Baum EZ, Bebernitz GA, Hulmes JD, Muzithras VP, Jones TR, Gluzman Y (1993) Expression and analysis of the human cytomegalovirus UL80-encoded protease: identification of autoproteolytic sites. J Virol 67:497–506

Beaulieu BL, Sullivan JL (1997) Epstein-Barr virus. In: Richman DD, Whitley RJ, Hayden FG (eds) Clinical virology. Churchill Livingstone, New York, p 485

Bonneau PR, Grand-Maitre C, Greenwood DJ, Lagace L, LaPlante SR, Massariol MJ, Ogilvie WW, O'Meara JA, Kawai SH (1997) Evidence of a conformational change in the human cytomegalovirus protease upon binding of peptidyl-activated carbonyl inhibitors. Biochemistry 36:12644–12652

Braun DK, Dominguez G, Pellett PE (1997) Human herpesvirus 6. Clin Microbiol Rev 10:521–567

Burck PJ, Berg DH, Luk TP, Sassmannshausen LM, Wakulchik M, Smith DP, Hsiung HM, Becker GW, Gibson W, Villarreal EC (1994) Human cytomegalovirus maturational proteinase: expression in *Escherichia coli*, purification, and enzymatic characterization by using peptide substrate mimics of natural cleavage sites. J Virol 68:2937–2946

Chen P, Tsuge H, Almassy RJ, Gribskov CL, Katoh S, Vanderpool DL, Margosiak SA, Pinko C, Matthews DA, Kan CC (1996) Structure of the human cytomegalovirus protease catalytic domain reveals a novel serine protease fold and catalytic triad. Cell 86:835–843

Chu M, Mierzwa R, Truumees I, King A, Patel M, Pichardo J, Hart A, Dasmahapatra B, Das PR, Puar MS (1996) Tetrahedron Lett 37:3943–3946

Cole JL (1996) Characterization of human cytomegalovirus protease dimerization by analytical centrifugation. Biochemistry 35:15601–15610

Dabrowski CE, Ashman SM, Fernandez AV, Gorczyca M, Lavery P, Parratt MJ, Serafinowska HT, Sternberg EJ, Tew DG, West A, Jarvest RJ (1998) Inhibition of herpesvirus proteases by novel thieno[2,3-d]oxazinones: demonstration of inhibition of virus protein processing and selective antiviral activity in cell culture. Antimicrob Agents Chemother (submitted)

Darke PL, Chen E, Hall DL, Sardana MK, Veloski CA, LaFemina RL, Shafer JA, Kuo LC (1994) Purification of active herpes simplex virus-1 protease expressed in *Escherichia coli*. J Biol Chem 269:18708–18711.

Darke PL, Cole JL, Waxman L, Hall DL, Sardana MK, Kuo LC (1996) Active human cytomegalovirus protease is a dimer. J Biol Chem 271:7445–7449

Deckman IC, Hagen M, McCann PJ III (1992) Herpes simplex virus type 1 protease expressed in *Escherichia coli* exhibits autoprocessing and specific cleavage of the ICP35 assembly protein. J Virol 66:7362–7367

DiIanni CL, Drier DA, Deckman IC, McCann III PJ, Liu F, Roizman B, Colonno RJ, Cordingley MG (1993a) Identification of the herpes simplex virus-1 protease cleavage sites by direct sequence analysis of autoproteolytic cleavage products. J Biol Chem 268:2048–2051

DiIanni CL, Mapelli C, Drier DA, Tsao J, Natarajan S, Riexinger D, Festin SM, Bolgar M, Yamanaka G, Weinheimer SP, Meyers CA, Colonno RJ, Cordingley MG (1993b) In vitro activity of the herpes simplex virus type 1 protease with peptide substrates. J Biol Chem 268:25449–25454

DiIanni CL, Stevens JT, Bolgar M, O'Boyle DR II, Weinheimer SP, Colonno RJ (1994) Identification of the serine residue at the active site of the herpes simplex virus type 1 protease. J Biol Chem 269:12672–12676

Eizuru Y (1998) Multidrug resistance in human cytomegalovirus. Int Antivir News 6:61–63

Flynn DL, Becker DP, Dilworth VM, Highkin MK, Hippenmeyer PJ, Houseman KA, Levine LM, Li M, Moormann AE, Rankin A, Toth MV, Villamil CI, Wittwer AJ, Holwerda BC (1997) The herpes virus protease: mechanistic studies and discovery of inhibitors of the human cytomegalovirus protease. Drug Des Discov 15:3–15

Gao M, Matusick-Kumar L, Hurlburt W, DiTusa SF, Newcomb WW, Brown JC, McCann PJ III, Deckman I, Colonno RJ (1994) The protease of herpes simplex virus type 1 is essential for functional capsid formation and viral growth. J Virol 68:3702–3712

Gershon AA, Silverstein SJ (1997) Varicella-zoster virus. In: Richman DD, Whitley RJ, Hayden FG (eds) Clinical virology. Churchill Livingstone, New York, p 421

Griffiths PD (1995) Progress in the clinical management of herpesvirus infections. Antivir Chem Chemother 6:191–209

Griffiths PD, Emery VC (1997) Cytomegalovirus. In: Richman DD, Whitley RJ, Hayden FG (eds) Clinical virology. Churchill Livingstone, New York, p 445

Hall DL, Darke PL (1995) Activation of the herpes simplex virus type 1 protease. J Biol Chem 270:22697–22700

Holskin BP, Bukhtiyarova M, Dunn BM, Baur P, de Chastonay J, Pennington MW (1995) A continuous fluorescence-based assay of human cytomegalovirus protease using a peptide substrate. Anal Biochem 227:148–155

Holwerda BC (1997) Herpesvirus proteases: targets for novel antiviral drugs. Antivir Res 35:1–21

Hoog SS, Smith WW, Qiu X, Janson CA, Hellmig B, McQueney MS, O'Donnell K, O'Shannessy D, DiLella AG, Debouck C, Abdel-Meguid SS (1997) Active site cavity of herpesvirus proteases revealed by the crystal structure of herpes simplex virus protease/inhibitor complex. Biochemistry 36:14023–14029

Jarvest RL, Parratt MJ, Debouck CM, Gorniak JG, Jennings LJ, Serafinowska HT, Strickler JE (1996) Inhibition of HSV-1 protease by benzoxazinones. Bioorg Med Chem Lett 6:2463–2466

Jarvest RL, Connor SC, Gorniak JG, Jennings LJ, Serafinowska HT, West A (1997) Potent selective thienoxazinone inhibitors of herpes proteases. Bioorg Med Chem Lett 7:1733–1738

LaFemina RL, Bakshi K, Long WJ, Pramanik B, Veloski CA, Wolanski BS, Marcy AI, Hazuda DJ (1996) Characterization of a soluble stable human cytomegalovirus protease and inhibition by M-site peptide mimics. J Virol 70: 4819–4824

Liang PH, Doyle ML, Brun KA, O'Donnell K, Green SM, Baker AE, Feild JA, Blackburn MN, Abdel-Meguid SS (1998) Site-directed mutagenesis probing the catalytic role of arginines 165 and 166 of human cytomegalovirus protease. Biochemistry 37:5923–5929

Liebowitz D, Kieff E (1993) Epstein-Barr virus. In: Roizman B, Whitley RJ, Lopez C (eds) The human herpesviruses. Raven, New York, p 107

Liu F, Roizman B (1991a) The promoter, transcriptional unit, and coding sequence of herpes simplex virus 1 family 35 proteins are contained within and in frame with the UL26 open reading frame. J Virol 65:206–212

Liu F, Roizman B (1991b) The herpes simplex virus 1 gene encoding a protease also contains within its coding domain the gene encoding the more abundant substrate. J Virol 65:5149–5156

Liu F, Roizman B (1992) Differentiation of multiple domains in the herpes simplex virus 1 protease encoded by the UL26 gene. Proc Natl Acad Sci USA 89:2076–2080

Liu F, Roizman B (1993) Characterization of the protease and other products of amino-terminus-proximal cleavage of the herpes simplex virus 1 UL26 protein. J Virol 67:1300–1309

Lopez C (1993) Human herpesviruses 6 and 7. In: Roizman B, Whitley RJ, Lopez C (eds) The human herpesviruses. Raven, New York, p 309

Margosiak SA, Vanderpool DL, Sisson W, Pinko C, Kan C-C (1996) Dimerization of the human cytomegalovirus protease: kinetic and biochemical characterization of the catalytic homodimer. Biochemistry 35:5300–5307

Moore PS, Chang Y (1997) Kaposi's sarcoma-associated herpesvirus. In: Richman DD, Whitley RJ, Hayden FG (eds) Clinical virology. Churchill Livingstone, New York, p 509

Moore PS, Gao S-J, Dominguez G, Cesarman E, Lungu O, Knowles DM, Garber R, Pellett PE, McGeoch DJ, Chang Y (1996) Primary characterization of a herpesvirus agent associated with kaposi's sarcoma. J Virol 70:549–558

Patil A, Freyer AJ, Killmer L, Breen A, Johnson RK (1997) A cycloartanol sulfate from the green alga *Tuemoya sp.*: an inhibitor of VZV protease. Bioorg Med Chem Lett 7:1733–1738

Perona JJ, Craik CS (1995) Structural basis of substrate specificity in the serine proteases. Protein Sci 4:337–360

Person S, Laquerre S, Desai P, Hempel J (1993) Herpes simplex virus type 1 capsid protein, VP21, originates within the UL26 open reading frame. J Gen Virol 74: 2269–2273

Pinko C, Margosiak SA, Vanderpool DL, Gutowski JC, Condon B, Kan CC (1995) Single-chain recombinant human cytomegalovirus protease. J Biol Chem 270: 23634–23640

Pinto IL, West A, Debouck CM, DiLella AG, Gorniak JG, O'Donnell KC, O'Shannessy DJ, Patel A, Jarvest RL (1996) Novel, selective mechanism-based inhibitors of the herpes proteases. Bioorg Med Chem Lett 6:2467–2472

Preston VG, Coates JAV, Rixon FJ (1983) Identification and characterization of a herpes simplex virus gene product required for encapsidation of virus DNA. J Virol 45:1056–1064

Preston VG, Rixon FJ, McDougall IM, McGregor M, Al Kobaisi MF (1992) Processing of the herpes simplex virus assembly protein ICP35 near its carboxy terminal end requires the product of the whole of the UL26 reading frame. Virol 186:87–98

Qiu X, Culp JS, DiLella AG, Hellmig B, Hoog SS, Janson CA, Smith WW, Abdel-Meguid SS (1996) Unique fold and active site in cytomegalovirus protease. Nature 383: 275–279

Qiu X, Janson CA, Culp JS, Richardson SB, Debouck C, Smith WW, Abdel-Meguid SS (1997) Crystal structure of varicella-zoster virus protease. Proc Natl Acad Sci USA 94:2874–2879

Radhakrishnan R, Presta LG, Meyer EF Jr, Wildonger R (1987) Crystal structures of the complex of porcine pancreatic elastase with two valine-derived benzoxazinone inhibitors. J Mol Biol 198:417–424

Roizman B (1993) The family herpesviridae. In: Roizman B, Whitley RJ, Lopez C (eds) The human herpesviruses. Raven, New York, p 1

Roizman B, Furlong D (1974) The replication of herpes viruses. In: Fraenkel-Conrat H, Wagner RR (eds) Comprehensive virology. (Vol 3) Plenum, New York, p 229

Sardana VV, Wolfgang JA, Veloski CA, Long WJ, LeGrow J, Wolanski B, Emini EA, LaFemina RL (1994) Peptide substrate cleavage specificity of the human cytomegalovirus protease. J Biol Chem 269:14337–14340

Schmidt U, Darke PL (1997) Dimerization and activation of the herpes simplex virus type 1 protease. J Biol Chem 272:7732–7735

Shieh HS, Kurumbail RG, Stevens AM, Stegeman RA, Sturman EJ, Pak JY, Wittwer AJ, Palmier MO, Wiegand RC, Holwerda BC, Stallings WC (1996) Three-dimensional structure of human cytomegalovirus protease. Nature 383:279–282

Shu YZ, Ye Q, Kolb JM, Huang S, Veitch JA, Lowe SE, Manly SP (1997) Bripiodionen, a new inhibitor of human cytomegalovirus protease from *Streptomyces* sp. WC76599. J Nat Prod 60:529–532

Steffy KR, Schoen S, Chen C-M (1995) Nucleotide sequence of the herpes simplex virus type 2 gene encoding the protease and capsid protein ICP35. J Gen Virol 76: 1069–1072

Teshima T, Griffin JC, Powers JC (1982) A new class of heterocyclic serine protease inhibitors. Inhibition of human leukocyte elastase, porcine pancreatic elastase, cathepsin G, and bovine chymotrypsin A alpha with substituted benzoxazinones, quinazolines, and anthranilates. J Biol Chem 257:5085–5091

Tigue NJ, Matharu PJ, Roberts NA, Mills JS, Kay J, Jupp R (1996) Cloning, expression, and characterization of the proteinase from human herpesvirus 6. J Virol 70: 4136–4141

Tong L, Qian C, Massariol MJ, Bonneau PR, Cordingley MG, Lagace L (1996) A new serine-protease fold revealed by the crystal structure of human cytomegalovirus protease. Nature 383:272–275

Unal A, Pray TR, Lagunoff M, Pennington MW, Ganem D, Craik CS (1997) The protease and the assembly protein of Kaposi's sarcoma-associated herpesvirus (human herpesvirus 8). J Virol 71:7030–7038

Weinheimer SP, McCann III PJ, O'Boyle II DR, Stevens JT, Boyd BA, Drier DA, Yamanaka GA, DiIanni CL, Deckman IC, Cordingley MG (1993) Autoproteolysis of herpes simplex virus type 1 protease releases an active catalytic domain found in intermediate capsid particles. J Virol 67:5813–5822

Welch AR, McNally LM, Gibson W (1991a) Cytomegalovirus assembly protein nested gene family: four 3'-coterminal transcripts encode four in-frame, overlapping proteins. J Virol 65:4091–4100

Welch AR, Woods AS, McNally LM, Cotter RJ, Gibson W (1991b) A herpesvirus maturational proteinase, assemblin: identification of its gene, putative active site domain, and cleavage site. Proc Natl Acad Sci USA 88:10792–10796

Welch AR, McNally LM, Hall MRT, Gibson W (1993) Herpesvirus proteinase: site-directed mutagenesis used to study maturational, release, and inactivation cleavage sites of precursor and to identify a possible catalytic site serine and histidine. J Virol 67:7360–7372

Welch AR, Villarreal EC, Gibson W (1995) Cytomegalovirus protein substrate are not cleaved by the herpes simplex virus type 1 proteinase. J Virol 69:341–347

Whitley RJ, Roizman B (1997) Herpes simplex viruses. In: Richman DD, Whitley RJ, Hayden FG (eds) Clinical virology. Churchill Livingstone, New York, p 375

The 3C Proteinases of Picornaviruses and Other Positive-Sense, Single-Stranded RNA Viruses

E.M. BERGMANN and M.N.G. JAMES

A. Introduction

Picornaviruses are a family of viruses which belong to the large group of positive-sense, single-stranded RNA viruses (RUECKERT 1996). It was realized 30 years ago that the product of the translation of the RNA genome of these viruses is proteolytically processed to yield the mature viral proteins (SUMMERS and MAIZEL 1968; KORANT 1972). Subsequently, it could be shown for two different picornaviruses that the processing enzyme is a specific, virally encoded proteinase (PELHAM 1978; GORBALENYA et al. 1979; KORANT et al. 1979; PALMENBERG et al. 1979). Once the amino-acid sequences of the viral proteinases became available, predictions were made concerning the structure of the picornaviral 3C proteinases (GROBALENYA et al. 1986; BAZAN and FLETTERICK 1988; GORBALENYA et al. 1989). These predictions were remarkable. Based on an analysis of several conserved sequence motifs within the amino-acid sequence of the 3C proteinases, it was suggested that these proteinases are structurally related to the chymotrypsin-like proteinases but with a cysteine residue as the active-site nucleophile. Crystal structures of 3C proteinases from two picornaviruses confirmed this prediction (ALLAIRE et al. 1994; MATTHEWS et al. 1994). At present, crystal structures of 3C proteinases from viruses, belonging to three different genera of the picornaviruses, have been published (MATTHEWS et al. 1994; BERGMANN et al. 1997; MOSIMANN et al. 1997).

The chymotrypsin-like cysteine proteinases have so far been found only in positive-sense, single-stranded RNA viruses (GORBALENYA and SNIJDER 1996; RYAN and FLINT 1997; BERGMANN and JAMES 1999). 3C or 3C-like proteinases are found in all picornaviruses, many related plant viruses and at least two other important families of animal viruses. As these enzymes are distinct from cellular enzymes and their function is essential for viral replication, the 3C proteinases constitute an obvious target for the design of anti-viral drugs (KRÄUSSLICH and WIMMER 1988).

Some positive-sense, single-stranded RNA viruses also carry genes coding for chymotrypsin-like serine proteinases or papain-like cysteine proteinases (KRÄUSSLICH and WIMMER 1988; PALMENBERG 1990; DOUGHERTY and SEMLER 1993; RYAN and FLINT 1997). For other proteinases from these viruses, the enzyme classes to which they belong are not yet established. It is very likely that there are more novel classes of proteinases awaiting discovery in the positive-sense, single-stranded RNA viruses.

Table 1. The picornaviruses

Genus	Number of serotypes	Viruses	Diseases caused in humans
Enterovirus	>90	Polio virus, coxsackievirus, echovirus	Intestinal infections, poliomyelitis, myocarditis, meningitis, encephalitis, hand, foot and mouth disease, herpangina, myalgia, pleurodynia
Rhinovirus	>100	Rhinovirus	Common cold
Aphthovirus	7	FMDV, equine rhinovirus	Foot and mouth disease of cloven-hoofed animals
Cardiovirus	2	Encephalomyocarditis virus, Theiler's murine encephalitis virus	None known
Hepatovirus	1	Hepatitis A virus	Infectious hepatitis
Parechovirus	2 (3?)	Parechovirus 1, parechovirus 2, Ljungan river virus (?)	Myocarditis, intestinal infections

FMDV, foot and mouth disease virus.

B. Picornaviridae

Picornaviruses constitute a very large family of positive-sense, single-stranded RNA viruses (Rueckert 1996), and some are among the oldest known and best-studied viruses (Landsteiner and Popper 1909; Loeffler and Frosch 1964). Picornaviruses cause a wide variety of different diseases. Presently, viruses of the family Picornaviridae are classified into six genera (Table 1).

The more than 100 serotypes of human rhinoviruses (HRVs) are responsible for most common colds in humans (Couch 1996; Makela et al. 1998). Foot-and-mouth disease virus, the prototype of the aphthoviruses, is the causative agent of one of the most important diseases of livestock (Belsham 1993).

Hepatitis A virus (HAV) is the only known member of the genus *Hepatovirus*, and causes an acute form of infectious hepatitis (Hollinger and Ticehurst 1996). Hepatitis A is still fairly widespread in those parts of the world that do not have safe drinking water supplies. Isolated cases or mini-epidemics of hepatitis A still occur regularly in the developed world and are usually attributed to contaminated food (Pebody et al. 1998). Safe and effective vaccines against hepatitis A have recently become available (Thiel 1998), but their widespread use appears unlikely. Whilst acute HAV infections are, in most cases, relatively harmless, co-infection of patients with chronic hepatitis is often more dangerous (Sjogren 1998; Vento et al. 1998). An increase in the number of chronic hepatitis infections may therefore change the significance of hepatitis A as an infectious disease.

The genus *Enterovirus* consists of the polio-, coxsackie- and echoviruses (MELNICK 1996). These viruses cause a wide variety of illnesses in humans, ranging from mild respiratory tract and intestinal infections to meningitis, myocarditis, encephalitis and poliomyelitis (Table 1). Poliovirus (PV), the major cause of poliomyelitis in humans, has been targeted for world-wide eradication by the turn of the millennium (COCHI et al. 1997; CENTER FOR DISEASE CONTROL 1998). In spite of the success of polio vaccination, it is not clear whether this goal can be achieved (SUTTERS and COCHI 1997; TAYLOR et al. 1997). Non-vaccine related cases of poliomyelitis are very rare in most parts of the world.

Enteroviruses remain a serious health problem. A recent epidemic in Asia provided a grim reminder of this (CHANG et al. 1998). In many clinical settings, the majority of cases of viral meningitis and myocarditis are caused by enteroviruses. The enteroviruses have also been implicated as triggers of autoimmune diseases such as multiple sclerosis, myocarditis and diabetes (ANDREOLETTI et al. 1997; CARTHY et al. 1997; STEINMANN and CONLON 1997; NIKLASSON et al. 1998; ROIVAINEN et al. 1998). Definite proof of a link between enteroviral infections and the onset of autoimmune diseases is still not established. Recently, an animal model of a demyelinating disease that resembles multiple sclerosis and is caused by the picornavirus Theiler's murine encephalitis virus, has provided evidence for a mechanism whereby viruses can trigger autoimmune diseases (MILLER et al. 1997).

Two of the echoviruses (EV22 and EV23) have recently been reclassified into a new genus, the parechoviruses. The establishment of the new genus was based partly on observed differences in the mechanism of the proteolytic processing of the polyprotein (SCHULTHEISS et al. 1995).

Theiler's murine encephalitis virus is a member of the genus *Cardiovirus*. It causes the above-mentioned demyelinating disease in mice and constitutes an important model system for these diseases. None of the cardioviruses has been linked to any known disease in humans.

It is very likely that the taxonomy of the Picornaviridae will be further modified in the future. As has happened with the parechoviruses, elucidation of details of the viral replication mechanism may lead to reclassification of individual viruses or the establishment of new genera.

C. Other Families of Positive-Sense, Single-Stranded RNA Viruses

I. Caliciviridae

The caliciviruses were discovered relatively recently and were initially considered to be picornaviruses. Elucidation of details of their structure and replication mechanism made it clear that they constitute a different family of viruses (Clarke and Lambden 1997; Kapikian et al. 1996). They derive their name (calix is Latin for cup) from cup-like indentations of their capsid that are visible in the electron microscope. Their genome structure is also different from that of the picornaviruses, e.g., the structural proteins are found at the carboxy-terminus of the viral polyprotein.

Several caliciviruses cause intestinal infections in humans. They are now considered to be one of the leading causes of what is often described as a "stomach flu" in humans (Green 1997). The proteolytic processing enzyme of the caliciviruses is a chymotrypsin-like cysteine proteinase (Wirblich et al. 1995; Martín-Alonso et al. 1996).

II. Coronaviridae

The largest of the positive-sense, single-stranded RNA viruses are the coronaviruses. They are enveloped viruses and derive their name from their star-like appearance in the electron microscope. Coronaviruses have developed a more complex replication mechanism than other positive-sense, single-stranded RNA viruses, including several mRNA species (Holmes and Lai 1996). The gene products from the major RNA species are nevertheless produced by specific proteolytic cleavage of the translated polyprotein. The viral proteinase that is responsible for most cleavages is a chymotrypsin-like cysteine proteinase (Tibbles et al. 1996; Liu et al. 1997; Seybert et al. 1997; Lu et al. 1998; Schiller et al. 1998). The 3C-like proteinase of the coronavirus avian-infectious-bronchitis virus cleaves following a glutamine residue, similar to the 3C proteinase of the picornaviruses (Ng and Liu 1998).

Coronaviruses cause upper respiratory tract and intestinal infections in humans and animals and are considered the second major cause of the common cold in humans (Makela et al. 1998). They have also been implicated as a cause of viral diarrhea (Gonzalez et al. 1997).

III. Others

All known positive-sense, single-stranded RNA viruses utilize the strategy of specific proteolytic processing of polyproteins to express their genomes. Some have developed additional strategies, such as subgenomic mRNAs or multiple open reading frames. The specific proteolytic processing of polyproteins by a viral proteinase remains an important part of their replication strategy. Therefore, all known positive-sense, single-stranded RNA viruses carry at least one and often several genes that code for proteolytic enzymes (Kräusslich and Wimmer 1988; Dougherty and Semler 1993).

At least three different classes of proteinases are found in these viruses, and there are probably others awaiting discovery. The chymotrypsin-like cysteine proteinases are so far uniquely found in the three families of viruses discussed above and in related plant viruses. The Nsp4 proteinase of the arteriviruses is a serine proteinase. Analysis of the enzyme and its sequence led SNIJDER et al. (1996) to propose that the enzyme is more closely related to the picornaviral 3C proteinases than to any serine proteinase.

D. Functions of Viral Proteinases in Positive-Sense, Single-Stranded RNA Viruses

I. The Picornaviral Life Cycle

Figure 1 shows a simplified scheme of the life cycle of hepatitis A virus, a typical picornavirus. It serves to illustrate the significance of the functions of the viral proteinases during viral replication (RUECKERT 1996).

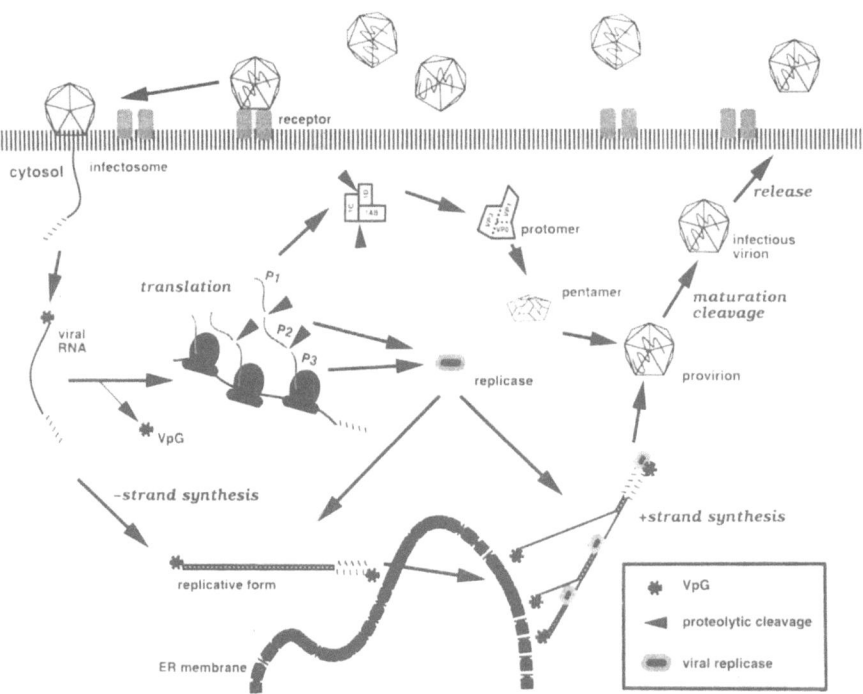

Fig. 1. A simplified scheme of the life-cycle of hepatitis A virus, a typical picornavirus. Viral replication takes place in the cytosol of the host cell. The RNA replication is performed by a viral replicase complex and is localized at modified intracellular membrane structures. Translation of the viral RNA and co-translational, proteolytic processing of the resulting polyprotein is the initial event of a picornaviral infection. The P1 gene products are the structural proteins and are further proteolytically processed to allow capsid assembly. The P2 and P3 gene products are further proteolytically processed and assemble into the viral replicase. In other genera of the Picornaviridae, the P1|P2 cleavage is not performed by the 3C proteinase but by the 2A proteinase or through a completely different mechanism. In the enteroviruses, the cleavages within the structural proteins P1 require the precursor of the 3C proteinase, 3CD

The virus attaches to a specific cell-surface receptor and undergoes some conformational changes that allow it to release its genome into the cytosol of the host cell. A small protein (VPg), which is covalently attached to the 5′ terminus of the picornaviral RNA genome, is cleaved by a host factor, and the resulting RNA is translated into a large polyprotein. The viral polyprotein is co-translationally processed by one or several specific viral proteinases (Palmenberg 1990). The first cleavage typically separates the structural from the non-structural proteins of the virus (Ryan and Flint 1997). The non-structural proteins of the picornaviruses are further proteolytically processed and assemble to form the viral replicase.

Viral replicase complexes perform both negative-sense and positive-sense RNA replication (Porter 1993; Wimmer et al. 1993). The exact composition of the complexes is not clear in even the best-studied picornaviruses (Harris et al. 1994; Xiang et al. 1998). The viral RNA polymerase (the 3D gene product in picornaviruses), the putative RNA helicase (the 2C gene product) and the 3C proteinase form part of this complex. There is also good evidence that some cellular proteins are recruited to form part of the picornaviral-replication complex (Andino et al. 1993; Xiang et al. 1995; Gamarnik and Andino 1997; Parsley et al. 1997; Roehl et al. 1997). Viral replication takes place on modified intracellular-membrane structures. Modification of the intracellular-membrane structures is a common feature of picornaviral infection and is mediated, at least in part, by the 2B and 2C gene products (Bienz et al. 1983; Bienz et al. 1990; Teterina et al. 1997a,b).

The 3C proteinase also has an RNA-binding site and plays a part in the binding of the RNA during the initiation of RNA replication. Small RNA viruses are under evolutionary pressure to maintain the small size of their genome and a limited number of genes. Therefore, many of their gene products, e.g., 3C, have multiple functions. The exact function of the 3C proteinase within the picornaviral replicase complex is not clear. It is possible that some proteolytic cleavages are performed within the replicase complex. For example, the proteolytic cleavage between 3A and 3B could be performed within the replicase complex. 3A is a hydrophobic protein which presumably serves to anchor the replicase complex to modified, intracellular-membrane structures. 3B is the small protein, VPg, which remains covalently attached to the viral RNA (Wimmer 1982).

In a typical picornavirus infection, the ratio of positive-sense to negative-sense RNA is about 50 to 1. Most of the negative-sense RNA exists in the form of a double-stranded replicative form. The VPg-associated positive-sense RNA genome is packaged into provirions, which are transformed into infectious virions by a non-enzymatic proteolytic cleavage of one of the capsid proteins. This is referred to as the maturation cleavage (Palmenberg 1990).

After the structural proteins have been cleaved from the viral polyproteins, their assembly into procapsids is regulated by successive proteolytic cleavages. Only after the proteolytic cleavages have been performed by the

3C proteinase can the capsid proteins undergo the conformational changes that will allow them to assemble into the precursors of the procapsid. Final assembly into the provirion requires the presence of the VPg-associated RNA, but the details of this process are not clear (RUECKERT 1996).

There are three distinct functions of viral proteinases during the life cycle of the picornaviruses (KRÄUSSLICH and WIMMER 1988). The first is the specific, co-translational, proteolytic processing of the viral polyprotein. The second is the processing of the precursors of the viral capsid. The proteolytic cleavages of the capsid precursor regulate the assembly of the procapsid and should therefore be considered a distinct function of the viral proteinases. A third function of viral proteinases, at least in some picornaviruses, is the cleavage of some host cell proteins (HAGHIGHAT et al. 1996; ROEHL et al. 1997; RYAN and FLINT 1997 and references therein; YALAMANCHILI et al. 1997). This serves either to downregulate cellular processes that compete with the viral replication or to recruit cellular proteins to become part of the viral-replication machinery.

II. Proteolytic Processing of the Viral Polyprotein

The full-length polyprotein is not detectable under normal conditions because it is already co-translationally processed. The proteolytic cleavages are performed by one or two specific viral proteinases that are themselves part of the polyprotein. The polyprotein processing is performed sequentially. Some of the cleavage sites are cleaved faster than others. Usually, the P1|P2 cleavage, which separates the structural and nonstructural proteins, is the first cleavage event. The mechanism of this primary P1|P2 cleavage is different in the different genera of the picornaviruses (RYAN and FLINT 1997).

In the viruses that belong to the genera *Entero-* and *Rhinovirus*, the primary cleavage is performed by a separate 2A proteinase. The 2A proteinase is also a chymotrypsin-like cysteine proteinase and cleaves at its own amino-terminus (RYAN and FLINT 1997 and references therein). In aphtho- and cardioviruses, the primary cleavage occurs at the carboxy-terminus of 2A by a non-enzymatic mechanism (PALMENBERG et al. 1992; DONNELLY et al. 1997). In HAV and, presumably, also in the parechoviruses, the primary cleavage is a 3C-mediated proteolytic cleavage at the amino-terminus of the 2B gene product (JIA et al. 1993; SCHULTHEISS et al. 1994; MARTIN et al. 1995; SCHULTHEISS et al. 1995).

In all picornaviruses, the majority of the proteolytic cleavages within the polyprotein are performed by the 3C proteinase. The 3C proteinases cleave specifically following a glutamine residue. Additional residues around the scissile bond contribute to the recognition of the cleavage sites (NICKLIN et al. 1988; LONG et al. 1989; PALLAI et al. 1989; CORDINGLEY et al. 1990; WEIDNER and DUNN 1991; JEWELL et al. 1992). These are 4 to 5 residues that precede the scissile bond and 2 to 3 residues that follow it [P_5–P_3' in the nomenclature of SCHECHTER and BERGER (1967)].

The 3C-like proteinase of the coronaviruses also specifically cleaves following a glutamine residue (NG and LIU 1998). The corresponding enzyme in the calicivirus rabbit-hemorrhagic-disease virus specifically recognizes a glutamate residue in the P_1 position of a substrate (WIRBLICH et al. 1995; MARTÍN-ALONSO et al. 1996). The specificities of the 3C and 3C-like proteinases of positive-sense, single-stranded RNA viruses are unique and distinct from those of known mammalian proteinases.

III. Regulation of Capsid Assembly by Proteolytic Cleavages of the Capsid-Protein Precursors

Viral-structural proteins are designed to assemble into large symmetrical structures. They are usually synthesized as precursors to prevent premature assembly or aggregation. The precursor is subsequently covalently modified. The most common form of covalent modification of the capsid precursor in viruses is proteolytic processing (KRÄUSSLICH and WIMMER 1988). This is one reason that proteolytic enzymes are common gene products of viruses, even in large DNA viruses.

Two successive proteolytic cleavages by the 3C proteinase are required in picornaviruses to allow capsid assembly (Fig. 1; RUECKERT 1996). In the enteroviruses, these cleavages are performed by the precursor 3CD (YPMA-WONG et al. 1988). The RNA-polymerase domain has an effect on the substrate specificity or on the catalytic efficiency, but in the absence of structural information for 3CD it is not clear how this is accomplished. Following the proteolytic cleavages between VP0|VP3 and VP3|VP1, the capsid precursors undergo a conformational change and assemble into pentameric structures. The final assembly of the provirion requires the presence of the RNA (RUECKERT 1996).

IV. Inhibition of Cellular Functions by Proteolytic Cleavages of Host Cell Proteins

The third function of proteolytic enzymes in Picornaviruses is to cleave specific cellular proteins. The best-studied example is the cleavage of eIF4G by the 2A proteinase of entero- and rhinoviruses or by the L proteinase of the aphthoviruses (RYAN and FLINT 1997). This cleavage impairs translation of capped, cellular mRNAs and therefore improves the translational efficiency of the viral RNA. Not all picornaviruses inhibit host cell translation by this mechanism. Hepatitis A virus carries only a single gene coding for a proteolytic enzyme, 3C (SCHULTHEISS et al. 1994). HAV has no equivalent to either the enteroviral 2A proteinase or the aphthoviral L proteinase. In HAV-infected cells, there is no evidence of a cleavage of eIF4G, and HAV translation even depends on the intact cellular eIF4G. Cleavage of eIF4G by an enteroviral 2A proteinase inhibits HAV replication (BORMAN and KEAN 1997).

There are reports of other cellular proteins that are cleaved by the picornaviral 3C proteinase (ROEHL et al. 1997; RYAN and FLINT 1997 and references therein; YALAMANCHILI et al. 1997). The function of these cleavages in vivo is not completely clear. Some cleavages appear to impair host cell transcription; others may modify cellular proteins to become part of the viral-replication machinery (ROEHL et al. 1997).

E. The 3C Proteinase

I. Structure

Refined crystal structures of the 3C proteinases from three of the six genera of the Picornaviruses have been published (BERGMANN et al. 1997; MATTHEWS et al. 1994; MOSIMANN et al. 1997). Figure 2 (s. appendix, page 400/401) shows a ribbon representation of the three-dimensional structures of the 3C proteinase from HAV and PV. The two enzymes represent two different classes of the 3C proteinases (GORBALENYA and SNIJDER 1996). The HAV 3C proteinase is larger (219 residues); the enzymes from enteroviruses represent a smaller type (183 residues).

The fold of the two-domain structure of the 3C proteinases is similar to that of the chymotrypsin-like serine proteinases (ALLAIRE et al. 1994). The structure consists of two β-barrel domains with identical topology. The proteolytic active site is in a cleft between the two domains, and residues from both domains contribute to the catalytic mechanism and substrate binding (PERONA and CRAIK 1995). There are two alternative descriptions for the structure of the β-barrel domains. They can be described either as a six-stranded β-barrel or a barrel formed by two orthogonal, four-stranded β-sheets in which the edge strands are a part of both sheets (CHOTIA 1984). We feel that the latter description is more appropriate for the larger HAV 3C proteinase (Fig. 2a; BERGMANN et al. 1997). The second and fifth β-strands of each domain are interrupted by a β-bulge or, in the case of β-strand eI, by a single turn of a helix. This introduces a bend into the edge strands (bI, eI, bII and eII) that allows them to continue from one β-sheet to the other. In the smaller enteroviral enzyme, the edge strands are less bent and more continuous (Fig. 2b; MOSIMANN et al. 1997). It is simply the intrinsic twist of the edge β-strands which allows them to wrap around the whole barrel. Each domain of the smaller enteroviral enzyme can be adequately described as a six-stranded β-barrel.

In spite of the differences, the 3C proteinases from HAV and PV show remarkable conservation of their structures. The core of the β-barrel domains superimpose well. The diameter of the two β-barrels, their relative orientation and the direction of the individual β-strands are very similar. A structural superposition of the two structures reveals 30 identical residues in the sequence. The differences between the two 3C proteinases manifest primarily in the turns and loops which connect the β-strands and in the lengths of the individual secondary-structure elements.

The residues of both the amino- and carboxy-termini of the 3C proteinases form helices. An amino-terminal α-helix is a unique feature of the 3C proteinase. The amino-terminal helix of the 3C proteinase packs against the surface of one of the β-sheets of the carboxy-terminal β-barrel, and the carboxy-terminal helix packs against the amino-terminal domain (Figs. 2, 3, s. appendix, page 400/401). The two helices act like latches in stabilizing the structure.

The proteolytic active site of the 3C proteinase is less accessible than the active site of most chymotrypsin-like serine proteinases. This is primarily due to another structural feature that distinguishes the 3C proteinases from the mammalian chymotrypsin-like enzymes: two β-strands from the carboxy-terminal domain extend past the β-barrel and twist back toward the active site. They form an isolated β-ribbon, with hydrogen bonds formed only between the two β-strands (colored in *light gray* in Figs. 2, 3). This extension of β-strands bII and cII is quite long in HAV 3C, and contributes residues to the active site (Bergmann et al. 1997); it is nine residues shorter in PV 3C (Mosimann et al. 1997).

The molecular surface of 3C on the side opposite from the proteolytic active site is formed by the part of the polypeptide chain that connects the two domains (Fig. 3). The domain connection is flanked by the amino- and carboxy-terminal helices. This region of 3C is important for a function of 3C that is distinct from its proteolytic activity (Andino et al. 1990; Hämmerle et al. 1992; Andino et al. 1993; Leong et al. 1993; Kusov and Gauss-Müller 1997).

II. Specificity and Substrate Binding

Chymotrypsin-like serine proteinases bind their cognate substrates and protein inhibitors in a canonical conformation (Read and James 1986; Bode and Huber 1992). The proteinases typically bind four to five residues preceding the scissile peptide bond and two to three residues following the site of cleavage [P_5 to P_1 and P_1' to P_3' in the nomenclature of Schechter and Berger (1967)]. Most of the residues of the peptide substrate adopt a β-strand conformation. The P_1 residue adopts a conformation that represents a tight 3_{10} helix. This causes the carbonyl of the scissile peptide bond to point into the so-called oxyanion hole of the enzyme. The two neighboring substrate residues, P_2 and P_1', adopt a more twisted β-strand conformation to accommodate this. Binding of the peptide substrate involves main-chain hydrogen-bond interactions between the substrate and β-strands of the enzyme, which resembles an anti-parallel β-sheet. As a result of the substrate conformation, side chains of the peptide substrate point into specificity pockets on the surface of the enzyme (Perona and Craik 1995).

There is now evidence, from cocrystal structures of 3C proteinases with bound, peptide-mimetic inhibitors, that peptide substrates bind to the 3C proteinases in a similar mode (DRAGOVICH et al. 1998a, b; WEBBER et al. 1998; BERGMANN and JAMES, unpublished observations). Before these cocrystal structures became available, models were built of the enzyme-substrate interactions (BERGMANN et al. 1997; MATTHEWS et al. 1994; MOSIMANN et al. 1997; WEBBER et al. 1998). The models built utilized the same conformation of the bound substrate and could successfully rationalize the specificity of the 3C proteinases.

The residues from P_5 to P_2 of a substrate form anti-parallel β-sheet interactions with β-strand eII of the proteinase. This interaction is a common feature of enzyme-substrate interactions in chymotrypsin-like proteinases (PERONA and CRAIK 1995). The β-strand b2II in HAV 3C, part of the unique anti-parallel β-ribbon, could form an additional, parallel β-sheet interaction with the P_4 through P_2 residues of a substrate.

The amino-terminal domain of a chymotrypsin-like proteinase provides the majority of the interactions with the substrate residues following the scissile peptide bond (PERONA and CRAIK 1995). In HAV 3C, the P_1' and P_2' residues would interact with the edge of β-strand bI. This β-strand is interrupted by a β-bulge that causes several carbonyl groups of the peptide bonds of this strand to point into the active site. Presumably, they can act as hydrogen-bond acceptors for the binding of a substrate. The β-strand bI of the entero- and rhinoviral 3C proteinases is continuous (Fig. 2b), and there are fewer possible interactions with a substrate in these enzymes. Rhino- and enteroviral 3C proteinases prefer a glycine as the P_1' residue of a substrate (NICKLIN et al. 1988; LONG et al. 1989; PALLAI et al. 1989; CORDINGLEY et al. 1990; WEIDNER and DUNN 1991). It has been suggested that the main chain of the substrate of these enzymes turns at the P_1' residue (MATTHEWS et al. 1994). There is no significant sequence preference for the P_1' residue of a substrate of HAV 3C. We believe that the difference in the conformations of β-strand bI between the two different classes of enzymes results in different conformations of the P_n' residues of the bound substrates.

The cleavage sites for the picornaviral 3C proteinases within the polyprotein are distinguished by the residues in the P_4, P_2, P_1 and P_1' positions (reviewed by BERGMANN 1998; SKERN 1998). All piconaviral 3C proteinases require a glutamine in the P_1 position of a substrate, but the sequence preferences of the enzymes from different viruses for the other positions are distinct. For example, the sequence preference for the residue in the P_4 position of a substrate is different among the various 3C proteinases. The 3C proteinases from entero- and rhinoviruses prefer a small, hydrophobic residue in the P_4 position of a substrate (NICKLIN et al. 1988; LONG et al. 1989; PALLAI et al. 1989; CORDINGLEY et al. 1990; WEIDNER and DUNN 1991). The HAV 3C proteinase prefers a larger, hydrophobic residue (Leu or Ile) in this position (JEWELL et al. 1992). The model of substrate binding places the side chain of the P_4 residue into a hydrophobic cleft formed by β-strands eII, fII and b2II from the carboxy-terminal domain (BERGMANN et al. 1997). The hydrophobic S_4 binding cleft of the entero- and rhinoviral enzymes is smaller than that of the HAV 3C proteinase (MATTHEWS et al. 1994; MOSIMANN et al. 1997; WEBBER et al. 1998). This is due to the fact that several of the residues which form this hydrophobic cleft are larger in the entero- and rhinoviral enzymes (e.g., PV 3C Leu125 and Phe170 correspond to Ala141 and Val 200 in HAV 3C).

All the models of substrate recognition agree that the glutamine residue in the P_1 position of a substrate probably forms a hydrogen bond between the carbonyl oxygen atom of its side chain and the N^ε atom of the imidazole ring of a conserved histidine in the S_1 pocket of the 3C proteinases (MATTHEWS et al. 1994; BERGMANN et al. 1997; MOSIMANN et al. 1997). This role of the conserved histidine (His191 in HAV 3C, His160 in HRV, His161 in PV) had been proposed prior to the elucidation of the first crystal structure (GORBALENYA et al. 1989). In the entero- and rhinoviral enzymes, a threonine residue (Thr142) forms an additional hydrogen bond to the side-chain carbonyl of the P_1 glutamine (WEBBER et al. 1998). There are no suitable groups on the enzyme that can interact with the amide nitrogen atom of the side chain of the P_1 glutamine in the crystal structures of the 3C proteinases. This correlates well with experimental results that show that inihibitors with N-substituted glutamine isosteres are effective inhibitors of the 3C proteinases (MALCOLM et al. 1995; MORRIS et al. 1997; DRAGOVICH et al. 1998b; WEBBER et al. 1998)

How can the picornaviral 3C proteinases distinguish the invariant glutamine residue in the P_1 position of a substrate from glutamate? A mechanism for this distinction has been suggested based on the details of the structure of the S_1 pocket of the HAV 3C proteinase (BERGMANN et al. 1997). One edge of the imidazole side chain of the conserved His191 forms part of the S_1 specificity pocket of HAV 3C (Fig. 4a). The N^ε atom of the imidazole ring provides a hydrogen-bond donor to recognize the P_1 glutamine side chain. The other edge of the imidazole ring interacts with two buried water molecules in the core of the carboxy-terminal domain of HAV 3C. The water, in turn, interacts with the side-chain carboxyl group of Glu132. Because the side chain of Glu132 is buried inside the hydrophobic environment of the carboxy-terminal β-barrel domain of HAV 3C, it is very likely uncharged. Deprotonation and charging of the side chain of Glu 132, in its hydrophobic environment, would be energetically unfavorable (QASIM et al. 1995 and references therein). Because the two residues, His 191 and Glu132, interact inside the core of the carboxy-terminal β-barrel domain, protonation of His191 would also be energetically unfavorable, much more so than simply having a protonated, positively-charged histidine residue in this environment (QASIM et al. 1995; BERGMANN et al. 1997). These interactions thus ensure that His191 of HAV 3C is neutral. A tyrosine residue (Tyr138 in PV 3C) plays a role similar to that of Glu 132 in HAV 3C in the smaller entero- and rhinoviral 3C proteinases (MOSIMANN et al. 1997).

The available structural information concerning the 3C proteinases can explain the specific recognition of the proteolytic cleavage sites within the viral polyprotein. It is not possible, with the available structural information, to explain why some cleavage sites within the viral polyprotein are preferably

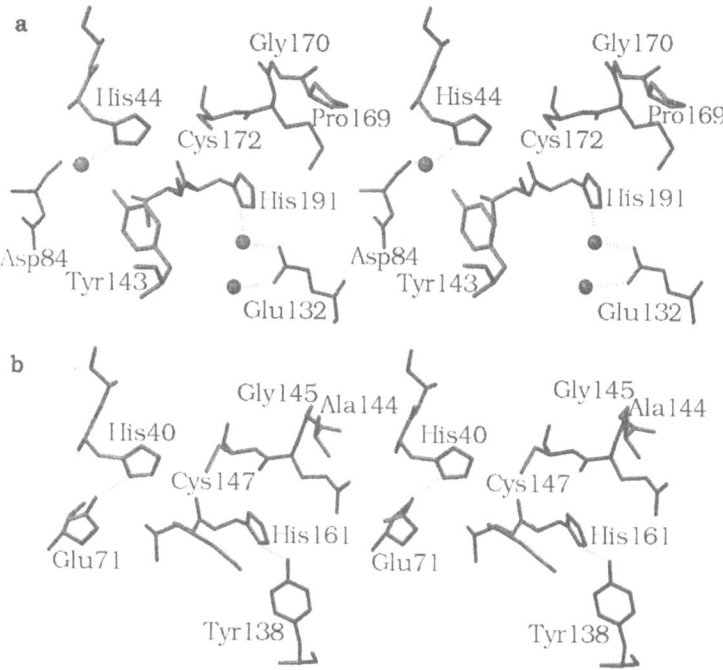

Fig. 4a,b. An all-atom representation of the active sites of the 3C proteinases from (**a**) hepatitis-A virus (HAV) and (**b**) poliovirus in stereo. The views in **a** and **b** are similar. The residues of the oxyanion hole are on the *right*; the residues of the S_1 specificity pocket are *below* it. Water molecules in the structure of HAV 3C are represented by *spheres*

cleaved during the sequential processing of the polyprotein. Presumably, there are other factors that influence the sequence of cleavage events, such as the accessibility or conformation of the cleavage sites.

III. Enzymatic Mechanism

The structure of the active site of the two classes of 3C proteinases is shown in Fig. 4a, b. The three main chemical groups that contribute to the catalytic reaction are in an arrangement which resembles the one in the active site of the chymotrypsin-like serine proteinases (JAMES 1993). The S^γ atoms of Cys172 and Cys147 in HAV 3C and PV 3C, respectively, act as the nucleophiles. They are assisted by general acid-base catalysts (His 44 in HAV 3C and His40 in PV) and an electrophilic oxyanion hole. To accommodate the cysteine nucleophile, the active site of the 3C proteinases is larger than that of the chymotrypsin-like serine proteinases. The distance from the S^γ atom of the nucleophilic cysteine to the N^ε atom of the histidine general acid-base catalyst is 3.7–4.0 Å in the various crystal structures of 3C proteinases. This is 0.7–

1.0 Å longer than the analogous distance between the O^γ of the nucleophilic serine and the N^ε atom of the histidine in the serine proteinases.

The oxyanion hole of the chymotrypsin-like proteinases is formed by the NH groups of two peptide bonds, which are in an orientation facilitating donation of hydrogen bonds to the carbonyl of the scissile peptide bond of a substrate (Whiting and Peticolas 1994). The conformation of the oxyanion hole is not the lowest-energy conformation. In chymotrypsin-like serine proteinases, the main-chain conformation of the residues which form the oxyanion hole is maintained by interactions of the peptide bonds with other parts of the structure. There are no such interactions in the structures of the 3C proteinases. What, then, maintains the conformation of the oxyanion hole in the 3C proteinases?

Mutation of the nucleophilic cysteine to alanine in the HAV 3C proteinase causes the collapse of the oxyanion hole (Allaire et al. 1994). A similar orientation of the oxyanion hole is observed in the cocrystal structure of a peptide-aldehyde-derived hemithioacetal inhibitor of the HRV 3C proteinase (Webber et al. 1998). Apparently, the nucleophile itself has a role in maintaining the conformation of the active site. This provides evidence for the existence of a thiolate-imidazolium ion pair in the active site of the 3C proteinases. The negative charge of the thiolate of the nucleophilic cysteine would assist in orienting the dipole of the peptide bonds of the oxyanion hole. The generally accepted mechanism for other cysteine proteinases assumes a thiolate-imidazolium ion pair in the active site (Brocklehurst et al. 1998; Storer and Ménard 1994).

Additional groups in the active site of the 3C proteinases, besides the nucleophile, oxyanion hole and general acid-base catalyst, are also important for the enzymatic activity. An aspartate or glutamate residue, corresponding topologically to the third member of the catalytic triad in chymotrypsin-like serine proteinases, is conserved throughout the 3C proteinases (Gorbalenya et al. 1989; Gorbalenya and Snijder 1996; Ryan and Flint 1997). Nevertheless, a true catalytic triad does not exist in these cysteine proteinases (Fig. 4). In the entero- and rhinoviral 3C proteinases, the conserved Glu71 interacts with the imidazole of the histidine general acid-base catalyst in an unusual way (Matthews et al. 1994; Mosimann et al. 1997). It forms a hydrogen bond with His 40 through the *anti* lone-pair electrons of its carboxylate (Fig. 4b). This is generally assumed to be a weaker interaction than the more common hydrogen bond through the *syn* lone-pair electrons.

In HAV 3C, the corresponding residue, Asp84, does not interact with the imidazole of His44 at all (Bergmann et al. 1997; Fig. 4a). A water molecule occupies the position of the carboxylate of a third member of the catalytic triad and is hydrogen bonded to the N^δ of His 44. A tyrosine residue interacts with His44. Tyr143 of HAV 3C is perpendicularly above the plane of the imidazole of His44 and, therefore, cannot form a hydrogen bond to the imidazole. Its interaction with His44 must be electrostatic.

The additional groups in the active site of the 3C proteinases, which interact with the histidine general acid-base catalyst, most likely have two functions: they maintain the proper orientation of the active-site residues and they stabilize the charges of those residues. That these interactions appear to be mostly electrostatic and do not form typical hydrogen bonds could be taken as further evidence for an enzymatic mechanism of the 3C proteinases, involving a thiolate-imidazolium ion pair. The details of the actual enzymatic mechanism of the chymotrypsin-like cysteine proteinases remain to be established.

IV. Autocatalytic Excision of the 3C Proteinase

The 3C proteinases can cleave themselves out of the respective viral polyprotein in *cis* when they are part of the polyprotein, or in *trans* when they are expressed separately (HARMON et al. 1992). PALMENBERG and RUECKERT (1982) were the first to provide evidence that the autocatalytic cleavage of the 3C proteinases could be an intramolecular event. Further evidence was provided by HANECAK et al. (1984). Nevertheless, the available experimental evidence cannot distinguish between a truly intramolecular cleavage and a proteolytic cleavage within a tight dimer or larger oligomer of 3C proteinase precursors.

The three crystal structures of the 3C proteinases suggest a possible model for an intramolecular cleavage event at the amino-terminus of 3C (MATTHEWS et al. 1994; BERGMANN et al. 1997; MOSIMANN et al. 1997). In this model, the unique, amino-terminal α-helix of the 3C proteinases is folded only after cleavage at the amino-terminus of 3C. Prior to the cleavage, only the last turn of helix A exists as a reverse turn, and the amino-terminal residues reach into the active site along β-strand bI in an extended conformation. The last turn of helix A is formed by a conserved sequence motif, R/K-R/K-N-I/L. After the amino-terminus is cleaved, the folding of the stable helix A removes the P_n' residues from the active site to prevent intramolecular product inhibition.

It is very difficult to imagine a similar intramolecular cleavage at the carboxy-terminus of 3C. The authors of all the published crystal structures instead favor an intermolecular proteolytic cleavage within a tight polymer of 3C precursors (Matthews et al. 1994; Bergmann et al. 1997; Mosimann et al. 1997). All three crystal structures are made up of at least two independent molecules. In all the structures, the carboxy-terminus of one molecule of a dimer is within reach of the proteolytic active site of another molecule. However, the dimers in the three different crystal structures are different, and there is no independent experimental evidence for the formation of a tight dimer of the 3C proteinases in solution. Structural work on larger precursors of 3C will be required to resolve the mechanism of autocatalytic cleavage.

V. Other Functions of the Picornaviral 3C Gene Product

The most conserved motif in the sequence of the picornaviral 3C proteinases is not part of the proteolytic active site (GORBALENYA et al. 1989; RYAN and FLINT 1997). It was first shown for poliovirus 3C that these residues are important for a function of the 3C gene product which is distinct from the proteolytic activity (Andino et al. 1990; Hämmerle et al. 1992; Andino et al. 1993). That the few gene products of small RNA viruses perform multiple functions in viral replication is not an uncommon situation. The picornaviral 3C proteinase is recruited as part of the viral replicase complex and has an RNA-binding site. The RNA-binding activity of 3C is important for the recognition of the non-translated regions of the viral RNA during RNA replication (Andino et al. 1990; Hämmerle et al. 1992; Andino et al. 1993; Leong et al. 1993; Harris et al. 1994; Kusov and Gauss-Müller 1997; Walker et al. 1995). The exact function of 3C during viral RNA replication is not known. It is possible that the RNA-binding activity of 3C simply serves to recruit the proteinase to the replicase complex in order to perform essential proteolytic cleavages within these complexes (Xiang et al. 1998).

The conserved RNA-binding site of 3C is on the surface of the molecule opposite from the proteolytic active site (Matthews et al. 1994; Bergmann et al. 1997; Mosimann et al. 1997). The conserved sequence motif KFRDI forms part of the connection between the two domains of the 3C proteinase (Fig. 3). The domain connection is in a partly helical and partly extended conformation and is flanked by the amino- and carboxy-terminal helices. Several of the β-turns that connect the strands of the two β-barrel domains also contribute to this surface. This surface of the molecule is highly charged (Bergmann et al. 1997).

The RNA-binding site of 3C is on the opposite site of the proteolytic active site. Therefore, the structures suggest that the two activities could be completely independent. Both the amino- and the carboxy-terminal helices do, however, contribute to the RNA-binding site. Therefore, RNA binding could have an influence on the processing of the termini of 3C (Ryan and Flint 1997). Similarly, a larger precursor of 3C, such as 3ABC or 3CD, would most likely have different RNA-binding activity. The RNA-binding activity presents another possible target for the design of antiviral inhibitors. Because little is known about the molecular details of the RNA-binding activity of 3C, there has been little effort directed against this function to date.

F. Inhibition of the 3C Proteinase

I. Effect of 3C Proteinase Inhibitors on Viral Replication

The 3C proteinase performs an important and indispensable function during the viral life cycle. The chymotrypsin-like cysteine proteinases also represent a unique class of proteolytic enzymes, with a specificity that is distinct from all known cellular proteinases (Gorbalenya and Snijder 1996; Ryan and Flint 1997; Bergmann and James 1999). As such, the picornaviral 3C proteinases represent ideal targets for the design of proteinase inhibitors with antiviral activity (Kräusslich and Wimmer 1988). Some 3C proteinase inhibitors effectively inhibit viral replication and reduce viral load when tested in cell cultures (MORRIS et al. 1997; DRAGOVICH et al. 1998a, b; KONG et al. 1998; WEBBER et al. 1998).

Even for the best-reported proteinase inhibitors, the ex vivo inhibition of viral replication in cell cultures is usually significantly less potent than the in vitro proteinase inhibition. For very good proteinase inhibitors with nanomolar inhibition constants (K_is), the doses producing a response in 50% of animals are typically micromolar or slightly below. The antiviral activity of some good proteinase inhibitors was disappointing when tested ex vivo; these inhibited viral replication in cell culture not at all or only at concentrations that are toxic for the cells (WEBBER et al. 1996). Effective antivirals need to be tight-binding proteinase inhibitors, possess low toxicity and be able to reach sufficiently high intracellular levels.

II. Strategies for the Design of 3C Proteinase Inhibitors

It is important to keep in mind that the recent successful development of HIV-proteinase inhibitors has benefited tremendously from the understanding of the mechanism of aspartic proteinases, which was derived from many years of experimental work on other aspartic proteinases, such as renin. The chemical functionality at the core of all the new anti-HIV drugs, which are HIV-proteinase inhibitors, is very similar to classical aspartic proteinase inhibitors (HOETELMANS et al. 1997; KORANT and RIZZO 1997; see also the introduction to this volume). This illustrates the point that a detailed understanding of enzymatic structure, function and mechanism is invaluable for the design of effective inhibitors. While there exists a considerable amount of information about specific intermolecular interactions between 3C proteinases and their substrates or inhibitors, little is known about the catalytic mechanisms of these enzymes.

The most commonly applied approach to the development of 3C proteinase inhibitors combines a reactive chemical functionality with groups that satisfy the known specificity determinants of the proteinases. The reactive chemical functionalities are usually groups that are known to react covalently with the active-site thiol nucleophile of cysteine proteinases (RASNICK 1996). It has been shown that several inhibitors react covalently with the active-site thiol of the 3C proteinases. To achieve specificity, these functionalities are combined with groups that mimic the specificity determinants of a peptide substrate of the proteinase. Several sources of experimental information contributed information about the specificity requirements of the 3C proteinases.

Analysis of the sequence of the cleavage sites within the natural substrate, the viral polyprotein, usually reveals sequence preferences (reviewed by BERGMANN 1998 and SKERN 1998). Kinetic studies with small peptide substrates are also informative in identifying the substrate preferences of the proteinase (NICKLIN et al. 1988; ORR et al. 1989; PALLAI et al. 1989; CORDINGLEY et al. 1990; WEIDNER and DUNN 1991; JEWELL et al. 1992). Most of these studies found that the 3C proteinases prefer certain amino acids in the P_4, P_2 and P_1 positions of the substrate. The entero- and rhinoviral enzymes also require glycine and proline in the P_1' and P_2' positions of a substrate. The sequence preferences for substrates of the 3C proteinases from individual viruses differ. It is also noteworthy that the optimal cleavage sequence found in the context of a hexapeptide substrate can be different from the consensus sequence of the cleavage sites in the viral polyprotein. It is generally assumed that the cleavage sequence preferences derived from kinetics studies with small peptide substrates are more useful for the design of specific inhibitors.

The third source which contributes experimental information about the specific interactions between the 3C proteinases and their cognate substrates are the crystal structures of 3C proteinases (Matthews et al. 1994; Bergmann et al. 1997; Mosimann et al. 1997) and, more recently, proteinase-inhibitor complexes (Webber et al. 1996; Dragovich et al. 1998 a, b; Webber et al. 1998). Initially, the crystal structures of the free enzymes were used to model the binding of a substrate or inhibitor. Crystal structures of inhibitor complexes proved the general validity of the binding modes utilized in these models. There were, however, differences in the details of the enzyme-substrate interactions between the models and the corresponding cocrystal structures (Webber et al. 1998). While theoretical models of enzyme-substrate interactions are useful in the absence of experimental structures, they rarely predict all details of intermolecular interactions correctly or as accurately as an actual cocrystal structure of a complex.

The detailed understanding of the specific interactions between the 3C proteinases and their preferred substrates has led to the development of potent inhibitors of the enzymes. Most of these inhibitors mimic the specific interactions of a P_4 to P_1 tetrapeptide substrate and combine this with a chemical functionality which reacts covalently with the active-site thiol (Kaldor et al. 1995; Malcolm et al. 1995; Morris et al. 1997; Dragovich et al. 1998 a,b; Kong et al. 1998; Webber et al. 1998). The chemical functionalities utilized are usually classic cysteine proteinase inhibitors that react covalently with the enzyme. In the case of the rhinovirus 3C proteinase, these inhibitors have been further improved by optimization of the individual groups that target the specific subsites of the enzyme (Dragovich et al. 1998a,b; Kong et al. 1998; Webber et al. 1998). This has resulted in some potent proteinase inhibitors having sub-nanomolar inhibition constants.

There have also been alternative approaches to the discovery and design of 3C proteinase inhibitors. An interesting method used to identify specific inhibitors of the HAV 3C proteinase has been employed by McKendrick et al. (1998). HAV 3C proteinase was incubated with a mixture of peptide-based covalent inhibitors, and the inhibited enzyme was analyzed by mass spectrometry. The analysis showed that the enzyme was able to select from the mixture one inhibitor that best fit its specific interactions. This approach could be used generally to optimize enzyme inhibitors that reacted covalently. In this case, it also identified peptide-mimetic inhibitors that targeted the specific S′ subsites of the enzyme.

A very labor-intensive method is the screening of large libraries of natural compounds or cultures of microorganisms for enzyme inhibitors (SINGH et al. 1991; KADAM et al. 1994; McCALL et al. 1994; BRILL et al. 1997; JUNGHEIM et al. 1997). When successful, this approach can identify completely new and unexpected classes of compounds. The resulting compounds are rarely very potent inhibitors; none of the proteinase inhibitors that resulted from these screening procedures have been developed into potent inhibitors.

III. Inhibitors of the Chymotrypsin-Like Cysteine Proteinases

The most effective 3C proteinase inhibitors combine a chemical group that interacts covalently with the active-site cysteine nucleophile with other groups that interact non-covalently with the specificity determinants of the enzyme. The chemical functionalities that react covalently with the nucleophilic thiol of the enzyme are classical cysteine proteinase inhibitors (RASNICK 1996). Among the reactive groups are aldehydes (KALDOR et al. 1995; MALCOLM et al. 1995; SHEPHERD et al. 1996), iodoacetylpeptidyl amides (McKENDRICK et al. 1998), β-lactams (SKILES et al. 1990), halomethyl ketones (ORR et al. 1989; SHAM et al. 1995; MORRIS et al. 1997), isatins (WEBBER et al. 1996) and vinylogous sulfones and esters. The best, presently known, inhibitors of the 3C proteinases are the vinylogous esters (DRAGOVICH et al. 1998 a,b; KONG et al. 1998). These compounds react with the nucleophilic thiol of the enzyme via a Michael addition. This was confirmed by experimental evidence, including cocrystal structures of proteinase-inhibitor complexes (DRAGOVICH et al. 1998a,b). Other chemical functionalities also provide potent proteinase inhibitors but are far less promising as antivirals. No results of experiments regarding the toxicity and bioavailability of 3C proteinase inhibitors in animals or humans have been published at this time.

The most commonly encountered problems, when proteinase inhibitors were tested for their antiviral activity in cell cultures, were toxicity and poor intracellular availability. Some very promising proteinase inhibitors, such as the isatins, showed no antiviral efficacy at concentrations below their toxicity levels in cell culture (WEBBER et al. 1998). Presumably, problems with toxicity are, at least in part, due to the reactivity of the covalent 3C proteinase inhibitors, but no potent, non-covalent inhibitors of the 3C proteinases are known at this point. Other potent proteinase inhibitors are significantly less effective as antivirals in cell cultures. Among those inhibitors are the fluoromethyl ketones and vinylogous sulfones. Presumably, these inhibitors do not achieve sufficiently high intracellular concentrations.

The minimum size of the effective 3C proteinase inhibitors corresponds to the equivalent of a tetrapeptide which mimics the P_4 to P_1 residues of a substrate. Smaller inhibitors are significantly less effective because of the reduced number of specific, intermolecular interactions they can form with the enzyme. The problem of poor intracellular availability, which is encountered with some proteinase inhibitors, can therefore not be overcome by reducing the size of the inhibitors.

G. Summary and Outlook

The 3C proteinases perform an essential function during the life cycle of three large families of animal viruses (Picornaviridae, Caliciviridae and Coronaviridae). Members of these virus families are responsible for a large number of respiratory and intestinal infections and also cause more serious viral infections. The chymotrypsin-like cysteine proteinases of positive-sense, single-stranded RNA viruses constitute a distinct class of enzymes, and the specificity of the 3C proteinases is unique. They are, therefore, an ideal target for the design of specific antiviral drugs. Extensive kinetic studies and crystal structures of the enzymes from three genera of the Picornaviridae have provided important insights into the structure–function relationships of these enzymes, but little is known about the enzymatic mechanism of the 3C proteinases. So far, all effective inhibitors of the 3C proteinase react covalently with the active-site cysteine nucleophile.

Rhinoviruses and some of the coronaviruses are together responsible for the vast majority of common colds. Both could potentially be inhibited by effective 3C proteinase inhibitors. Because many viruses from different families cause upper-respiratory-tract infections, which are essentially indistinguishable by their clinical symptoms alone, effective treatment would also require simple analytical procedures to identify the causative agent. Members of the family Caliciviridae and Coronaviridae cause a large number of intestinal infections, often referred to as "stomach flu". These viruses should, therefore, be considered attractive drug-design targets. A treatment for the rarer but often serious enteroviral infections would be of great value. Whether or not a treatment of enteroviral infection could be beneficial for the prevention of autoimmune diseases is not clear.

Acknowledgements. The research of the authors on the picornaviral proteinases was supported by the National Sciences and Engineering Research Council of Canada and the National Institute of Allergies and Infectious Diseases of the USA. We are grateful to Tim Skern (Vienna) for critical comments.

References

Allaire M, Chernaia MM, Malcolm BA, James MNG (1994) Picornaviral 3C cysteine proteinases have a fold similar to chymotrypsin-like serine proteinases. Nature 369:72–76

Andino R, Rieckhof GE, Trono DE, Baltimore D (1990) Substitutions in the protease (3C^{Pro}) can suppress a mutation in the 5′ noncoding region. J Virol 64:607–612

Andino R, Rickhof GE, Achacoso PL, Baltimore D (1993) Poliovirus RNA synthesis utilizes an RNP complex formed around the 5'-end of viral RNA. EMBO J 12:3587–3598

Andreoletti L, Hober D, Hober-Vandenberghe C, Belaich S, Vantyghem MC, Lefebvre J, Wattre P (1997) Detection of Coxsackie B virus RNA sequences in whole blood samples from adult patients at the onset of type I diabets mellitus. J Med Virol 52:121–127

Bazan JF, Fletterick RJ (1988) Viral cysteine proteinases are homologous to the trypsin-like family of serine proteinases: structural and functional implications. Proc Natl Acad Sci U S A 85:7872–7876

Belsham GJ (1993) Distinctive features of foot-and-mouth disease virus, a member of the picornavirus family; aspects of virus protein synthesis, protein processing and structure. Prog Biophys Mol Biol 60:241–260

Bergmann EM (1998) Hepatitis A virus picornain 3C. In: Barrett AD, Rawlings NJ, Woesner F (eds) Handbook of proteolytic enzymes. Academic Press, London

Bergmann EM, James MNG (1999) Proteolytic enzymes of the viruses of the family Picornaviridae. In: Dunn B (ed) Proteinases of infectious agents. Academic Press, San Diego

Bergmann EM, Mosimann SC, Chernaia MM, Malcolm BA, James MNG (1997) The refined crystal structure of the 3C gene product from hepatitis A virus: specific proteinase activity and RNA recognition. J Virol 71:2436–2448

Bienz K, Egger D, Rasser Y, Bossart W (1983) Intracellular distribution of poliovirus proteins and the induction of virus-specific cytoplasmic structures. Virology 131:39–48

Bienz K, Egger D, Troxler M, Pasamontes L (1990) Structural organization of poliovirus RNA replication is mediated by viral proteins of the P2 genomic region. J Virol 64:1156–1163

Bode W, Huber R (1992) Natural protein proteinase inhibitors and their interactions with proteinases. Eur J Biochem 204:433–451

Borman AM, Kean KM (1997) Intact eukaryotic initiation factor 4G is required for hepatitis A virus internal initiation of translation. Virology 237:129–136

Brill BM, Kati WM, Montgomery D, Karwowski JP, Humphrey PE, Jackson M, Clement JJ, Kadam S, Chen RH, McAlpine JB (1997) Novel triterpene sulfates from fusarium compactum using a rhinovirus 3C protease inhibitor screen. J Antibiot (Tokyo) 49:541–546

Brocklehurst K, Watts AB, Patel M, Verma C, Thomas EW (1998) Cysteine proteinases. In: Sinnott ML (ed) Comprehensive biological catalysis. Academic Press, London

Carthy CM, Yang D, Anderson DR, Wilson JE, McManus BM (1997) Myocarditis as systemic disease: new perspectives on pathogenesis. Clin Exp Pharmacol Physiol 24:997–1003

Center for Disease Control (1998) Progress towards global eradication of poliomyelitis 1997. MMWR Morb Mort Wkly Rep 47:414–419

Chang L-Y, Huang Y-C, Lin, TY (1998) Fulminant neurogenic pulmonary oedema with hand, foot and mouth diease. Lancet 352:367–368

Chotia C (1984) Principles that determine the structure of proteins. Annu Rev Biochem 53:537–572

Clarke IN, Lambden PR (1997) Viral zoonoses and food of animal origin: caliciviruses and human disease. Arch Virol Suppl 13:141–152

Cochi SL, Hull HF, Sutter RW, Wilfert CM, Katz SL (eds) (1997) Global poliomyelitis eradication initiative: status report. J Infect Dis 175[Suppl 1]

Cordingley MG, Callahan PL, Sardana VV, Garsky VM, Colonno RJ, (1990) Substrate requirements of human rhinovirus 3C protease for peptide cleavage in vitro. J Biol Chem 265:9062–9065

Couch RB (1996) Rhinoviruses. In: Fields BN, Knipe DM, Howley PM, Channock RM, Melnick JL, Monath TP, Roizmann BE, Straus SE (eds) Fields' virology. Lippincott-Raven, Philadelphia

Donnelly MLL, Gani D, Flint M, Monaghan S, Ryan MD (1997) The cleavage activity of aphtho and cardiovirus 2A proteins. J Gen Virol 78:13–21

Dougherty WG, Semler BL (1993) Expression of virus-encoded proteinases: functional and structural similarities with cellular enzymes. Microbiol Rev 57:781–822

Dragovich PS, Webber SE, Babine, RE, Fuhrman SA, Patick AK, Matthews DA, Lee CA, Reich SH, Prins TJ, Marakovits JT, Littlefield ES, Zhou R, Tikhe J, Ford CE, Wallace MB, Meador JW III, Ferre RA, Brown EL, Binford Sl, Harr JE, DeLisle DM, Worland ST (1998a) Structure-based design, synthesis and biological evaluation of irreversible human rhinovirus 3C protease inhibitors. 1. Michael acceptor structure-activity studies. J Med Chem 41:2806–2818

Dragovich PS, Webber SE, Babine, RE, Fuhrman SA, Patick AK, Matthews DA, Reich SH, Marakovits JT, Prins TJ, Zhou R, Tikhe J, Littlefield ES, Bleckman TM, Wallace MB, Little TL, Ford CE, Meador JW III, Ferre RA, Brown EL, Binford Sl, DeLisle DM, Worland ST (1998b) Structure-based design, synthesis and biological evaluation of irreversible human rhinovirus 3C protease inhibitors. 2. Peptide structure-activity studies. J Med Chem 41:2819–2834

Gamarnik AV, Andino R (1997) Two functional complexes formed by KH domain containing proteins with the 5′ noncoding region of poliovirus RNA. RNA 3:882–892

Gonzalez P, Sanches A, Rivera P, Jimenez C, Hernandez F (1997) Rotavirus and Coronavirus outbreak: etiology of annual diarrhea in Costa Rican children. Rev Biol Trop 45:898–991

Gorbalenya AE, Snijder EJ (1996) Viral cysteine proteinases. Perspect Drug Discovery Design 6:64–86

Gorbalenya AE, Svitkin YV, Kazachkov YA, Agol VI (1979) Encephalomyocarditis virus-specific polypeptide p22 is involved in the processing of the viral precursor polypeptide. FEBS Lett 108:1–5

Gorbalenya AE, Blinov VM, Donchenko AP (1986) Poliovirus-encoded proteinase 3C: a possible evolutionary link between cellular serine and cysteine proteinase families. FEBS Lett 194:253–257

Gorbalenya AE, Donchenko AP, Blinov VM, Koonin EV (1989) Cysteine proteinases of positive strand RNA viruses and chymotrypsin-like serine proteinases: a distinct protein superfamily with a common structural fold. FEBS Lett 243:103–114

Green KY (1997) The role of human caliciviruses in epidemic gastroenteritis. Arch Virol Suppl 13:153–165

Haghighat A, Svitkin Y, Novoa I, Küchler E, Skern T, Sonnenberg N (1996) The eIF4G-eIF4E complex is the target for direct cleavage by the rhinovirus 2A proteinase. J Virol 70:8444–8450

Hämmerle T, Molla A, Wimmer E (1992) Mutational analysis of the proposed FG loop of poliovirus proteinase 3C identified amino acids that are necessary for 3CD cleavage and might be determinants of a function distinct from proteolytic activity. J Virol 66:6028–6034

Hanecak R, Semler BL, Ariga H, Anderson CW, Wimmer E (1984) Expression of a cloned gene segment of poliovirus in E. coli: evidence for autocatalytic production of the viral proteinase. Cell 37:1063–1073

Harmon SA, Updike W, Xi-Ju J, Summers DF, Ehrenfeld E (1992) Polyprotein processing in cis and in trans by hepatitis A virus 3C protease cloned and expressed in E. coli. J Virol 66:5242–5247

Harris KS, Xiang W, Alexander LS, Lane WS, Paul AV, Wimmer E (1994) Interactions of poliovirus polypeptide 3CD^Pro with the 5′ and 3′ termini of the poliovirus genome. J Biol Chem 269:27004–27014

Hoetelmans RMW, Meenhorst PL, Mulder JW, Burger DM, Koks CHW, Beijnen JH (1997) Clinical pharmacology of HIV protease inhibitors: focus on saquinavir, inidinavir and ritonavir. Pharm World Sci 19:159–175

Hollinger FB, Ticehurst JR (1996) Hepatitis A virus. In: Fields BN, Knipe DM, Howley PM, Channock RM, Melnick JL, Monath TP, Roizmann BE, Straus SE (eds) Fields' virology. Lippincott-Raven, Philadelphia

Holmes KV, Lai MC (1996) Coronaviridae: the viruses and their replication. In: Fields BN, Knipe DM, Howley PM, Channock RM, Melnick JL, Monath TP, Roizmann BE, Straus SE (eds) Fields' virology. Lippincott-Raven, Philadelphia

James MNG (1993) Convergence of active-centre geometries among the proteolytic enzymes. In: Bond JS, Barrett AJ (eds) Proteolysis and protein turnover. Portland, London

Jewell DA, Swietnicki AW, Dunn, BM, Malcolm BA (1992) Hepatitis A virus 3C proteinase substrate specificity. Biochemistry 31:7862–7869

Jia X-Y, Summers DF, Ehrenfeld E (1993) Primary cleavage of the HAV capsid protein precursor in the middle of the proposed 2A coding region. Virology 193:515–519

Jungheim LN, Cohen JD, Johnson RB, Villareal EC, Wakulchik M, Loncharich RJ, Wang QM (1997) Inhibition of human rhinovirus 3C protease by homophthalimides. Bioorg Med Chem Lett 7:1589–1594

Kadam S, Poddig J, Humphrey P, Karwowski J, Jackson M, Tennent S, Fung L, Hochlowski J, Rasmussen R, McAlpine J (1994) Citrinin hydrate and radicinin: human rhinovirus 3C protease inhibitors discovered in a target-directed microbial screen. J Antibiot (Tokyo) 47:836–839

Kaldor SW, Hammond M, Dressman BA, Labus JM, Chadwell FW, Kline AD, Heinz BA (1995) Glutamine-derived aldehydes for the inhibition of human rhinovirus 3C proteinase. Bioorg Med Chem Lett 5:2021–2026

Kapikian AZ, Estes MK, Chanock RM (1996) Norwalk group of viruses. In: Fields BN, Knipe DM, Howley PM, Channock RM, Melnick JL, Monath TP, Roizmann BE, Straus SE (eds) Fields' virology, Lippincott-Raven, Philadelphia

Kong JS, Venkatraman S, Furness K, Nimkar S, Shepherd TA, Wang QM, Aubé J, Hanzlik R (1998) Synthesis and evaluation of peptidyl Michael acceptors that inactivate human rhinovirus 3C protease and inhibit virus replication. J Med Chem 41:2579–2587

Korant BD (1972) Cleavage of viral precursor proteins in vivo and in vitro. J Virol 10:751–759

Korant BD, Rizzo CJ (1997) The HIV protease and therapies for AIDS. Adv Exp Med Biol 421:279–284

Korant BD, Chow N, Lively M, Powers J (1979) Virus-specified protease in poliovirus-infected HeLa cells. Proc Natl Acad Sci USA 76:2992–2995

Kräusslich H-G, Wimmer E (1988) Viral proteinases. Annu Rev Biochem 57:701–754

Kusov YY, Gauss-Müller V (1997) In vitro RNA binding of the hepatitis A virus proteinase 3C (HAV 3C^{Pro}) to secondary structure elements within the 5' terminus of the HAV genome. RNA 3:291–302

Landsteiner K, Popper E (1909) Übertraqung der Poliomyelitis acuta auf Affen. Z Immunitätsforschung Orig 2:377–390

Leong LEC, Walker PA, Porter AG (1993) Human rhinovirus 14 protease 3C (3C^{Pro}) binds specifically to the 5'-noncoding region of the viral RNA. J Biol Chem 268:25735–25739

Liu DX, Xu HY, Brown TD (1997) Proteolytic processing of the coronavirus infectious bronchitis virus 1a polyprotein: identification of a 10 kilodalton polypeptide and determination of its cleavage sites. J Virol 71:1814–1820

Loeffler F, Frosch P (1964) Report of the Commission for Research on Foot-and-Mouth Disease. In: Hahon N (ed) Selected papers on virology. Prentice-Hall, Englewood Cliffs

Long LA, Orr DC, Cameron JM, Dunn BM, Kay J (1989) A consensus sequence for substrate hydrolysis by rhinovirus 3C proteinase. FEBS Lett 258:75–78

Lu XT, Sims AL, Denison MR (1998) Mouse hepatitis virus 3C-like proteinase cleaves a 22 kilodalton protein from the ORF 1a polyprotein in virus-infected cells and in vitro. J Virol 72:2265–2279

Makela MJ, Puhakka T, Runskanen O, Leinonen M, Saikku AP, Kimpimaki M, Blomquist S, Hyppia T, Arstilla P (1998) Viruses and bacteria in the etiology of the common cold. J Clin Microbiol 36:539–542

Malcolm BA, Lowe C, Shechosky S, McKay RT, Yang CC, Shah VJ, Simon RJ, Vederas JC, Santi DV (1995) Peptide aldehyde inhibitors of hepatitis A virus 3C proteinase. Biochemistry 34:8172–8179

Martín-Alonso JM, Casais R, Boga JA, Parra F (1996) Processing of rabbit hemorrhagic disease virus polyprotein. J Virol 70:1261–1265

Martin A, Escriou N, Chao SF, Girard M, Lemon SM, Wychowski C (1995) Identification and site-directed mutagenesis of the primary (2A/2B) cleavage site of the hepatitis A virus polyprotein: functional impact on the infectivity of HAV RNA transcripts. Virology 213:213–222

Matthews DA, Smith WW, Ferre RA, Condon B, Budahazi G, Sisson W, Villafranca JE, Janson CA, McElroy HE, Gribskov CL, Worland S (1994) Structure of human rhinovirus 3C protease reveals a trypsin-like polypeptide fold, RNA-binding site and means for cleaving precursor polyprotein. Cell 77:761–771

McCall JO, Kadam S, Katz L (1994) A high capacity microbial screen for inhibitors of human rhinovirus protease 3C. Biotechnology 12:1012–1016

McKendrick JE, Frormann S, Luo C, Semchuck P, Vederas JC, Malcolm BA (1998) Rapid mass spectrometric determination of preferred irreversible proteinase inhibitors in combinatorial libraries. Int J Mass Spectrom 176:113–124

Melnick JL (1996) Enteroviruses: Polioviruses Coxsackie viruses, echoviruses, and newer enteroviruses. In: Fields BN, Knipe DM, Howley PM, Channock RM, Melnick JL, Monath TP, Roizmann BE, Straus SE (eds) Fields' virology. Lippincott-Raven, Philadelphia

Miller SD, Vanderlugt CL, Smith-Begolka W, Pao W, Yauch RL, Neville KL, Katz-Levy Y, Carrizosa A, Kim BS (1997) Persistent infection with Theiler's virus leads to CNS autoimmunity via epitope spreading. Nat Med 3:1133–1136

Morris TS, Frormann S, Shechosky S, Lowe C, Lall MS, Gauss-Müller V, Purcell RH, Emerson SU, Vederas JC, Malcolm BA (1997) In vitro and ex vivo inhibition of hepatitis A virus 3C proteinase by a peptidyl monofluoromethyl ketone. Bioorg Med Chem Lett 5:797–807

Mosimann SC, Chernaia MM, Sia S, Plotch S, James MNG (1997) Refined X-ray crystallographic structure of the poliovirus 3C gene product. J Mol Biol 273:1032–1047

Ng LFP, Liu DX (1998) Identification of a 24-kDa polypeptide processed from the coronavirus infectious bronchitis virus 1a polyprotein by the 3C-like proteinase and determination of its cleavage sites. Virology 243:388–395

Nicklin MJ, Harris KS Pallai PV, Wimmer E (1988) Poliovirus proteinase 3C: large-scale expression purification and specific cleavage activity on natural and synthetic substrates in vitro. J Virol 62:4586–4593

Niklasson B, Hornfeldt B, Lundman B (1998) Could myocarditis, insulin-dependent diabetes mellitus and Guillain-Barre syndrome be caused by one or more infectious agents carried by rodents? Emerg Infect Dis 4:187–193

Orr DC, Long AC, Kay J, Dunn BM, Cameron JM (1989) Hydrolysis of a series of synthetic peptide substrates by the human rhinovirus 14 3C proteinase cloned and expressed in E. coli. J Gen Virol 70:2931–2992

Pallai PV, Burkhardt F, Shoog M, Schreiner K, Bax P, Cohen KA, Hansen G, Palladino DE, Harris KS, Nicklin MJ, Wimmer E (1989) Cleavage of synthetic peptides by purified poliovirus 3C proteinase. J Biol Chem 264:9738–9741

Palmenberg AC (1990) Proteolytic processing of picornaviral polyprotein. Annu Rev Microbiol 44:602–623

Palmenberg AC, Rueckert RR (1982) Evidence for intramolecular self-cleavage of picornaviral replicase precursors. J Virol 41:244–249

Palmenberg AC, Pallansch MA, Rueckert RR (1979) Protease required for processing picornaviral coat protein residues resides in the viral replicase gene. J Virol 32:770–778

Palmenberg AC, Parks GD, Hall DJ, Ingraham RH, Seng TW, Pallai PV (1992) Proteolytic processing of the cardioviral P2 region: primary 2A/2B cleavage in clone derived precursors. Virology 190:754–762

Parsley TB, Towner JS, Blyn LB, Ehrenfeld E, Semler BL (1997) Poly (rC) binding protein 2 forms a ternary complex with the 5'-terminal sequences of poliovirus RNA and the viral 3CD proteinase. RNA 3:1124–1134

Pebody RG, Leino T, Ruuutu P, Kinnunen L, Davidkin I, Nohynek H, Leinikki P (1998) Foodborne outbreaks of hepatitis A in a low endemic country: an emerging problem? Epidemiol Infect 120:55–59

Pelham HRB (1978) Translation of encephalomyocarditis virus RNA in vitro yields an active proteolytic processing enzyme. Eur J Chem 85:457–462

Perona JJ, Craik CS (1995) Structural basis of substrate specificity in the serine proteinases. Protein Sci 4:337–360

Porter AG (1993) Picornavirus nonstructural proteins: emerging roles in virus replication and inhibition of host cell functions. J Virol 67:6917–6921

Qasim MA, Ranjbar MR, Wynn S, Anderson S, Laskowski M Jr (1995) Ionizable P_1 residues in serine proteinase inhibitors undergo large pK shifts on complex formation. J Biol Chem 270:27419–27422

Rasnick D (1996) Small synthetic inhibitors of cysteine proteinases. Perspect Drug Discov Design 6:47–63

Read RJ, James MNG (1986) Introduction to the protein inhibitors: X-ray crystallography. In: Barrett AJ, Salvesen G (eds) Proteinase inhibitors. Elsevier Science, Amsterdam

Roehl HH, Parsley TB, Ho TV, Semler BL (1997) Processing of a cellular polypeptide by 3CD proteinase is required for poliovirus ribonucleoprotein comlex formation. J Virol 71:578–585

Roivainen M, Knip M, Hyöty H, Kulmala P, Hiltunen M, Vähäsalo P, Hovi T, Åkerblom HK (1998) Several different enterovirus serotypes can be associated with prediabetic autoimmune episodes and onset of overt IDDM. Childhood Diabetes in Finland (DiMe) Study Group J Med Virol 56:74–78

Rueckert RR (1996) Picornaviridae: the viruses and their replication. In: Fields BN, Knipe DM, Howley PM, Channock RM, Melnick JL, Monath TP, Roizmann BE, Straus SE (eds) Fields' virology. Lippincott-Raven, Philadelphia

Ryan MD, Flint M (1997) Virus-encoded proteinases of the picornavirus super-group. J Gen Virol 78:699–723

Schechter I, Berger A (1967) On the size of the active site in proteases. I. Papain. Biochem Biophys Res Commun 27:157–162

Schiller JJ, Kanjanahaluethai A, Baker SC (1998) Processing of the coronavirus MHV-JHM polymerase polyprotein: identification of precursors and proteolytic products spanning 400 kilodaltons of ORF 1a. Virology 242:288–302

Schultheiss T, Kusov YY, Gauss-Müller V (1994) Proteinase 3C of hepatitis A virus (HAV) cleaves the HAV polyprotein P2-P3 at all sites including VP1/2A and 2A/2B. Virology 198:275–281

Schultheiss T, Emerson SU, Purcell RH, Gauss-Müller V (1995) Polyprotein processing in echovirus 22 – a first assessment. Biochem Biophys Res Commun 219: 1120–1127

Seybert A, Ziebuhr J, Siddell SG (1997) Expression and characterization of a recombinant murine coronavirus 3C-like proteinase. J Gen Virol 78:71–75

Sham JL, Rosenbrook W, Kari W, Betebenner DA, Wideburg NE, Saldivar A, Plattner JJ, Norbeck DW (1995) Potent inhibitors of the human rhinovirus 3C protease containing a backbone modified glutamine. J Chem Soc Perkin Trans 1:1081–1082

Shepherd TA, Cox GA, McKinney A, Tang J, Wakulchik M, Zimmermann RE, Villareal EC (1996) Small peptidic aldehyde inhibitors of human rhinovirus 3C protease. Bioorg Med Chem Lett 6:2893–2896

Singh SB, Cordingley MG, Ball RG, Smith JL, Dombrowski AW, Goetz MA (1991) Structure and stereochemistry of thysanone: A novel human rhinovirus 3C protease inhibitor from *Thysanophora penicilloides*. Tetrahedron Lett 32:5279–5282

Sjogren MH (1998) Preventing acute liver disease in patients with chronic liver disease. Hepatology 27:887–888

Skern T (1998) Picornain 3C. In: Barrett AD, Rawlings NJ, Woesner F (eds) Handbook of proteolytic enzymes. Academic Press, London

Skiles JW, McNeil D (1990) Spiro indolinone β-lactams inhibitors of poliovirus and rhinovirus 3C-proteinases. Tetrahedron Lett 31:7277–7280

Snijder EJ, Wassenar ALM, van Dinten LC, Spaan WJM, Gorbalenya AE (1996) The arterivirus Nsp4 protease is the prototype of a novel group of chymotrypsin-like enzymes, the 3C-like serine proteases. J Biol Chem 271:4864–4871

Steinmann L, Conlon P (1997) Viral damage and the breakdown of self-tolerance. Nat Med 3:1085–1087

Storer AC, Ménard R (1994) Catalytic mechanism in papain family of cysteine peptidases. Methods Enzymol 244:486–500

Summers DF, Maizel JV (1968) Evidence for large precursor proteins in poliovirus synthesis. Proc Natl Acad Sci USA 59:966–971

Sutters RW, Cochi SL (1997) Ethical dilemmas in worldwide polio eradication programs. Am J Public Health 87:913–916

Taylor CE, Taylor ME, Cutts F (1997) Ethical dilemmas in current planning for polio eradication. Am J Public Health 87:922–925

Teterina NL, Bienz K, Egger D, Gorbalenya AE, Ehrenfeld E (1997a) Induction of intracellular membrane rearrangements by HAV proteins 2C and 2BC. Virology 237:66–77

Teterina NL, Gorbalenya AE, Egger D, Bienz K, Ehrenfeld E (1997b) Poliovirus 2C protein determinants of membrane binding and rearrangements in mammalian cells. J Virol 71:8962–8972

Thiel TK (1998) Hepatitis A vaccination. Am Fam Physician 57:1500–1501

Tibbles KW, Brierley I, Cavanagh D, Brown TDK (1996) Characterization in vitro of an autocatalytic processing activity associated with the predicted 3C-like proteinase domain of the Coronavirus avian infectious bronchitis virus. J Virol 70:1923–1930

Vento S, Garofano T, Renzini C, Cainelli F, Casali F, Ghironzi G, Ferraro T, Conaia E (1998) Fulminant hepatitis associated with hepatitis A virus superinfection in patients with chronic hepatitis C. N Engl J Med 338:286–290

Walker PA, Leong LEC, Porter AG (1995) Sequence and structural determinants of the interaction between the 5'-noncoding region of picornavirus RNA and rhinovirus protease 3C. J Biol Chem 270:14510–14516

Webber SE, Tikhe J, Worland ST, Fuhrmann SA, Hendrickson TF, Matthews DA, Love RA, Patick AK, Meador JW, Ferre PA, Brown EL, Delisle DM, Ford CE, Binford SL (1996) Design synthesis and evaluation of nonpeptide inhibitors of human rhinovirus 3C proteinase. J Med Chem 39:5072–5882

Webber SE, Okano K, Little TL, Reich SH, Xin Y, Fuhrman SA, Matthews DA, Love RA, Hendrickson TF, Patick AK, Meador JW III, Ferre RA, Brown EL, Ford CE, Binford SL, Worland ST (1998) Tripeptide aldehyde inhibitors of human rhinovirus 3C protease: design, synthesis, biological evaluation and cocrystal structure solution of P1 glutamine isosteric replacements. J Med Chem 41:2786–2805

Weidner JR, Dunn BM (1991) Development of synthetic peptide substrates for the poliovirus 3C proteinase. Arch Biochem Biophys 286:402–408

Whiting AK, Peticolas WL (1994) Details of the acyl-enzyme intermediate and the oxyanion hole in serine protease catalysis. Biochemistry 33:552–561

Wimmer E (1982) Genome linked proteins of viruses. Cell 28:199–201

Wimmer E, Hellen CUT, Cao X (1993) Genetics of poliovirus. Annu Rev Genet 27:353–436

Wirblich C, Sibilia M, Boniotti MB, Rossi C, Thiel H-J, Meyers G (1995) 3C-like protease of rabbit hemorrhagic disease virus: identification of cleavage sites in the ORF1 polyprotein and analysis of cleavage specificity. J Virol 69:7159–7169

Xiang WS, Harris KS, Alexander L, Wimmer E (1995) Interaction between the 5'-terminal cloverleaf and 3AB/3CD^{Pro} of poliovirus is essential for RNA application. J Virol 69:3658–3667

Xiang W, Cuconati A, Hope D, Kirkegaard K, Wimmer E (1998) Complete protein
    linkage map of poliovirus P3 proteins: interaction of polymerase 3D^{pol} with Vpg
    and with genetic variants of 3AB. J Virol 72:6732–6741
Yalamanchili D, Weidman K, Dasgupta A (1997) Cleavage of transcriptional activator
    Oct-1 by poliovirus encoded protease 3C^{pro}. Virology 239:176–185
Ypma-Wong MF, Dewalt PG, Johnson VH, Lamb JG, Semler BL (1988) Protein 3CD
    is the major poliovirus proteinase responsible for cleavage of the P1 capsid pre-
    cursor. Virology 166:265–270

CHAPTER 8

Adenovirus Proteinase-Antiviral Target for Triple-Combination Therapy on a Single Enzyme: Potential Inhibitor-Binding Sites

W. F. MANGEL, D.L. TOLEDO, M.T. BROWN, J. DING, R.M. SWEET, D.L. BARNARD, and W.J. McGRATH

A. Virus-Coded Proteinases as Targets for Antiviral Therapy

Virus-coded proteinases are attractive targets for antiviral therapy. These enzymes are essential for the synthesis of infectious virus and perform a wide variety of tasks at different times and places during an infection. Among the medically important virus families with members known to encode proteinases (KRAUSSLICH and WIMMER 1988; BEBÉ and CRAIG 1997) are: the picornaviruses, which include polio, rhino and hepatitis A; the retroviruses, which include human immunodeficiency virus (HIV); the flaviviruses, which include hepatitis C; the orthomyxoviruses, which include influenza; the herpes viruses, which include cytomegalovirus; and the adenoviruses. Viral proteinases are extremely specific enzymes, and there has been expectation that equally specific inhibitors may be effective antiviral agents. This expectation has recently been realized with the advent of HIV-proteinase inhibitors that have been shown to be extremely specific biochemically and effective clinically.

Human adenovirus presents a good model system to study the exploitation of virus-coded proteinases as targets for antiviral therapy. The adenovirus proteinase contains three potential sites for antiviral therapy – the active site and the sites to which two cofactors, the viral DNA and the 11 amino acid peptide pVIc, bind to activate the enzyme. Thus human adenovirus can be used as a model system to test the hypothesis that the probability of generating a virus resistant to three different inhibitors directed against three different sites on the same virus-coded protein is much lower than to three different inhibitors directed against three different virus-coded proteins.

I. Adenovirus and Its Proteinase in the Virus Life Cycle

There are 47 serotypes of adenovirus, and they can cause acute infections of the respiratory and gastrointestinal tracts and of the eye (HORWITZ 1990; HIERHOLZER et al. 1991). Adenovirus is a nonenveloped virus that contains 34,000–36,000bp of double-stranded (ds) DNA, with a potential coding capacity for more than 50 proteins (HORWITZ 1990; HIERHOLZER et al. 1991). The human adenovirus proteinase (AVP) is activated late in infection. After formation of empty capsid shells, the viral DNA, along with capsid components, enters the shells to form immature virions. The virus-coded proteinase is then activated; it cleaves six virion precursor proteins, thereby rendering the virus particles infectious (MIRZA and WEBER 1980; HANNAN et al. 1983). There are about 50 proteinase molecules per virion (BROWN et al. 1996), and they cleave the multiple copies of six different virion precursor proteins 2500 times in each virus particle. AVP may play a role in virus entry into cells (COTTEN and WEBER, 1995; GREBER et al. 1996); inhibition of the viral proteinase blocked the degradation of the capsid-stabilizing protein VI and prevented virus uncoating – and thereby release of the viral DNA – at the nuclear membrane (GREBER et al. 1996).

II. The AVP as a Model System for Antiviral Agents

The AVP is an ideal enzyme system within which to test proteinase inhibitors as antiviral agents. A wide range of animals can be infected by different strains of adenovirus, including mice, chickens and monkeys, so that once good inhibitors of the enzyme are found they can be tested as antiviral agents in several animal systems. At a minimum, the AVP utilizes three geographically distinct sites for optimal enzyme activity: an active site and two cofactor-binding sites. These sites have been characterized biochemically, and the crystal structure of the enzyme with one of the cofactors bound is known at resolutions of 2.6 Å (DING et al. 1996) and 1.6 Å (McGRATH et al. unpublished observations). Here, we describe these three sites and discuss the design of different types of inhibitors to bind to these sites and function as antiviral agents. We also address the issue of resistance and how this enzyme system may be used to study the efficacy of different variations of combination therapy.

B. Biochemistry of the AVP

I. Cloning of the Gene and Development of an Assay for the Adenovirus-2 Proteinase

The gene for the AVP has been identified (YEH-KAI et al. 1983), cloned and expressed in *Escherichia coli* (ANDERSON 1993), and the resultant 204-amino-acid protein purified (MANGEL et al. 1993; TIHANYI et al. 1993; WEBSTER et al. 1993; MANGEL et al. 1996). The enzyme was shown to be highly specific; analysis by WEBSTER et al. (1989a, b) of the cleavage sites in the six virion precursor proteins processed by AVP indicated requirements of either Leu, Ile or Met in the P4 position and Gly in the P2, followed by Gly-Xxx or Xxx-Gly. In this notation (SCHECHTER and BERGER 1967), P1 is the amino acid in a substrate that is cleaved at its C-terminus and P2 is the adjacent amino acid N-terminal to P1 (SCHECHTER and BERGER 1967). Xxx is any amino acid residue. We synthesized (Leu-Arg-Gly-Gly-NH)$_2$-rhodamine and showed it to be an extremely specific, sensitive and selective substrate for AVP within disrupted wild-type adenovirus serotype 2 (Ad2) virions (MANGEL et al. 1993; MCGRATH et al. 1996). However, purified recombinant AVP exhibited no proteinase activity (MANGEL et al. 1993; WEBSTER et al. 1993; MANGEL et al. 1996). Eventually, cofactors were discovered.

II. Discovery and Characterization of Two Cofactors

One cofactor is the viral DNA (MANGEL et al. 1993). If disrupted wild-type Ad2 virus is treated with DNase and then assayed, proteinase activity is lost but can be restored upon addition of Ad2 DNA. A second cofactor is a plasmin-sensitive virion protein, which turned out to be the 11-amino-acid peptide, pVIc, from the C-terminus of the precursor to virion protein VI (MANGEL et al. 1993; WEBSTER et al. 1993). Its sequence is GVQSLKRRRCF.

The cofactors affect the macroscopic kinetic constants of the interaction of AVP with the rhodamine-based fluorogenic substrates (MANGEL et al. 1996). AVP alone has a small amount of activity. By incubating Ad2 DNA with AVP, the Michaelis constant (K_m) increases twofold and the catalytic rate constant (k_{cat}) threefold. By incubating pVIc with AVP, K_m increases twofold and k_{cat} increases 350-fold. With all three components together, AVP plus Ad2 DNA plus pVIc, K_m increases twofold and k_{cat} increases 6000-fold relative to those with AVP alone. Thus, the cofactors increase proteinase activity by increasing the k_{cat}, not decreasing the K_m.

III. Binding Interactions among the Cofactors

1. AVP Binding to pVIc in the Absence and Presence of DNA

The binding interactions among the cofactors were characterized by titration curves and by fluorescence polarization. For the interaction of AVP with pVIc, we incubated a constant amount of AVP with increasing amounts of pVIc and assayed for enzyme activity. When the amount of enzyme activity was plotted on the ordinate versus the concentration of pVIc, a hyperbola was obtained. If one assumes that at the plateau all the AVP had been titrated with pVIc, then the data could be transformed into bound pVIc versus free pVIc (Fig 1A) and, from this, an equilibrium dissociation constant (K_d) of 682 $\pm$ 121 nM could be calculated. When this titration was repeated in the presence of an excess of T7 DNA, and with AVP varied as opposed to pVIc, the K_d dropped to 54.7 $\pm$ 0.6 nM (Fig. 1B).

2. AVP–pVIc-Complex Binding to DNA

In order to characterize the interaction of AVP–pVIc complexes with DNA, we titrated a constant amount of ds 36-mer DNA with increasing amounts of AVP–pVIc complexes and assayed for enzyme activity (Fig. 2). Here, tight binding, characterized by two straight lines, was observed; below saturation,

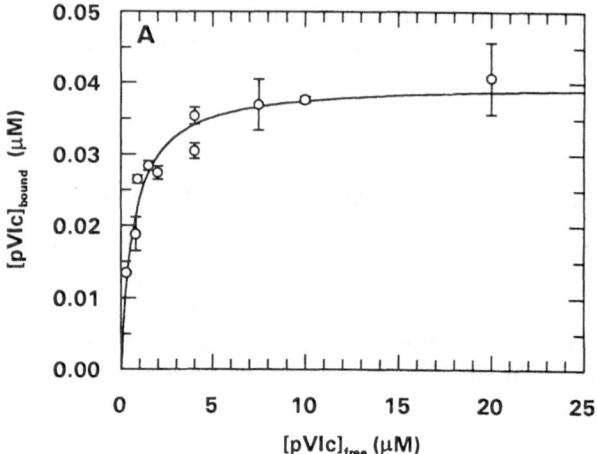

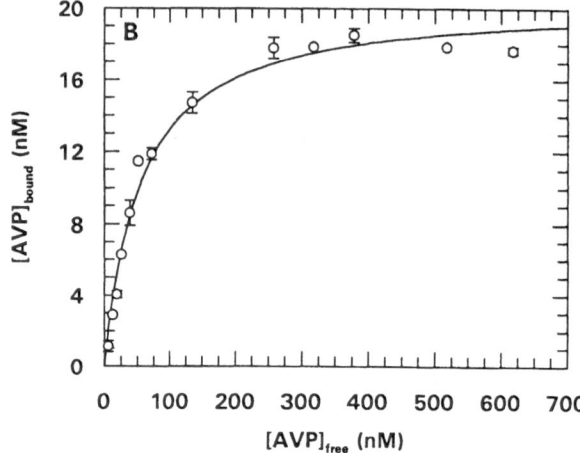

Fig. 1A,B. Binding of pVIc to adenovirus proteinase (AVP) in the absence (**A**) and presence (**B**) of DNA. **A** Reactions (1 ml) that contained 0.01 M tris(hydroxymethyl)aminomethane (Tris, pH 8.0), 5 mM octylglucoside, 40 nM AVP and concentrations of pVIc that ranged from 0–20 μM were incubated at 37°C for 5 min. Then (Leu-Arg-Gly-Gly-NH)$_2$-Rhodamine was added to a concentration of 10 μM and the increase in fluorescence was monitored as a function of time. The change in fluorescence was calculated by subtracting the fluorescence of the sample without pVIc from the fluorescence of samples containing pVIc. The change in fluorescence was plotted versus time and the resulting rates were transformed to concentrations of bound and free pVIc. **B** Reactions (1 ml) that contained 0.01 M Tris (pH 8.0), 5 mM octylglucoside, 1.5 μg/ml T7 DNA, 20 nM pVIc and concentrations of AVP that ranged from 0–600 nM were incubated at 37°C for 3 min. Then (Leu-Arg-Gly-Gly-NH)$_2$-Rhodamine was added to a concentration of 2 μM and the increase in fluorescence was monitored as a function of time. The change in fluorescence was calculated by subtracting the fluorescence of the sample without AVP from the fluorescence of samples containing AVP. The change in fluorescence was plotted against time and the resulting rates transformed to concentrations of bound and free AVP

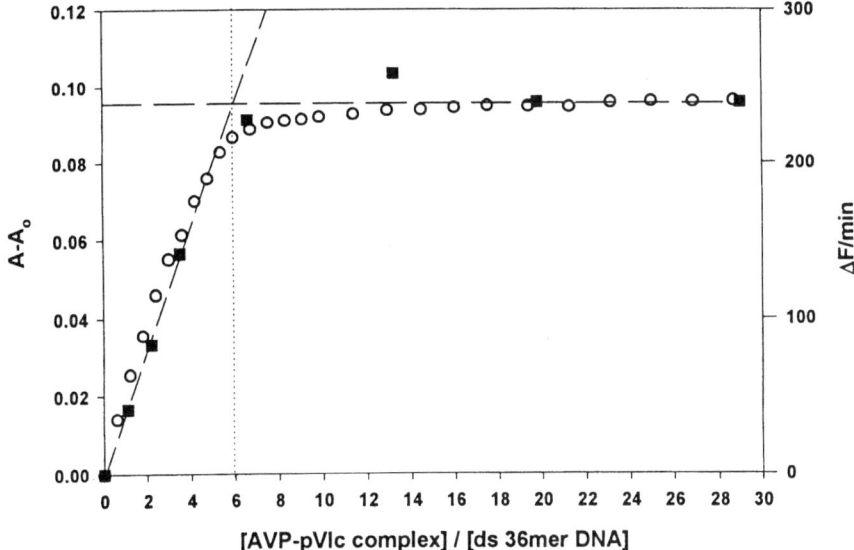

Fig. 2. Binding of the adenovirus proteinase (AVP)-pVIc complexes to DNA is coincident with stimulation of enzyme activity by DNA. AVP–pVIc complexes were formed by incubating a 1.5-M excess of pVIc with AVP in 0.01 M tris(hydroxymethyl)aminomethane (Tris, pH 8.0), 5 mM NaCl and 1 mM ethylenediaminetetraacetic acid (EDTA) on ice for 30 min. Fluorescence anisotropy measurements (*open circle*) were performed with 12.5-nM 5′-fluorescein-labeled double-stranded (ds) 36-mer DNA in 10 mM Tris (pH 8.0), 0.1 mM EDTA and 0.0125% NP-40 at 20°C. Aliquots of AVP-pVIc complexes were added to the DNA; the solutions were mixed and, 30 s after each addition, the change in anisotropy (A–A$_0$) was measured using an excitation wavelength of 490 nm and a 520-nm bandpass filter. Activity measurements (*closed square*) were performed with the AVP–pVIc complex in the presence or absence of 12.5-nM ds 36-mer DNA in 10 mM Tris (pH 8.0), 5 mM octylglucoside and 5 μM (Leu-Arg-Gly-Gly-NH)$_2$-Rhodamine at 20°C. The change in fluorescence (ΔF) was monitored as a function of time using an excitation wavelength of 492 nm and an emission wavelength of 523 nm. The units of the abscissa are moles of AVP–pVIc complex divided by moles of ds 36-mer DNA. The *dashed lines* are least-squares fits to the data points. The *vertical dotted line* defines the intersection of the two dashed lines at a molar ratio of 6

there was no unbound AVP–pVIc complex. Saturation, the intersection of the two straight lines, was achieved at 6 AVP–pVIc complexes per ds 36-mer DNA. If the experiment was repeated with fluorescein-labeled ds 36-mer DNA and if, instead of assaying for enzyme activity, the change in anisotropy was measured, the two curves were superimposable. These two sets of data indicate that the enzyme binds to DNA and that binding to DNA is coincident with enzyme stimulation.

IV. Roles of AVP Cofactors in Virus Maturation

The functions of the cofactors might be to regulate the temporal and spatial activity of the enzyme (MANGEL et al. 1997). Our working hypothesis is that the enzyme is initially synthesized with negligible activity. Presumably, if it were active before virion assembly, it would cleave virion precursor proteins, thereby preventing virion assembly. Late in infection, virion proteins, including precursor proteins, assemble into "empty capsids" (BHATTI and WEBER 1978). Then the core proteins and AVP bound to the viral DNA are encapsidated, generating "young virions." Binding to DNA increases the k_{cat} of AVP threefold. It is in these "top components" that the proteinase is activated and the precursor proteins are processed to yield mature, infectious virus. This could occur by pVI binding to the viral DNA (RUSSELL and PRECIOUS 1982) such that AVP already bound to the viral DNA can excise pVIc. The released pVIc could then bind AVP. The AVP–pVIc complex next binds to the viral DNA, and this ternary complex, AVP–pVIc–DNA, is a fully active proteinase.

How can 50 fully-activated proteinases (BROWN et al. 1996) bound to the viral DNA inside the virion cleave 2500 peptide bonds in precursor proteins to render a virus particle infectious? Perhaps the viral DNA serves as a guidewire, next to which are the 2500 processing sites that must be cleaved. The proteinase complex could then slide along the viral DNA, cleaving the precursor proteins. The binding of AVP–pVIc complexes to DNA is not sequence specific (MANGEL et al. 1993), a property that allows the proteinase to move along the viral DNA. This would be analogous to the binding of the *E. coli* RNA polymerase holoenzyme to nonpromoter DNA sequences (HINKLE and CHAMBERLIN 1972). RNA polymerase binds to nonpromoter DNA sequences with a K_d of 100 nM, and the polymerase slides along the DNA via one-dimensional diffusion (SINGER and WU 1988) until it locates a promoter. In the case of AVP, it slides along the viral DNA, encountering precursor cleavage sites.

C. Crystal Structure of the Adenovirus-2 Proteinase Complexed with pVIc

The crystal structure of an AVP–pVIc complex has been solved to 2.6-Å resolution by means of X-ray-crystallographic analysis using single isomorphous replacement supplemented with anomolous scattering (DING et al. 1996). The AVP–pVIc complex is ovoid, with α-helices forming the wide end (Fig. 3). The narrow end contains another α-helix, and the region between comprises one central and two peripheral α-helices that interact with a β-sheet. The β-sheet consists of five β-strands from AVP; a sixth β-strand originates from the last eight amino-acid residues of pVIc.

AVP appears to represent a new type of proteinase. The sequence of the gene for the proteinase is not related to any gene sequences in the databases. Inhibitor profiles of enzyme activity give ambiguous results in revealing the type of proteinase. Comparing the structure of AVP–pVIc with all unique protein molecules in the Brookhaven Protein Data Bank (BERNSTEIN et al.

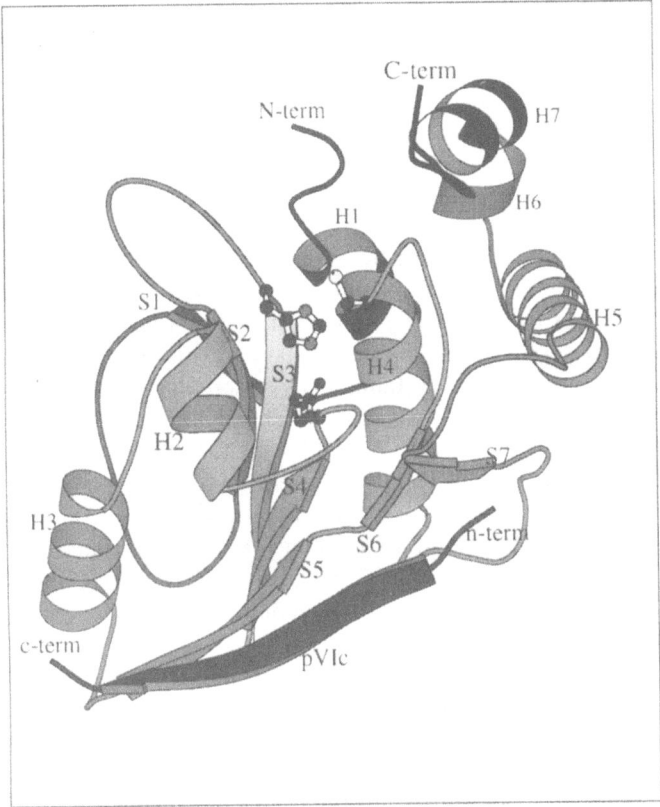

Fig. 3. The secondary structure of the adenovirus proteinase–pVIc complex. α-Helices are labeled H1 through H7; β-strands S1 through S7 from the N- to C-terminus. pVIc, is the nearest β-strand of the figure. Side chains are shown only for the active-site residues Cys122, His54 and Glu71

1977) revealed no equivalent structure, suggesting that AVP represents a new family of protein molecules.

However, we noticed some structural similarities between papain and the AVP–pVIc complex; a helix and several β-strands within the central region of papain appear to be in similar positions in the AVP–pVIc complex. When we superimposed those structural similarities, we found that the nucleophilic Cys25 of papain superimposed over Cys122 of AVP (Fig. 4, s. appendix, page 398/399). Furthermore, His159 and Asn175 of papain superimposed over His54 and Glu71 of AVP. Even the major component of the oxyanion hole, Gln19 of papain, superimposed over Gln115 of AVP. Thus, AVP is the first member of a new class of cysteine proteinases, C5 (RAWLINGS and BARRETT 1994); it is an example of convergent evolution. Despite the similarities with papain in the positions of the amino-acid residues involved in catalysis, the sequential order of these amino-acid residues in the polypeptide chain is different. In AVP, the triad involved in catalysis is His54, Glu71 and Cys122, whereas in papain the order is Cys25, His159 and Asn175.

D. Potential Inhibitor-Binding Sites

There are general "rational" and "irrational" ways to obtain inhibitors of AVP. One "rational" way is "structure-based drug design". This approach utilizes computer graphics to display the topography of a potential drug-binding site so that one can design drugs complementary to this site. A computer can be used both automatically and systematically to screen structurally diverse compounds for prototypes that fit. Searches of chemical data bases using the coordinates of AVP deposited in the Brookhaven Data Base (accession number 1AVP) via programs such as DOCK (Kuntz 1992; Shoichet et al. 1992; Shoichet and Kuntz 1993), CAVEAT (Lauri and Bartlett 1994) or SYSDOC, a supercomputing-based dimeric analog approach for drug optimization (Pang and Kollman 1995; Pang and Brimijoin 1998), should aid in the identification of lead compounds. A general "irrational" approach to drug design is the "combinatorial-library" method. Here, many thousands of structurally diverse compounds are generated by combinatorial chemistry and the products evaluated with automated, high-throughput assays. Once a large number of active structures is ascertained, their common structural motifs are identified by deconvolution (Lam et al. 1997).

Lead compounds, identified "rationally" or "irrationally," are then refined to become even better inhibitors. First, one measures the inhibitory equilibrium dissociation constant, expecting it to be micromolar or lower. Then, the lead compound is co-crystallized with the enzyme. Are the contacts as predicted? Based upon the structure, a second generation inhibitor is designed, synthesized, tested and co-crystallized with the enzyme. The process is repeated until a selective inhibitor with a low equilibrium dissociation constant is obtained.

I. Active Site

The active site of AVP is on the surface of the molecule. It lies within a 25-Å-long, bent groove that is ~8-Å wide. Cys122 and His54 lie in the middle of the groove. There are several areas within the active site to which inhibitors can be designed to bind.

AVP is both different from and similar to papain, and this can be exploited in the design of unique inhibitors of AVP. For example, there are many different low-molecular-weight inhibitors of papain; their contacts within the active site of papain are known because they have been co-crystallized with papain and their structures determined by X-ray diffraction (Kim et al. 1992). Regions of papain where inhibitors bind can be compared with similar regions in AVP. Then, the inhibitors of papain can be redesigned to take into account the differences. For example, E-64 (N-[N-(L-3-trans-carboxirane-2-carbonyl)-L-leucyl]-agmatine) is an excellent inhibitor of papain that does not inhibit AVP (MCGRATH et al. 1996). One may be able to redesign E-64, retaining common contacts with papain and AVP, removing contacts unique to papain, and inserting contacts unique to AVP, such that the resultant compound becomes an inhibitor of AVP and not of papain.

There are several ways in which the substrate specificity of AVP can be exploited in inhibitor design. Competitive inhibitors can be substrate like. We know the specificity determinants in substrates for AVP are the P4 and P2 amino acids. Thus, compounds that reflect this specificity and have the potential to inhibit the enzyme can be designed, synthesized and tested. For example, these compounds could contain a nonhydrolyzable reduced peptide bond (HOLSKIN et al. 1995), an aldehyde (MACKENZIE et al. 1986) or a monofluoroketone (RASNICK 1985; MCGRATH et al. 1995) at the P1 position.

The substrate specificity of AVP can be taken advantage of to deliver specific inhibitors to the active-site groove. We have performed modeling studies using the five amino acids at the site in the virus-coded protein pVI where cleavage by AVP liberates pVIc-Ile-Val-Gly-Leu-Gly (Fig. 5, s. appendix, page 398/399). Energy-minimization studies indicate the substrate binds to the active-site groove as a β-strand. The P2 amino acid binds in a very small hydrophobic groove. The P4 amino acid binds in a larger pocket that can best accommodate a Leu, Ile or Met residue. The side chains of the P1 and P3 amino acids point away from the surface of AVP, which is why they are not specificity determinants. Thus, one can attach inhibitory groups to the P1 and/or P3 amino-acid side chains or even a putative P5 side chain, and the resultant peptide should still bind specifically in the active-site groove.

II. DNA-Binding Sites

We don't know where DNA binds on the surface of the AVP–pVIc complex. However, the molecular surface of the AVP–pVIc complex has four large clusters of positive-charge density ranging in area from $45 \, \text{Å}^2$ to $65 \, \text{Å}^2$; these are potential DNA-binding sites (Fig. 6, s. appendix, page 398/399). The shortest distance between clusters (~24 Å) is commensurate with the rise of a single turn of ds helical DNA. We are trying to obtain X-ray diffraction-quality crystals of an AVP–pVIc–DNA complex. Once the structures of the DNA-binding sites are known, compounds will be designed to bind to these sites. Although DNA binding is not sequence specific, the DNA-binding sites may be unique, and it is to these sites that specific inhibitors may be designed. Our presumption is that a compound is a potential antiviral agent if it prevents AVP from binding to the viral DNA.

III. pVIc-Binding Sites

The pVIc-binding site is another place inhibitory molecules may bind. Surprisingly, pVIc, which exerts powerful control over the rate of catalysis, binds quite far from the active-site residues involved in catalysis. The cysteine residue of pVIc forms a disulfide bond with Cys104 of AVP, which is 32 Å away from the active-site nucleophile Cys122. The residue of pVIc closest to the active site is Val2', whose side chain is 14.5 Å away from Cys122. There is extensive contact between pVIc and AVP: 28 hydrogen bonds, 4 ionic bonds and a disulfide bond (Fig. 7, s. appendix, page 402/403). pVIc directly interacts with two different regions of AVP, and this suggests both how pVIc may increase k_{cat} and how to design potential proteinase inhibitors. pVIc appears as a "strap" that spans a region from the conserved Cys104 near the narrow end of the structure across the back of the molecule, i.e., away from the active-site groove, to the turn at the end of helix 4 near the active-site groove. One possible function of pVIc may be that, upon binding, it brings two regions of AVP together: the formation of the β-sheet on one end of pVIc with the central β-structure of the molecule and the extensive contact of pVIc's amino-terminal region with residues that interact near the C-terminal end of helix 4. The functional consequences of bringing two regions of AVP together may be that this alters the geometry around the putative Cys–His ion pair, thereby increasing the pair's catalytic power. If pVIc really functions as a strap, then molecules that prevent pVIc from bringing two regions of AVP together are potential proteinase inhibitors.

Different ligands that bind to the pVIc-binding site in AVP in preference to pVIc can have different effects. Some ligands may increase the k_{cat} for substrate hydrolysis, like pVIc. The binding of other ligands may prevent the activation of AVP. Since we don't yet know how the binding of pVIc to AVP activates the enzyme, we cannot predict whether a specific ligand will stimulate or prevent the activation of the enzyme. However, both types of ligands may be antiviral agents. A ligand that stimulates the activation of AVP could be used to activate AVP before virion assembly. Then, that complex would cleave virion precursor proteins, thereby preventing the formation of nascent virus particles. Alternatively, the ligand that prevents activation of AVP would prevent maturation of precursor proteins within virions, thus preventing the virus particle from becoming infectious.

If pVIc is a "strap" that holds together two domains of AVP, then peptides that interfere with the binding of native pVIc and prevent a "strap" from forming should inhibit the activation of AVP. For example, the N- and C-terminal amino-acid residues of pVIc – GVQ and SLKRRRCF, respectively – should bind to AVP. Although GVQ should bind to the C-terminal end of helix 4, binding alone it should not be able to bring both domains together. Similarly, although SLKRRRCF may be able to form a β-strand with the central β-sheet structure, binding alone it should not be able to bring that domain closer to the C-terminal end of helix 4.

Another potential inhibitor is the peptide GVAALAARACA. It is a mutant in which the amino-acid residues of pVIc whose side chains interact with AVP have been replaced by an alanine residue. Side-chain interactions are very specific, possibly more so than the interactions of AVP with the peptide backbone of pVIc. This mutant peptide may prevent a series of subtle positional changes throughout the AVP–pVIc complex that result in the 350-fold increase in enzyme activity.

The crystal structure of the AVP–pVIc complex revealed that amino-acid residues $3'$, $7'$, $9'$ and $11'$ in pVIc form salt bridges with AVP. Thus, GVASLKA-RAC is a peptide that should not be able to form any of the salt bridges with AVP. Testing of this mutant will show whether the salt bridges are required for pVIc stimulation of enzyme activity. If they are required, then additional experiments can be done to determine which salt bridges are important. This approach may lead to an abbreviated form of pVIc that can stimulate enzyme activity.

E. Summary and Prospects

Human adenovirus is a good model system with which to study the exploitation of virus-coded proteinases as targets for antiviral therapy. Because activation of AVP utilizes two cofactors, the viral DNA and the 11-amino-acid peptide pVIc, this is a good model system with which to test the hypothesis that the probability of generating a virus resistant to three different inhibitors directed against three different sites in the same virus-coded protein is much lower than to three different inhibitors directed against three different virus-coded proteins. Once we have compounds that inhibit the enzyme in vitro, we shall test them with viruses and cells in culture. A wide range of animals, including mice, chickens and monkeys, can be infected with different strains of adenovirus so that we can test the more promising inhibitors in several animal systems.

Although we present human adenovirus as a model system within which to study the use of proteinase inhibitors as antiviral agents, the results of our experiments are directly applicable to other medically important proteinases, because human adenovirus is "less unique" than it used to be. The Sindbis virus serine proteinase has a very basic N-terminal segment responsible for association with the viral RNA (Tong et al. 1993). The NS3 protein of hepatitis-C virus is a serine proteinase whose activity is enhanced by a cofactor, the 54-amino-acid residue NS4A protein. A 12-residue synthetic peptide, comprising amino acids 12–33 of NS4A, forms a complex with the NS3 proteinase domain and activates the enzyme so that it can cleave at certain processing sites (Butkiewicz et al. 1996; Kim et al. 1996).

There are numerous sites on AVP where ligands can bind and, in doing so, should inhibit enzyme activity. Within the active site, there are several different regions. The DNA-binding site will soon be revealed, and polyanions designed to bind to it should be good antiviral agents. Although we still don't know how the binding of pVIc to AVP increases k_{cat} for substrate hydrolysis, we do know how pVIc binds to AVP; this has enabled us to design several potential antiviral agents.

Acknowledgements. The research was supported by National Institute of Allergy and Infectious Diseases Grant AI41599, the Office of Health and Environmental Research of the United States Department of Energy, the National Science Foundation and the Department of Energy's Office of Science Education and Technical Information. The work on testing pVIc as an antiviral agent on virus-infected cells in culture was funded by Contract N01-AI-35178 from the Virology Branch, Division of Microbiology and Infectious Diseases, National Institute of Allergy and Infectious Diseases (NIAID), National Institutes of Health. We also thank S. Klein and M. Garner for help with the manuscript.

References

Anderson CW (1993) Expression and purification of the adenovirus proteinase polypeptide and of a synthetic proteinase substrate. Protein Expr Purif 4:8–15

Bebé LM, Craig CS (1997) Viral proteases: evolution of diverse structural motifs to optimize function. Cell 91:427–430

Bernstein FC, Koetzle TF, Williams, GJB, Meyer EF, Brice MD, Rodgers JR, Kennard O, Shimanouchi T, Tasumi M (1977) The protein data bank: A computer-based archival file for macromolecular structure. J Mol Biol 112:535–542

Bhatti AR, Weber J (1978) Protease of adenovirus type 2. In vitro processing of core protein. Biochem Biophys Res Commun 81:973–979

Brown MT, McGrath WJ, Toledo DL, Mangel WF (1996) Different modes of inhibition of human adenovirus proteinase, probably a cysteine proteinase, by bovine pancreatic trypsin inhibitor. FEBS Lett 388:233–237

Butkiewicz NJ, Wendel M, Zhang R, Jubin R, Pichardo J, Smith EB, Hart AM, Ingram R, Durkin J, Mui PW, Murray MG, Ramanathan L, Dasmahapatra B (1996) Enhancement of hepatitis C virus NS3 proteinase activity by association with NS4A-specific synthetic peptides: identification of sequence and critical residues of NS4A for the cofactor activity. Virology 225:328–338

Cotten M, Weber J (1985) The adenovirus protease is required for virus entry into host cells. Virology 213:494–502

Ding J, McGrath WJ, Sweet RM, Mangel WF (1996) Crystal structure of the human adenovirus proteinase with its 11 amino-acid cofactor. EMBO J 15:1778–1783

Greber UF, Webster P, Weber J, Helenius A (1996) The role of the adenovirus protease in virus entry into cells. EMBO J 15:1766–1777

Hannan C, Raptis LH, Dery CV, Weber J (1983) Biological and structural studies with adenovirus type 2 temperature-sensitive mutant defective for uncoating. Intervirology 19:213–223

Hinkle D, Chamberlin M (1972) Studies of the binding of E. coli RNA polymerase to DNA. II. The kinetics of the binding reaction. J Mol Biol 70:187–196

Hierholzer JC, Stone YO, Broderson JR (1991) Antigenic relationships among the 47 human adenoviruses determined in reference horse antisera. Arch Virol 121:179–197

Holskin BP, Bukhtiyarova M, Dunn BM, Baur P, de Chastonay J, Pennington MW (1995) A continuous fluorescence-based assay of human cytomegalovirus protease using a peptide substrate. Anal Biochem 226:148–155

Horwitz MS (1990) Adenoviruses. In: Fields BN, Knipe DM (eds) Fields' virology (vol 2), 2nd edn. Raven, New York, pp 1723–1742

Kim M-J, Yamamoto D, Matsumoto K, Inoue M, Ishida T, Mizuno H, Sumiya S, Kitamura K (1992) Crystal structure of papain-E64-c complex: binding diversity of E64-c to papain S2 and S3 subsites. Biochem J 287:797–803

Kim JL, Morgenstern KA, Lin C, Fox T, Dwyer MD, Landro JA, Chambers SP, Markland W, Lepre CA, O'Malley ET, Haberson SL, Rice CM, Murcko MA,

Caron PR, Thomson JA (1996) Crystal structure of the hepatitis C virus NS3 protease domain complexed with a synthetic NS4A peptide cofactor. Cell 87:343–355

Krausslich H-G, Wimmer E (1998) Vital proteinases. Annu Rev Biochem 57:701–754

Kuntz ID (1992) Structure-based strategies for drug design and discovery. Science 257:1078–1082

Lam KS, Lebl M, Krchnak V (1997) The "one bead-one compound" combinatorial library method. Chem Rev 97:411

Lauri G, Bartlett PA (1994) A program to facilitate the design of organic molecules. J Comput Aided Mol Des 8:51–66

Mackenzie NE, Grant SK, Scott AI, Malthouse PG (1986) ^{13}C NMR study of the stereospecificity of the thiohemiacetals formed on inhibition of papain by specific enantiomeric aldehydes. Biochemistry 25:2293–2298

Mangel WF, McGrath WJ, Toledo DL, Anderson CW (1993) Viral DNA and a viral peptide can act as cofactors of adenovirus virion proteinase activity. Nature 361:274–275

Mangel WF, Toledo DL, Brown MT, Martin JH, McGrath WJ (1996) Characterization of three components of human adenovirus proteinase activity in vitro. J Biol Chem 271:536–543

Mangel WF, Toledo DL, Ding J, Sweet RM, McGrath WJ (1997) Temporal and spatial control of the adenovirus proteinase by both a peptide and the viral DNA. Trends Biochem Sci 22:393–398

McGrath ME, Eakin AE, Engel JC, McKerrow JH, Craik CS, Fletterick RJ (1995) The crystal structure of cruzain: a therapeutic target for Chagas' disease. J Mol Biol 247:251–259

McGrath WJ, Abola AP, Toledo DL, Brown MT, Mangel WF (1996) Characterization of human adenovirus proteinase activity in disrupted virus particles. Virology 217:131–138

Mirza A, Weber J (1980) Infectivity and uncoating of adenovirus cores. Intervirology 13:307–311

Pang Y-P, Kollman PA (1995) Applications of free energy derivatives to analog design. Perspect Drug Discov Des 3:106–122

Pang Y-P, Brimijoin S (1998) Supercomputing-based dimeric analog approach for drug optimization. Parallel Computing 24:1557–1566

Rasnick D (1985) Synthesis of peptide fluoromethylketones and the inhibition of human cathepsin B. Anal Biochem 149:461–465

Rawlings ND, Barrett AJ (1994) Families of cysteine peptidases. Meth Enzymol 244:461–486

Russell WC, Precious B (1982) Nucleic acid-binding properties of adenovirus structural polypeptides. J Gen Virol 63:69–79

Schechter I, Berger A (1967) On the size of the active site in proteases. I. Papain. Biochem Biophys Res Commun 27:157–162

Shoichet BK, Bodian DL, Kuntz ID (1992) Molecular docking using shape descriptors. J Comp Chem 13:380–397

Shoichet BK, Kuntz ID (1993) Matching chemistry and shape in molecular docking. Protein Eng 6:723–732

Singer PT, Wu C-W (1988) Kinetics of promoter search by *Escherichia coli* RNA polymerase: effets of monovalent and divalent cations and temperature. J Biol Chem 263:4208–4214

Tihanyi K, Bourbonniere M, Houde A, Rancourt C, Weber J M (1993) Isolation and properties of adenovirus type 2 proteinase. J Biol Chem 268:1780–1785

Tong L, Wengler G, Rossmann MG (1993) Refined structure of Sindbis virus core protein and comparison with other chymotrypsin-like serine proteinase structures. J Mol Biol 230:228–247

Webster A, Russell S, Talbot P, Russell WC, Kemp GD (1989a) Characterization of the adenovirus proteinase: substrate specificity. J Gen Virol 70:3225–3234

Webster A, Russell W C, Kemp GD (1989b) Characterization of the adenovirus pro-
    teinase; development and use of a specific peptide assay. J Gen Virol 70: 3215–
    3223
Webster A, Hay RT, Kemp G (1993) The adenovirus protease is activated by a virus-
    coded disulphide-linked peptide. Cell 72:97–104
Yeh-Kai L, Akusjarvi G, Alestrom P, Petterson U, Tremblay M, Weber J (1983) Genetic
    identification of an endopeptidase encoded by the adenovirus genome. J Mol Biol
    167:217–222

Proteinases as Virulence Factors in Bacterial Diseases and as Potential Targets for Therapeutic Intervention with Proteinase Inhibitors

J. POTEMPA and J. TRAVIS

A. Introduction

In recent years a marked increase in the resistance of many bacterial pathogens to conventional antibiotics has been observed. The most dramatic example of this process has been the appearance of bacterial strains that were susceptible to only a single clinically available antibiotic, vancomycin. More importantly, examples of bacterial resistance to this antibiotic are being reported. During the same time, a sharp decline occurred in numbers of new or modified antibiotics which have recently become available for medical practice (BAX 1997), raising the gloomy prognosis of an end to the antibiotic era. Thus, although this declaration is certainly premature, we consider it prudent to suggest other mechanisms for the development of alternative antibacterial strategies, particularly because of the progress made in the sequencing of bacterial genomes. This has opened the unparalleled possibility for designing new anti-infective therapies by providing an opportunity to identify molecular targets indispensable for bacterial growth and/or survival and pathogenesis (KNOWLES 1997).

B. Common Themes in Bacterial Virulence

I. Host Defenses Against Bacterial Pathogens

The human body is in continuous contact with a myriad of microorganisms, many of which are potential pathogens. Nevertheless, thanks to formidable defense mechanisms, infective diseases are rare. Unless a disease-causing bacterium is introduced by an insect bite or through a wound, it first comes into contact with either skin or mucosal membranes. In most types of bacterial infections, colonization of these surfaces is the initial step in disease development, and several defensive mechanisms have evolved to protect these tissues against colonization (SALYERS and WHITT 1994). They include, among others, the resident microflora of the skin, lysozyme and bactericidal peptides, lactoferrin, and secretory immunoglobulin A (sIgA), the last playing an important function in preventing bacterial attachment to mucosal cells and trapping invading organisms in mucin (KILIAN et al. 1988).

Microorganisms that overcome surface defenses and reach underlying tissue or blood encounter an array of interior host defenses, which can be roughly divided into two categories. The first line of defense is constitutive and nonspecific, predominantly comprised of professional phagocytes, enforced iron limitation, and ready activation of the complement cascade. In tissues, the combination of complement activation and phagocyte attack on invading microbes produces inflammation, whose major function is to limit the spreading of infection, thus buying time for the host to develop an invader-specific, induced defense line composed of antibodies, activated macrophages and cytotoxic T cells (Salyers and Whitt 1994). Once this line of defense is breached, systemic infection will occur leading to bacteremia and/or colonization of internal organs and, ultimately, life threatening conditions.

II. Virulence Factors

Virulence, or pathogenicity, is generally delineated as the ability of a bacterium to cause infection, and virulence factors represent either bacterial products or strategies that contribute to virulence or pathogenicity. According to this definition, any bacterial trait, structure or molecule that helps a pathogen to accomplish colonization, evade host defense mechanisms, facilitate dissemination and cause host damage may be recognized as a virulence factor (Isenberg 1988; Mekalanos 1992). In many respects proteolytic enzymes produced by several pathogenic bacterial species fit into the category of virulence factors and, therefore, may be suitable targets for therapeutic intervention with specific inhibitors (Goguen et al. 1995; Travis et al. 1995; Maeda 1996; Lantz 1997).

C. Bacterial Proteinases as Potential Virulence Factors

Taking into account how precisely and tightly host proteinases are regulated and the fact that disturbance of the balance between endogenous proteinases and their natural inhibitors lies at the foundation of many diseases, one may argue that exogenous enzymes of invading microbes have a high potency to contribute to pathogenesis. This contention is gaining further support from the unnoticed but important fact that, in most cases, microbial enzymes are not controlled by endogenous inhibitors and, potentially, can reek havoc in host-regulated proteolytic systems. For this reason, this concept is reviewed in several of the following sections of this chapter.

I. Distribution of Proteinases among Pathogens

Only proteinases that reside on the bacterial surface or are released into the environment can act directly on host proteins. As presented in Table 1, these kinds of enzymes belong to three catalytic classes and are broadly distributed

Table 1. Bacterial pathogens known to produce endopeptidases with a potential to be considered as virulence factors

Pathogen	Endopeptidase[a]	Catalytic type and proteinase family affiliation[b]
Bacteroides fragilis	*B. fragilis* (entero)toxin, fragilysin	M10
Clostridium histolyticum	Clostridiopeptidase A, collagenase A, microbial collagenase, clostridial collagenase	M9
Clostridium perfringens<?2>	1) Clostridiopeptidase A, collagenase A, microbial collagenase, clostridial collagenase	M9
	2) Clostridiopeptidase B, clostripain	C11
	3) Lambda toxin[c]	M4
C. tetani	Tetanus neurotoxin, tetanus toxin, tentoxilysin	M27
C. botulinum	Botulinum neurotoxin, botulinum toxin, bontoxilysin	M27
Escherichia coli	OmpT, protein a, protein 3b, protease VII, omptin	S18
Haemophilus influenzae, *Neisseria gonorrhaea,* *N. meningitidis*	IgA protease, IgA1 protease, Igase, IgA-specific serine endopeptidase	S6
Legionella pneumophila	Major secretory protein, thermolysin homologue (Legionella), Legionella metalloendopeptidase	M4
Listeria monocytogenes	*L. monocytogenes* metalloproteinase, thermolysin homologue (Listeria), *L. monocytogenes* Mpl protease	M4
Porphyromonas gingivalis<?3>	1) Arg-gingipain, argingipain, gingipain-1, gingivain, prpR1, prtH gene product, gingipain R[c]	C25
	2) Lys-gingipain. porphypain, gingipain K	C25
	3) Periodontain	C10
	4) PrtT gene product, streptopain homologue	C10
Pseudomonas aeruginosa<?3>	1) Neutral metalloproteinase, LasB gene product, elastase, pseudolysin	M10
	2) Alkaline protease, aeruginolysin	M4
	3) LasA endopeptidase, staphylolytic endopeptidase, bacteriolytic enzyme, staphylolysin	M23
	4) Protease IV	S[d]
Salmonella typhimurium	Protein E, PrtA	S18
Serratia marcescens<?1>	1) 50kDa Metalloproteinase, *S. marcescens* extracellular proteinase, serralysin	M10
	2) Serratial serine protease, SSP, subtilisin extracellular homologue (Serratia)	S8

Table 1. *Continued*

Pathogen	Endopeptidase[a]	Catalytic type and proteinase family affiliation[b]
Staphylococcus aureus<?4>	1) Epidermolysin A, exfoliatin A, epidermolytic toxin A, exfoliative toxin A (ETA)	S2
	2) Epidermolysin B, exfoliatin B, epidermolytic toxin B, exfoliative toxin B (ETB)	S2
	3) V8 protease, endopeptidase Glu-C, Staphylococcal serine proteinase, glutamyl endopeptidase I	S2
	4) *S. aureus* metalloproteinase, protease II, aureolysin	M4
	5) *S. aureus* cysteine proteinase, protease III, staphylopain	C[d]
Staphylococcus epidermidis	sepP1; sepA gene products, elastase (staphylococcus)	M4
Streptococcus pyogenes<?1>	1) Streptococcal chemotactic factor inactivator, C5a peptidase, streptococcal C5a peptidase (SCP)	S8
	2) Streptococcal cysteine proteinase, streptococcal peptidase A, streptococcus pyrogenic exotoxin B (SpeB), interleukin-1β convertase, staphylopain	C10
Streptococcus mitis, S. oralis, S. sanguis, S. pneumoniae	IgA protease, immunoglobulin A1 proteinase, IgA1-specific metalloproteinase	M26
Treponema denticola	Chymotrypsin-like protease PtrP, dentilysin, trepolisin	S8
Vibrio cholerae	Hap gene product, haemagglutinin/protease, mucinase, cholera lectin, soluble hemagglutinin, HA/protease	M4
Vibrio vulnificus	Aeromonas proteolytica neutral proteinase, elastolytic protease, aeromonolysin, vibriolysin	M4
Yersinia pestis	Pla gene product, coagulase, Pla virulence factor	S18

[a] Names of proteinases most often used in the literature are listed with the name recommended by NC-IUMBM (Barrett et al. 1998) underlined.
[b] Indicates catalytic type of proteinase, while a number following refers to the family of evolutionary related enzymes to which a given bacterial proteinase belongs. For more information search the database. *M* metallo; *C* cysteine, *S* serine.
[c] Often mistaken as being microbial collagenase.
[d] Not assigned to a particular proteinase family.
[e] Gingipains R are product of two related genes, rgp1 and rgp2. These genes are almost identical in the part encoding the catalytic domain, but rgp2 gene is the missing nucleotide sequence encoding the C-terminal hemagglutinin/adhesin domain (Mikolajczyk-Pawlinska et al. 1998). In comparison with the rgp1 gene product, which is found predominantly as a complex of the catalytic and hemagglutinin domains referred to as 95-kDa gingipain R1, the second gene product, RGP-2, is obtained as a single-chain protein (POTEMPA et al. 1998).

among pathogenic bacteria. In same cases their primary importance is to aid in satisfying the nutritional needs of bacteria during infection, especially for microbes that are asaccharolytic. As a side effect, however, they can also cause direct tissue damage. However, more often such proteinases act to manipulate host antibacterial responses and other important proteinase-regulated host systems, in this way potentially contributing to the enhanced virulence of proteinase-producing pathogens.

II. Potential Targets for Bacterial Proteinases

1. Inactivation of Host Proteinase Inhibitors

Proteinase inhibitors constitute about 10% of the total protein content of human serum, and inhibitors present in tissues further supplement their levels. Except for α-2-macroglobulin, which inhibits all four catalytic classes of proteolytic enzymes, the other groups are specialized in inhibition of either cysteine-, metallo-, or serine proteinases. This last group of inhibitors is the largest and most diverse with Serpin (serine proteinase inhibitors) family members being key factors in the regulation of proteolytic cascades utilized for coagulation, fibrinolysis, complement activation and kinin generation (POTEMPA et al. 1994). Although it is commonly accepted that inactivation or degradation of host proteinase inhibitors by microbial proteinases may have significant pathological effects, data supporting this contention is sketchy due to a lack of systematic investigations. This is apparent from data recently reviewed by MAEDA (1996), where it was shown that only a small number of inhibitors have been studied for susceptibility to inactivation by a very limited number of proteinases from pathogenic bacteria. In addition, there is not a single case in which the kinetics of inhibitor inactivation was determined, making it difficult to evaluate whether such a reaction could take place in vivo.

The only human inhibitor with the ability to control the activity of bacterial proteinases is α-2-macroglobulin. Because of a change in $\alpha_2 M$ structure during inactivation of proteinases, the complex is rapidly cleared from the circulation. Many cells, including fibroblasts and macrophages, have $\alpha_2 M$ receptors that bind $\alpha_2 M$-proteinase complexes and internalize them. This allows neutralization and the rapid removal of various proteinases from the circulation or tissues. However, for *Serratia marcescens* metalloproteinase (serralysin), *Pseudomonas aeruginosa* alkaline proteinase (aeruginolysin), and a few other microbial proteolytic enzymes, this process is subverted to the advantage of the pathogen because the internalized proteinase is regenerated inside cells, escapes from the $\alpha_2 M$-complex into the cytoplasm and exerts a cytotoxic effect (MAEDA et al. 1987; MAEDA 1996).

2. Direct and Indirect Degradation of Connective Tissue

Several pathogenic bacteria release very high levels of proteinases with broad specificity and which are able to target many host tissue proteins, including those present in very high concentrations, such as the primary connective tissue proteins, collagen and elastin. The most prominent examples are enzymes from *Clostridium perfringens*, *C. histolyticum* and *P. aeruginosa* (Galloway 1991; Harrington 1996). *C. perfringens* is the most common pathogen of clostridial myonecrosis (gas gangrene) and is also implicated in other necrotizing diseases, including necrotizing enteropathy, gangrenous cholecystis, necrotizing pneumonia, and enteritis necroticans, as well as, crepitant cellulitis, a spreading infection associated with destruction of subcutaneous connective tissue within the skin (Lorber 1995). This bacterium produces several proteolytic enzymes, including a collagenase and kappa toxin, that can contribute to tissue damage. Similar proteinases are known to be produced by *C. histolyticum*, *P. aeruginosa*, *Vibrio vulnificus* (Smith and Merkel 1982) and some other human pathogens (Harrington 1996). In general, these broad-specificity bacterial proteinases are likely to serve one of two functions: the release of amino acids and small peptides necessary for bacterial growth, and/or the breakdown of natural tissue barriers in order to facilitate bacterial infiltration.

Tissue destruction and/or pathogen spread can also be achieved, indirectly, with a devious efficiency by many bacterial species that have not been endowed by nature with high levels of potent proteinases. First, inactivation of endogenous proteinase inhibitors can reduce tight control of the powerful neutrophil serine proteinases, elastase and cathepsin G, from the endogenous inhibitors α-1-proteinase inhibitor and α-1-antichymotrypsin, respectively. Second, it became apparent recently that the elastase of *P. aeruginosa* (pseudolysin), the *V. cholerae* hemagglutinin metalloproteinase and the cysteine proteinases of *Porphyromonas gingivalis* are able to activate host matrix metalloproteinases, including proMMP-1, proMMP-8 and proMMP-9 (Sorsa et al. 1992; Decarlo et al. 1997; Okamoto et al. 1997), thus enhancing local tissue destruction. Third, many pathogens have the ability to activate plasminogen and immobilize plasmin on the cell surface in a form refractory to inhibition by natural proteinase inhibitors but with a functionality to degrade both fibrin and fibrinogen (Lottenberg et al. 1994; Boyle and Lottenberg 1997). These acts of molecular piracy are examples of the sophisticated utilization of host proteinases for invasion and dissemination of the pathogen and are discussed in more details in Sect. C.II.3.c.

3. Dysregulation of Proteinase Cascades

a) Kallikrein–Kinin Cascade

From the pioneering work of Maeda's group, it has become apparent that bradykinin (or kinin) generation by pathogenic proteinases is a universal event occurring during most bacterial infections (Maeda et al. 1992). Normally, this pathway is controlled by limited proteolysis steps in which bradykinin is released from high-molecular-weight kininogen (HMWK) by plasma kallikrein which, in turn, is generated from prekallikrein by activated

Hageman factor (activated factor XII, XIIa). In another, probably less important pathway, kallidin (Lys-bradykinin) is generated from low-molecular-weight kininogen by tissue or glandular kallikreins. Significantly, endogenous proteinase inhibitors tightly regulate both pathways. In contrast, bacterial proteinases from pathogens such as *S. marcescens*, *Staphylococcus aureus*, *P. aeruginosa*, *Streptococcus pyogenes*, *V. vulnificus*, *V. cholerae* and *P. gingivalis* are impervious to regulation by host inhibitors and are able to generate bradykinin in human plasma either directly by degradation of HMWK or through activation of Hageman factor and/or plasma prekallikrein (MATSUMOTO et al. 1984; KAMATA et al. 1985; MOLLA et al. 1989; MARUO et al. 1993; IMAMURA et al. 1994a, 1995; HERWALD et al. 1996; MAEDA and YAMAMOTO 1996).

The released bradykinin exerts powerful biological activity, and at the site of infection/inflammation is responsible for pain and local extravasation leading to edema and, at a systemic level, the development of hypotension and shock. Recently, compelling evidence has been generated that triggering the kallikrein/kinin cascade by bacterial proteinases can greatly enhance pathogen dissemination from the local site of infection into the systemic circulation (MAEDA 1996; SAKATA et al. 1996). Taken together, these data indicate a potential strategy for the treatment of bacterial infections using appropriate inhibitors of either the bacterial proteinases, the kallikreins and/or kinin antagonists.

b) Blood Coagulation Cascade

Blood clotting is essentially a ubiquitous host response to bacterial invasion, and it plays an important function in the confinement of infection and enhancement of phagocytosis. In some cases, however, excessive clotting protects bacteria from phagocytosis by structurally obstructing the immigration of phagocytes. The best example is infective (bacterial) endocarditis caused by colonization of heart valves by bacteria, which then multiply inside the clot, referred to as vegetation. Several bacterial species have been implicated as etiologic factors of infective endocarditis, but streptococci and staphylococci account for 80–90% of the cases (SCHELD and SANDE 1995). In spite of the fact that it is firmly established that local activation of the extrinsic clotting pathway is a major stimulus for vegetation formation (DRAKE et al. 1984), nothing is known as to whether bacterial proteinases may participate in this process, although such enzymes are produced by pathogenic strains, including *Viridans streptococci* (STRAUS 1982; HARRINGTON and RUSSELL 1994; MAYO 1995), which, otherwise, lack significant virulence factors (SALYERS and WHITT 1994). However, it is clear that once a vegetative clot is formed, other mechanisms must also be invoked to reduce clot lysis, a normal host response to maintain proper blood flow. This may involve inhibitor production by some organism, which has initially released proteinases to stimulate clot formation.

Uncontrolled activation of the clotting cascade can result in disseminated intravascular coagulation, the most serious consequence of the infectious disease. In this context it is surprising how little is known about the potential of bacterial proteinases to trigger coagulation. Early reports had indicated that culture supernatants of *Bacteroides melaninogenicus* caused clotting activity (PULVERER et al. 1977), and it was latter shown that purified proteinases from *P. aeruginosa* and *S. aureus* were also able to activate prothrombin (PULVERER et al. 1980; WEGRZYNOWICZ et al. 1981).

Recently KAMINISHI et al. (1994) determined that proteinases from *P. aeruginosa* and *S. marcescens* can trigger the coagulation cascade at the level of factor XII and X, respectively, as well as by direct activation of prothrombin, but no kinetic parameters of these reactions were determined. However, more details are known for factor X activation by two gingipains R (arginine-specific gingipains) from *P. gingivalis*, an organism believed to be involved in the pathogenesis of periodontitis and, possibly, in the development of related cardiovascular disease. Although both gingipains can activate factor X, kinetic parameters of the reaction indicate that, in vivo at the site of infection, gingipain R1, the complex of the catalytic and hemagglutinin/adhesin domains, is a more important activator than the single chain, 50-kDa gingipain R2 (RGP-2). Indeed, the k_{cat}/K_m value $(4.1 \times 10^6\,M^{-1}s^{-1})$ is similar to the value of Russell's viper venom factor X coagulant protein $(9.6 \times 10^6\,M^{-1}s^{-1})$, the strongest exogenous factor X activator outside of the coagulation cascade itself. Activation of factor X by high molecular mass (H)RGP mimics the physiological process by its dependence not only on the presence of calcium ions but also by stimulation by phospholipids (IMAMURA at al. 1997).

c) Fibrinolysis Cascade

As discussed in the previous section, clot formation is important for the confinement of a pathogen to the site of infection. In this context, it is not surprising that invasive pathogens have developed sophisticated systems to dissolve fibrin. It is highly significant, however, that in this process pathogens utilize the host system rather than their own fibrinolytic proteinases. Many of the mechanisms to subvert entrapment in the clot and, subsequently, invade tissues, and disseminate are examples of an astonishing adaptation of the pathogen to thrive on the host.

Several human pathogens, including *Borrelia burgdorferi*, *Escherichia coli*, *Haemophilus influenzae*, *Heliobacter pylori*, *Neisseria gonorrhoeae*, *Neisseria menningitidis*, *S. aureus* and Streptococcus species, group A, C and G are capable of binding plasmin(ogen) (BOYLE and LOTTENBERG 1997). In all cases, binding occurs via kringle domains of plasmin(ogen) and, in several cases, cell-surface-expressed glyceraldehyde-3-phosphate dehydrogenase was identified as the receptor. Active, cell-surface bound plasmin generated either by the action of bacterial activators such as staphylokinase, streptokinase, or host plasminogen activators is refractory to inhibition by α-2-antiplasmin. This mechanism bestows the pathogen with a surface proteinase that not only can release it freely from the clot but also degrade most proteins within basement membranes and the extracellular matrix, thus facilitating both invasion and dissemination. Such a process parallels the physiological mechanisms utilized for ovulation, trophoblast implantation and embryogenesis (VASSALLI et al. 1991), as well as in the pathology of metastasis of tumor cells (POLLANEN et al. 1991).

Recently, this attractive model of bacterial virulence by plasmin(ogen) piracy was confirmed. In an elegantly conducted study COLEMAN et al. (1997) have provided compelling evidence that plasminogen activation contributes to the virulence of *B. burgdorferi*, the tick-borne spirochetal etiologic agent of Lyme disease. Using plasminogen (Pla) knockout mice, these investigators have shown that the blood meal of the tick provides plasminogen and plasminogen activator activity, which is necessary for dissemination of *B. burgdorferi* in the tick vector and is important in the establishment of bacteremia in mice. A similar mechanism may function in the pathogenicity of streptococci, but the lack of a suitable animal model makes this hypothesis difficult to verify experimentally (LOTTENBERG 1997).

Host plasminogen activation, but without immobilization of plasmin on bacterial surface, is a crucial event in the pathogenicity of *Yersinia pestis*, the etiologic agent of plague. *Y. pestis* usually enters the host by a flea bite. To disseminate from the site of inoculation and cause fulminating and fatal infections in multiple organ sites, the functional product of a gene (*pla*) encoding plasminogen activator (coagulase), and located on a 9.5-kb virulence plasmid, is necessary (SODEINDE et al. 1992). In addition to plasminogen activation, this bacterial proteinase can cleave C3, thus interfering with complement activation and subsequent phagocyte chemoattractant C5a formation. However, this reaction is unlikely to be important in plague pathology because there is no correlation of C5a production with susceptibility to *Y. pestis* in $C5a^+$ and $C5a^-$ congenic mice (WELKOS et al. 1997). In addition, the lack of the *pla* gene does not affect *Y. pestis* virulence if the bacterium is inoculated intravenously or intraperitoneally (SODEINDE et al. 1992). It should be noted that homologues of pla have been found in *E. coli* (ompT, 50% identity) and *Salmonella typhimurium* (prtA, 70% identity) (SODEINDE and GOGUEN 1989); however, their function as virulence factors is still obscure. Taken together, it is apparent that plasmin activation is the primary mechanism utilized for breaching the mechanical barriers for the spreading of infection within the host organism.

d) Complement Cascade

Complement is one of the potent defense systems which, when activated, produces high humoral bactericidal activity, opsonizing bacteria and recruiting phagocytic cells to the infection locus through the chemotactic component, anaphylatoxin C5a. Pathogens have developed several countermeasures to defend themselves from complement and some of their strategies depend on proteolytic activity (JOINER 1988). *S. pyogenes* produces a subtilisin-type serine proteinase (C5 peptidase) which specifically cleaves the human serum C5a near its carboxyl terminus, destroying its ability to serve as a chemoattractant (HILL et al. 1988; CHEN and CLEARY 1990). C5a activity is also eliminated by treatment with serralysin, a metalloproteinase from *S. marcescens* (ODA et al. 1990). Similarly, *P. aeruginosa* elastase (pseudolysin) cleaves a number of complement molecules, inactivating complement-derived chemotactic and phagocytic factors (SCHULTZ and MILLER 1974; HONG and GHEBREHWET 1992). In this context *P. gingivalis* employs a seemingly suicidal tactic because instead of eliminating the complement dependent chemotactic activity gingipains R are very efficient in terms of the generation of C5a through direct cleavage of C5 (see Sect.E.IV). At the same time, however, gingipain degrades C3 eliminating in this way C3-derived opsonins (WINGROVE et al. 1992), thus rendering *P. gingivalis* resistant to phagocytosis (CUTLER et al. 1993). In addition, degradation of other components of complement by this organism hampers formation of the bactericidal membrane attack complex (SCHENKEIN 1988).

4. Degradation of Immunoglobulin Function

The fact that many of the important mucosal bacterial pathogens, including *N. meningitides*, *N. gonorrhoeae*, *H. influenzae*, *Streptococcus pneumoniae*, and successful members of the human resident flora have developed proteinases exclusively specific for the cleavage at the hinge region of IgA1 has been well documented. These proteases belong to three different catalytic classes (serine-, cysteine- and metalloproteinases) and are a striking example of convergent evolution of putative bacterial virulence factors (KILIAN et al. 1996). All cleave peptide bonds at a P1 proline residue within the hinge region of IgA1, separating the antigen binding Fab fragments from the Fc fragment, the effective domain of the immunoglobulins for binding to phagocyte receptors. This mode of cleavage of IgA1 not only eliminates its protective ability, but also can camouflage bacteria with Fab fragments, which mask epitopes recognized by intact, functional antibodies. The characteristic cleavage fragments of sIgA1 or IgA1 can be detected in nasopharyngeal, oral, and intestinal secretions of subjects colonized or infected with IgA1 protease producing bacteria, as well as in the cerebrospinal fluid of patients with bacterial meningitis caused by either *N. meningitidis*, *H. influenzae*, or *S. pneumoniae* (METHA et al. 1973; KILIAN 1981; KILIAN et al. 1983; AHL and REINHOLDT 1991). Despite all of this incriminating evidence denoting IgA1 proteases as important virulence factors, the exact role of these enzymes in bacterial pathogenesis is still somewhat unclear due to the lack of an appropriate animal model and the unique specificity of these enzymes for only human, gorilla or chimpanzee IgA1 molecules (REINHOLDT and KILIAN 1991).

Several other bacterial proteinases may cleave human IgA, as well as other immunoglobulins, but none show the unique specificity limited exclusively to the hinge region of IgA1. Indeed, they usually digest other proteins. Nonetheless, antibody cleavage by bacterial proteinases may contribute to subversion of host defense mechanisms, especially if functional, intact antigen-binding Fab fragments are dissected from the Fc fragment (MOLLA et al. 1988).

5. Dysregulation of Cytokine Networking Systems

The interaction between bacteria and cytokine networks is an emerging theme of microbial pathogenicity (WILSON et al. 1998). Cytokines are indispensable for the maintenance of both innate and acquired immunity, and an understanding of how bacteria can affect cytokine function at the site of infection may explain the pathological outcome of some diseases. One mode of dysregulating cytokine networking can occur through proteolytic degradation of these molecules and, without surprise, cytokines have been shown to be susceptible to degradation by bacterial proteinases. *P. aeruginosa* elastase and alkaline protease, as well as a metalloproteinase from *Legionella pneumophila*, degraded interleukin (IL)-2 abrogating the biological activity of this cytokine (MINTZ et al. 1993; THEANDER et al. 1988). In contrast, however, IL-1α and IL-1β were resistant to degradation by both proteinases from the former bacterium (PARMELY et al. 1988), and only the alkaline proteinase was capable of cleaving and inactivating gamma interferon (IFN-γ) (HORVAT and PARMELY 1988).

Supernatants from cultures of *P. gingivalis*, as well as bacterial biofilms containing this organism, have been shown to degrade IL-1β, IL-6 and IL-1 receptor antagonist (IL-1ra), the activity most likely being contributed to by the presence of bacterial proteinases (FLETCHER et al. 1997, 1998). Purified gingipains, both lysine- and arginine-specific, have been shown to efficiently degrade tumor necrosis factor-α (TNF-α) abolishing its biological activity (CALKINS et al. 1998). TNF-α is also susceptible to degradation by proteinases from *P. aeruginosa* (PARMELY et al. 1990) but highly refractory to cleavage by host proteinases from neutrophils (CALKINS et al. 1998), indicating that some cytokines can work in the host-derived, highly proteolytic environment of an inflammatory locus unless bacterial proteinases are present. Thus, inactivation of cytokines by pathogenic proteinases is very likely to disrupt cell-to-cell communication and influence the course of both the inflammatory reaction and the elimination of infection.

The message carried by cytokines is transmitted inside the target cells by specific cell-surface receptors. Such receptors, including those binding to complement factors, are involved in the regulation of the cellular response to the presence of bacteria and, therefore, may constitute a perfect target for bacterial proteinases. Proteolytic inactivation of these receptors may virtually blind the cells and hinder their response to infection. Indeed, there is a growing body of evidence that some pathogens do utilize this strategy.

Metalloproteinases from *S. marcescens*, *S. aureus*, *P. aeruginosa* and *Listeria monocytogenes* were shown to liberate soluble IL-6 receptor (sIL-6R) from human monocytes which can then act as an agonist by binding to bystander cells (fibroblasts or epithelial cells) and rendering them sensitive to the action of IL-6 (Vollmer et al. 1996). In this way, receptor shedding can provoke long-range biological effects in the host organism. In a similar manner, but acting as antagonists, proteolytic liberation of urokinase plasminogen activator (u-PA) receptor (u-PAR) from monocytes by streptococcal pyrogenic exotoxin B (streptopain) may also have a longer-range effect than just decreasing the level of functional cell surface u-PAR since a soluble form of u-PAR can compete at a distance with the cellular receptor for ligand (Wolf et al. 1994).

In comparison with the shedding of functional soluble receptors, proteolytic inactivation of neutrophil receptors involved in chemotaxis and bacteria phagocytosis is limited in range but of great potential advantage for the invading pathogen. This strategy is employed by *P. aeruginosa* elastase, which cleaves the formyl-Met-Leu-Phe receptor on neutrophils (Ijiri et al. 1994). A similar effect seems to be exerted by cysteine proteinases from *P. gingivalis* (Lala et al. 1994). In addition, it was well documented that gingipain K and another yet uncharacterized serine proteinase from this bacterium were able to inactivate the C5a receptor on neutrophils (Jagels et al. 1996).

The discovery of a family of proteinase activated receptors (PARs) on mammalian cells (Vu et al. 1991) has also opened a new dimension for possible targets for bacterial proteinases. Apparently, this pathway is utilized by gingipains of *P. gingivalis* which activate platelets in a proteinase-dependent manner because enzymes with blocked active sites do not affect platelets (Curtis et al. 1993). All of these data again support the importance of developing inhibitors against proteinases, which are able to dysregulate important biochemical pathways in an uncontrolled manner.

6. Virus Activation

For a long time it has been known that influenza virus-bacterial co-infections result in serious clinical consequences, such as severe life threatening pneumonia (Schwarzmann et al. 1971). It was generally assumed that viral infection in a given tissue favors growth conditions for bacteria, resulting in a bacterial secondary infection (Babiuk et al. 1988). From more recent investigations, however, it is apparent that parallel active interplay between viruses and bacteria are responsible for the severity of viral infections. The proteolytic cleavage of the hemagglutinin (HA) of influenza A virus is a prerequisite for the formation of an infectious virus and, thereby, for spreading of the virus in the organism (Klenk and Rott 1987). There is compelling evidence that bacteria can take part in virus activation by this mechanism either directly or indirectly. It was shown that *Aerococcus viridans* and some strains of *S. aureus* secrete proteases that were capable of virus activation by HA proteolytic cleavage which, in the case of co-infection, led to replication of the influenza virus in the respiratory track and the development of fetal pneumonia (TASHIRO et al. 1987a; SCHEIBLAUER et al. 1992). Proteinases of *S. marcescens*, *P. aeruginosa* and *S. pyogenes*, although unable to cleave HA, still considerably enhance viral pathogenicity, apparently due to activation of host proteolytic cascades (AKAIKE et al. 1989; SCHEIBLAUER et al. 1992) and release of human proteinases, including plasmin and kallikrein, which proteolytically activate the virus (LAZAROWITZ et al. 1973). In this respect, it is not surprising that proteinase inhibitors, including aprotinin and leupeptin, have a protective effect on the development of viral pneumonia mediated by concomitant bacterial infection (ZHIRNOV et al. 1984; TASHIRO et al. 1987b; HAYASHI et al. 1991).

7. Proteolytic Activity of Bacterial Toxins

a) Clostridium Neurotoxins

Neurotoxins produced by *C. botulinum* (BoNT types A to G) and *C. tetanum* (TeNT) are the most extremely potent natural toxins known to date, exerting their effects at the subfemtomolar level. All have a similar structural organization and are synthesized as single polypeptide chains of 150 kDa. They are released into the environment upon bacterial cell lysis and must undergo proteolytic cleavage for their activity to be manifested (GORDON and LEPPLA 1994). This generates two chain neurotoxins composed of a heavy chain (100 kDa) and a light chain (50 kDa) and held together by a single disulfide bridge. The heavy chain is responsible for the specific binding of the toxin to presynaptic membranes and translocation of the light chain into the neuron. The light chains of toxins represent a new group of zinc-dependent endopeptidases with a unique specificity limited to a very small subset of proteins which play a central role in synaptic signal transduction (MONTECUCCO and SCHIAVO 1993). Once in the neuronal cytoplasm, the light chain of TeNT and BoNT types B, D, F and G cleave synaptobrevin while SNAP-25 is a target for BoTN types A and E. Cleavage of these proteins leads directly to the neuroparalysis associated with tetanus and botulism.

b) Anthrax Lethal Factor

Anthrax toxin, the most deadly natural toxin, is produced by bacterium *Bacillus anthracis*. It is composed of three proteins: protective antigen (PA), edema factor (EF) and lethal factor (LF). PA binds to specific cell-surface receptors and, upon proteolytic activation, forms a membrane channel that mediates entry of EF and LF into the cell. Based on the amino acid sequence, the presence of a consensus zinc-binding site (-His-Glu-Phe-Gly-Phe-) and mutagenesis experiments, LF was classified as a homologue of zinc-metalloproteinases (KIMPEL et al. 1994). Recently, it was shown that, indeed, LF is a very specific proteinase that cleaves the amino terminus of mitogen activated protein kinase kinases 1 and 2 (MAPKK1 and MAPKK2). This cleavage inactivates MAPKK1 and inhibits the MAPK signal transduction pathway, which plays a fundamental role in the intracellular signaling network contributing in this way to the pathogenesis of anthrax (Duesbery et al. 1998).

c) Epidermolytic (Exfoliative) Toxins of S. Aureus

Epidermolytic toxins (ETs; synonyms exfoliative toxin, epidermolysin, exfoliatin) are protein toxins secreted by S. aureus. Two serotypes (ETA and ETB) of ET have been identified, purified and implicated as etiologic factors of the staphylococcal scaled skin syndrome (SSSS) (Bailey et al. 1995). ETs are homologues of serine proteinases. Substitution of the catalytic serine residue (Ser-195) abolishes activity against the synthetic substrate Boc-L-Glu-OPhenyl as well as toxin biological activity when tested in the newborn mouse-skin model. Both activities are also inhibited by diisopropylfluorophosphate (DFP), indicating that proteolytic activity of ETs is responsible for the pathological features of SSSS. Nevertheless, ETs lack proteolytic activity against several tested proteins and peptides, and the natural substrate of these enzymes is still unknown (Bailey et al. 1995) indicating an extraordinary specificity of these unusual proteinases. Recently, the tertiary structure of ETA was solved, confirming the fact that ETs are certainly serine proteinases which cleave after acidic residues and with catalytic activity regulated through a specific flip of a main-chain, peptide bond. The structure also explains the narrow specificity of the enzyme, which is probably modulated through a specific interaction of the unique N-terminal domain and one of the surface loops with another molecule (Cavarelli et al. 1997; Vath et al. 1997).

D. Dilemmas in Considering Bacterial Proteinases as Target for Antibacterial Chemotherapy

The astonishing success with the designing of anti-HIV therapy based on proteinase inhibitors encourages the suggestion of a similar approach for the treatment of bacterial diseases. By definition, such therapy should be aimed at the essential virulence factors. However, despite all of the destructive potential of bacterial proteinases, as presented in previous sections, there is an inherited problem with defining these enzymes as virulence factors (Lantz 1997). Goguen and his colleagues (1995) have outlined the ideal scheme for pinpointing the role of proteinases in bacterial pathogenicity. First, pathogen mutants with scrupulously defined knockout lesions of the proteinase gene should be tested using the best available model of experimental infection. Second, the proteinase in question should be precisely characterized with regard to biochemical properties, particularly specificity, and the main enzyme target(s) in the infected host should be defined. Third, pathogenicity of the bacterium should be tested in the host strain, constructed in such way that the main target molecule(s) is modified so that its retains physiological activity but is resistant to the pathogen proteinase. To accomplish all these goals is not easy because very often it is difficult to mimic natural infection in an appropriate animal model. The most perplexing example is the potential pathogenic importance of IgA1 proteinases, which cannot be verified due to the lack of a convenient animal model. Fortunately, the progress in transgenic-animal technology should make it eventually possible to apply the rigorous guidelines outlined above to determine whether a given proteinase is a virulence factor, as was already demonstrated for *B. burgdorferi* (see Sect. C.II.3.c).

Despite this lack of certainty about proteinases as being the *sine qua non* of bacterial virulence, they still can be considered as an attractive target for therapeutic intervention with inhibitors, and some examples are discussed below.

E. Paradigms for Testing Proteinase Inhibitors as Therapeutic Agents

I. *P. Aeruginosa* Infections

P. aeruginosa is a gram-negative opportunistic pathogen of compromised hosts, including patients with burns, cancer, trauma, and cystic fibrosis. It is also one of the leading causes of corneal infection prevalent among individuals wearing contact lenses and the most frequent gram-negative bacterium implicated in nosocomial pneumonia. In cancer, burn, and pneumonia patients, *Pseudomonas* infections are often invasive and, despite advances in burn therapy, the mortality associated with *P. aeruginosa* wound sepsis is 78%, while ventilator-associated pneumonia caused by this bacterium has a mortality rate of 40–68% (POLLACK 1995). In all of these pathologies, proteolytic enzymes secreted by *Pseudomonas* have been implicated as the most important virulence factors, and the beneficial effect of inhibitor treatment has been documented in several animal models. However, the results presented were sometimes contradictory, apparently due to the fact that the exact number of proteinases produced by *P. aeruginosa* was unknown. It is now apparent that this organism secretes at least five different proteinases: elastase (pseudolysin), alkaline proteinase (aeruginolysin), LasA (staphylolytic endopeptidase), protease IV, and a lysine-specific endopeptidase. Although recent evidence suggests that LasA and protease IV also contribute to the pathology of pseudomonal infections (GALLOWAY 1991; ENGEL et al. 1997, 1998), the bulk of the information has been focussed on elastase and the alkaline protease.

Non-proteolytic-enzyme-producing-strains of *P. aeruginosa* exhibit low pathogenicity in burned mice infection, but the mortality increased significantly when proteolytic enzymes were applied to the infected wound together with the bacterium; this effect was reversed by treatment with $\alpha_2 M$

(HOLDER and HAIDARIS 1979). Also, the treatment of burned mice infected with a proteinase-producing strain of *P. aeruginosa* together with α_2M substantially enhanced animal survival (HOLDER 1983). Ovomacroglobulin, a bird equivalent of α_2M, has shown high therapeutic value in abrogating cornea destruction in experimental keratitis (MIYAGAWA et al. 1991a, 1994). In addition, a synthetic inhibitor of metalloproteinases was shown to be effective in treatment of chronic suppurative otitis media caused by *P. aeruginosa* in chinchillas (COTTER et al. 1996). Apparently, local application of broad-spectrum inhibitors not only protects tissues against the deleterious, direct and/or indirect action of bacterial proteinases (MOLLA et al. 1987), but also may suppress growth of pathogens at the infection site (MIYAGAWA et al. 1991b) and prevent bacterial dissemination, as was shown in pseudomonal septic shock in the guinea-pig model (KHAN et al. 1994, 1995).

Summarizing, pseudomonal infections are an attractive target for antibacterial therapy with proteinase inhibitors. First, involvement of proteinases in pathogenicity of this opportunistic pathogen is well established. Second, appropriate models of infection in compromised animals are available. Third, the protective effect of proteinase inhibitors has already been well documented.

II. *S. Pyogenes* Infections

S. pyogenes (group-A streptococcus) is one of the most important bacterial pathogens of humans. This ubiquitous organism is the most frequent bacterial cause of acute pharyngitis (streptococcal sore throat and scarlet fever), the etiologic factor of localized purulent infections of the skin (pyoderma), and the cause of a variety of invasive infections of skin and soft tissues, including erysipelas, cellulitis, necrotizing fasciitis, myositis, and toxic-shock syndrome. Complications of the upper respiratory-tract infections with group-A streptococci include acute rheumatic fever, a disease characterized by non-suppurative inflammatory lesions involving primarily the heart, joints, subcutaneous tissues, and the central nervous system. In addition, pharyngeal or cutaneous infections with certain nephritogenic group-A streptococcal strains may lead to post-streptococcal acute glomerulonephritis (BISNO 1995).

The list of proven group-A Streptococcus virulence factors is relatively short and includes M protein, M-related protein, hyaluronic acid capsule, a fibrinogen-binding protein, streptococcal pyrogenic exotoxin A (SpeA), C5a peptidase, and an extracellular cysteine proteinase, streptopain, known as streptococcal pyrogenic exotoxin B (SpeB) (MUSSER et al. 1991; WESSELS et al. 1991; MUSSER and KRAUSE, 1997). While C5a peptidase activity is limited to leukocyte chemotactic complement fragment C5a inactivation, streptopain possesses a broader spectrum of specificities relevant for bacterium pathogenicity. In addition to the ability to degrade extracellular matrix proteins, such as fibronectin and vitronectin (KAPUR et al. 1993a), and activate matrix metalloproteinase 2 (BURNS et al. 1996), the enzyme is involved in proteolytic shedding of the urokinase plasminogen activator receptor from monocytes (WOLF et al. 1994) and the release of biologically active fragments of streptococcal surface proteins including fibrinogen-binding fragment of M1 protein, IgG-binding fragment of protein H and C5a peptidase (BERGE and BJORCK 1995). Furthermore, the enzyme can profoundly accelerate an inflammatory reaction by both the direct liberation of bradykinin from high-molecular-weight kininogen (HERWALD et al. 1996) and activation of interleukin-1β precursor (KAPUR et al. 1993b).

Unlike many other bacterial proteolytic enzymes, two streptococcal proteinases have been firmly confirmed to participate in virulence. First, the isogenic, C5a peptidase-deficient mutant constructed by insertional mutagenesis was efficiently cleared by neutrophils while the wild-type parental strain avoided phagocytosis by retarding the inflow of inflammatory cells to the focus of infection (JI et al. 1996). Second, the importance of streptopain as a major virulence factor was proven by comparison of the pathogenicity of two pairs of *S. pyogenes* isogenic strains of M3 and M49 serotypes in an interperitoneal infection mouse model, in which it was shown that proteinase-deficient isogenic mutants had a significantly lower ability to kill the host (LUKOMSKI et al. 1997). Apparently, in both cases, decreased virulence of mutant strains was due to the fact that, in comparison with the wild-type parental strain, the mutants were phagocytosed and cleared more efficiently by neutrophils. This led to lower level of bacteremia, dissemination to organs, and subsequent host death (LUKOMSKI et al. 1998). These results are in line with the observation that immunization with streptopain protects mice against challenge with heterologous group-A streptococci (KAPUR et al. 1994) and suggests a similar opportunity for the use of C5a peptidase as an immunogen.

Laboratory investigations on the importance of streptopain in streptococcal infections correlate very well with clinical observations. First, streptopain is expressed in vivo during the course of a diverse range of streptococcal invasive disease episodes because patients infected with a variety of distinct M serotypes developed antibodies against this proteinase (GUBBA et al. 1998). Second, individuals suffering from invasive episodes with low acute-phase levels of serum antibody to streptopain are more likely to die or have serious debilitating, clinical outcomes than patients with high antibody titer (HOLM et al. 1992). Third, group-A streptococci strains recovered from patients with severe soft-tissue infections produced higher levels of protease than did organisms isolated from less severe infections.

All together, results of the laboratory and clinical research on streptopain clearly suggest that the growth and virulence of group A streptococci can be arrested by specific proteinase inhibitors. In grossly undervalued investigations by BJORCK et al. (1989), it was shown that, in vitro, the streptopain inhibitor, N-benzoxycarbonyl-leucyl-valyl-glycine diazomethane (Z-LVG-CHN$_2$), specifically blocked growth of several strains of group-A streptococci with an efficiency comparable to that of well-established, anti-streptococcal antibiotics such as tetracycline. Furthermore, in vivo, a single injection of the inhibitor cured mice inoculated with lethal doses of bacteria. Obviously, these results, together with a fuller understanding of the pathological consequences of streptococcal proteinase activity in the infected host, should be a foundation for the development of specific inhibitors against streptopain and C5a peptidase which, one day, may replace antibiotics in the treatment of diseases caused by group-A streptococci.

III. Diseases Caused by Proteolytic Toxins

The recent advances in understanding the molecular structure and the proteolytic mode of action of clostridial neurotoxins, anthrax LF and staphylococcal ET have opened a new avenue for a potential therapeutic approach based on proteinase inhibitors. Utilizing the unique specificity of bacterial proteolytic toxins, which should make it possible to design inhibitory compounds exclusively specific for target enzymes yet inert against host proteinases, can tremendously facilitate this approach. Despite the fact that antibiotics can easily kill *B. anthracis* and *C. botulinum*, such antibiotic treatment is often unsuccessful because by the time of administration there is enough toxin in the host system to cause death or other severe deleterious effects. This, however, could be prevented by treatment with specific proteinase inhibitors blocking the enzymatic activity of botulinum neurotoxin or anthrax LF.

In developed countries, *B. anthracis* infections are very rare, but there are continuous fears that terrorists will attempt to wage germ warfare with *anthrax bacillus*. This serious threat could be defused if the proteolytic activity of the lethal factor of anthrax toxin could be neutralized with specific inhibitors. Such a fact justifies intense research and development of such compounds.

In contrast to anthrax, botulism, a disease caused by ingestion of *C. botulinum* toxin, is fairly frequent, and 140 cases were reported in the USA from 1983 to 1987. In many cases, the treatment failed because it came too late after the toxin had entered neurons and begun to exert its deleterious effect. At this stage toxin-neutralizing antibodies were ineffective because they were unable to enter neurons, but it can be envisioned that the administration of specific inhibitors which could follow toxins into neurons might have a beneficial effect in botulism treatment.

Proteinase inhibitors may also be extremely useful in treatment of the SSSS. The development of such compound is justified by the fact that exfoliate toxin is a unique serine proteinase clearly implicated in disease pathogenicity. Clearly, utilization of such compounds in an anti-staphylococcal armament may be very important in the near future, since rampant antibiotic resistance among *S. aureus* strains is becoming more common.

IV. Periodontal Disease

Periodontitis is the most common infectious disease inflicting human beings. The clinical hallmarks of the disease, including massive accumulation of neutrophils, bleeding on probing, increased flow of gingival crevicular fluid (edema), bone resorption, loss of tooth attachment and formation of periodontal pockets are consequences of a persistent inflammatory reaction triggered by specific bacterial species in the subgingivial plaque. Several distinct clinical categories of periodontitis can be distinguished, but the most common form is adult periodontitis with *Porphyromonas gingivalis*, *Treponema denticola* and *Bacteroides forsythus* being generally recognized as the major pathogens involved in the induction and/or progression of this disease.

The common feature of these three established periodontopathogens is production of a significant amount of proteolytic enzymes, among which proteinases cleaving Arg-Xaa and/or Lys-Xaa peptide bounds are predominant. Because activation of several host proteolytic cascade systems, including coagulation, fibrinolysis, complement activation and kinin release is based on limited proteolysis after specific arginine residues, bacterial proteinases mimicking this activity in an unregulated manner can have a devastating effect on both host tissue destruction and host defense systems (see Sect. C.II.3). Indeed, a wealth of data supports the contention that, at the very least, *P. gingivalis* arginine-specific gingipains R work in this way and may contribute significantly to the pathogenesis of periodontitis.

First, gingipains R are potent vascular permeability enhancement factors, which readily activate plasma kallikrein and cause the subsequent release of bradykinin (IMAMURA et al 1994). In addition, gingipains R, in concert with gingipain K, can induce vascular permeability by cleaving bradykinin directly from high-molecular-weight kininigen (IMAMURA et al. 1995a). In this way gingipains may contribute to generation of gingival crevicular fluid at periodontitis sites infected with *P. gingivalis*.

Second, gingipains R cleave complement factor C5, liberating a potent chemotactic factor C5a, while at the same time destroying C3, and eliminating formation of C3-derived opsonins (WINGROVE et al. 1992). The release of the proinflammatory C5a-like molecule is further augmented if C5 is oxidized before proteinase digestion. In these conditions, which may likely occur at inflammatory sites, not only is gingipain R more potent, but also gingipain K becomes efficient in releasing the active fragment from oxidized C5 (DISCIPIO et al. 1996). These data implicate a mechanism for recruiting neutrophils to the periodontitis site, which may contribute to the massive accumulation of phagocytes in inflamed periodontal tissue. However, they also abrogate neutrophil function, apparently through cleavage of surface receptors (LALA et al. 1994; JAGELS et al. 1996).

Third, gingipains R can efficiently activate the coagulation cascade. In addition to the phospholipids and Ca^{++} dependent activation of factor X by 95-kDa gingipain R1 (IMAMURA et al. 1997), clotting activity can be generated through direct activation of prothrombin. At the same time gingipains, especially gingipain K, abrogate the clotting potential of fibrinogen (IMAMURA 1995b). These pro- and anti-coagulant activities of gingipains may not only contribute to the bleeding tendency but also to persistent inflammation in peri-

odontitis sites because the thrombin released has powerful proinflammatory activities including stimulation of the local production of IL-1 and
prostaglandins, the major mediators of bone resorption.

Fourth, gingipains may contribute to the pathology of periodontitis
through their ability to activate matrix metalloproteinases (DECARLO et al.
1997). Finally, gingipains R seems to be very important factors in the processing of bacterial proproteins such as the 75-kDa major outer membrane
protein, profimbrilin and progingipain K (OGAWA et al. 1994; NAKAYAMA et al.
1996).

As listed above, all of these apparent key roles of gingipains R in both *P.
gingivalis* house-keeping functions and in pathological events associated with
infection qualifies this enzyme(s) as a suitable target for the development of
inhibitors which may be used in the treatment of periodontitis (POTEMPA et al.
1995; POTEMPA and TRAVIS 1996). The credibility of this approach is further
heightened by the fact that such enzymes are expressed in vitro because their
activity can be detected in gingival crevicular fluid collected from periodontitis sites infected with *P. gingivalis* (WIKSTROM et al. 1994), and anti-gingipain
antibodies are present in serum from infected patients. Moreover, it has
already been shown in a murine model that *P. gingivalis* virulence was compromised if bacterial cells were treated with proteinase inhibitors before inoculation (FEUILLE et al. 1996; KESAVALU et al. 1996; GENCO et al. 1998). Also,
immunization with gingipains or gingipain-derived peptide fragments yields
full protection (GENCO et al. 1999b). These data, as well as results from experiments which showed that isogenic strains deficient in one or both gingipains
R (FLETCHER et al. 1995; NAKAYAMA et al. 1995; TOKUDA et al. 1998) have compromised virulence, indicate that such enzymes are the obvious major pathogenic factor of *P. gingivalis*.

In summary, periodontitis may represent a useful model for testing the
therapeutic potential of bacterial proteinase inhibitors because (1) gingipains
R are well-characterized proteinases playing a central role in the virulence of
P. gingivalis, a major periodontopathogen; (2) animal models of periodontitis
are available; (3) a drug can be applied locally at periodontitis sites using time-
release devices (KORNMAN 1993); (4) application of tetracycline-related compounds inhibiting matrix metalloproteinases have been shown to reduce
periodontal tissue destruction (RIFKIN et al. 1993). The last observation
confirms that host proteinases are directly responsible for much of the periodontal tissue damage and implies that future local treatment of periodontitis with proteinase inhibitors must be aimed at both bacterial and host
proteinases.

V. Plague

Infections of experimental animals with *Y. pestis* may also constitute an interesting model to study the therapeutic use of proteinase inhibitors. As was mentioned before (Sect. C.II.3.c), plasminogen activator, an unique serine

proteinase related to omptin, the product of the ompT gene of *E. coli*, seems to be a single factor involved in dissemination of some virulent strains of *Y. pestis* from an initial subcutaneous infection site (SODEINDE et al. 1992; WELKOS et al. 1997). Due to the uniqueness of the *Y. pestis* plasminogen activator, it is likely that very specific inhibitors of this proteinase can be designed and tested using existing relevant animal models of the disease. Such experiments may not only provide further understanding of the pathogenesis of plague but also furnish an alternative treatment for this terrible disease which still looms in many parts of the world, including the USA (BUTLER 1995).

F. Bacterial Proteinases to the Rescue

Microorganisms, including resident bacterial species and pathogens, compete among themselves for ecological niches, and in this warfare they use a multitude of compounds with adversary or even lethal effects on other microbes. Although the best examples are antibiotics produced mainly be *Streptomyces* spp. and fungi, the content of this armory is much more diverse and includes toxins, hemolysins, bacteriophages, byproducts of primary metabolic pathways, peptide antibiotics, bacteriocins and bacteriolytic enzymes (JACK et al. 1995). Bacteriolytic enzymes are primarily zinc metalloproteinases specific for cleavage after Gly, especially in -Gly-Gly ↓ Xaa sequences. They lyse the cell walls of gram-positive bacteria in which the peptidoglycan cross-links contain Gly or multiple Gly residues. There are two families of bacteriolytic metalloproteinases. One family of homologous extracellular metalloproteinases consists of *Lysobacter enzymogenes* β-lytic endopeptidases, *Achromobacter lyticus* β-lytic endopeptidases, *Areomonas hydrophilia* proteinase (AhP) and *P. aeruginosa* staphylolytic endopeptidase referred to as LasA (KESSLER 1995). The second family contains a staphylolytic proteinase (lysostaphin) produced by certain Staphylococci strains (SCHINDLER and SCHUHARDT 1965; RAMADURAI and JAYASWAL 1997; SUGAI et al. 1997; THUMM and GOTZ 1997). Recently, enzyme homologues to lysostaphin produced by *Streptococcus zooepidemicus* have been described and shown to be lytic against *S. pyogenes* strains (SIMMONDS et al. 1996, 1997).

Because of the anti-staphylococcal activity of LasA and lysostaphin, both enzymes have a potential to be used as a treatment against pathogenic *S. aureus* strains. Indeed, in several in vitro and in vivo studies, lysostaphin alone or in combination with antibiotics was shown to be an effective anti-staphylococcal agent (GUNN and HENGESH 1969; BRAMLEY and FOSTER 1990; POLAK et al. 1993) being used with success in therapy for bovine mastitis (OLDHAM and DALEY 1991) and in the topical treatment of persistent nasal carriage of *S. aureus* in humans (QUICKEL et al. 1971). With the advent of molecular biology techniques, it is logical to believe that, in the future, we will use recombinant strains of the resident bacterial flora expressing bacteriolytic proteinases to protect the skin and mucosal surfaces from colonization by pathogenic microorganisms. The legacy of such therapy already exists in the successful application of the relatively avirulent *S. aureus* 502A in the prevention of serious staphylococcal diseases in neonates and in the treatment of furunculosis (ALY et al. 1982).

G. Conclusions

The data reviewed above indicate that bacterial proteinases represent an attractive target for development of inhibitors which may be used for therapeutic intervention in some contiguous bacterial diseases. This approach has not yet been explored at the clinical scale. However, because the pharmaceutical industry today has the unprecedented ability to produce highly specific proteinase inhibitors with superior pharmacological properties, it is certain that in the near future inhibitors of bacterial proteinases will become common antibacterial drugs. In comparison with antibiotics, such drugs will have their disadvantages and advantages. First, their spectrum of activity will be narrow and, therefore, any application will require very precise identification of the pathogen. However, this feature will eliminate the drawback of broad-spectrum antibiotics, which not only attack the pathogen but also reduce the number of resident, protective microflora that sometimes can cause secondary serious infections. In addition, in contrast to antibiotics, proteinase inhibitors are quite unlikely to show bactericidal or bacteriostatic effects but, instead, would slow the progress of infection and require help from additional drugs. However, it is possible that, as in cases of human immunodeficiency virus (HIV) infection, inhibitors may induce development of resistance, although acquired immunodeficiency syndrome (AIDS) is a chronic condition while most bacterial infections are acute in nature and can be rapidly eliminated.

In summary, it is apparent that inhibitors of bacterial proteinases are an obvious class of compounds which may prove useful in the future treatment of such diseases as *P. aeruginosa* burn and corneal infections, periodontitis, group-A streptococcal invasive episodes, botulism and other maladies caused by bacterial pathogens.

References

Ahl T, Reinholdt J (1991) Detection of immunoglobulin A1 protease induced Fab fragments on dental plaque bacteria. Infect Immun 59:563–569

Akaike T, Molla A, Ando M, Araki S, Maeda H (1989) Molecular mechanism of complex infection of bacteria and virus analyzed by a model using serratial protease and influenza virus in mice. J Virol 63:2252–2259

Aly R, Shinefield H, Maibach H (1982) Bacterial interference among Staphylococcus aureus strains. In: Aly R, Shinefield H (eds) Bacterial interference. CRC Press, Boca Raton, pp 13–23

Babiuk LA, Lawmen MJP, Bielefeldt OH (1988) Viral-bacterial synergistic interaction in respiratory disease. Adv Virus Res 35:219–249

Bailey CJ, Lockhart BP, Redpath MB, Smith TP (1995) The epidermolytic (exfoliative) toxins of Staphylococcus aureus. Med Microbiol Immunol 184:53–61

Barrett AJ, Rawlings ND, Woessner JF (eds) (1998) Handbook of proteolytic enzymes. Academic Press, San Diego

Bax R (1997) Antibiotic resistance: a view from the pharmaceutical industry. Clin Infect Dis 24[Suppl 1]:S151–S153

Berge A, Bjorck L (1995) Streptococcal cysteine proteinase releases biologically active fragments of streptococcal surface proteins. J Biol Chem 270:9862–9867

Bisno AL (1995) Streptococcus pyogenes. In: Mandel GL, Bennett JE, Dolin R (eds) Principles and practice of infectious diseases, 4th edn. Churchill Livingstone, Edinburgh, pp 1786–1799

Bjorck L, Akesson P, Bohus M, Trojaner J, Abrahamson M, Olafsson I, Grubb A (1989) Bacterial growth blocked by a synthetic peptide based on the structure of a human proteinase inhibitor. Nature 337:385–386

Boyle MDP, Lottenberg R (1997) Plasminogen activation by invasive human pathogens. Thromb Haemost 77:1–10

Bramley AJ, Foster R (1990) Effects of lysostaphin on Staphylococcus aureus infections of the mouse mammary gland. Res Vet Sci 49:120–121

Burns EH Jr, Marciel AM, Musser JM (1996) Activation of a 66-kilodalton human endothelial cell matrix metalloproteinase by Streptococcus pyogenes extracellular cysteine protease. Infect Immun 64:4744–4750

Butler T (1995) Yersinia species (including plague). In: Mandel GL, Bennett JE, Dolin R (eds) Principles and practice of infectious diseases, 4th edn. Churchill Livingstone, Edinburgh, pp 2070–2078

Calkins CC, Platt K, Potempa J, Travis J (1998) Inactivation of tumor necrosis factor-α by proteinases from periodontal pathogen, Porphyromonas gingivalis. J Biol Chem 273:6611–6614

Cavarelli J, Prevost G, Bourguet W, Moulinier L, Chevrier B, Delagoutte B, Bilwes A, Mourey L, Rifai S, Piemont Y, Moras D (1997) The structure of Staphylococcus aureus epidermolytic toxin A, an atypical serine protease, at 1.7A resolution. Structure 5:813–824

Chen CC, Cleary PP (1990) Complete nucleotide sequence of the streptococcal C5a peptidase gene of Streptococcus pyogenes. J Biol Chem 265:3161–3167

Coleman JL, Gebbia JA, Piesman J, Degen JL, Bugge TH, Benach JL (1997) Plasminogen is required for efficient dissemination of B. burgdorferi in ticks and for enhancement of spirochetemia in mice. Cell 89:1111–1119

Cotter CS, Avidano MA, Stringer SP, Achultz GS (1996) Inhibition of proteases in Pseudomonas otitis media in chinchillas. Otolaryngol Head Neck Surg 115: 342–351

Curtis MA, Macey M, Slaney JM, Howells GL (1993) Platelet activation by protease I of Porphyromonas gingivalis W83. FEMS Microbiol Lett 110:167–174

Cutler CW, Arnold RR, Scheinkein HA (1993) Inhibition of C3 and IgG proteolysis enhances phagocytosis of Porphyromonas gingivalis. J Immunol 151:7016–7029

DeCarlo AA Jr, Windsor LJ, Bodden MK, Harber GJ, Birkedal-Hansen B, Birkedal-Hansen H (1997) Activation and novel processing of matrix metalloproteinases by thiol-proteinases from the oral anaerobe Porphyromonas gingivalis. J Dent Res 76:1260–1270

Discipio RG, Daffern PJ, Kawahara M, Pike R, Travis J, Hugli TE (1996) Cleavage of human complement component C5 by cysteine proteinases from Porphyromonas (Bacteroides) gingivalis. Prior oxidation of C5 augments proteinase digestion of C5. Immunology 87:660–667

Drake TA, Rodgers GM, Sande MA (1984) Tissue factor is a major stimulus for vegetation formation in enterococcal endocarditis in rabbits. J Clin Invest 73:1750–1753

Duesbery NS, Webb CP, Leppla SH, Gordon VM, Klimpel KR, Copeland TD, Ahn NG, Oskarsson MK, Fukasawa K, Paull KD, Vande Woude GF (1998)

Proteolytic inactivation of MAP-kinase-kinase by anthrax lethal factor. Science 280:734–737

Engel LS, Hobden JA, Moreau JM, Collagen MC, Hill JM, O'Callaghan RJ (1997) Pseudomonas deficient in protease IV has significantly reduced corneal virulence. Invest Ophthamol Vis Sci 38:1535–1542

Engel LS, Hill JM, Moreau JM, Green LC, Hobden JA, O'Callaghan RJ (1998) Pseudomonas aeruginosa protease IV produces corneal damage and contributes to bacterial virulence. Invest Ophthamol Vis Sci 39:662–665

Feuille F, Ebersole JL, Kesavalu L, Stephan MJ, Holt SC (1996) Mixed infections with Porphyromonas gingivalis and Fusobacterium nucleatum in a murine lesion model: potential synergistic effect on virulence. Infect Immun 64:2094–2100

Fletcher HM, Schenkein HA, Morgan RM, Bailey KA, Berry CR, Macrina FL (1995) Virulence of Porphyromonas gingivalis W83 mutant defective in the prtH gene. Infect Immun 63:1521–1528

Fletcher J, Reddi K, Poole S, Nair S, Henderson B, Tabona P, Wilson M (1997) Interaction between periodontopathogenic bacteria and cytokines. J Periodontal Res 32:200–205

Fletcher J, Nair S, Poole B, Henderson B, Wilson M (1998) Cytokine degradation by biofilms of Porphyromonas gingivalis. Curr Microbiol 4:216–219

Galloway DR (1991) Pseudomonas aeruginosa elastase and elastolysis: recent developments. Mol Microbiol 5:2315–2321

Genco CA, Odusanya BM, Potempa J, Mikolajczyk-Pawlinska J, Travis J (1998) A peptide domain on gingipain R which confers immunity against Porphyromonas gingivalis infection in mice. Infect Immun 66:4108–4114

Genco CA, Potempa J, Mikolajczyk-Pawlinska J, Travis J (1999) Role of gingipains R in Porphyromonas gingivalis pathogenesis. Clin Infect Dis (in press)

Goguen JD, Hoe NP, Subrahmanyam YVBK (1995) Proteases and bacterial virulence: a view from the trenches. Infect Agents Dis 4:47–54

Gordon VM, Leppla SH (1994) Proteolytic activation of bacterial toxins: role of bacterial and host cell proteinases. Infect Immun 62:333–340

Gubba A, Low DE, Musser JM (1998) Expression and characterization of group A streptococcus extracellular cysteine proteinase recombinant mutant proteins and documentation of seroconversion during human invasive disease episodes. Infect Immun 66:765–770

Gunn LC, Hengesh J (1969) The use of lysostaphin in treatment of staphylococcal wound infections. Rev Surg 26:214

Harrington DJ (1996) Bacterial collagenases and collagen-degrading enzymes and their potential role in human disease. Infect Immun 64:1885–1891

Harrington DJ, Russell RRB (1994) Identification and characterization of two extracellular proteases of Streptococcus mutans. FEMS Microbiol Lett 121:237–242

Hayashi T, Hotta H, Itoh M, Homma M (1991) Protection of mice by protease inhibitor, aprotonin, against lethal Sendai virus pneumonia. J Gen Virol 72:979–982

Herwald H, Collin M, Muller-Esterl W, Bjorck L (1996) Streptococcal cysteine proteinase releases kinins: a virulence mechanism. J Exp Med 184:665–673

Hill HR, Bohnsuck JF, Morris EZ, Augustine NH, Parker CJ, Cleary P, Wu JT (1988) Group B Streptococci inhibits the chemotactic activity of the fifth component of complement. J Immunol 141:3551–3556

Holder IA (1983) Experimental studies of the pathogenesis of infections due to Pseudomonas aeruginosa: effect of treatment with protease inhibitors. Rev Infect Dis 5[Suppl 5]:S914–S921

Holder IA, Haidaris CG (1979) Experimental studies of the pathogenesis of infections due to Pseudomonas aeruginosa: extracellular protease and elastase as in vivo virulence factors. Can J Microbiol 25:939–599

Holm SE, Norrby A, Berghold A-M, Norgren M (1992) Aspects of pathogenesis of serious group A streptococcal infection in Sweden, 1988–1989. J Infect Dis 166:31–37

Hong YQ, Ghebrehwet B (1992) Effect of Pseudomonas aeruginosa elastase and alkaline protease on serum complement and isolated components C1q and C3. Clin Immunol Immunopathol 62:133–138

Horvat RC, Parmely MJ (1988) Pseudomonas alkaline protease degrades human gamma interferon and inhibits its bioactivity. Infect Immun 56:2925–2932

Ijiri Y, Matsumoto K, Kamata R, Nishino N, Okamura R, Kambara T, Yamamoto T (1994) Suppression of polymorphonuclear leucocyte chemotaxis by Pseudomonas aeruginosa elastase in vitro: a study of the mechanism and the correlation with ring abscess in pseudomonal keratitis. Int J Exp Pathol 75:441–541

Imamura T, Pike RN, Potempa J, Travis J (1994) Pathogenesis of periodontitis: a major arginine-specific cysteine proteinase from Porphyromonas gingivalis induces vascular permeability enhancement through activation of the kallikrein/kinin pathway. J Clin Invest 94:361–367

Imamura T, Pike RN, Potempa J, Travis J (1995a) Dependence of vascular permeability enhancement on cysteine proteinases in vesicles of Porphyromonas gingivalis. Infect Immun 63:1999–2003

Imamura T, Potempa J, Pike RN, Moore JN, Barton MH, Travis J (1995b) Effect of free and vesicle-bound cysteine proteinases of Porphyromonas gingivalis on plasma clot formation: implications for bleeding tendency at periodontitis sites. Infect Immun 63:4877–4882

Imamura T, Potempa J, Tanase S, Travis J (1997) Activation of blood coagulation factor X by arginine specific cysteine proteinases (gingipain-Rs) from Porphyromonas gingivalis. J Biol Chem 272:16062–16067

Isenberg H (1988) Pathogenicity and virulence: another view. Clin Microbiol Rev 1:40–53

Jack RW, Tagg JR, Ray B (1995) Bacteriocins of gram-positive bacteria. Microbiol Rev 59:171–200

Jagels MA, Travis J, Potempa J, Pike R, Hugli TE (1996) Proteolytic inactivation of the leukocyte C5a receptor by proteinases derived from Porphyromonas gingivalis. Infect Immun 64:1984–1991

Ji Y, McLandsborough L, Kondagunta A, Cleary PP (1996) C5a peptidase alters clearance and trafficking of group A streptococci by infected mice. Infect Immun 64:503–510

Joiner KA (1988) Complement evasion by bacteria and parasites. Annu Rev Microbiol 42:201–230

Kamata R, Yamamoto T, Matsumoto K, Maeda H (1985) A serratial protease causes vascular permeability reaction by activation of the Hageman factor dependent pathway guinea pigs. Infect Immun 48:747–753

Kaminishi H, Hamatake T, Cho T, Tamaki T, Suenaga N, Fujii T, Hagihara Y, Maeda H (1994) Activation of blood clotting factors by microbial proteinases. FEMS Microbiol Lett 121:327–332

Kapur V, Topouzis S, Majesky W, Li LL, Hamrick MR, Hamill RJ, Patti JM, Musser JM (1993a) A conserved Streptococcus pyogenes extracellular cysteine proteinase cleaves human fibronectin and degrades vitronectin. Microb Pathog 15:327–346

Kapur V, Majesky MW, Li LL, Black RA, Musser JM (1993b) Cleavage of interleukin 1β (IL-1β) precursor to produce active IL-1 beta by conserved extracellular cysteine protease from Streptococcus pyogenes. Proc Natl Acad Sci USA 90:7676–7680

Kapur V, Maiffei JT, Greer RS, Li LL, Adams GJ, Musser JM (1994) Vaccination with streptococcal extracellular cysteine proteinase (interlukin-1β convertase) protects mice against challenge with heterologous group A streptococci. Microbiol Pathog 16:443–450

Kesavalu L, Holt SC, Ebersole JL (1996) Trypsin-like protease activity of Porphyromonas gingivalis as a potential virulence factor in a murine lesion model. Microb Pathog 20:1–10

Kessler E (1995) β-lytic endopeptidases. Methods Enzymol 248:740–756

Khan MMH, Shibuya Y, Nakagaki T, Kambara T, Yamamoto T (1994) Alpha-2-macroglobulin as the major defense in acute pseudomonal septic shock in guinea-pig model. Int J Exp Pathol 75:285–293

Khan MMH, Shibuya Y, Kambara T, Yamamoto T (1995) Role of alpha-2-macroglobulin and bacterial elastase in guinea-pig pseudomonal septic shock. Int J Exp Pathol 76:21–28

Kilian M (1981) Degradation of immunoglobulins A1, A2, and A3 by suspected principal pathogens. Infect Immun 34:757–765

Kilian M, Reinholdt J, Mortensen SB, Sorensen CH (1983) Perturbation of mucosal immune defense mechanisms by bacterial IgA proteases. Bull Eur Physiopathol Respir 19:99–104

Kilian M, Mestecky J, Russell MW (1988) Defense mechanisms involving Fc-dependent functions of immunoglobulin A and their subversion by bacterial immunoglobulin A proteases. Microbiol Rev 52:296–303

Kilian M, Reinholdt J, Lomholt H, Poulsen K, Frandsen EVG (1996) Biological significance of IgA1 proteases in bacterial colonization and pathogenesis: critical evaluation of experimental evidence. Acta Pathol Microbiol Immunol Scand 104:321–338

Klimpel KR, Arora N, Leppla SH (1994) Anthrax toxin lethal factor contains a zinc metalloproteinase consensus sequence which is required for lethal toxin activity. Mol Microbiol 13:1093–1100

Klenk HD, Rott R (1987) The molecular biology of influenza virus pathogenicity. Adv Virus Res 34:247–281

Knowles DJC (1997) New strategies for antibacterial drug design. Trends Microbiol 5:379–383

Lala A, Amano A, Sojar HT, Radel SJ, De Dardin E (1994) Porphyromonas gingivalis trypsin-like protease: a possible natural ligand for the neutrophil formyl peptide receptor. Biochem Biophys Res Commun 199:1489–1496

Lantz MS (1997) Are bacterial proteases important virulence factors? J Periodont Res 32:126–132

Lazurovitz SG, Goldberg AR, Choppin PW (1973) Proteolytic cleavage by plasmin of the HA polypeptide of influenza virus. Host cell activation of serum plasminogen. Virology 56:172–180

Lorber B (1995) Gas gangrene and other clostridium-associated diseases. In: Mandel GL, Bennett JE, Dolin R (eds) Principles and practice of infectious diseases, 4th edn. Churchill Livingstone, Edinburgh, pp 2182–2195

Lottenberg R (1997) A novel approach to explore the role of plasminogen in bacterial pathogenesis. Trends Microbiol 5:466–467

Lottenberg R, Minning-Wenz D, Boyle MDP (1994) Capturing host plasmin(ogen): a common mechanism for invasive pathogens? Trends Microbiol 2:20–24

Lukomski S, Sreevatsan S, Amberg C, Reichardt W, Woischnik M, Podbielski A, Musser JM (1997) Inactivation of Streptococcus pyogenes extracellular cysteine protease significantly decreases mouse lethality of serotype M3 and M49 strains. J Clin Invest 99:2574–2580

Lukomski S, Burns EH Jr, Wyde PR, Podbielski A, Rurangirwa J, Moore-Poveda DK, Musser JM (1998) Genetic inactivation of an extracellular cysteine protease (SpeB) expressed by Streptococcus pyogenes decreases resistance to pathogenesis and dissemination to organs. Infect Immun 66:771–776

Maeda H (1996) Role of microbial proteases in pathogenesis. Microbiol Immunol 40:685–699

Maeda H, Yamamoto T (1996) Pathogenic mechanisms induced by microbial proteases in microbial infections. Biol Chem 377:217–226

Maeda H, Molla A, Oda T, Katsuki T (1987) Internalization of serratial protease into cells as an enzyme inhibitor complex with α2-macroglobulin and regeneration of protease activity and cytotoxicity. J Biol Chem 262:10946–10950

Maeda H, Maruo K, Akaike T, Kaminishi H, Hagiwara Y (1992) Microbial proteinases as an universal trigger of kinin generation in microbial infections. In: Bönner G, Fritz H, Schoelkens B, Dietze G, Luppertz K (eds) Recent progress on kinins, agents and action, Suppl 38/III. Birkhäuser Verlag, Basel, pp 362–368

Maruo K, Maeda H, Akaike T, Imada Y, Ohkubo I, Ono T. (1993) Effect of microbial and mite proteases on low- and high-molecular-weight kininogens: generation of kinin and inactivation of thiol-protease inhibitory activity. J Biol Chem 268: 17711–17715

Matsumoto K, Yamamoto T, Kamata R, Maeda H (1984) Pathogenesis of serratial infection: activation of the Hageman factor–prekallikrein cascade by serratial protease. J Biochem 96:739–749

Mayo JA, Zhu H, Harty DW, Knox KW (1995) Modulation of glycosidase and protease activities by chemostat growth conditions in an endocarditis strain of Streptococcus sanguis. Oral Microbiol Immunol 10:342–348

Mekalanos JJ (1992) Environmental signals controlling expression of virulence determinants in bacteria. J Bacteriol 174:1–7

Metha S, Plaut AG, Calvanico NJ, Tomasi TB (1973) Human immunoglobulin A: production of an Fc fragment by an enteric microbial proteolytic enzyme. J Immunol 111:1274–1276

Mikolajczyk-Pawlinska J, Kordula T, Pavloff N, Pemberton PA, Chen WCA, Travis J, Potempa J (1998) Genetic variation of Porphyromonas gingivalis genes encoding gingipains, cysteine proteinases with arginine or lysine specificity. Biol Chem 379:205–211

Mintz CS, Miller RD, Gutgsell NS, Malek T (1993) Legionella pneumophila inactivates interleukin-2 and cleaves CD4 on human T-cells. Infect Immun 61:3416–3421

Miyagawa S, Kamata R, Matsumoto K, Okamura R, Maeda H (1991a) Inhibitory effects of ovomacroglobulin on bacterial keratitis in rabbits (1991) Graefe's Arch Clin Exp Ophthamol 229:291–286

Miyagawa S, Matsumoto K, Kamata R, Okamura R, Maeda H (1991b) Effects of protease inhibitors on growth of Serratia marcescens and Pseudomonas aeruginosa: chicken egg white ovomacroglobulin is potent growth suppressor. Microb Pathog 11:137–141

Miyagawa S, Kamata R, Matsumoto K, Okamura R, Maeda H (1994) Therapeutic intervention with chicken egg white ovomacroglobulin and a new quinolone on experimental Pseudomonas keratitis. Graefe's Arch Clin Exp Ophthamol 232: 488–493

Molla A, Matsumura Y, Yamamoto T, Okamura R, Maeda H (1987) Pathogenic capacity of proteases from Serratia marcescens and Pseudomonas aeruginosa and suppression by chicken egg white ovomacroglobulin. Infect Immun 55:2509–2517

Molla A, Kagimoto T, Maeda H (1988) Cleavage of immunoglobulin G (IgG) and IgA around hinge region by proteases from Serratia marcescens. Infect Immun 56:916–920

Molla A, Yamamoto T, Akaike T, Miyoshi S, Maeda H (1989) Activation of Hageman factor and prekallikrein and generation kinin by various microbial proteinases. J Biol Chem 264:10589–10595

Montecucco C, Schiavo G (1993) Tetanus and botulism neurotoxins: a new group of zinc proteases. Trends Biochem Sci 18:324–327

Musser JM, Krause RM (1998) The revival of group A streptococcal diseases, with a commentary on staphylococcal toxic shock syndrome. In: Krause RM (ed) Emerging infections. Academic Press, San Diego, pp 185–218

Musser JM, Hauser AR, Kim MH, Schlievert PM, Nelson K, Selander RK (1991) Streptococcus pyogenes causing toxic-shock-like syndrome and other invasive diseases: clonal diversity and pyrogenic exotoxin expression. Proc Natl Acad Sci USA 88:2668–2672

Nakayama K, Kadowaki T, Okamoto K, Yamamoto K (1995) Construction and characterization of arginine-specific cysteine proteinase (Arg-gingipain)-deficient mutants of Porphyromonas gingivalis. Evidence for significant contribution of Arg-gingipain to virulence. J Biol Chem 270:23619–23626

Nakayama K, Yoshimura F, Kadowaki T, Yamamoto K (1996) Involvement of arginine specific cysteine proteinase (Arg-gingipain) in fimbriation of Porphyromonas gingivalis. J Bacteriol 178:2818–2824

Oda T, Kojima Y, Akaike T, Ijiri S, Molla A, Maeda H (1990) Inactivation of chemotactic activity of C5a by serratial 56-kilodalton protease. Infect Immun 58:1269–1272

Ogawa T, Mori H, Yasuda K, Hasegawa M (1994) Molecular cloning and characterization of the genes encoding the immunoreactive major cell-surface proteins of Porphyromonas gingivalis. FEMS Microbiol Lett 120:23–30

Okamoto T, Akaike T, Suga M, Tanase S, Horie H, Miyajima S, Ando M, Ichinose Y, Maeda H (1997) Activation of human matrix metalloproteinases by various bacterial proteinases. J Biol Chem 272:6059–6066

Oldham ER, Daley MJ (1991) Lysostaphin: use of a recombinant bactericidal enzyme as a mastitis therapeutic. J Dairy Sci 74:4175–4182

Parmely M, Gale A, Clabaugh M, Horvat R, Zhou WW (1990) Proteolytic inactivation of cytokines by Pseudomonas aeruginosa. Infect Immun 58:3009–3014

Polak J, Della Latta P, Blackburn P (1993) In vitro activity of recombinant lysostaphin-antibiotic combinations toward methicillin-resistant Staphylococcus aureus. Diagn Microbiol Infect Dis 17:265–270

Pollack M (1995) Pseudomonas aeruginosa. In: Mandel GL, Bennett JE, Dolin R (eds) Principles and practice of infectious diseases, 4th edn. Churchill Livingstone, Edinburgh, pp 1980–2003

Pollanen J, Stephens RW, Vaheri A (1991) Directed plasminogen activation at the surface of normal and malignant cells. Adv Cancer Res 57:273–328

Potempa J, Travis J (1996) Porphyromonas gingivalis proteinases in periodontitis, a review. Acta Biochim Pol 43:455–466

Potempa J, Korzus E, Travis J (1994) The serpin superfamily of proteinase inhibitors: structure, function, and regulation. J Biol Chem 269:15957–15960

Potempa J, Pike R, Travis J (1995) Host and Porphyromonas gingivalis proteinases in periodontitis: a biochemical model of infection and tissue destruction. Prospect Drug Discovery Design 2:445–458

Potempa J, Mikolajczyk-Pawlinska J, Brassell D, Nelson D, Thogersen IB, Enghild JJ, Travis J (1998) Comparative properties of two cysteine proteinases (gingipains R), the products of two related but individual genes of Porphyromonas gingivalis. J Biol Chem 273:21648–21657

Pulverer G, Wegrzynowicz Z, Ko HL, Jeljaszewicz J (1977) Clotting and fibrinolytic activity of Bacteroides melaninogenicus. Zbl Bakt Hyg, I Abt Orig A 248:99–109

Pulverer G, Wegrzynowicz Z, Ko HL, Jeljaszewicz J (1980) Influence of Pseudomonas aeruginosa proteases on prothrombin, plasminogen and fibrinogen. Zbl Bakt Hyg, I Abt Orig A 255:368–379

Quickel KE Jr, Selden R, Caldwell JR, Nora NF, Schaffner W (1971) Efficacy and safety of topical lysostaphin treatment of persistent nasal carriage of Staphylococcus aureus. Appl Environ Microbiol 22:446–450

Ramadurai L, Jayaswal RK (1997) Molecular cloning, sequencing, and expression of lytM, a unique autolytic gene of Staphylococcus aureus. J Bacteriol 179:3625–3631

Reinholdt J, Kilian M (1991) Lack of cleavage of immunoglobulin A (IgA) from rhesus monkeys by bacterial IgA1 proteases. Infect Immun 59:2219–2221

Rifkin BR, Vernillo AT, Golub LM (1993) Blocking periodontal disease progression by inhibiting tissue-destructive enzymes: a potential therapeutic role for tetracyclines and their chemically-modified analogs. J Periodontol 64:819–827

Sakata Y, Akaike T, Suga M, Ijiri S, Ando M, Maeda H (1996) Bradykinin generation triggered by Pseudomonas proteases facilitates invasion of the systemic circulation by Pseudomonas aeruginosa. Microbiol Immunol 40:415–423

Salyers AA, Whitt DD (1994) Bacterial pathogenesis. A molecular approach. ASM Press, Washington DC, pp 3–62

Scheiblauer H, Reinacher M, Tashiro M, Rott R (1992) Interactions between bacteria and influenza A virus in the development of influenza pneumonia. J Infect Dis 166:783–791

Schenkein HA (1988) The effect of periodontal proteolytic Bacteroides species on proteins of human complement system. J Periodont Res 23:187–192

Scheld WM, Sande MA (1995) Endocarditis and intravascular infections. In: Mandel GL, Bennett JE, Dolin R (eds) Principles and practice of infectious diseases, 4th edn. Churchill Livingstone, Edinburgh, pp 740–783

Schnidler CA, Schuhardt VT (1965) Purification and properties of lysostaphin – a lytic agent for Staphylococcus aureus. Biochim Biophys Acta 97:242–250

Schultz DR, Miller KD (1974) Elastase of Pseudomonas aeruginosa: inactivation of complement components and complement-derived chemotactic and phagocytic factors. Infect Immun 10:128–135

Schwarzmann SW, Adler JL, Sullivan RJ Jr, Marine WM (1971) Bacterial pneumonia during the Hong Kong influenza epidemic of 1968–1969. Arch Intern Med 127:1037–1041

Simmonds RS, Pearson L, Kennedy RC, Tagg JR (1996) Mode of action of lysostaphin-like bacteriolytic agent produced by Streptococcus zooepidemicus 4881. Appl Environ Microbiol 62:4536–4541

Simmonds RS, Simpson WJ, Tagg JR (1997) Cloning and sequence analysis of zooA, a Streptococcus zooepidemicus gene encoding a bacteriocin-like inhibitory substance having a domain structure similar to that of lysostaphin. Gene 189:255–261

Smith GC, Markel JR (1982) Collagenolytic activity of Vibrio vulnificus: potential contribution to its invasiveness. Infect Immun 35:1155–1156

Sodeinde OA, Goguen JD (1989) Nucleotide sequence of the plasminogen activator gene of Yersinia pestis: relationship to ompT of Escherichia coli and gene E of Salmonella typhimurium. Infect Immun 57:1517–1523

Sodeinde OA, Subrahmanyam YVBK, Stark K, Quan T, Bao Y, Goguen JD (1992) A surface protease and the invasive character of plague. Science 258:1004–1007

Sorsa T, Ingman T, Soumalainen K, Haapasalo M, Konttinen Y, Lindy O, Saari H, Uitto V-J (1992) Identification of proteases from periodontopathogenic bacteria as activators of latent human neutrophil and fibroblast-type interstitial collagenases. Infect Immun 60:4491–4495

Straus DC (1982) Protease production by Streptococcus sanguis associated with subacute bacterial endocarditis. Infect Immun 38:1037–1045

Sugai M, Fujiwara T, Akiyama T, Ohara M, Komatsuzawa H, Inoue S, Suginaka H. (1997) Purification and molecular characterization of glycylglycine endopeptidase produced by Staphylococcus capitis EPK1. J Bacteriol 179:1193–1202

Tashiro M, Ciborowski P, Reinacher M, Klenk HD, Pulverer G, Rott R (1987a) Synergistic role of staphylococcal proteases in the induction of influenza virus pathogenicity. Virology 157:421–430

Tashiro M, Klenk HD, Rott R (1987b) Inhibitory effect of a proteinase inhibitor, leupeptin, on the development of influenza pneumonia, mediated by concomitant bacteria. J Gen Virol 68:2039–2041

Theander TG, Kharazmi A, Pedersen BK, Christensen LD, Tvede N, Poulsen LK, Odum N, Svenson M, Bendtzen K (1988) Inhibition of human lymphocyte proliferation and cleavage of interlukin-2 by Pseudomonas aeruginosa proteases. Infect Immun 56:1673–1677

Thumm G, Gotz F (1997) Studies on prolysostaphin processing and characterization of lysostaphin immunity factor (Lif) of Staphylococcus simulans biovar staphylolyticus. Mol Microbiol 23:1251–1265

Tokuda M, Karunakaran T, Duncan M, Hamada N, Kuramitsu H (1998) Role of Arg-gingipain A in virulence of Porphyromonas gingivalis. Infect Immun 66:1159–1166

Travis J, Potempa J, Maeda H (1995) Are bacterial proteinases pathogenic factors? Trends Microbiol 3:405–407

Vassalli JD, Sappino AP, Belin D (1991) The plasminogen activator/plasmin system. J Clin Invest 88:1067–1072

Vath GM, Earhart CA, Rabo JV, Kim MH, Bohach GA, Schlievert PM, Ohlendorf DH (1997) The structure of the superantigen exfoliative toxin A suggests a novel regulation as a serine protease. Biochemistry 36:1559–1566

Vollmer P, Walev I, Rose-John S, Bhakdi S (1996) Novel pathogenic mechanism of microbial metalloproteinases: liberation of membrane-anchored molecules in biologically active form exemplified by studies with the human interleukin-6 receptor. Infect Immun 64:3646–3651

Vu T-K, Hung DT, Wheaton VI, Coughlin SR (1991) Molecular cloning of a functional thrombin receptor reveals a novel proteolytic mechanism of receptor activation. Cell 64:1057–1068

Wegrzynowicz Z, Heczko PB, Drapeau GR, Jeljaszewicz J, Pulverer G (1981) Prothrombin activation by a metalloproteinase from Staphylococcus aureus. J Clin Microbiol 12:123–139

Welkos SL, Friedlander AM, Davis KJ (1997) Studies on the role of plasminogen activator in systemic infection by virulent Yersinia pestis strain C092. Microbiol Pathog 23:211–223

Wessels MR, Moses AE, Goldberg JB, DiCesare TJ (1991) Hyaluronic acid capsule is a virulent factor for mucoid group A streptococci. Proc Natl Acad Sci USA 88:8317–8321

Wikstrom M, Potempa J, Polanowski A, Travis J, Renvert S (1994) Detection of Porphyromonas gingivalis in gingival exudate by a dipeptide-enhanced trypsin-like activity. J Periodontol 65:47–55

Wilson M, Seymour R, Henderson B (1998) Bacterial perturbation of cytokine networks. Infect Immun 66:2401–2409

Wingrove JA, DiScipio RG, Chen Z, Potempa J, Travis J, Hugli TE (1992) Activation of complement components C3 and C5 by a cysteine proteinase (Gingipain-1) from Porphyromonas (Bacteroides) gingivalis. J Biol Chem 267:18902–18907

Wolf BB, Gibson CA, Kapur V, Hussaini IM, Musser JM, Gonias SL (1994) Proteolytically active streptococcal pyrogenic exotoxin B cleaves monocytic cell urokinase receptor and release an active fragment of the receptor from the cell surface. J Biol Chem 269:30682–30687

Zhirnov OP, Ovcharenko AV, Bukrinskaya AG (1984) Suppression of influenza virus replication in infected mice by proteinase inhibitors. J Gen Virol 65:191–106

CHAPTER 10

Parasite Proteases as Targets for Therapy

J.H. McKerrow, C.R. Caffrey, and J.P. Salter

A. Introduction

Parasitic diseases represent some of the world's greatest health problems. As many as 500 million people are infected with malaria, 250 million with schistosomiasis, and 40 million with various species of trypanosomes (World Health Organization 1993). The problem is now compounded by the emergence of parasite resistance to traditional chemotherapy. New drugs are desperately needed but because these are primarily diseases of poor people in poor countries, investment in research and drug development by industry has been minimal or absent.

One promising avenue of new drug development for parasitic diseases is the development of inhibitors targeting parasite proteases. Research over several decades has uncovered a panoply of proteases that play key roles in parasite life cycles or the diseases the parasites produce. The important "proof of concept" that protease inhibitors can, in fact, selectively arrest parasite replication in a mammalian host has recently been shown (Engel et al. 1998b; Rosenthal 1993). Renewed efforts in this endeavor have also been spurred by the success of anti-human-immunodeficiency-virus (HIV) protease inhibitors as drugs.

An important consideration in the development and evaluation of inhibitors of parasite proteases is that the treatment course is expected to be relatively short (as with an antibiotic). For example, a murine model of *Trypanosoma cruzi* infection was cured by a 20-day course of inhibitor treatment, and murine malaria can be cured with an even shorter treatment (Engel et al. 1998b; Rosenthal et al. 1998). Time dependent, slow off-rate, or covalent irreversible inhibitors may be practical in this setting, whereas there might be concern about their use in more chronic settings like cancer or osteoporosis.

This review will not be a comprehensive listing of known parasite proteases but rather a selective discussion of classes and examples of parasite proteases that have been more extensively evaluated or represent illustrative examples of suitable targets.

B. Metalloproteases

Metalloproteases represent a large and diverse group of enzymes, comprising as many as 30 families (Rawlings and Barrett 1995). A signature motif for most metalloproteases is the active-site sequence HEXXH (McKerrow 1987), which forms a key part of the metal-binding site (usually binding a zinc atom). The specificity of zinc metalloproteases resides in van der Waals and ionic interactions between the enzyme binding pockets (generally extending from S2 to S2'), and the corresponding side chains of amino-acid residues in the peptide or protein substrate. The recognition and visualization of these binding specificities using computer graphics has led to the successful design of relatively specific inhibitors of metalloproteases (Schwartz and van Wart 1992; Lovejoy et al. 1994).

To date, parasite metalloproteases have been divided into three major groups: the first are aminopeptidases associated with egg hatching or molting; the second, extracellular-matrix-degrading proteases reminiscent of mammalian collagenases; and the third, unusual proteases with a relatively specific substrate specificity like gp63. A general discussion using the more biochemically characterized examples will be given below. Parasite-genome projects are daily uncovering new members of each of these general categories, so this review is by no means an exhaustive discussion of all known parasite metalloproteases. The reader is referred to parasite and sequence-database websites for additional and more current examples.

I. Parasite Aminopeptidases

Aminopeptidases are metalloproteases with the capacity to cleave one or two amino acids from the amino terminus of proteins or peptides. This restricted substrate specificity and catalytic action correlates with the known biologic activity of better-characterized members of this family. For example, mammalian enzymes in this category generally process peptide hormones or other relatively small (less than 3000 Da) substrates involved in metabolic or neural signaling pathways (Erdös and Skidgel 1989). Leucine-aminopeptidase activity has been associated with the hatching of schistosome eggs (Xu and Dresden 1988) and may also be present on the surface of helminth parasites (Xu et al. 1990). While the specific function of these proteases is not known, a role in protein or peptide processing can be inferred from the mammalian homologues. Other aminopeptidases with specificity for larger hydrophobic amino acids at the amino terminus have been associated with the molting of helminth larvae. For example, filarial parasites have a burst of aminopeptidase activity detectable in "molting fluid" released at the transition between the L3 and L4 larval stages (Hong et al. 1993). Inhibition of this activity by metal-chelating inhibitors blocks larval molting. Further biochemical characterization of these enzymes, and the important link between biochemically characterized activity and specific genes, should be forthcoming as genome projects reveal more filarial nematode genes. Little effort has gone into further development of inhibitors for these proteases because the molting step in and of itself is not the most logical target for antiparasitic chemotherapy. However, further analysis of these types of enzymes may reveal similar activity in adult worms more suitable for a protease-inhibitor drug-development program (DAY and CHEN 1998).

II. Parasite Metalloproteases and Tissue Invasion

Invasive larval forms of several parasites secrete metalloproteases (McKERROW et al. 1990; HOTEZ et al. 1985). Some of these proteases have significant activity against extracellular-matrix macromolecules such as collagen and elastin. While, at the time of this review, there was no information on the structural characteristics of these enzymes, their inhibitor profile and macromolecular-substrate specificity is reminiscent of the matrix metalloproteases (MMPs) of vertebrates (RAWLINGS and BARRETT 1995). Their association with invasive larval forms and, in one case, the demonstration that invasion of skin by *Strongyloides* L3 larvae can be inhibited by metalloprotease inhibitors, suggests at least one biologic role for this group of enzymes. These metalloproteases are more "promiscuous" in their substrate activity versus the aminopeptidases or gp63. Combinatorial libraries of MMP inhibitors from pharmaceutical efforts aimed at MMPs in cancer or inflammatory disease could be exploited for their potential antiparasitic value (HODGSON 1995). While they may not necessarily prove to be useful antiparasitic drugs, these inhibitors would have immense value in helping to define the biologic function of these enzymes via "chemical knockout."

III. The Protease gp63

At a density of 500,000 molecules/cell, gp63 (also called promastigote surface protease or leishmanolysin) is the major surface protein of *Leishmania* promastigotes (BOUVIER et al. 1995). In the promastigote stage, it is glycosylphosphatidylinositol (GPI)-anchored to the parasite surface and is active at a relatively broad pH range. gp63 is a zinc metalloprotease with the signature zinc-binding active-site sequence, but otherwise little tertiary-structure similarity to other metalloproteases (BOUVIER et al. 1995). Unlike the more promiscuous metalloproteases (like those from *Strongyloides*), gp63 has a relatively specific substrate-sequence preference. Sequence analysis of cleavage sites generated by gp63 in a variety of proteins and peptides has led to a consensus substrate sequence of LIAY//LKKAT where "//" indicates the site of peptide cleavage (BOUVIER et al. 1990). A family of seven genes in *L. major* codes for gp63. Genes 1–6 encode promastigote forms, including the GPI-anchored gp63, while gene 7 is expressed exclusively in the infectious metacyclic stage and in amastigotes (JOSHI et al. 1998; VOTH et al. 1998). Site-directed mutagenesis has confirmed that glutamic acid 265 in the HEXXH zinc-binding motif is indispensable for catalytic activity of gp63 (McGWIRE and CHANG 1996). There are three potential *N*-linked-glycosylation sites, and mutation of asparagine 577 causes extracellular release of gp63, suggesting that it is the gGPI-addition site for membrane anchoring of gp63 (McGwire and Chang 1996).

While the exact biologic roles of gp63 have not been definitively identified, there are several clues from biochemical and genetic analyses that suggest experimentally testable hypotheses.

1. The relatively restricted substrate specificity of gp63 suggests that its role is probably not in parasite nutrition, but is more likely cleavage of a specific substrate or substrates involved in key processing or activation steps in the parasite life cycle.

2. Targeted gene disruption of gp63 genes 1–6 resulted in loss of expression in promastigotes and increased sensitivity to complement-mediated lysis (Joshi et al. 1998). These deletion mutants grew normally in the midgut of sandfly vectors and underwent normal stage differentiation from amastigotes to infectious metacyclics. They were also capable of producing lesions in BALB/c mice.

3. In *L. mexicana mexicana*, gp63 was implicated in parasite uptake by macrophages. Antibodies against gp63 blocked the interaction between promastigotes and macrophages by inhibiting attachment (Russell and Wilhelm 1986). Macrophages rapidly phagocytized gp63-containing liposomes. Uptake of gp63 by lysosomes was suppressed by anti-gp63-antibody fragments. Nevertheless, deletion of gp63 genes 1–6 yielded organisms still capable of infecting macrophages and differentiating into amastigotes (Joshi et al. 1998).

4. Insect trypanosomatids, including *Crithidia fasciculata* and *Herpetomonas samuelpessoai* (which do not infect vertebrates), feature gp63 homologues. This suggests that at least one of the functions of gp63 is involved in specific parasite-insect interactions (Bouvier et al. 1995).

5. Survival of amastigotes within host macrophages also seems to involve gp63. Cloning and comparison of virulent and attenuated strains of *L. mexicana amazonensis* showed that failure of the attenuated parasites to survive inside macrophages was associated with a 20- to 50-fold reduction in surface gp63 protein (Seay et al. 1996). Coating of attenuated cells or liposomes with proteolytically active gp63 protected them from degradation inside macrophage phagolysosomes.

6. A homologue of gp63 has also been found in African trypanosomes that do not have an intracellular stage. Bloodstream trypanosomes activate the alternative pathway of complement, but are not lysed because the complement cascade does not progress beyond the association of C3 convertase with the parasite surface (Donelson et al. 1998). Therefore, the homologue of gp63 in African trypanosomes has been hypothesized to play a role in protection against complement-mediated lysis, as was previously proposed for *Leishmania* gp63 (Brittingham et al. 1995).

7. Finally, a potential function of gp63 may be inferred from reports of the function of the zinc metalloprotease of *Listeria monocytogenes*. Transposon

insertion has been used to isolate 5000 mutants of *L. monocytogenes* (RAVENAU et al. 1992). One transposon insertion took place within the gene *mpl*, which encodes a zinc metalloprotease. It was subsequently shown that this specific mutation was associated with a loss of phospholipase activity, specifically due to a phosphatidylcholine-specific phospholipase C (PC-PLC). The results strongly suggest that the zinc metalloprotease of *Listeria* plays a role in the maturation of *Listeria* PC-PLC. Although the uptake and intracellular growth of bacteria were not affected, the virulence of the mutant was strongly impaired in mouse models of infection. It would be worthwhile to evaluate whether gp63 is involved in processing or activation of specific *Leishmania* or bloodstream trypanosome proteins or enzymes like phospholipase C.

IV. Is gp63 a Logical Target for Development of Protease Inhibitors as Therapy?

Surface localization of gp63 on *Leishmania* promastigotes places it in a vulnerable location for inhibition. At the same time, the promastigote stage is relatively transient in the mammalian host, and the copy number of the protease on the parasite surface is large. A more promising target for drug development may be the amastigote form of the protease. Targeted delivery of the zinc chelator 1,10-phenanthroline selectively eliminated intracellular *Leishmania* amastigotes (SEAY et al. 1996). Analysis of the effect of new generations of metalloprotease inhibitors on amastigotes and lesion development in mice is therefore warranted.

V. Future Development of Metalloprotease Inhibitors Targeting Parasite Proteases

While development of aspartyl and cysteine protease inhibitors targeting parasite enzymes is ongoing, very little work has been done on metalloprotease targets. This is due, in part, to the unavailability of structural information on parasite metalloproteases, and, until recently, the pharmaceutical industry's greater interest in aspartyl and cysteine protease inhibitors. However, a recent surge in development of metalloprotease inhibitors targeting mammalian proteases involved in inflammation or tumor cell invasion has provided an opportunity for new research efforts in parasitic disease. Several metalloprotease structures have been solved, representing family members with probable homologues in the parasite genome (BODE et al. 1993). Small-molecular-weight metalloprotease inhibitors suitable for use in animal models of parasite infection, and for which substantial pharmacokinetic and bioavailability data is available, have been produced (HODGSON 1995). It may be a propitious time for investigators of parasitic diseases to form strategic collaborations with peers in the pharmaceutical or biotechnology industries to evaluate combinatorial libraries of metalloprotease inhibitors that affect helminth and protozoan parasites.

C. Cysteine Proteases

The preferred characteristic for any synthetic protease inhibitor is a combination of oral bioavailability, low toxicity and selectivity for the target protease. These goals were pursued when focused research on the use of small, synthetic cysteine protease inhibitors (CPIs) used to target parasite cysteine proteases (CPs) began in the late 1980s. An unexpected boost to this approach was the recognition of the relative lack of redundancy in parasite proteases and the lower intracellular concentrations of parasite CPs compared with mammalian cells. In addition, some parasites appear to concentrate CPIs, thus increasing the potency of these compounds (McGrath et al. 1995; McKerrow 1995). Thus, even with relatively non-selective CPIs, it has been possible to kill parasites in vitro and in vivo (see below). Attempts to induce resistance to CPIs in vitro indicate that the mechanisms of induced resistance are often different from the mechanisms of resistance to the commonly applied chemotherapeutic agents (McKerrow et al. 1998). Thus, CPIs may provide an alternative therapy for drug-resistant parasites. A brief synopsis of the current state of the discipline with regard to the use of irreversible and reversible CPIs is outlined below.

I. Irreversible Inhibitors

Most research on the parasiticidal action of CPIs analyzed peptide-based inhibitors in which a peptidyl moiety dictates enzyme selectivity and a chemically reactive group forms a covalent, irreversible bond with the active site. Initial studies evaluated dipeptidyl diazomethyl ketones [DMKs; e.g., benzyloxycarbonyl (Z)-Phe-Ala-DMK] or fluoromethyl ketones (FMKs; e.g., Z-Phe-Arg-FMK) against *Plasmodium falciparum* (Rosenthal et al. 1989, 1991), *T. cruzi* (Ashall 1990; Harth et al. 1992; Meirelles et al. 1992), and *T. brucei* (Robertson et al. 1990) in vitro. An association between inhibition of CP activity and killing of the parasite was confirmed, but the specific effects on parasite metabolism, morphology or life cycle varied with species.

The first demonstration that CPIs could be detrimental to parasite survival in vivo was the intraperitoneal use of the peptidomimetic Mu-F-homophenylalanine (hF)-FMK (Fig. 1, Table 1), which cured 80% of mice infected with the model for falciparum malaria, *P. vinckeii* (Rosenthal et al. 1993). Killing was associated with an almost complete inhibition of the target CP falcipain. This inhibitor represented an advance over the earlier peptidyl compounds (Table 2). The introduction of the "non-natural" amino acid derivative homoPhe (hF) for Phe at P1 (nomenclature of Schechter and Berger

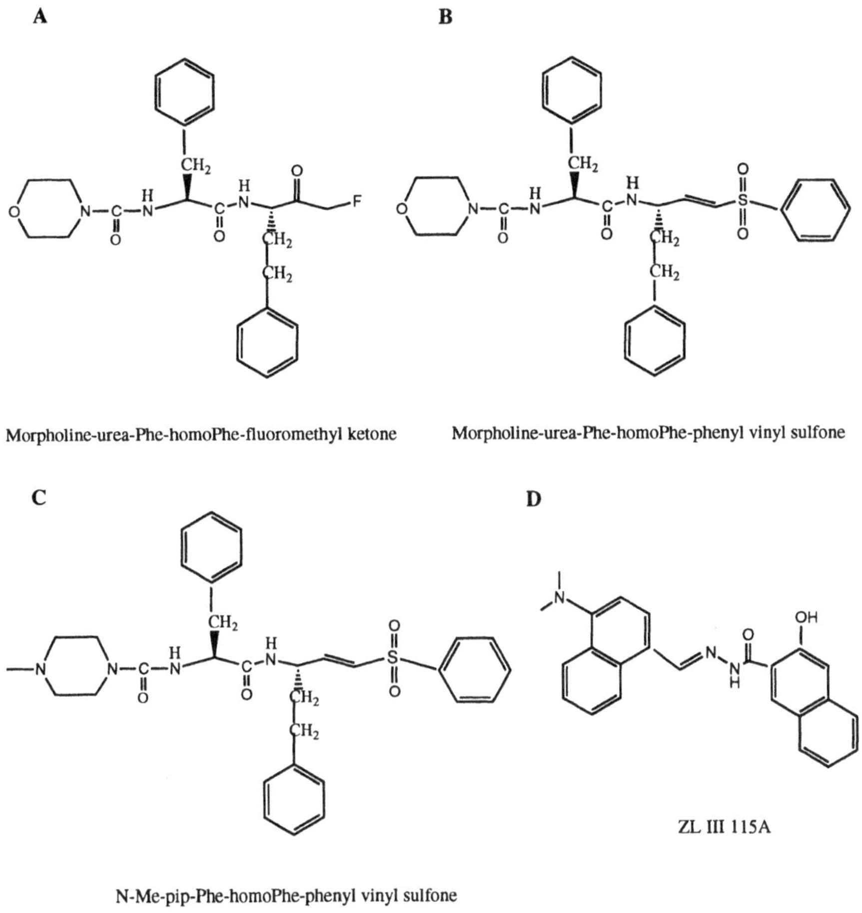

A

Morpholine-urea-Phe-homoPhe-fluoromethyl ketone

B

Morpholine-urea-Phe-homoPhe-phenyl vinyl sulfone

C

N-Me-pip-Phe-homoPhe-phenyl vinyl sulfone

D

ZL III 115A

Fig. 1A–D. Structures of the irreversible peptidomimetic inhibitors (**A**) morpholine-urea-Phe-homoPhe-fluoromethyl ketone, (**B**) morpholine-urea-Phe-homoPhe-phenyl vinyl sulfone, (**C**) *N*-methyl-piperazine-Phe-homoPhe-phenyl vinyl sulfone and (**D**) of the reversible non-peptidyl hydrazide ZL III 115A

1967) increased the half life and reduced the toxicity of the inhibitor. Secondly, the substitution of the morpholine urea (Mu) protecting group for Z improved aqueous solubility. Mu-F-hF-FMK was also effective against *S. mansoni* infection in mice by decreasing both the worm burden and egg output (WASILEWSKI et al. 1996).

Despite this success in vivo, it was clear that further development of inhibitors incorporating either DMK or FMK was not optimal due to concerns over the toxicity of the fluorine group (EICHHOLD et al. 1997). Also, the poor stability of DMKs at low pH would have likely limited their development for use orally (RASNICK 1996). The introductions of peptidomimetic vinyl sulfone (VS) inhibitors (PALMER et al. 1995; RASNICK 1996) and heterocyclic

Table 1. Effect of cysteine protease inhibitors (CPIs) against experimental parasitic infections in mice

Parasite	Target	CPI tested	Dosing regime	Outcome	Reference
Plasmodium vinckeii<?3>	Homolog of falcipain	Mu-F-hF-FMK	S.c. 400 mg/kg/day for 4 days	Cure in 80% of mice	Rosenthal et al. 1993
			In food 100 mg/kg/day for 7 days	Slowed development of parasitemia	Olson et al. 1998
		N-Me-pip-L-hF-VSPh	Gavage 100 mg/kg/day for 7 days	Slowed development of parasitemia	Olson et al. 1998
		N-Me-pip-L-hF-VS2Np	Gavage 100 mg/kg/day for 7 days	Cure in 40% of mice	Olson et al. 1998
Trypanosoma cruzi<?2>	Cruzain	Mu-F-hF-FMK	I.p. 100 mg/kg/day for 18 days (at end of experiment)	All mice survived a lethal infection	Engel et al. 1998b
		Mu-F-hF-VSPh	I.p. 100 mg/kg/day	All mice survived for 14 days (until end of experiment)	Engel et al. 1998b
		N-Me-pip-hF-VSPh	I.p. 100 mg/kg/day for 24 days	60% of mice were alive at 240 days (end of experiment); parasitological cure	Engel et al. 1998b
Leishmania major	cpB, cpL	Mu-F-hF-VSPh	I.p. 100 mg/kg/day for 28 days	At 4 weeks, 83% reduction in lesion size	Selzer et al. 1998
Trypanosoma brucei	Trypanopain	Z-F-A-DMK	I.p. 200 mg/kg/day for 5 days	Decreased parasitemia; mice survived longer	Scory et al. unpublished observations
Schistosoma mansoni	Sm31	Mu-F-hF-FMK	I.p. 100 mg/kg/day for 8 days	Decreased worm burden; decreased egg production	Wasilewski et al. 1996

DMK, diazomethyl ketone; *FMK*, fluoromethyl ketone; *hF*, homophenylalanine; *i.p.*, intraperitoneally; *Mu*, morpholine urea; *N-Me-pip*, N-methyl piperazine; *s.c.*, subcutaneously; *VS2Np*, 2-naphthyl vinyl sulfone; *VSPh*, phenyl vinyl sulfone; *Z*, benzyloxycarbonyl.

Table 2. Evolution in design of irreversible peptide-based inhibitors used to target parasite cysteine proteases

Inhibitor type	Examples	Advantages/drawbacks
Peptidyl fluoromethyl ketones	Z-F-R-FMK, Z-F-A-FMK	Readily metabolized, toxic metabolite
Peptidyl diazomethyl ketones	Z-F-A-DMK	Readily metabolized, mutagenic
Peptidomimetic fluoromethyl ketones	Mu-F-hF-FMK	Increased solubility due to Mu, longer half-life due to hF
Peptidomimetic phenyl vinyl sulfones<?1>	Mu-F-hF-VSPh, N-Me-pip-F-hF-VSPh	Relative non-toxicity of VSPh, Increased solubility and oral bioavailability due to N-Me-pip

DMK, diazomethyl ketone; *FMK*, fluoromethyl ketone; *hF*, homophenylalanine; *Mu*, morpholine urea; *N-Me-pip*, N-methyl piperazine; *VSPh*, phenyl vinyl sulfone; *Z*, benzyloxycarbonyl.

oxygen-containing "trapping groups" (KRANTZ 1991; Zimmerman, Prototek Inc., personal communication) were significant advances in that they are relatively well tolerated in experimental animals (ENGEL et al. 1998b; ROSENTHAL et al. 1998). Following successful application against parasites in culture (ROSENTHAL et al. 1996; ENGEL et al. 1998a), these compounds, or modifications thereof (Fig. 1, Table 1), cured in vivo infections of *P. vinckeii* (OLSON et al. 1998), *T. cruzi* (ENGEL et al. 1998b), and *L. major* (SELZER et al., manuscript submitted). Radiolabeled vinyl sulfone inhibitors incubated with whole cells or cell extracts indicated that the sole target of the inhibitor was the relevant CP. With *P. vinckeii*, an improvement in the inhibition of falcipain and parasiticidal properties has been realized with the incorporation of a 2-naphthyl (2Np) group on the VS (OLSON et al. 1998). Most importantly, these latter authors have shown that both peptidomimetic FMKs and VSs are effective as oral agents in delaying the clinical signs of malaria and curing infection, respectively.

II. Reversible Inhibitors

Only recently has the development of reversible inhibitors against parasite CPs begun to approach the success of the irreversible inhibitors. Two technological advances have provided momentum to this endeavor, namely computer-assisted drug design and combinatorial chemistry. A seminal study by RING et al. (1993) screened a computer model of falcipain against the Available Chemical Database (ACD; formerly the Fine Chemicals Directory; Molecular Design Ltd., San Leandro, Calif.) of 55,313 commercially available small molecules using the DOCK 3.0 software. Thirty-one compounds were chosen for experimental testing and, of these, three displayed IC_{50} values (concentration at which enzyme activity is inhibited by 50%) against falcipain of under $10\,\mu M$. One compound, oxalic bis(2-hydroxy-1-naphthylmethylene)hydrazide, was later used as a lead by Selzer et al. (1997) to screen potential anti-leishmanials in vitro using computer models of *L. major* cathepsins B and L. From this lead, second-generation compounds with hydrazide scaffolds, such as ZL III 115A (see Fig. 1) were synthesized, with IC_{50} values under $10\,\mu M$ against *Leishmania* cathepsin B. These compounds killed *L. major* promastigotes and arrested the growth of amastigotes in vitro (Selzer et al. 1997). In vivo, ZL III 115A reduced the size of footpad lesions in mice by 50% compared with controls, indicating an arrest of parasite replication (Selzer et al., manuscript submitted).

Compounds incorporating a second scaffold, the chalcone, have been designed by Cohen and coworkers using a computer model of falcipain. The most active chalcone, 1-(2,5-dichlorophenyl)-3-(4-quinolinyl)-2-propen-1-one, had an IC_{50} value of 200nM against *P. falciparum* in vitro (Li et al. 1995). Importantly, both chloroquine-sensitive and resistant parasites were equally susceptible to the chalcone.

Spurred on by the success of the development of HIV protease inhibitors, renewed interest in both academic and industrial laboratories has led to the construction of large combinatorial libraries of CPIs. For example, a new type of reversible inhibitor has been developed against cathepsin K, a homologue of some parasite CPs, and implicated in the resorption of bone by osteoclasts (Drake et al. 1996; Votta et al. 1997). These inhibitors contain a 1,3-diamino-propanone scaffold and span both the S and S' subsites of the protease active site (Thompson et al. 1997; Yamashita et al. 1997). Selectivity for cathepsin K over cathepsin L, B and S has been attained for these inhibitors with inhibition constants versus cathepsin K in the nanomolar and subnanomolar ranges.

In summary, the concept of targeting essential parasite CPs with small synthetic inhibitors to mitigate against parasite survival has been well demonstrated both in vitro and in vivo. Improvements in both the chemistry and pharmacokinetics of such inhibitors have been brought to the point where cures of parasitic infections in experimental animals are possible, including the use of these compounds orally. Further development can now focus on improved oral bioavailability and new reversible-inhibitor scaffolds.

D. Serine Proteases

Serine proteases are members of the largest of the four protease families. Their name is derived from the highly reactive serine that is used to attack the scissile peptide bond. The versatility of this catalytic mechanism is demonstrated by multiple examples of convergent evolution, and available sequences exceeding 500 entries (Perona and Craik 1997). Serine proteases also represent the largest number of protease crystal structures solved. This allows for accurate three-dimensional modeling, creating the opportunity for the rapid generation of lead inhibitors, and then providing direction for their refinement (RING et al. 1993).

I. Cercarial Elastase, an Example of a Parasite Larval Serine Protease

Waterborne cercariae of the blood fluke *S. mansoni* rapidly penetrate human skin and, during this process, secrete proteases to facilitate invasion. The identification of a serine protease within these secretions that degrades elastin and other skin macromolecules provided an opportunity to use homology-based modeling to predict active-site structure and computationally screen for inhibitor leads (NEWPORT et al. 1988). A structural model was made using the known structures of six different mammalian templates and the primary sequence of cercarial elastase. Using this information, a series of synthetic inhibitors were developed with inhibition characteristics that matched the predictions of the model. Additionally, in vitro skin inhibition assays also demonstrated biologic activity that corresponded to the inhibitor kinetic-data profile (COHEN et al. 1991). This work was extended to the identification of nonpeptidic inhibitors through a docking algorithm that predicted the best fit of publicly available compounds into the active site of the enzyme (RING et al. 1993).

II. Other Potential Serine-Protease Targets

Trypanosoma brucei is known to release a serine oligopeptidase into the blood stream of its host. Although its biologic role is unknown, the fact that none of the mammalian-protein proteinase inhibitors tested had inhibitory effects suggests that its activity may be important in the pathogenesis of African sleeping sickness (TROEBERG et al. 1996).

A serine protease from *Schistosoma mansoni* has been shown to cleave complement, thereby preventing lysis of parasites in vitro (FISHELSON 1995). Protease inhibitors targeting this enzyme might interrupt the schistosome's defense and allow normal immune clearance. The mechanism of *P. falciparum* merozoite invasion into red blood cells has also been shown to involve a serine protease, which is hypothesized to restructure the red cell cytoskeleton (ROGGWILLER et al. 1996).

E. Aspartyl Proteases

Aspartyl proteases are named for the two catalytic aspartic acids at the active site. Renin and the HIV aspartyl protease are the best known members of the group, and the current success of the HIV protease inhibitors as drugs has revived interest in many other protease-inhibitor-therapy projects.

I. Plasmepsins I and II

Hemoglobin degradation by the malaria parasite *P. falciparum* involves at least two aspartic (plasmepsin I and II) and one cysteine protease (falcipain). Inhibition studies with all three enzymes have validated the hypothesis that these proteases are potential targets for the development of antimalarial chemotherapy. The development of a recombinant expression system for both plasmepsin I (Moon et al. 1997) and plasmepsin II (Hill et al. 1994) opens the door for crystallization studies, already carried out for plasmepsin II (Silva et al. 1996), and the screening of a large number of inhibitors.

The fact that the repertoire of mammalian aspartyl proteases is small should aid in the development of selective inhibitors, because any questions about reactivity with other host enzymes can be tested directly. This fact has already been exploited by Silva et al. (1996). Five derivatives were synthesized and tested, using pepstatin A as a lead compound, against plasmepsin II and human cathepsin D (Cat D). These five derivatives were made using data from the structure of Cat D and had roughly equal inhibition characteristics between the two enzymes. Three more compounds were then designed using the structural data from plasmepsin II. Each of these new compounds had, on

Table 3. Aspartic and serine protease inhibitors used with parasites

Parasite	Class	Target	Inhibitor	Reference	Details
Plasmodium falciparum<?3>	Aspartic	Plasmepsin I	SC-50083	Francis et al. 1994	In vitro; IC_{50} 5 μM
			Ro 40-4383	Moon et al. 1997	In vitro; IC_{50} 250–300 nM
		Plasmepsin II	Compound 7	Silva et al. 1996	In vitro; IC_{50} 20 μM
	Serine	gp76	S-12-P	Roggwiller et al. 1996	In vitro; invasion assay 10 μM
Schistosoma mansoni	Serine	Cercarial elastase	AAPL-CMK	Cohen et al. 1991	In vitro; invasion assay 50 μM
Cryptosporidium parvum	Serine	Unknown	α-1-Antitrypsin	Forney et al. 1996	In vitro; invasion assay 5 μg/ml
Enterocytozoon bieneusi	Aspartic	Unknown	Retroviral inhibitors	Conteas et al. 1998	In vivo; indirect evidence

IC50, concentration at which enzyme activity is inhibited by 50%.

average, 20 times more selectivity for plasmepsin II than Cat D, confirming the predictive power of structural data.

Finally, plasmepsin I and II do not seem to be redundant (HILL et al. 1994; MOON et al. 1997; SILVA et al. 1996). Their expression patterns, immunoreactivity, and inhibitor specificity are all unique. This may allow the design of distinct inhibitors that, used in conjunction, could provide synergistic growth inhibition. More importantly, the use of two inhibitors would greatly reduce the chance that the parasite could develop resistance – a problem always one step away from any chemotherapeutic strategy.

II. The Indirect Discovery of an Antiparasitic Protease Inhibitor

The growing use of HIV aspartic protease inhibitors led to the discovery that their effects may include clearance of *Enterocytozoon bieneusi*. The strongest correlation with parasite clearance in a clinical trial was the concurrent use of HIV protease inhibitors (CONTEAS et al. 1998). Increasing global use of these inhibitors may provide an additional opportunity to observe their effects on other parasitic diseases in which parasite-derived aspartyl proteases are critical (Table 3).

References

Ashall F (1990) Characterisation of an alkaline peptidase of *Trypanosoma cruzi* and other trypanosomatids, Mol Biochem Parasitol 38:77–88

Bode W, Gomis-Ruth FX, Stockler W (1993) Astacins, serralysins, snake venom and matrix metalloproteinases exhibit identical zinc-binding environments (HEXXHXXGXXH and Met-turn) and topologies and should be grouped into a common family, the "metzincins". FEBS Lett 331:134–140

Bouvier J, Schneider P, Etges RJ (1990) Peptide substrate specificity of the membrane-bound metalloprotease of *Leishmania*. Biochemistry 29:10113–10119

Bouvier J, Schneider P, Etges RJ (1995) Leishmanolysin: surface metalloproteinase of *Leishmania*. Methods Enzymol 248:614–633

Brittingham A, Morrison CJ, McMaster WR, McGwire BS, Chang K-P (1995) Role of the *Leishmania* surface protease gp63 in complement fixation, cell adhesion, and resistance to complement-mediated lysis. J Immunol 155:3102–3111

Cohen FE, Gregoret LM, Amiri P, Aldape K, Railey J, McKerrow JH (1991) Arresting tissue invasion of a parasite by protease inhibitors chosen with the aid of computer modeling. Biochemistry 30:11221–11229

Conteas CN, Berlin OG, Speck CE, Pandhumas SS, Lariviere MJ, Fu C (1998) Modification of the clinical course of intestinal microsporidiosis in acquired immunodeficiency syndrome patients by immune status and anti-human immunodeficiency virus therapy. Am J Trop Med Hyg 58:555–558

Day TA, Chen GZ (1998) The metalloprotease inhibitor 1,10-phenanthroline affects *Schistosoma mansoni* motor activity, egg laying and viability. Parasitology 116: 319–325

Donelson JE, Hill KL, El-Sayed NMA (1998) Multiple mechanisms of immune evasion by African trypanosomes. Mol Biochem Parasitol 91:51–66

Drake FH, Dodds RA, James IE, Connor JR, Debouck C, Richardson S, Lee-Rykaczewski E, Coleman L, Rieman D, Barthlow R, Hastings G, Gowen M (1996)

Cathepsin K, but not cathepsins B, L, or S, is abundantly expressed in human osteoclasts, J Biol Chem 271:12511–12516

Eichhold TH, Hookfin EB, Taiwo YO, De B, Wehmeyer KR (1997) Isolation and quantification of fluoroacetate in rat tissues, following dosing of Z-Phe-Ala-CH2-F, a peptidyl fluoromethyl ketone protease inhibitor. J Pharm Biomed Anal 16:459–467

Engel JC, Doyle PS, Palmer J, Hsieh I, Bainton DF, McKerrow JH (1998a) Cysteine protease inhibitors alter Golgi complex ultrastructure and function in *Trypanosoma cruzi*. J Cell Sci 111:597–606

Engel JC, Doyle PS, Hsieh I, McKerrow JH (1998b) Cysteine protease inhibitors cure an experimental *Trypanosoma cruzi* infection. J Exp Med 188:725–734

Erdös EG, Skidgel RA (1989) Neutral endopeptidase 24.11 (enkephhalinase) and related regulators of peptide hormones. FASEB J 3:145–151

Fishelson Z (1995) Novel mechanisms of immune evasion by *Schistosoma mansoni*. Mem Inst Oswaldo Cruz 90:289–292

Forney JR, Yang S, Du C, Healey MC (1996) Efficacy of serine protease inhibitors against *Cryptosporidium parvum* infection in a bovine fallopian tube epithelial cell culture system. 82:638–640

Francis SE, Gluzman IY, Oksman A, Knickerbocker A, Mueller R, Bryant ML, Sherman DR, Russell DG, Goldberg DE (1994) Molecular characterization and inhibition of a Plasmodium falciparum aspartic hemoglobinase. EMBO J 13:306–317

Hill J, Tyas L, Phylip LH, Kay J, Dunn BM, Berry C (1994) High level expression and characterisation of Plasmepsin II, an aspartic proteinase from *Plasmodium falciparum*. FEBS Lett 352:155–158

Hodgson J (1995) Remodeling MMPIs. Biotechnology 13:554–557

Hong X, Bouvier J, Wong M, Lee G, McKerrow JH (1993) *Brugia pahangi*: identification and characterization of an aminopeptidase associated with larval molting. Exp Parasitol 76:127–133

Hotez P, Trang N, McKerrow JH, Cerami A (1985) Isolation and characterization of a proteolytic enzyme from the adult hookworm. J Biol Chem 260:7343–7348

Joshi PB, Sacks DL, Modi G, McMaster WR (1998) Targeted gene deletion of *Leishmania major* genes encoding developmental stage-specific leishmanolysin (GP63). Mol Microbiol 27:519–530

Krantz A, Copp LJ, Coles PJ, Smith RA, Heard SB (1991) Peptidyl (acloxy)methyl ketones and the quiescent affinity label concept: the departing group as a variable structural element in the design of inactivators of cysteine proteinases. Biochemistry 30:4678–4687

Li R, Kenyon GL, Cohen FE, Chen X, Gong B, Dominguez JN, Davidson E, Kurzban G, Miller RE, Nuzum EO, Rosenthal PJ, McKerrow JH (1995) In vitro antimalarial activity of chalcones and their derivatives. J Med Chem 38:5031–5037

Lovejoy B, Cleasby A, Hassell AM, Longley K, Luther MA, Weigl D, McGeehan G, McElroy AB, Drewry D, Lambert MH, Jordan SR (1994) Structure of the catalytic domain of fibroblast collagenase complexed with an inhibitor. Science 263:375–377

McGrath ME, Eakin AE, Engel JC, McKerrow JH, Craik CS, Fletterick RJ (1995) The crystal structure of cruzain: a therapeutic target for Chagas' disease. J Mol Biol 247:251–259

McGwire BS, Chang KP (1996) Posttranslational regulation of a *Leishmania* HEXXH metalloprotease (gp63) – the effects of site-specific mutagenesis of catalytic, zinc binding, *N*-glycosylation, and glycosyl phosphatidylinositol addition sites on *N*-terminal end cleavage, intracellular stability, and extracellular exit. J Biol Chem 271:7903–7909

McKerrow JH (1987) Human fibroblast collagenase contains an amino acid sequence homologous to the zinc-binding site of *Serratia* protease. J Biol Chem 262:5943

McKerrow JH (1995) Cysteine proteases of parasites: a remarkable diversity of function. Perspect Drug Discov Design 2:437–444

McKerrow JH, Brindley P, Brown M, Gam AA, Neva F (1990) *Strongyloides stercoralis*: identification of a protease that facilitates penetration of skin by the infective larvae. Exp Parasitol 70:134–143

McKerrow JH, Engel JC, Caffrey CR (1998) Cysteine protease inhibitors as chemotherapy for parasitic infections. (submitted)

Moon RP, Tyas L, Certa U, Rupp K, Bur D, Jacquet C, Matile H, Loetscher H, Grueninger-Leitch F, Kay J, Dunn BM, Berry C, Ridley RG (1997) Expression and characterisation of plasmepsin I from *Plasmodium falciparum*. Eur J Biochem 244:552–560

Newport GR, McKerrow JH, Hedstrom R, Petitt M, McGarrigle L, Barr PJ, Agabian N (1988) Cloning of the proteinase that facilitates infection by schistosome parasites. J Biol Chem 263:13179–13184

Olson JE, Lee GK, Semenov A, Rosenthal PJ (1998) Antimalarial effects in mice of orally administered peptidyl cysteine protease inhibitors. Bioorg Med Chem (in press)

Palmer JT, Rasnick D, Klaus JL, Bromme D (1995) Vinyl sulfones as mechanism-based cysteine proteases inhibitors. J Med Chem 38:3193–3196

Perona JJ, Craik CS (1997) Evolutionary divergence of substrate specificity within the chymotrypsin-like serine protease fold. 272:29987–29990

Rasnick D (1996) Small synthetic inhibitors of cysteine proteases. Perspect Drug Discov Design 6:47–63

Raveneau J, Geoffroy C, Beretti J-L, Gaillard J-L, Alouf JE, Berche P (1992) Reduced virulence of a *Listeria monocytogenes* phospholipase-deficient mutant obtained by transposon insertion into the zinc metalloprotease gene. Infect Immun 60:916–921

Rawlings ND, Barrett AJ (1995) Evolutionary families of metallopeptidases. Methods Enzymol 248:183–272

Ring CS, Sun E, McKerrow JH, Lee GK, Rosenthal PJ, Kuntz ID, Cohen FE (1993) Structure-based inhibitor design using protein models for the development of antiparasitic agents. Proc Natl Acad Sci USA 90:3583–3587

Robertson CD, North MJ, Lockwood BC, Coombs GH (1990) Analysis of the proteinases of *Trypanosoma brucei*. J Gen Microbiol 136:921–925

Roggwiller E, Betoulle ME, Blisnick T, Braun Breton C (1996) A role for erythrocyte band 3 degradation by the parasite gp76 serine protease in the formation of the parasitophorous vacuole during invasion of erythrocytes by *Plasmodium falciparum*. 82:13–24

Rosenthal PJ, McKerrow JH, Rasnick D, Leech JH (1989) *Plasmodium falciparum*: Inhibitors of lysosomal cysteine proteinases inhibit a trophozoite proteinase and block parasite development. Mol Biochem Parasitol 35:177–184

Rosenthal PJ, Wollish WS, Palmer JT, Rasnick D (1991) Antimalarial effects of peptide inhibitors of a *Plasmodium falciparum* cysteine proteinase. J Clin Invest 88:1467–1472

Rosenthal PJ, Lee GK, Smith RE (1993) Inhibition of a *Plasmodium vinckei* cysteine proteinase cures murine malaria. J Clin Invest 91:1052–1056

Rosenthal PJ, Olson JE, Lee GK, Palmer JT, Klaus JL, Rasnick D (1996) Antimalarial effects of vinyl sulfone cysteine proteinase inhibitors. Antimicrob Agents Chemother 40:1600–1603

Rosenthal PJ, McKerrow JH, Rasnick D, Leech JH (1998) A cysteine proteinase of *Plasmodium falciparum* is a potential target of antimalarial chemotherapy. (manuscript submitted)

Russell DG, Wilhelm H (1986) The involvement of the major surface glycoprotein (gp63) of *Leishmania* promastigotes in attachment to macrophages. J Immunol 136:2613–2620

Schechter I, Berger A (1967) On the size of the active site in proteases. I. Papain. Biochem Biophys Res Commun 27:157–162

Schwartz MA, van Wart HE (1992) Synthetic inhibitors of bacterial and mammalian interstitial collagenases. Prog Med Chem 29:271–334

Seay MB, Heard PL, Chaudhuri G (1996) Surface Zn-proteinase as a molecule for defense of *Leishmania mexicana amazonensis* promastigotes against cytolysis inside macrophage phagolysosomes. Infect Immun 64:5129–5137

Selzer PM, Chen X, Chan VJ, Cheng M, Kenyon GL, Kuntz ID, Sakanari JA, Cohen FE, McKerrow JH (1997) *Leishmania major*: molecular modeling of cysteine proteases and prediction of new nonpeptide inhibitors. Exp Parasitol 87:212–221

Selzer PM, Pingel S, Hsieh I, Chan VJ, Engel JC, Russell DG, Sakanari JA, McKerrow JH (1998) Cysteine protease inhibitors arrest growth of *Leishmania major* in vitro and in vivo. (submitted)

Silva AM, Lee AY, Gulnik SV, Maier P, Collins J, Bhat TN, Collins PJ, Cachau RE, Luker KE, Gluzman IY, Francis SE, Oksman A, Goldberg DE, Erickson JW (1996) Structure and inhibition of plasmepsin II, a hemoglobin-degrading enzyme from *Plasmodium falciparum*. Proc Natl Acad Sci USA 93:10034–10039

Thompson SK, Halbert SM, Bossard MJ, Tomaszek TA, Levy MA, Zhao B, Smith WW, Abdel-Meguid SS, Janson CA, D'Alessio KJ, Briand J, Sarkar SK, Huddleston MJ, Ijames CF, Carr SA, Garnes KT, Shu A, Heys JR, Bradbeer J, Zembryki D, Lee-Rykaczewski L, James IE, Lark MW, Drake FH, Gowen M, Gleason JG, Veber DF (1997) Design of potent and selective human cathepsin K inhibitors that span the active site. Proc Natl Acad Sci USA 94:14249–14254

Troeberg L, Pike RN, Morty RE, Berry RK, Coetzer THT, Lonsdale-Eccles JD (1996) Proteases from *Trypanosoma brucei brucei*. Purification, characterisation and interactions with host regulatory molecules. Eur J Biochem 238:728–736

Voth BR, Kelly BL, Joshi PB, Ivens AC, McMaster WR (1998) Differentially expressed *Leishmania major gp63* genes encode cell surface leishmanolysin with distinct signals for glycosylphosphatidylinositol attachment. Mol Biochem Parasitol 93:31–41

Votta BJ, Levy MA, Badger A, Bradbeer J, Dodds RA, James IE, Thompson S, Bossard MJ, Carr T, Connor JR, Tomaszek TA, Szewczuk L, Drake FH, Veber DF, Gowen M (1997) Peptide aldehyde inhibitors of cathepsin K inhibit bone resorption both in vitro and in vivo. J Bone Miner Res 12:1396–1406

Wasilewski MM, Lim KC, Phillips J, McKerrow JH (1996) Cysteine protease inhibitors block schistosome hemoglobin degradation in vitro and decrease worm burden and egg production in vivo. Mol Biochem Parasitol 61:179–189

World Health Organization (1993) Tropical disease research: progress 1991–1992. Eleventh programme report. In: UNDP/World Bank/WHO (eds) Special programme for research and training in tropical disease. 8. Leishmaniasis

Xu QY, Shively JE (1988) Microsequence analysis of peptides and proteins. VIII. Improved electroblotting of proteins onto membranes. Anal Biochem 170:19–30

Xu YZ, Shawar SM, Dresden MH (1990) *Schistosoma mansoni*: Purification and characterization of membrane-associated leucine aminopeptidase, Exp Parasitol 70:124–133

Yamashita DS, Smith WW, Zhao B, Janson CA, Tomaszek TA, Bossard MJ, Levy MA, Oh H-J, Carr TJ, Thompson SK, Ijames CF, Carr SA, McQueney M, D'Alessio KJ, Amegadzie BY, Hanning CR, Abdel-Meguid S, DesJarlais RL, Gleason JG, Veber DF (1997) Structure and design of potent and selective cathepsin K inhibitors. J Am Chem Soc 119:11351–11352

Section III
(Non-Viral) Proteases Involved in Diseases

CHAPTER 11

Host Proteinases as Targets for Therapeutic Intervention

J.C. Cheronis

A. Introduction

Proteinases (also called proteases or peptidases) are classified on the basis of their mechanisms of peptide-bond cleavage and are referred to in this regard by the nucleophile used to attack the carbonyl carbon of the scissile bond. Since proteins are one of the fundamental building blocks of biological systems, proteinases are essential regulators of biological activity. For example, they are directly involved in such diverse physiologic processes as fertility and reproduction, cellular proliferation, tissue remodeling, wound healing, clotting, clot dissolution, digestion, and protein turnover. Similarly, an increased understanding of the causative role proteinases play in a number of disease processes (such as cancer invasion and metastasis, the inflammatory response, antigen processing and presentation, degenerative diseases, and cardiovascular/pulmonary disorders) has led to an increased effort to exploit the potential of proteinase inhibition as a therapeutic alternative for a variety of diseases. Previous chapters in this book have focused on the potential for novel therapies based on the inhibition of exogenous, pathogen-derived proteinases for important human diseases ranging from human immunodeficiency-virus (HIV) infection to periodontal disease. The following chapters are intended to give the reader a general understanding and appreciation of the potential that inhibition of endogenous or host proteinases may have for the treatment of a variety of human disorders or conditions.

While it is beyond the scope of this chapter to review the entire biochemical, physiological, pathophysiological, and medicinal history of the proteinase/proteinase-inhibitor field as it applies to the modification of host proteinases and their activities, it is appropriate to give the reader a sense of how this area developed. To date, the only two therapeutic areas that have been successfully addressed by the use of selective synthetically derived proteinase inhibitors are that of hypertension [angiotensin-converting-enzyme (ACE) inhibitors] and the treatment of HIV (HIV proteinase inhibitors). The history of the former has been the subject of numerous reviews (EHLERS and RIORDAN 1989; EHLERS and RIORDAN 1990; CORVOL and WILLIAMS 1998), and the latter is the subject of several chapters in this volume. However, if one were to widen the definition of proteinase-directed therapies beyond the use of unnatural synthetic inhibitors, clearly the field of thrombosis and thrombolysis has been a fertile ground for developing an understanding of the importance of proteinases in human physiology/pathophysiology. More specifically, the use of heparin for augmenting the activity of a natural pro-teinase inhibitor (antithrombin III) and warfarin for decreasing the produc-tion of clotting factors (pro-proteinases) clearly underscores the potential for regulating proteolytic activity in vivo. Similarly, the use of streptokinase and, more recently, tissue-plasminogen activator illustrates the potential for aug-menting proteolytic processes in the treatment of clinically important diseases. For an encyclopedic review of the known universe of proteolytic enzymes please refer to *Handbook of Proteolytic Enzymes* (BARRETT et al. 1998).

Table 1 is a sampling of potential therapeutic targets covering the five known classes of proteinases (serine, cysteine, metallo, aspartyl, and threo-nine), and the general clinical areas in which inhibition of specific proteinases might be of therapeutic value. The table tries to incorporate both the accepted nomenclature for the proteinases in question as well as references for further reading. The single exception to the above, with respect to the identification of a therapeutic area, listed in this table, is acrosin, the trypsin-like proteinase of the acrosome of the sperm head responsible for sperm penetration of the zona pellucida of the ovum (SRIVASTAVA et al.1965; STAMBAUGH and BUCKLEY 1968). In this case, the "therapeutic" area for acrosin inhibition does not fall within the general categories used to organize Table 1, but may be used in the development of a male (or female) non-hormonal birth-control agent, further emphasizing the tremendous breadth of potential proteinase-based therapeu-tics may have. However, this breadth of potential is a double-edged sword, in that it carries with it the alternative potential of important toxicities or side effects.

Any time a therapy is directed toward the modification of endogenous mediators (in this case, host proteinases), there may well be attendant mech-anism-based toxicities associated with the inhibition of the target proteinase(s) that cannot be avoided by changing the structural class of the compound or the mechanism of inhibition. An obvious example of this dilemma is the inhibition of thrombin or factor X which would be of tremendous therapeu-tic value for patients with a history of pulmonary embolism or deep venous thrombosis and many patients with advanced atherosclerotic diseases. However, thrombin inhibition carries with it the potential for bleeding com-plications that can be as dangerous as the disease for which the patient is being treated. This is dramatically different from side effects or toxicities associated with the targeting of exogenous pathogen-derived (e.g. virally encoded) pro-teinases, which tend to have unique specificities and/or structures. In these latter situations, side effects associated with one class of compound are inher-ent to the drug and may be categorically different, both in quality and kind, than those of a different class of compound targeting the same enzyme. As a result, the development of therapies based on the modification of host enzymes is inherently more difficult than that of therapies described in the preceding chapters, in that it must take into consideration both the potential

Table 1. Host proteinases as targets for therapeutic interventions

Target enzyme	Protease class	Therapeutic application					References
		Cardiovascular	Autoimmune	Cancer	Degenerative	Inflammatory	
Acrosin	S						URCH (1991); ADHAM et al. (1998)
Caspase-1 (ICE)	C		X			X	THORNBERRY (1994); THORNBERRY and MOLINEAUX (1995); THORNBERRY (1998)
Cathepsin B	C			X			BARRETT and KIRSCHKE (1981); BERQUIN and SLOANE (1996); MORT and BUTTLE (1997); MORT (1998); SHRIDHAR et al. (Chap. 16)
Cathepsin G	S					X	SALVESEN (1998)
Cathepsin K	C				X		BOSSARD et al. (1996); BRÖMME et al. (1996a,b, 1998)
Cathepsin S	C		X				RIESE et al. (1996); KIRSCHKE (1998)
Chymase	S		X			X	SCHWARTZ (1990); CAUGHEY (1995); CAUGHEY (1998)
Collagenase 3 (MMP-13)	Zn			X			FREIJE et al. (1994); MITCHELL (1996); JEFFREY (1998)
Endothelin-converting enzyme	Zn	X					TURNER and MURPHY (1996); AHN (1998)
Factor X	S	X					JAMES (1994); STENFLO (1998)
Gelatinase A (MMP-2)	Zn			X			BIRKEDAL-HANSEN (1995); MURPHY and CRABBE (1995); MURPHY (1998)
Gelatinase B (MMP-9)	Zn			X		X	BIRKEDAL-HANSEN (1995); MURPHY and CRABBE (1995); COLLIER and GOLDBERG (1998)
Granzyme B	S		X				SITKOVSKY and HENKART (1993); SALVESEN and FROELICH (1998)
Guanidino-benzotase	S			X			STEVEN (1998)

Table 1. Continued

Target enzyme	Protease class	Therapeutic application					References
		Cardiovascular	Autoimmune	Cancer	Degenerative	Inflammatory	
Interstitial collagenase (MMP-1)	Zn			X	X		Woessner (1991); Birkedal-Hansen et al. (1993); Dioszegi et al. (1995); Cawston (1998)
Macrophage elastase (MMP-12)	Zn	X			X	X	Banda et al. (1987); Shapiro et al. (1992, 1993); Senior and Shapiro (1998)
Matrilysin (MMP-7)	Zn			X			Woessner (1998)
Membrane-type MMPs (MT-MMPs 1, 2, 3, and 4)	Zn			X		X	Seiki (1998)
Mu-Calpain	C				X		Murachi (1989); Mellgren and Murachi (1990); Sorimachi et al. (1996); Sorimachi and Suzuki (1998)
Neutrophil collagenase (MMP-8)	Zn	X				X	Tschesche et al. (1992); Van Wart (1992); Tschesche (1995); Tschesche and Pieper (1998)
Neutrophil elastase	S	X			X	X	Bieth (1998); Cheronis and Rabinovitch (Chap.14)
Non-ICE caspases	C		X		X		Salvesen and Dixit (1997); Thornberry and Lazebnik (1998); Deveraux et al. (Chap.17)
Peptidyl-dipeptidase A (ACE)	Zn	X					Ehlers and Riordan (1989, 1990); Bhoola et al. (1992); Corvol and Williams (1998)

Proteinase	Class					References
Plasma kallikrein	S	X				Bhoola et al. (1992); Wachtfogel et al. (1993); Colman (1998)
Proteinase-3 (myeloblastin)	S	X				Hoidal et al. (1995); Hoidal (1998); Cheronis and Rabinovitch (Chap. 14)
Proteasome	T	X		X		Rubin and Finley (1995); Coux et al. (1996); De Martino (1998); Tanaka and Kawahara (1998) (Chap. 18)
Renin	D				X	Inagami (1989); Fukamizu and Murakami (1995); Suzuki et al. (1998)
Stomelysin 1 and 2 (MMPs-3, 10)	Zn	X	X	X		Windsor et al. (1993); Nagase (1995); Matrisian (1998); Nagase (1998)
Stomelysin 3 (MMP-11)	Zn			X		Rouyer et al. (1995); Pei and Weiss (1998)
Thrombin	S				X	Stubbs and Bode (1993); Grand et al. (1996); Stone and Le Bonniec (1998); Schmaier (Chap. 15)
Tissue kallikrein	S	X			X	Bhoola et al. (1992); Madeddu (1993); Lintz et al. (1995); Chao (1998)
TNF-α-converting enzyme (TACE)	Zn	X				Black and Becherer (1998)
Tryptase	S	X		X		Schwartz (1990, 1994); Caughey (1995); Johnson (1998)
Urokinase plasminogen activator	S			X		Danø et al. (1985, 1994); Ellis and Danø (1998)

S, serine; C, cysteine; D, aspartyl; Zn, metallo; T, threonine; ICE, interleukin-1β-converting enzyme; MMP, matrix metalloproteinase; ACE, angiotensin-converting enzyme; TNF, tumor necrosis factor.

unique toxicities of the specific compound being evaluated and the attendant problems associated with inhibiting the target enzyme.

A further complication of the targeting of endogenous proteinases is that there is a significant degree of functional redundancy associated with mammalian systems, which does not appear to be the case with viral and other microbial systems. For example, there is only one HIV-processing proteinase, and inhibition of that proteinase will prevent viral protein processing and, thereby, replication. However, the processing of mammalian proteins [such as the cytokines, interleukin (IL)-1β and tumor necrosis factor (TNF)α, see 14] can be accomplished by a variety of proteinases, even though there may be a hierarchical or predominant mechanism for this biological process. With therapies directed toward microbial proteinases, the efficacy of the therapy can be compromised by any number of different problems, such as viral mutation or strain heterogeneity, bioavailability, intracellular penetration, etc. However, the probability of success (assuming that the inhibitor is of adequate potency) is far greater than the probability of a clinically meaningful effect produced by the inhibition of an endogenous proteinase, regardless of the potency of the compound. In other words, there is the potential for a significant divergence between the potency and efficacy of a potential therapeutic agent when inhibiting an endogenous proteinase relative to that for the inhibition of an exogenous proteinase, all other things (bioavailability, penetration into protected microenvironments, clearance, etc.) being equal. All of the foregoing aside, however, the opportunity for novel therapeutic interventions based on proteinase inhibition, both of exogenous and endogenous proteinases, is a promising area of research in the biopharmaceutical industry today.

B. History

Not surprisingly, the field of proteinase research initially developed out of studies concerning digestive processes in the gut, and pepsin, the principal acid protease of the stomach, is generally considered to be the first enzyme to be "discovered" (sometime in the 18th century). Further advances in this field did not occur before the late 19th century, when the pancreatic juices of cattle were shown to have the ability to digest proteins and that this activity was different from that of pepsin by virtue of a significantly higher optimum pH for "trypsin" relative to that of pepsin (KÜHNE 1876). Despite these early efforts, it was not until the 1930s that protein digestion by pancreatic juices was found to be due to more than one enzyme (trypsin and chymotrypsin) and that these enzymes were initially secreted as proenzymes undergoing activation upon exposure to extracts of the small intestine or slightly acidic solutions (NORTHRUP and KUNITZ 1932; NORTHRUP et al. 1939). The third major serine proteinase of the digestive system, pancreatic elastase, was not discovered until 1949 (BALÓ and BANGA 1949).

Other proteolytic processes, such as those associated with white blood cells, primarily neutrophils (leukoprotease activity), were also described in the early part of this century but required several decades before the enzymes responsible for this activity were identified [neutrophil elastase: JANOFF and SCHERER (1968); DEWALD et al. (1975); RINDLER-LUDWIG and BRAUNSTEINER (1975); cathepsin G: BAUGH and TRAVIS (1976); STARKEY and BARRETT (1976a, b); proteinase-3: BAGGIOLINI et al. (1978); KAO et al. (1988); neutrophil collagenase: LAZARUS et al. (1968); MURPHY et al. (1977); gelatinase B: HIBBS et al. (1985)]. Similarly, tissue kallikrein was first described as a substance in human urine that produced profound but transient hypotension when injected intravenously into dogs (FREY 1926; FREY and KRAUT 1928). Initially this substance was thought to be hormone-like, but through the efforts of Werle and his associates over several decades the proteolytic qualities of this substance became better understood (FREY et al. 1950a,b; FREY et al. 1968). Complementing this work on leukocyte enzymes and the kallikrein–kinin system are studies dating from well back into the 1940s that investigate the proteases involved in the coagulation (clotting factors) and complement systems and the interaction of the contact system with the initiation of both the coagulation and inflammatory pathways [complement system – ROSS (1986); MORGAN (1990); WHALEY et al. (1992); KERR (1994); KERR (1998a,b); contact system and factor XII – PIXLEY and COLMAN (1993); RATINOFF (1998); coagulation system, factor X and thrombin – STUBBS and BIODE (1993); BLOOM et al. (1994); JAMES et al. (1994); GRAND et al. (1996); STENFLO (1998); STONE and LE BONNIEC (1998)].

Continued effort in the laboratories of Hans Fritz and Marianne Jochum at the Ludwigs Maximillians University in Munich, as well as others throughout the world extending the pioneering efforts of these early researchers, have demonstrated the importance of inflammatory and coagulation-associated proteinases in a variety of pathophysiologic conditions, most notably in sepsis and shock (BONE 1992; WAYDHAS et al. 1992; JOCHUM et al. 1993). More recently, these researchers have extended this work into the clinical arena by supplementing patients with sepsis or severe trauma with high-dose antithrombin III (INTHORN et al. 1998; WAYDHAS et al. 1998).

Finally, it is important to recognize that the endogenous proteinase field, while venerable in many respects, is still an area that is actively evolving. One of the most exciting areas in the fields of proteinase research and the importance of proteinase inhibitors as therapeutics is that of the caspases. This family of enzymes, which includes IL-1-converting enzyme (caspase-1) and the apoptosis-related proteinases (caspases 3, 8 and 9), were only recognized as a distinct family of enzymes in the early 1990s, and other members of this family are still being identified (THORNBERRY and MOLINEAUX 1995; THORNBERRY 1998; THORNBERRY and LAZEBNIK 1998) (see Chap. 14).

C. Section Overview

Any book (or portion thereof) that is intended to deal with endogenous/host proteinases as potential therapeutic targets has the difficulty of selecting topics and establishing the organization of the chapters and their relationship to the field in general. Topics can be segregated on the basis of enzyme class (aspartyl, serine, cysteine, threonine or metallo) or on the basis of clinical indications (inflammation, cancer, degenerative diseases, etc.). Either approach is somewhat arbitrary, and both have their own drawbacks. In this volume we have chosen to select topics that are more of a sampling of therapeutic areas, some well defined and some still being investigated, that will give the reader a general appreciation of the breadth and scope proteinase-inhibitor-based therapeutic approaches may have for the future of medicine. In the cases in which the targets are "old" (thrombin, Chap. 15 or the elastases, Chap. 14, for example) we have tried to emphasize new data and concepts and their implications for clinical medicine rather than to restate the well-established ideas described in the past. In other cases, such as with the proteolytic processing of amyloid proteins in the central nervous system (Chap. 19), the enzymes responsible for these activities have yet to be isolated and identified (hence they are not included in Table 1), but are the subjects of intensive research efforts in a number of different laboratories as they are considered to be critical components of the pathophysiology of Alzheimer's disease. In addition to these two extremes, we have also included chapters on a variety of different enzymes that are clearly of interest to both the academic and pharmaceutical industries. These include the cysteine and metalloproteinases associated with tumor cell invasion and metastasis (Chaps. 12 and 16), the proteasome, and the cysteine proteinases associated with cell death or apoptosis. We hope that you find these chapters both informative and provocative.

References

Adham IM, Schlösser M, Engel E (1998) Acrosin. In: Barrett AJ, Rawlings ND, Woessner JE (eds) Handbook of proteolytic enzymes. Academic, San Diego, pp 90–93

Ahn K (1998) Endothelin-converting enzyme 1. In: Barrett AJ, Rawlings ND, Woessner JE (eds) Handbook of proteolytic enzymes. Academic, San Diego, pp 1085–1089

Baggiolini M, Bretz U, Dewald B (1978) The polymorphonuclear leukocyte. Inflamm Res 8:3–10

Baló J, Banga I (1949) Elastase and elastase inhibitor. Nature 164:491

Banda MJ, Werb Z, McKerrow JH (1987) Elastin degradation. Methods Enzymol 144:288–305

Barrett AJ, Kirschke H (1981) Cathepsin B, cathepsin H, and cathepsin L. Methods Enzymol 80:535–561

Barrett AJ, Rawlings ND, Woessner JE (eds) Handbook of proteolytic enzymes. Academic, San Diego

Baugh R, Travis J (1976) Human leukocyte granule elastase: rapid isolation and characterization. Biochemistry 15:836–841

Berquin IM, Sloane BF (1996) Cathepsin B expression in human tumors. Adv Exp Med Biol 389:281–294

Bhoola KD, Figueroa CD, Worthy K (1992) Bioregulation of kinins: kallikreins, kininogens, and kininases. Pharmacol Rev 44:1–80

Bieth JG (1998) Leukocyte elastase. In: Barrett AJ, Rawlings ND, Woessner JE (eds) Handbook of proteolytic enzymes. Academic, San Diego, pp 54–60

Birkedal-Hansen H (1995) Proteolytic remodeling of extracellular matrix. Curr Opin Cell Biol 7:728–735

Birkedal-Hansen H, Moore WGI, Bodden MK, Windsor LJ, Birkedal-Hansen B, DeCarlo A, Engler JA (1993) Matrix metalloproteinases: a review. Crit Rev Oral Biol Med 4:197–250

Black RA, Becherer JD (1998) Tumor necrosis factor alpha-converting enzyme. In: Barrett AJ, Rawlings ND, Woessner JE (eds) Handbook of proteolytic enzymes. Academic, San Diego, pp 1315–1317

Bloom AL, Forbes CD, Thomas DP, Tuddenham EGD (eds) (1994) Haemostasis and thrombosis, 3rd edn. Churchill Livingstone, Edinburgh

Bone RC (1992) Modulators of coagulation: a critical appraisal of their role in sepsis. Arch Intern Med 152:1381–1389

Bossard MJ, Tomaszek TA, Thompson SK, Amegadszie BY, Hanning CR, Jones C, Kurdyla JT, McNulty DE, Drake FH, Gowen M, Levy MA (1996) Proteolytic activity of human osteoclast cathepsin K. Expression, purification, activation, and substrate specificity. J Biol Chem 271:12517–12524

Brömme D (1998) Cathepsin K. In: Barrett AJ, Rawlings ND, Woessner JE (eds) Handbook of proteolytic enzymes. Academic, San Diego, pp 624–628

Brömme D, Okamoto K, Wang BB, Biroc S (1996a) Human cathepsin O2, a matrix protein-degrading cysteine protease expressed in osteoclasts. Functional expression of human cathepsin O2 in *Spodoptera frugiperda* and characterization of the enzyme. J Biol Chem 271:2126–2132

Brömme D, Klaus JL, Okamoto K, Rasnick D, Palmer JT (1996b) Peptidyl vinyl sulphone: a new class of potent and selective cysteine protease inhibitors-S 2P 2 specificity of human cathepsin O2 in comparison with cathepsins S and L. Biochem J 315:85–89

Caughey GH (1995) Mast cell chymases and tryptases: phylogeny, family relations, and biogenesis. In: Caughey GH (ed) Mast cell proteases in immunology and biology. Marcel Dekker, New York, pp 305–329

Caughey GH (1998) Chymase. In: Barrett AJ, Rawlings ND, Woessner JE (eds) Handbook of proteolytic enzymes. Academic, San Diego, pp 66–70

Cawston TE (1998) Interstitial collagenase. In: Barrett AJ, Rawlings ND, Woessner JE (eds) Handbook of proteolytic enzymes. Academic, San Diego, pp 1155–1162

Chao J (1998) Human tissue kallikrein. In: Barrett AJ, Rawlings ND, Woessner JE (eds) Handbook of proteolytic enzymes. Academic, San Diego, pp 97–100

Collier IE, Goldberg GI (1998) Gelatinase B. In: Barrett AJ, Rawlings ND, Woessner JE (eds) Handbook of proteolytic enzymes. Academic, San Diego, pp 1205–1210

Colman RW (1998) Plasma prekallikrein and kallikrein. In: Barrett AJ, Rawlings ND, Woessner JE (eds) Handbook of proteolytic enzymes. Academic, San Diego, pp 147–153

Corvol P, Williams TA (1998) Peptidyl-dipeptidase A/angiotensin I-converting enzyme. In: Barrett AJ, Rawlings ND, Woessner JE (eds) Handbook of proteolytic enzymes. Academic, San Diego, pp 1066–1076

Coux O, Tanaka K, Goldberg AL (1996) Structure and functions of the 20 S and 26 S proteasomes. Annu Rev Biochem 65:801–847

Danø K, Andreasen PA, Grøndahl-Hansen J, Kristensen P, Nielsen LS, Skriver L (1985) Plasminogen activators, tissue degradation and cancer. Adv Cancer Res 44:139–266

Danø K, Behrendt N, Brünner N, Ellis V, Ploug M, Pyke C (1994) The urokinase receptor: protein structure and role in plasminogen activation and cancer invasion. Fibrinolysis 8[Suppl 1]:189–203

De Martino GN (1998) 26 S Proteasome. In: Barrett AJ, Rawlings ND, Woessner JE (eds) Handbook of proteolytic enzymes. Academic, San Diego, pp 505–512

Dewald B, Rindler-Ludwig R, Bretz U, Baggiolini M (1975) Subcellular localization and heterogeneity of neutral proteases in neutrophilic polynuclear leukocytes. J Exp Med 141:709–723

Dioszegi M, Cannon P, Van Wart HE (1995) Vertebrate collagenases. Methods Enzymol 248:413–431

Ehlers MRW, Riordan JF (1989) Angiotensin-converting enzyme: new concepts concerning its biological role. Biochemistry 28:5311–5318

Ehlers MRW, Riordan JF (1990) Angiotensin-converting enzyme. Biochemistry and molecular biology. In: Laragh JH, Brenner BM (eds) Hypertension pathophysiology, diagnosis and management. Raven, New York, pp 1217–1231

Ellis V, Danø K (1998) u-Plasminogen activator. In: Barrett AJ, Rawlings ND, Woessner JE (eds) Handbook of proteolytic enzymes. Academic, San Diego, pp 177–184

Freije JMP, Díez-Itza I, Balbín M, Sánchez LM, Blasco R, Tolivia J, López-Otin C (1994) Molecular cloning and expression of collagenase 3, a novel human matrix metalloproteinase produced by breast carcinomas. J Biol Chem 269:16766–16773

Frey EK (1926) Zusammenhänge zwischen Herzabeit und Nierentätigkeit. Arch Klin Chir 142:663

Frey EK, Kraut H (1928) Ein neues Kreislaufhormon und seine Wirkung. Arch Exp Pathol Pharmakol 133:1

Frey EK, Kraut H, Werle E (1950a) Zusammenhange Zwischen Herzarbeit und Nierentatigkeit. Arch Klin Chir p 142

Frey EK, Kraut H, Werle E (1950b) Kallikrein (Padutin). Enke, Stuttgart

Frey EK, Kraut H, Werle E, Vogel R, Zickgraf-Ruedel G, Trautschold I (1968) Das Kallikrein-kinin-system und seine Inhibitoren. Enke, Stuttgart

Fukamizu A, Murakami K (1995) New aspects of the renin-angiotensin system in blood pressure regulation. TEM 6:279–284

Grand RJ, Turnell AS, Grabham PW (1996) Cellular consequences of thrombin-receptor activation. Biochem J 313:353–368

Hibbs MS, Hasty KA, Seyer JM, Kang AH, Mainardi CL (1985) Biochemical and immunological characterization of the secreted forms of human neutrophil gelatinase. J Biol Chem 260:2493–2500

Hoidal JR (1998) Myeloblastin. In: Barrett AJ, Rawlings ND, Woessner JE (eds) Handbook of proteolytic enzymes. Academic, San Diego, pp 62–65

Hoidal JR, Rao NV, Gray B (1995) Myeloblastin: leukocyte proteinase 3. Methods Enzymol 244:61–67

Inagami T (1989) Structure and function of renin. J Hypertens Suppl 7:S3–S8

Inthorn D, Hoffmann JN, Hartl WH, Mòhlbayer D, Jochum M (1998) Effect of antithrombin III supplementation on inflammatory responses in patients with sever sepsis. Shock 10:90–96

James HL (1994) Physiology and biochemistry of factor X. In: Bloom AL, Forbes CD, Thomas DP, Tuddenham EGD (eds) Haemostasis and thrombosis, 3rd edn. Churchill Livingstone, Edinburgh, pp 439–464

Janoff A, Scherer I (1968) Mediators of inflammation in leukocyte lysosomes. IX. Elastinolytic activity in granules of human polymorphonuclear leukocytes. J Exp Med 128:1137–1140

Jeffrey JJ (1998) Collagenase 3. In: Barrett AJ, Rawlings ND, Woessner JE (eds) Handbook of proteolytic enzymes. Academic, San Diego, pp 1167–1170

Jochum M, Machleidt W, Fritz H (1993) Phagocyte proteinases in multiple trauma and sepsis: pathomechanisms and related therapeutic approaches. In: Neugebauer EA, Holaday JW (eds) Handbook of mediators in septic shock. CRC, Boca Raton, pp 335–361

Johnson DA (1998) Tryptase. In: Barrett AJ, Rawlings ND, Woessner JE (eds) Handbook of proteolytic enzymes. Academic, San Diego, pp 70–74

Kao RC, Wehner NG, Skubitz KM, Gray BH, Hoidal JR (1988) A distinct human polymorphonuclear leukocyte proteinase that produces emphysema in hamsters. J Clin Invest 82:1963–1973

Kerr MA (1994) Complement. In: Kerr MA, Thorpe R (eds) LABFAX immunochemistry. Bios Scientific, Oxford, pp 211–233

Kerr MA (1998a) Complement component C2 and the classical pathway C3/C5 convertase. In: Barrett AJ, Rawlings ND, Woessner JE (eds) Handbook of proteolytic enzymes. Academic, San Diego, pp 130–134

Kerr MA (1998b) Factor B and the alternative pathway C3/C5 convertase. In: Barrett AJ, Rawlings ND, Woessner JE (eds) Handbook of proteolytic enzymes. Academic, San Diego, pp 135–140

Kirschke H (1998) Cathepsin S. In: Barrett AJ, Rawlings ND, Woessner JE (eds) Handbook of proteolytic enzymes. Academic, San Diego, pp 621–624

Kühne WF (1876) über die Verdauung der Eiweisstoffe durch den Pankreassaft. Virchows Arch 39:130

Lazarus GS, Brown RS, Daniels JR, Fullmer HM (1968) Human granulocyte collagenase. Science 159:1483–1485

Lintz W, Wiemer G, Gohlke P, Unger T, Scholkens BA (1995) Contribution of kinins to the cardiovascular actions of angiotensin-converting enzyme inhibitors. Pharmacol Rev 47:25–49

Madeddu P (1993) Receptor antagonists of bradykinin: a new tool to study the cardiovascular effects of endogenous kinins. Pharmacol Rev 28:107–128

Matrisian LM (1998) Stomelysin 2. In: Barrett AJ, Rawlings ND, Woessner JE (eds) Handbook of proteolytic enzymes. Academic, San Diego, pp 1178–1180

Mellgren RL, Murachi T (1990) Intracellular calcium-dependent proteolysis. CRC Press, Boston

Mitchell PG, Magna HA, Reeves LM, Lopresti-Morrow LL, Yocum SA, Rosner PJ, Geoghegen KF, Hambor JE (1996) Cloning, expression and type II collagenolytic activity of matrix metalloproteinase-13 from human osteoarthritic cartilage. J Clin Invest 97:761–768

Morgan BP (1990) Complement: clinical aspects and relevance to disease. Academic Press, London

Mort JS (1998) Cathepsin B. In: Barrett AJ, Rawlings ND, Woessner JE (eds) Handbook of proteolytic enzymes. Academic, San Diego, pp 609–617

Mort JS, Buttle DJ (1997) Molecules in focus. Cathepsin B. Int J Biochem Cell Biol 29:715–720

Murachi T (1989) Intracellular regulatory system involving calpain and calpastatin. Biochem Int 18:263–294

Murphy G (1998) Gelatinase A. In: Barrett AJ, Rawlings ND, Woessner JE (eds) Handbook of proteolytic enzymes. Academic, San Diego, pp 1199–1205

Murphy G, Crabbe T (1995) Gelatinase A and B. Methods Enzymol 248:470–495

Murphy G, Reynolds JJ, Bretz U, Baggiolini M (1977) Collagenase is a component of the specific granules of human neutrophil leukocytes. Biochem J 162:195–197

Nagase H (1995) Stromelysins 1 and 2. Methods Enzymol 248:449–470

Nagase H (1998) Stromelysin 1. In: Barrett AJ, Rawlings ND, Woessner JE (eds) Handbook of proteolytic enzymes. Academic, San Diego, pp 1172–1178

Northrop JH, Kunitz M (1932) Crystalline trypsin. I. Isolation and tests of purity. J Gen Physiol 16:267

Northrop JH, Kunitz M, Herriott RM (1939) Crystalline enzymes. Columbia University, New York

Pei D, Weiss SJ (1998) Stromelysin 3. In: Barrett AJ, Rawlings ND, Woessner JE (eds) Handbook of proteolytic enzymes. Academic, San Diego, pp 1187–1190

Pixley RA, Colman RW (1993) Factor XII: Hageman factor. Methods Enzymol 222:51–65

Ratnoff OD (1998) Stromelysin 3. In: Barrett AJ, Rawlings ND, Woessner JE (eds) Handbook of proteolytic enzymes. Academic, San Diego, pp 144–146

Riese RJ, Wolf PR, Brömme D, Natkin LR, Villadangos JA, Ploegh HL, Chapman HA (1996) Essential role for cathepsin S in MHC class II-associated invariant chain processing and peptide loading. Immunity 4:357–366

Rindler-Ludwig R, Braunsteiner H (1975) Cationic proteins from human neutrophil granulocytes (evidence for their chymotrypsin-like properties). Biochim Biophys Acta 379:606–617

Ross GD (1986) Immunobiology of the complement system. Academic, Orlando

Rouyer N, Wolf C, Chenard MP, Rio MC, Chambon P, Bellocq JP, Basset P (1995) Stromelysin 3 gene expression in human cancer: an overview. Invasion Metastasis 14:269–275

Rubin DM, Finley D (1995) The proteasome: a protein-degrading organelle? Curr Biol 5:854–858

Salvesen GS (1998) Cathepsin G. In: Barrett AJ, Rawlings ND, Woessner JE (eds) Handbook of proteolytic enzymes. Academic, San Diego, pp 60–62

Salvesen GS, Dixit VM (1997) Caspases: intracellular signaling by proteolysis. Cell 91:443–446

Salvesen GS, Froelich CJ (1998) Granzyme B. In: Barrett AJ, Rawlings ND, Woessner JE (eds) Handbook of proteolytic enzymes. Academic, San Diego, pp 81–83

Schwartz LB (1990) Neutral proteases of mast cells. Karger, Basel

Schwartz LB (1994) Tryptase: a mast cell serine protease. Methods Enzymol 244:88–100

Seiki M (1998) Membrane-type matrix metalloproteinase 1. In: Barrett AJ, Rawlings ND, Woessner JE (eds) Handbook of proteolytic enzymes. Academic, San Diego, pp 1192–1195

Senior RM, Shapiro SD (1998) Macrophage elastase. In: Barrett AJ, Rawlings ND, Woessner JE (eds) Handbook of proteolytic enzymes. Academic, San Diego, pp 1180–1183

Shapiro SD, Griffin GL, Gilbert DJ, Jenkins NA, Copeland NG, Welgus HG, Senior RM, Ley TJ (1992) Molecular cloning, chromosomal localization, and bacterial expression of a murine macrophage metalloelastase. J Biol Chem 267:4664–4671

Shapiro SD, Kobayashi DK, Ley TJ (1993) Cloning and characterization of a unique elastolytic metalloproteinase produced by human alveolar macrophages. J Biol Chem 268:23824–23829

Sitkovsky MV, Henkart P (eds) (1993) Cytotoxic cells: recognition, effector function, generation and methods. Birkhäuser, Boston

Sorimachi H, Kimura S, Kinbara K, Kazama J, Takahashi M, Yajima H, Ishiura S, Sasagawa N, Nonaka I, Sugita H, Maruyama K, Suzuki K (1996b) Structure and physiological functions of ubiquitous and tissue-specific calpain species: muscle-specific calpain, p94 interacts with connectin/titin. Adv Biophys 33:101–122

Sorimachi H, Suzuki K (1998) Mu-Calpain. In: Barrett AJ, Rawlings ND, Woessner JE (eds) Handbook of proteolytic enzymes. Academic, San Diego, pp 643–649

Srivastava PN, Adams CE, Hartree IF (1965) Enzymatic action of acrosomal preparations on the rabbit ovum in vitro. J Reprod Fertil 10:61–67

Stambaugh R, Buckley J (1968) Zona pellucida dissolution enzymes of the rabbit sperm head. Science 161:585–586

Starkey PM, Barrett AJ (1976a) Neutral proteinases of human spleen. Purification and criteria for homogeneity of elastase and cathepsin G. Biochem J 155:255–263

Starkey PM, Barrett AJ (1976b) Human cathepsin G: catalytic and immunological properties. Biochem J 155:273–278

Stenflo J (1998) Coagulation Factor X. In: Barrett AJ, Rawlings ND, Woessner JE (eds) Handbook of proteolytic enzymes. Academic, San Diego, pp 163–167

Steven F (1998) Guanidinobenzotase. In: Barrett AJ, Rawlings ND, Woessner JE (eds) Handbook of proteolytic enzymes. Academic, San Diego, pp 230–232

Stone SR, Le Bonniec B F (1998) Thrombin. In: Barrett AJ, Rawlings ND, Woessner JE (eds) Handbook of proteolytic enzymes. Academic, San Diego, pp 168–174

Stubbs MT, Bode W (1993) A player of many parts: the spotlight falls on thrombin's structure. Thromb Res 69:1–58

Suzuki F, Murakami K, Nakamura Y, Inagami T (1998) Renin. In: Barrett AJ, Rawlings ND, Woessner JE (eds) Handbook of proteolytic enzymes. Academic, San Diego, pp 851–856

Thornberry NA (1994) Interleukin-1b converting enzyme. Methods Enzymol 244:615–631

Thornberry NA (1998) Caspase-1. In: Barrett AJ, Rawlings ND, Woessner JE (eds) Handbook of proteolytic enzymes. Academic, San Diego, pp 732–737

Thornberry NA, Lazebnik Y (1998) Caspases: enemies within. Science 281:1312–1316

Thornberry NA, Molineaux SM (1995) Interleukin-1b converting enzyme: a novel cysteine protease required for IL-1b production and implicated in programmed cell death. Protein Sci 4:3–12

Tschesche H (1995) Human neutrophil collagenase. Methods Enzymol 248:431–449

Tschesche H, Pieper M (1998) Neutrophil collagenase. In: Barrett AJ, Rawlings ND, Woessner JE (eds) Handbook of proteolytic enzymes. Academic, San Diego, pp 1162–1167

Tschesche H, Knäuper V, Krämer S, Michaelis J, Oberhoff R, Reinke H (1992) Latent collagenase and gelatinase from human neutrophils and their activation. Matrix Suppl 1:245–255

Turner AJ, Murphy LJ (1996) Molecular pharmacology of endothelin converting enzymes. Biochem Pharmacol 51:91–102

Urch UA (1991) Biochemistry and function of acrosin. In: Wassarman PM (ed) Elements of mammalian fertilization. basic concepts, vol 1. CRC, Boston, pp 233–248

Van Wart HE (1992) Human neutrophil collagenase. Matrix Suppl 1:31–36

Wachtfogel YT, DeLa Cadena RA, Colman RW (1993b) Structural biology, cellular interactions and pathophysiology of the contact system. Thromb Res 72:1–21

Waydhas C, Nast-Kolb D, Jochum M, Trupka A, Lenk S, Fritz H, Duswald K-H, Schweiberer L (1992) Inflammatory Mediators, Infection, Sepsis, and Multiple Organ Failure After Severe Trauma. Arch Surg 127:460–467

Waydhas C, Nast-Kolb D, Gippner-Steppert C, Trupka A, Pfundstein C, Schweiberer L, Jochum M (1998) High-dose antithrombin III treatment of severely injured patients: results of a prospective study. J Truama 45:931–940

Whaley K, Loos M, Weiler J (1992) Complement in health and disease, 2nd edn. Kluwer, Amsterdam

Windsor LJ, Grenett H, Birkedal-Hansen B, Bodden MK, Engler JA, Birkedal-Hansen H (1993) Cell type-specific regulation of SL-1 and SL-2 genes. Induction of the SL-2 gene but not the SL-1 gene by human keratinocytes in response to cytokines and phorbolesters. J Biol Chem 268:17341–17347

Woessner JF (1998) Matrilysin. In: Barrett AJ, Rawlings ND, Woessner JE (eds) Handbook of proteolytic enzymes. Academic, San Diego, pp 1183–1187

Woessner JF Jr (1991) Matrix metalloproteinases and their inhibitors in connective tissue remodeling. FASEB J 5:2145–2154

The Role of Metalloprotease Inhibitors in Cancer and Chronic Inflammatory Diseases

H.S. Rasmussen and K.P. Lynch

A. Introduction

Matrix metalloproteinases (MMPs) are a family of related enzymes that are secreted by connective tissue cells, inflammatory phagocytes, and a number of different transformed cells (Woessner 1991). They are termed metalloenzymes as each enzyme contains a zinc atom within a highly conserved active site. Under normal physiologic circumstances, MMPs are involved in turnover and remodeling of the extracellular matrix (ECM) and collectively are capable of breaking down the components of the ECM, including collagen, laminin, fibronectin, elastin and serpins (Matrisian and Hogan 1990; Woessner 1991). MMPs are also intimately involved in angiogenesis by a number of different mechanisms, including the breakdown of basement membranes and the facilitation of vascular invasion and tubule formation (Fischer et al. 1994; Wojtowicz-Praga et al. 1997).

The description of the human MMP family has been a dynamic process, beginning with the now well-characterized interstitial collagenase (MMP-1), which degrades fibrillar collagens, through to the more recent identification of membrane-type (MT-) MMPs, which are structurally linked to the cell surface and whose functions remain poorly defined. Presently, 15 different MMPs have been identified, varying in their substrate requirement and potency (Table 1) (Sato et al. 1994; Brown and Giavazzi 1995; Strongin et al. 1995; Takino et al. 1995; Will and Hinzman 1995; Cossins et al. 1996; Puente et al. 1996). As one would expect of enzymes with such destructive potential, their secretion and activity is tightly regulated through control of gene expression and their secretion as latent proenzymes which require modification of a 10-kDa amino terminal domain for the expression of enzyme activity (Kleiner and Stetler-Stevenson 1993). Once activated, MMPs are further regulated by both general protease inhibitors and a group of at least four specific MMP inhibitors known as tissue inhibitors of metalloproteinases (TIMPs) (Gomez et al. 1997).

Abnormal expression and regulation of MMP activity has been documented in a number of different diseases in which an increased breakdown of ECM or increased angiogenesis is believed to play a role, including progression of malignant tumors, certain inflammatory diseases [including rheumatoid arthritis (RA) and inflammatory bowel disease], periodontal disease, atherosclerosis, congestive heart failure, corneal disease, various fibrotic

Table 1. Matrix metalloproteinases (MMPs)

Enzyme	MMP number	Main substrate(s)
Interstitial collagenase	MMP-1	Fibrillar collagens
Gelatinase-A	MMP-2	Fibronectins, type-IV collagens
Stromelysin-1	MMP-3	Non-fibrillar collagen, laminin, fibronectin
Matrilysin	MMP-7	Laminin, fibronectin, non-fibrillar collagen
Neutrophil collagenase	MMP-8	Fibrillar collagens, types I, II and III
Gelatinase-B	MMP-9	Type-IV and -V collagens
Stromelysin-2	MMP-10	Laminin, fibronectin, non-fibrillar collagen
Stromelysin-3	MMP-11	Serpin
Metalloelastase	MMP-12	Elastin
Collagenase-3	MMP-13	Fibrillar collagens
MT-MMP	MMP-14	Progelatinase A
MT2-MMP	MMP-15	Not defined
MT3-MMP	MMP-16	Progelatinase A
MMT4-MMP	MMP-17	Not defined
–	MMP-18	Not defined

diseases and certain vascular abnormalities. Not surprisingly, treatment with inhibitors of MMPs (MMPIs) has been proposed for a number of these diseases, and, in particular, early clinical studies in cancer have attracted great interest.

B. MMP Expression in Disease

I. Cancer

A large body of experimental data indicates that excessive MMP activity is important in the invasion and metastasis of a range of solid tumors, including colorectal cancer (Urbanski et al. 1993; Zeng and Guillem 1995; Zeng et al. 1996), cervical cancer (Nuovo et al. 1995), non-small cell lung cancer (Urbanski et al. 1992; Brown et al. 1993), breast cancer (Davies et al. 1993b), bladder cancer (Davies et al. 1993c), skin cancer (Hamdy et al. 1994), brain cancer (Rao et al. 1993; Rutka et al. 1995) and non-Hodgkin malignant lymphoma (Kossakowska et al. 1993). Furthermore, a number of these studies have shown good correlation between the levels of MMPs and the aggressiveness or invasiveness of the tumor (Brown et al. 1993; Kossakowska et al. 1993; Rao et al. 1993; Hamdy et al. 1994; Rutka et al. 1995). Probably the most widely studied MMPs in human malignancy have been gelatinase A and B (MMPs 2 and 9), formerly known as type-IV collagenases. Most of these studies have not, however, clarified whether these enzymes are more or less important than any of the other human MMPs. It is hoped that future studies will identify individual MMPs or groups of MMPs that play a key role in the progression of particular cancer types. In the meantime, the enzymes clearly represent an attractive pharmacological target, although the optimum profile of enzyme inhibition required remains unclear. Both broad spectrum and more selective MMPIs have been shown to be effective in animal cancer models, and a number of different compounds are now being studied in clinical trials.

II. Arthritis

Elevated levels of MMP-9 and MMP-3 have been consistently demonstrated in the synovial fluid of patients with RA (KOOLWIJK et al. 1995), as well as patients with osteoarthritis (OA)(MANICOURT et al. 1994). Importantly, these MMPs have been localized at the cartilage-pannus junction near the site of the active joint destruction in RA (KOOLWIJK et al. 1995). MMP-9, produced by osteoclasts, plays a significant role in normal bone remodeling and pathologic bone resorption, and it has been hypothesized that excessive MMP-9 production contributes to the erosion of bone and articular cartilage, hallmarks of RA and OA. Indeed, MMPIs have been shown to slow development and progression of animal models of these diseases (ESSER et al. 1997; LEWIS et al. 1997).

III. Inflammatory Bowel Disease

In the normal intestine, MMP activity is negligible, whereas excessive amounts of MMP-9 have been detected in the inflammatory infiltrate in those with Crohn's disease (BAILEY et al. 1994). MMP-3 and MMP-9 have been demonstrated to be present in high concentrations in the lamina propria regions of inflamed mucosa in ulcerative colitis (BAILEY et al. 1994).

IV. Atherosclerosis

Recent studies have demonstrated the presence of MMPs in vulnerable human atherosclerotic plaques (HENNEY et al. 1991). It is thought that the enzymes are secreted by activated macrophages and are, in part, responsible for the ECM degradation which weakens the plaque's fibrous cap and subsequently leads to plaque rupture (SEPPO et al. 1995). The plaque disruption occurs most frequently at sites where the fibrous cap is the thinnest and most heavily infiltrated with so-called "foam cells" (ZORINA et al. 1995). It has therefore been suggested that MMP release by these cells plays an important role in the development of plaque vulnerability and rupture (ZORINA et al. 1995). Again, the use of MMPIs seems an attractive possibility in the treatment of this disease.

Excessive production of MMPs has also been demonstrated in pre-eclampsia (VETTRAINO et al. 1996), bronchiectasis (SEPPER et al. 1995), emphysema (SHAPIRO 1994), liver cirrhosis (TAKAHARA et al. 1995), graft-versus-host disease (HATTORI et al. 1997), and periodontal disease (INGMAN et al. 1996). The potential clinical utility of these experimental findings remain speculative.

C. General Considerations in the Development of MMPIs

As data have accumulated regarding the role of MMPs in a range of disease processes, so interest in the design and development of inhibitors of the enzymes has grown. The possible use of the endogenous MMPIs (TIMP-1 and TIMP-2) has been explored, although difficulties with pharmaceutical delivery of proteins have precluded their further development. Drug design programs have instead focused on the development of synthetic MMPIs.

Two particular issues have dominated thoughts in recent years regarding the design of synthetic inhibitors; first, the desire for compounds with good oral bioavailability and, second, the development of compounds with selective inhibitory activity against individual MMPs. X-ray crystallography data on the three-dimensional structure of the collagenase active site has greatly assisted drug design (GRAMS et al. 1996). Less fortunately, understanding of the roles and importance of individual MMPs lags well behind the drive to introduce "selective" inhibitors of these enzymes into the clinic. There remains a dichotomy of opinion as to whether a broad-spectrum or selective inhibitory approach is better. The latter has the potential to reduce the side effects seen with some broad-spectrum MMPIs. However, considering the significant overlap in substrate activity between the different enzymes (BECKETT et al. 1996) and the uncertainty as to which enzymes are most important to inhibit, more selective inhibitors may be less effective.

In practice, pharmaceutical companies have chosen to develop both broad-spectrum and selective inhibitors. More-selective inhibitors have been developed, usually by design of compounds with selective loss of activity against one or more enzymes. Several compounds have now entered clinical trials (Table 2) and results arising in the next half decade will have great influence on the design of future drug development programs. Most clinical data are available on the drugs batimastat and marimastat; so, this review concentrates largely on these agents.

Batimastat is a broad-spectrum MMPI with potent activity against most of the major MMPs, including interstitial collagenase (MMP-1) (IC_{50} 3 nM),

Table 2. Matrix metalloproteinases inhibitors in development

Drug	Company	Indication	Phase
Marimastat	British Biotech	Cancer	Phase III
Bay 12-9566	Bayer	Cancer	Phase I/II
CGS27023A	Novartis	Cancer	Phase I
AG3340	Agouron	Cancer	Phase I
OPB-3206	Otsuka	Cancer	Preclinical
KB 7785	Kanebo	Cancer	Preclinical
Ro32–3555	Roche	Arthritis	Phase I/II
Bryostatin-1	–	Cancer	Phase I
Metastat	Collagenex	Cancer	Preclinical
Squalamine	Magainin pharmaceuticals	Cancer	Preclinical

stromelysin-1 (MMP-3) (IC_{50} 20 nM), gelatinase A (MMP-2) (IC_{50} 4 nM), gelatinase B (MMP-9) (IC_{50} 4 nM), and matrilysin (MMP-7) (IC_{50} 6 nM). There is also evidence that batimastat is a potent inhibitor of progelatinase A (MMP-14) (unpublished observations). The molecular structure of batimastat is shown in Fig. 1. The molecule mimics the substrate of the MMPs, and works as a competitive, reversible inhibitor. Batimastat is poorly soluble and consequently shows minimal bioavailability when administered orally. Therefore, batimastat must be administered by direct injection into body spaces such as the peritoneal and pleural cavities. Intra-peritoneal injection of batimastat gives rise to elevated and sustained plasma concentrations with a half-life in humans of approximately 28 days, because the drug is gradually absorbed from the peritoneal cavity into the bloodstream. While this pharmacokinetic profile is less suitable for clinical trials in most cancers, it is a convenient administration schedule for rodent studies.

Marimastat (Fig. 2) is another broad-spectrum MMPI with an enzyme inhibitory spectrum very similar to its predecessor batimastat. The compound differs structurally from batimastat in the group adjacent to the hydroxamate moiety, and the group at the P2′ position. These changes render the compound soluble and, thus, more suitable for oral administration. The compound is, however, rapidly metabolized in rodents, undergoing a high first-pass effect, and therefore testing of marimastat in rodent models of cancer is difficult as sustained concentrations in this species are difficult to obtain by oral dosing. Consequently, much of the preclinical anti-tumor data have been generated

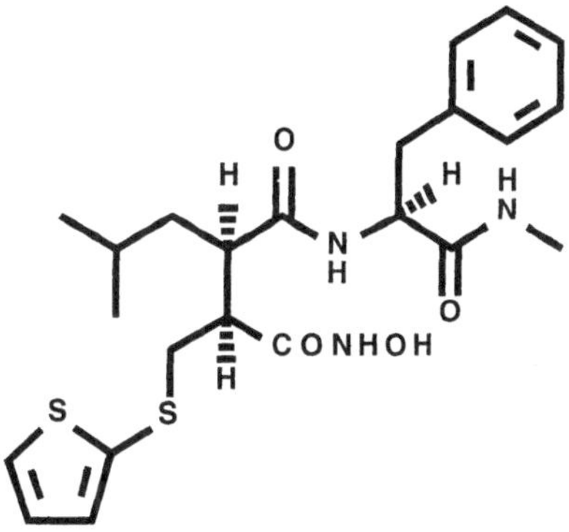

Fig. 1. The molecular structure of batimastat

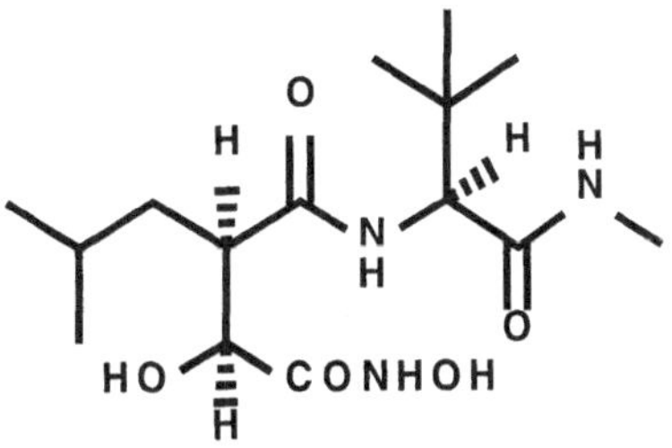

Fig. 2. The molecular structure of marimastat

with batimastat, although more recently delivery of marimastat via a mini osmotic pump has allowed effective study of the compound in animal models.

D. Preclinical Evidence of Anti-Tumor Activity of MMPIs

In a murine melanoma model, in which mouse B16 melanoma cells were injected subcutaneously in mice, batimastat significantly reduced the growth of the implanted tumors; when batimastat was administered from day 11–19 after tumor implantation, a 33% reduction in tumor growth was noted, whereas in animals administered batimastat from the day of tumor inoculation and up to day 19, a 58% reduction was detected (CHIRIVI et al. 1994). In the same series of experiments, batimastat also significantly inhibited the formation of spontaneous metastases (CHIRIVI et al. 1994). Animals were randomized to either batimastat (30 mg/kg daily for 18 days after surgical removal of the primary tumor) or control (saline); a 76% reduction in metastatic tumor burden was detected in the batimastat-treated animals ($P < 0.05$).

The anti-angiogenic activity of batimastat has been assessed in a murine End.1 hemangioma model (TARABOLETTI et al. 1995). End.1 is a virus-oncogene-transformed mouse endothelial cell line which forms vascular lesions similar to hemangiomas. End.1 tumors develop as they stimulate the host cell to create new blood vessels. This pro-angiogenic effect can be quantified and therefore represents a good model in which to assess anti-angiogenic activity of developmental compounds. Batimastat treatment resulted in a significant decrease in hemoglobin content (a quantitative measure for angiogenesis in this model) from 0.80 g/dl to 0.53 g/dl compared with control. These data indicate that batimastat possesses anti-angiogenic properties (TARABOLETTI et al. 1995).

In a xenograft model of human ovarian carcinoma, in which carcinoma cells were implanted in the peritoneum of nude mice, animals were randomly allocated to receive batimastat (30 mg/kg) or saline daily intra-peritoneally from day 7 to day 20 after tumor implantation. Batimastat significantly increased survival from 18 days in the control group to 105 days in the batimastat-treated group ($P < 0.001$). Also, histological analysis demonstrated a significant increase in both intra-tumoral and peri-tumoral fibrosis, and a subsequent reduction in tumor cells (DAVIES et al. 1993a).

In a xenograft model of human colorectal carcinoma, carcinoma cells were implanted in the intestinal wall of nude mice (WANG et al. 1994). Administration of intra-peritoneal batimastat resulted in a 50% reduction in tumor growth as well as a significant decrease in the incidence of local and regional invasion and distant metastases compared with saline control (WANG et al. 1994), leading to a significant increase in median survival ($P < 0.05$). Similar findings were observed in a xenograft breast cancer model, in which cancer cells were implanted in the mammary fat pad of nude mice. Nine weeks after implantation, the primary tumors were resected and the mice were allocated to receive batimastat or saline, administered intra-peritoneally daily from week 9 to week 16. On week 16, all animals were killed and autopsied for lung metastases (SLEDGE et al. 1995). In the batimastat-treated animals, significant reductions in local regrowth ($P = 0.035$), number of lung metastases (46% reduction, $P < 0.0001$), volume of lung metastases (60% reduction, $P < 0.0001$), and incidence of animals without any lung metastases (33% vs 9%, $P < 0.05$) were found compared with controls. Batimastat has also been studied in a xenograft model of pancreatic cancer. Cancer cells were implanted in nude mice (E. Zervos et al., unpublished observations), which were then allocated to receive batimastat or vehicle, starting 4 days prior to implantation and continuing until death or sacrifice on day 70. Batimastat resulted in a complete prevention of metastases from 20 in the control group to 0 in the batimastat group ($P < 0.05$); furthermore, a significant reduction in total tumor weight (from 0.65 g to 0.14 g, $P < 0.05$) as well as tumor volume (from 0.60 cc to 0.21 cc, $P < 0.05$) was detected.

It is important to note that several of these studies most closely reflect treatment in an adjuvant setting. In this context, a study using a syngeneic model of rat mammary carcinoma is of particular interest (ECCLES et al. 1996). These investigators examined the effects of short- and long-term administration of batimastat on metastasis and tumor growth rates. Batimastat was started immediately prior to removal of the primary mammary tumor, which had been implanted adjacent to the mammary fat pad, and continued for either 7 days (short course) or 58 days (long course). Short-term treatment resulted in fewer lung metastases than animals receiving saline control, although about half these animals developed metastatic disease. Conversely, only one animal treated with long-term batimastat developed metastases. In this animal, the study drug had been discontinued on day 45 because of peritoneal irritation. These results imply that as an adjuvant treatment, MMPIs may be best administered as a long-term or continuous therapy.

E. Clinical Studies with MMPIs

I. Design Considerations in Cancer

Designing a clinical trial program for the development of MMPIs in cancer is complex. Traditionally, when one develops a new cytotoxic agent, the purpose of the phase-I program is to define the maximum tolerated dose (MTD) and dosing schedule. As these drugs are cytostatic rather than cytotoxic in their actions, chronic administration may be required before anti-tumor activity can be evaluated. Moreover, in the absence of tumor cell death, the application of standard tumor reductive criteria for efficacy evaluation in this class of agents would be inappropriate. Finally, following complete enzyme inhibition, no further benefit would be expected when escalating the dose to levels that induce toxicity. Consequently, the purpose of the phase-I/II program with a drug of this class should be to identify an optimal biologic dose range rather than an MTD. After this is established, drug efficacy is best confirmed in randomized trials utilizing definitive primary end points such as survival or time to disease progression.

Such definitive trials are invariably large, of long duration, and costly. The challenge for drug companies involved in the development of these drugs is to identify a valid surrogate for use in a phase-II trial program that may confirm the therapeutic potential of the drug, as well as identifying an optimum dose range for investigation in these larger studies. One such approach is described below, and undoubtedly different approaches will be used as more of these agents enter clinical trials. Others have argued for a move directly from phase-I to phase-III trials, and at least one company is thought to have taken this bold strategy.

The special design problems described above are to a large extent unique for the development of MMPIs in cancer. In most other potential targets such as RA and inflammatory bowel disease relatively reliable surrogate markers do exist which can be employed in the phase-II program to help identify the optimal dose and get an indication of activity. Development in these indications should be somewhat simpler than in cancer.

II. Batimastat

Due to its pharmacokinetic limitations, batimastat has only been studied in selected diseases in which intra-cavity injection was appropriate. In patients with malignant ascites (BEATTIE et al. 1994), as well as malignant pleural effusion (MACAULAY et al. 1995), batimastat was injected directly into the peritoneal and pleural cavities. After intraperitoneal injection of marimastat, high and sustained plasma concentrations were achieved, detectable for up to 28 days after the injection, probably due to sustained release with subsequent absorption from the peritoneum (BEATTIE et al. 1994). In these patients, batimastat was well tolerated although some local pain was observed. The observation that intra-peritoneally administered batimastat led to sustained plasma concentration levels within the predicted therapeutic range led to a phase-I study in patients with advanced lung cancer (WOJTOWICZ-PRAGA et al. 1996). However, in those patients who did not have ascites, batimastat resulted in substantial local toxicity. No systemic toxicity was noted. Nonetheless, the insolubility of batimastat necessitating intra-peritoneal or intra-pleural administration, together with the development of an orally bioavailable drug with a similar inhibitory profile, resulted in cessation of further development of batimastat.

III. Marimastat

Like batimastat, marimastat is a broad-spectrum MMPI, although has a pharmacokinetic profile suitable for oral administration. It displays potent inhibitory activity against most of the major MMPs with IC_{50}s against MMP-1, MMP-2, MMP-3, MMP-7 and MMP-9 in the nanomolar range. Marimastat differs from batimastat in being a weaker inhibitor of stromelysin-1 (IC_{50} for marimastat is approximately 230 nM compared with 20 nM for batimastat). Whether this weaker activity against stromelysin-1 has any clinical implications is unknown.

As marimastat is not a cytotoxic agent, the initial phase-I work was done in healthy volunteers. The results of two of these studies have been published, showing a linear dose–plasma concentration relationship and a half-life of approximately 10 h (MILLAR et al. 1998). Plasma concentrations at all dose levels studied were well in excess of IC_{90} concentrations obtained in preclinical studies, indicating that oral administration of marimastat produces pharmacologically active blood levels.

Marimastat has been tested in more than 500 patients in phase-I/II studies in a number of different solid tumors in North America and Europe. In most of the studies, the effects of marimastat on cancer antigens (CA) [carcinogenic embryonic antigen (CEA) in colorectal cancer, CA 19/9 in pancreatic cancer, CA 125 in ovarian cancer and prostate-specific antigen (PSA) in prostate cancer] were assessed as a surrogate marker of biological activity.

Combined analysis of six of these studies indicated that marimastat treatment reduced the rates of rise of CAs in a dose-dependent fashion (NEMUNAITIS et al. 1998). It was concluded that a dose range of between 10 mg and 25 mg twice daily was appropriate for longer-term studies. These results were largely consistent with findings in each of the individual studies. A relationship was also observed between CA rate of rise and survival, indicating that such changes may be a valid surrogate end point for assessing potential drug activity in patients with cancer. It is noteworthy that these studies were conducted in patients with advanced, rapidly progressive, treatment refractory cancer, a group of patients which is notoriously difficult to treat.

A more optimal setting for the use of marimastat would be in patients with earlier stage disease and smaller tumor burdens. This strategy is currently being pursued in a number of randomized studies in which marimastat is being tested in patients with earlier stage disease, including a true adjuvant study in patients with resectable pancreatic cancer.

In clinical trials, marimastat has been generally well tolerated. The most common drug-related toxicity is a characteristic syndrome, consisting of musculo-skeletal pain and stiffness, often commencing in the small joints in the hand and, if dosing continues unchanged, tending to spread to other joints. At high and continuous dosages, the musculo-skeletal toxicity can be severe and may resemble "an inflammatory polyarthritis" (WOJTOWICZ-PRAGA et al. 1998). The symptomatology is similar to that observed in 6-month toxicology studies in the marmoset (unpublished data). The musculo-skeletal side effects occurred more commonly, were generally more severe, and developed more rapidly with higher doses of marimastat. At a dose of 10 mg twice daily, approximately 30% developed musculo-skeletal symptoms after 3–5 months of treatment. However, implementing a short dosage holiday of a few weeks, followed by dose reduction appears to make continued treatment possible in the majority of patients.

These encouraging phase-I/II data have provided the foundation for an extensive phase-III trial program; pivotal trials have been initiated in patients with advanced pancreatic cancer, gastric cancer, malignant glioblastoma, small cell lung cancer, non-small cell lung cancer, ovarian cancer and breast cancer. All of these trials have as their primary end points survival or disease progression and, when completed, should provide the first step towards defining the potential clinical use of this class of agents.

IV. Other MMP Inhibitors

A number of other MMP inhibitors are also in preclinical or clinical development for cancer. *AG3340* (Agouron Pharmaceuticals, Inc.) is a small synthetic, gelatinase "selective" MMPI. A phase-I study in healthy volunteers demonstrated that the drug was well tolerated and rapidly absorbed following oral dosages between 10 mg and 200 mg. A phase-I study in cancer patients has recently been initiated in the United States (Collier et al. 1997). *Bryostatins* are naturally occurring lactones derived from marine bryozoans with both cytotoxic and MMP inhibitory activity. Their MMP-inhibitory profile includes the enzymes MMP-1, -3, -9, -10 and 11 (Hornung et al. 1992). Bryostatin-1 has been tested in phase-I studies conducted by the NCI. So far, it has been well tolerated with myalgia being the principal toxicity. *CGS27023A* (Novartis) is an orally available broad-spectrum MMPI. A phase-I study in 36 patients with various tumors demonstrated that the drug was well tolerated, with myalgia/arthralgia and skin rashes being the major toxicities (Levitt et al. 1998). Like *AG3340, BAY 12-9566* is a gelatinase "selective" MMPI, with reduced influence on other MMPs (Rowinsky et al. 1998). So far, musculoskeletal toxicity has not been reported with BAY 12-9566, which may be due to its greater enzyme selectivity than the broader-spectrum compounds. Whether this greater selectivity reduces the clinical activity is an open question which will have to await comparative trials with a broad-spectrum inhibitor. *Ro32-3555* (Roche) is an MMPI with relatively weak activity against gelatinase A and stromelysin-1 but good activity against interstitial collagenase. The drug has activity in animal models of arthritis (Lewis et al. 1997), and good tolerability and oral bioavailability in healthy volunteers has been reported (Wood et al. 1996). Finally, *OPB-3206* (Otsuka) is another selective MMPI, inhibiting MMP-2 and MMP-9. The drug has shown anti-proliferative and anti-metastatic activity in preclinical models (Shono et al. 1997). OPB-3206 has not entered the clinic yet.

F. Clinical Studies in Non-Cancer Indications

Clinical studies in indications other than cancer have not yet been reported. However, most of the companies with MMPIs in clinical development are considering exploratory studies in indications such as arthritis, inflammatory bowel disease, periodontal disease, graft-versus-host reaction, aortic aneurysm and congestive heart disease; in all of these indications, there is a strong biological rationale for the use of MMPIs and, in some of the indications, there exists preclinical evidence suggesting that a MMPI might be effective in man. For most of these diseases, well-validated surrogate markers exist which would allow the use of relatively simple designs in early studies. It is anticipated that results of the use of MMPIs in one or more of these indications will start to emerge within the next few years.

G. Conclusions

There is a growing body of evidence confirming that excessive production of MMPs plays an important role in the growth and spread of malignant tumors, including colorectal, lung, breast, cervical and prostate cancers. Inhibitors of these enzymes have proven effective in a range of preclinical cancer models (ovarian, colorectal, brain, lung, pancreas, gastric, melanoma), slowing the growth of the tumor as well as reducing the incidence of metastases. Some data suggest that the optimal setting for drugs of this nature is in earlier-stage disease or tumors of low volume, and that longer-term treatment has advantages over short-term therapy. It is nevertheless clear that these agents represent a promising possibility for an additional weapon in the treatment of cancer. Phase-I/II studies in patients with advanced cancers have demonstrated that the drugs are generally well tolerated without the toxicity which characterizes traditional cytotoxic agents. Randomized clinical trials are now underway to establish their potential efficacy.

Theoretically, MMPIs may also be useful in the treatment of arthritis, inflammatory bowel disease, periodontal disease, graft-versus-host reaction, and some cardiological diseases; however, the research of these indications remains predominantly at the preclinical stage. Further preclinical and early exploratory clinical work is now required.

References

Bailey CJ, Hembry RM, Alaxander A, et al. (1994) Distribution of the matrix metalloproteinases stromelysin, gelatinases A and B, and collagenase in Crohn's disease and normal intestine. J Clin Pathol 47:113–116

Beattie GJ, Young HA, Smyth JF (1994) Phase I study of intra-peritoneal matrix metalloproteinase inhibitor in patients with malignant ascites (abstract). Proceedings of the 8th National Cancer Institute Canada–European Organization for the Research and Treatment of Cancer. Symposium on New Drug Development, Amsterdam

Beckett RP, Davidson AH, Drummond AH, et al. (1996) Recent advances in matrix metalloproteinase inhibitor research. Drug Dev Ther 1:16–26

Brown PD, Giavazzi R (1995) Matrix metalloproteinase inhibition: a review of antitumor activity. Ann Oncol 6:967–974

Brown PB, Bloxodge RE, Stuart NSA, et al. (1993) Association between expression of activated 72-kilodalton gelatinase and tumor spread in non-small cell lung carcinoma. J Natl Cancer Inst 85:574–578

Chirivi RGS, Garofalo A, Crimmin MJ, et al. (1994) Inhibition of the metastatic spread and growth of B-16-BL6 murine melanoma by a synthetic matrix metalloproteinase inhibitor. Int J Cancer 58:460–464

Collier MA, Yuen GJ, Bansal SK, et al. (1997) A phase I study of the matrix metalloproteinase inhibitor AG3340 given in single doses to healthy volunteers. In: Proceedings of the Am Assoc Cancer Res 38:221

Cossins J, Dudgeon TL, Catlin G, et al. (1996) Identification of MMP-18, a putative novel human matrix metalloproteinase. Biochem Biophys Res Commun 228:494–498

Davies B, Brown P, East N, et al. (1993a) A synthetic matrix metalloproteinase inhibitor decreases tumor burden and prolongs survival of mice bearing human ovarian carcinoma xenografts. Cancer Res 53:2087–2091

Davies B, Miles DW, Happerfield LC, et al. (1993b) Activity of type IV collagenases in benign and malignant breast disease. Br J Cancer 67:1126–1131

Davies B, Waxman J, Wasan H, et al. (1993c) Levels of matrix metalloproteinases in bladder cancer correlate with tumor grade and invasion. Cancer Res 53:5365–5369

Eccles SA, Box GM, Court WJ, et al. (1996) Control of lymphatic and hematogenous metastases of a rat mammary carcinoma by the matrix metalloproteinase inhibitor batimastat (BB-94). Cancer Res 56:2815–2822

Esser CK, Bugianesi RL, Caldwell CG, et al. (1997) Inhibition of stromelysin-1 (MMP-3) by P1'-biphenylylethyl carboxyalkyl dipeptides. J Med Chem 40:1026–1040

Fischer C, Gilbertson-Beadling S, Powers EA, et al. (1994) Interstitial collagenase is required for angiogenesis in vitro. Dev Biol 162:499–510

Gomez DE, Alonso DF, Yoshiji H, Thorgeirsson UP (1997) Tissue inhibitors of metalloproteinases – structure, regulation and biological functions. Eur J Cell Biol 74:111–122

Grams F, Crimmin M, Hinnes L, et al. (1996) Structure determination and analysis of human neutrophil collagenase complexed with a hydroxamate inhibitor. Biochemistry 34:14012–14020

Hamdy FC, Fadlon D, Cottam D, et al. (1994) Localization of messenger RNA for Mr 72,000 and 92,000 type IV collagenases in human skin cancers by in situ hybridization. Cancer Res 52:1336–1341

Hattori K, Hirano T, Ushiyama C, et al. (1997) A metalloproteinase inhibitor prevents acute lethal graft-versus-host disease in mice. Blood 90:542–548

Henney AM, Wakeley PR, Davies MJ, et al. (1991) Localization of stromelysin gene expression in atherosclerotic plaques by in situ hybridization. Proc Natl Acad Sci USA 88:8154–8158

Hornung RL, Pearson JW, Beckwith M, et al. (1992) Preclinical evaluation of bryostatin as an anticancer agent against several murine tumor lines: in vitro versus in vivo activity. Cancer Res 52:101–107

Ingman T, Tervahartiala T, Ding Y, et al. (1996) Matrix metalloproteinases and their inhibitors in gingival crevicular fluid and saliva of periodontitis patients. J Clin Periodontol 12:1127–1132

Kleiner DE, Stetler-Stevenson WG (1993) Structural biochemistry and activation of matrix metalloproteinases. Curr Opin Cell Biol 5:891–897

Koolwijk P, Miltenburg AM, Van Erck MG, et al. (1995) Activated gelatinase-B (MMP-9) and urokinase-type plasminogen activator in synovial fluids of patients with arthritis. Correlation with clinical and experimental variables of inflammation. J Rheumatol 22:385–393

Kossakowska AE, Urbanski SJ, Watson A, et al. (1993) Patterns of expression of matrix metalloproteinases and their inhibitors in human malignant lymphomas. Oncol Res 5:19–28

Levitt NC, Eskens F, Propper DJ, et al. (1998) A phase one pharmacokinetic study of CGS27023A, a matrix metalloproteinase inhibitor (abstract). Proceedings of Am Soc Clin Oncol 17:823

Lewis EJ, Bishop J, Bottomley KM, et al. (1997) Ro 32–3555, an orally active collagenase inhibitor, prevents cartilage breakdown in vitro and in vivo. Br J Pharmacol 121:540–546

Macaulay VM, O'Byrne KJ, Saunders MP, et al. (1995) Phase I study of the matrix metalloproteinase inhibitor batimastat in patients with malignant pleural effusion. Br J Cancer 71:11

Manicourt DH, Fujimoto N, Obata K, et al. (1994) Serum levels of collagenase, stromelysin-1, and TIMP-1. Age- and sex-related differences in normal subjects and relationship to the extent of joint involvement and serum levels of antigenic keratan sulfate in patients with osteoarthritis. Arthritis Rheum 37:1774–1783

Matrisian LM, Hogan BL (1990) Growth factor-regulated proteases and extracellular matrix remodeling during mammalian development. Curr Top Dev Biol 24:219–259

Millar AW, Brown PD, Moore J, et al. (1998) Results of single and repeat dose studies of the oral matrix metalloproteinase inhibitor marimastat in healthy male volunteers. Br J Clin Pharmacol 45:21–26

Nemunaitis J, Poole C, Primrose J, et al. (1998) Combined analysis of studies of the effects of the matrix metalloproteinase inhibitor marimastat on serum markers in advanced cancer: selection of a biologically active and tolerable dose for longer-term studies. Clin Cancer Res 4:1101–1109

Nuovo GJ, MacConnell PB, Simsir A, et al. (1995) Correlation of the in situ detection of polymerase chain reaction-amplified metalloproteinase complementary DNS and their inhibitors with prognosis in cervical carcinoma. Cancer Res 55:267–275

Puente XS, Pendas AM, Llano E, et al. (1996) Molecular cloning of a novel membrane-type matrix metalloproteinase from a human breast carcinoma. Cancer Res 56:944–949

Rao JS, Steck PA, Mohanam S, et al. (1993) Elevated levels of Mr 92,000 type IV collagenase in human brain tumors. Cancer Res 53:2208–2211

Rowinsky E, Hammond L, Aylesworth C, et al. (1998) Prolonged administration of BAY 12–9566, an oral non peptidic biphenyl matrix metalloproteinase inhibitor: a phase I and pharmacokinetic (PK) study (abstract). Proceedings of Am Soc Clin Oncol 17:836

Rutka JT, Matsuzawa K, Hubbard SL, et al. (1995) Expression of TIMP-1, TIMP-2, 72- and 92-kDa type IV collagenase transcripts in human astrocytoma cell lines: correlation with astrocytoma and invasiveness. Int J Oncol 6:877–884

Sato H, Takino T, Okada M, et al. (1994) A matrix metalloproteinase expressed on the surface of invasive tumor cells. Nature 37:61–65

Seppo T, Kevin D, Marina F, et al. (1995) Interstitial collagenase expression in human carotid atherosclerosis. Circulation 92:1393–1398

Sepper R, Konttinen YT, Ding Y, et al. (1995) Human neutrophil collagenase (MMP-8), identified in bronchieectasis BAL fluid, correlates with the severity of disease. Chest 107:1641–1647

Shapiro SD (1994) Elastolytic metalloproteinases produced by human mononuclear phagocytes. Potential roles in destructive lung disease. Am J Respir Crit Care Med 150:160–164

Shono T, Ono M, Jimi S, et al. (1997) A novel synthetic matrix metalloproteinase inhibitor OPB-3206: inhibition of tumor growth and metastases and angiogenesis (abstract). Proceedings of the Am Assoc Cancer Res 38:525

Sledge GW Jr., Qulali M, Goulet R, et al. (1995) Effect of the matrix metalloproteinase inhibitor batimastat on breast cancer regrowth and metastasis in athymic mice. J Natl Cancer Inst 87:1546–1550

Strongin A, Collier I, Bannikov G et al. (1995) Mechanism of cell surface activation of 72 kDa type IV collagenase. J Biol Chem 270:5331–5338

Takahara T, Furui K, Funaki J, et al. (1995) Increased expression of matrix metallo-proteinase-II in experimental liver fibrosis in rats. Hepatology 21:787–795

Takino T, Sato H, Shinagawa A, et al. (1995) Identification of the second membrane-type matrix metalloproteinase (MT-MMP2) gene from a human placenta cDNA library. MT-MMPs form a unique membrane-type subclass in the MMP family. J Biol Chem 270:23013–23020

Taraboletti G, Garofalo A, Belotti D, et al. (1995) Inhibition of angiogenesis and murine hemangioma growth by batimastat, a synthetic inhibitor of matrix metallopro-teinases. J Natl Cancer Inst 87:293–298

Urbanski SJ, Edwards DR, Maitland A, et al. (1992) Expression of metalloproteinases and their inhibitors in primary pulmonary carcinomas. Br J Cancer 66:1188–1194

Urbanski SJ, Edwards DR, Hershfield N, et al. (1993) Progression pattern of metallo-proteinases and other inhibitors changes with the progression of human sporadic colorectal neoplasia. Diagn Mol Pathol 2:81–89

Vettraino IM, Roby J, Tolley T, et al. (1996) Collagenase-I, stromelysin-I, and matrilysin are expressed within the placenta during multiple stages of human pregnancy. Placenta 17:557–563

Wang X Fu, Brown PD, Crimmin M, et al. (1994) Matrix metalloproteinase inhibitor BB-94 (batimastat) inhibits human colon cancer growth and spread in a patient-like orthotopic model in nude mice. Cancer Res 54:4726–4728

Will H, Hinzman B (1995) CDNA sequence and mRNA tissue distribution of a matrix metalloproteinase with a potential transmembrane segment. Eur J Biochem 231:602–608

Woessner JF, Jr. (1991) Matrix metalloproteinases and their inhibitors in connective tissue remodeling. FASEB 5:2145–2154

Wojtowicz-Praga S, Low J, Marshall J, et al. (1996) Phase I trial of a novel matrix met-alloproteinase inhibitor batimastat (BB-94) in patients with advanced cancer. Invest New Drugs 14:193–202

Wojtowicz-Praga SM, Dickson RB, Hawkins MJ (1997) Matrix metalloproteinase inhibitors. Invest New Drugs 15:61–75

Wojtowicz-Praga S, Torri J, Johnson M, et al. (1998) Phase I trial of marimastat, a novel matrix metalloproteinase inhibitor, administered orally to patients with advanced lung cancer. J Clin Oncol 16:2150–2156

Wood ND, Aitken M, Harris S, et al. (1996) The tolerability and pharmacokinetics of the cartilage protective agent (Ro32–3555) in healthy male volunteers. Br J Clin Pharmacol 42:676–677

Zeng ZS, Guillem JG (1995) Distinct pattern of matrix metalloproteinase 9 and tissue inhibitor of metalloproteinase 1 mRNA expression in human colorectal cancer and liver metastases. Br J Cancer 72:575–582

Zeng ZS, Huang Y, Cohen AM, et al. (1996) Prediction of colorectal cancer relapse and survival via tissue RNA levels of matrix metalloproteinase 9. J Clin Oncol 14:3133–3140

Zorina G, Galina S, Roger K, et al. (1995) Macrophage foam cells from experimental atheroma constitutively produce matrix-degrading proteinases. Proc Natl Acad Sci USA 92:402–406

CHAPTER 13

The Tumor Necrosis Factor-α Converting Enzyme

J.D. Becherer, M.H. Lambert, and R.C. Andrews

A. Biology of Tumor Necrosis Factor

I. Historical Perspective

Tumor necrosis factor (TNF)-α is a potent pro-inflammatory agent produced primarily by activated monocytes and macrophages (Vassalli 1992). Originally thought to be a selective anti-tumor agent (Old 1985) and a contributor to cachexia in cancer patients (Beutler and Cerami 1988), this protein, along with interleukin (IL)-1α, is recognized as a major inflammatory cytokine. Moreover, the production of TNF-α is not restricted to monocytes and macrophages. Other cellular sources of TNF-α include lymphocytes, mast cells, polymorphonuclear cells, astrocytes and microglial cells (Sung et al. 1988; Sawada et al. 1989; Gordon and Galli 1990; Djeu et al. 1990; Chung and Benveniste 1990). Increasing evidence suggests that overproduction of TNF-α is a major contributor to diverse pathologies such as septic shock, graft rejection, human immunodeficiency virus (HIV) infection and rheumatoid arthritis (RA) (Tracy et al. 1986; Piguet et al. 1987; Peterson et al. 1992; Williams et al. 1992).

The effects of TNF-α are mediated by either of two TNF receptors, TNFR1 or TNFR2. These receptors are type-II membrane proteins containing characteristic cysteine-rich repeat motifs in their extracellular domains. However, significant differences exist in their cytoplasmic domains. This accounts for the distinct and quite different effects induced by TNF upon binding, such as proliferation and apoptosis. Furthermore, since TNF receptors are present on most cell types in the body, the pleiotropic effects induced by TNF can be rationalized if not completely understood. Toward this end, mutant mice have been produced that lack TNFR1, TNFR2 or both. These mice clearly demonstrate the protective role of TNF in response to infections, as well as the potential toxicity associated with systemic exposure of TNF from uncontrolled infections. However, the most surprising finding is that these mutant mice develop and breed normally, suggesting that inhibition of TNF in disease states may have minimal side effects. These knockout mice have been thoroughly reviewed elsewhere (Bleuthmann 1998; Eugster et al. 1998; Shinbrot and Moore 1998).

II. The Role of TNF in Inflammatory Diseases

Characterization in the laboratory of the pro-inflammatory properties of TNF coincided with the identification of TNF in the synovial membrane and at the cartilage–pannus junction of patients with RA (CHU et al. 1991). Furthermore, cultured RA synovial cells spontaneously produced TNF, and neutralization of TNF led to a decrease in the production of other inflammatory cytokines such as IL-1 (BRENNAN et al. 1989). This was soon thereafter confirmed in animal models of arthritis (WILLIAMS et al. 1992), and intense efforts followed to prove this in the clinic. Initial success was demonstrated in patients with RA using a neutralizing antibody to TNF (ELLIOT et al. 1994). The continued clinical testing of several anti-TNF biological agents culminated in the approval of etanercept (ENBREL™ T) and infliximab (Remicade™ T). In clinical trials, both of these agents have successfully reduced the signs and symptoms experienced by RA patients (MORELAND et al. 1997; MAINI et al. 1998; WEINBLATT et al. 1999). The recent success of these biological agents that neutralize TNF has led to intense efforts to find small-molecule TNF antagonists that will mimic the efficacy, if not the mechanism, of these agents. Because TNF interacts at multiple contact points with either of its two receptors, researchers have struggled to find small-molecular-weight inhibitors that antagonize this interaction. Therefore, most efforts have focused on targets upstream of TNF synthesis or secretion and downstream of TNF receptor engagement, since these targets appear more amenable to modulation by small molecules (SHIRE and MULLER 1998).

B. Characterization of the TNF-α Converting Enzyme

I. Cell Secretion of TNF-α

After the cloning of the proteinase responsible for IL-1β secretion (see Chap. 17), a great deal of interest was directed toward understanding the mechanism by which TNF was synthesized and secreted. The translation of TNF-α messenger RNA (mRNA) produces a 26-kDa type-II transmembrane protein. Processing of this 26-kDa precursor by cleavage of the Ala^{76}–Val^{77} bond releases the 17-kDa carboxyl-terminal domain which is biologically active as a trimer. The amino-terminal 76-residue propeptide has a unique structure in that it lacks a typical signal peptide sequence (Fig. 1). Rather, it contains a central 25-residue transmembrane domain, while the Ala–Val processing site is approximately 20 residues removed from this region (KRIEGLER et al. 1988; CSEH et al. 1989; UTSUMI et al. 1993). By analogy with known processing enzymes, such as signal peptidase and KEX2/furins, it has been postulated that a serine-dependent peptidase may mediate the cleavage of the Ala–Val bond (SUFFYS et al. 1988; SCUDERI 1989; NIEHOERSTER et al. 1990; KIM et al. 1993). However, the 26-kDa precursor TNF-α is not co-translationally processed by

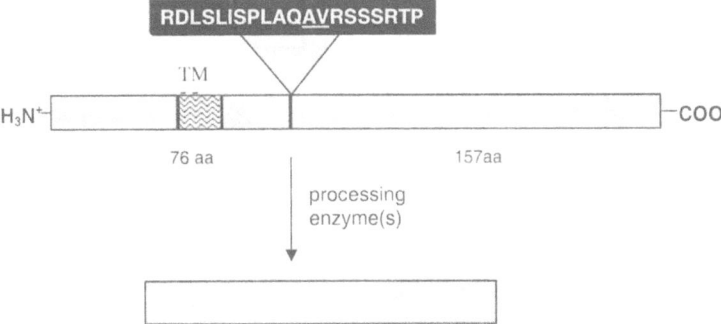

Fig. 1. Schema of tumor necrosis factor (TNF) processing. The Ala[76]–Val[77] scissile bond and surrounding sequence is highlighted. Cleavage of the 26-kDa precursor TNF molecule liberates the soluble 17-kDa secreted cytokine. The pro-domain is 76 amino acid residues in length and contains a transmembrane (TM) domain. The remaining molecule is comprised of 157 amino acid residues

signal peptidase, suggesting the presence of a novel pathway for TNF-α secretion (MULLER et al. 1986).

II. Purification and Cloning of TNF-α Converting Enzyme

The breakthrough findings, which eventually led to the identification of TNF-α converting enzyme (TACE), came in 1994 when several groups demonstrated that hydroxamate-based inhibitors of the matrix metalloprotease (MMP) family blocked TNF secretion from cells (GEARING et al. 1994; McGEEHAN et al. 1994; MOHLER et al. 1994). These inhibitors blocked TNF secretion from a variety of different cell types treated with various stimuli, indicating that a common target was being inhibited regardless of cell type. Furthermore, these inhibitors had no effect on TNF mRNA levels, unlike the serine protease inhibitors tested, indicating that they were not interfering with signal-transduction pathways or TNF transcription. The inhibitors also had a rapid onset of action. TNF secretion was inhibited more than 90% when cells were treated with the inhibitor GI 129471 90 min after lipopolysaccharide (LPS) stimulation of monocytes and well after the initial appearance of TNF in the media. Pulse-chase experiments in the presence of these hydroxamate inhibitors revealed that only the 26-kDa precursor form of TNF was present in the cell-associated fraction. Furthermore, this membrane-bound form of TACE existed only transiently and disappeared rapidly after 1 h (McGEEHAN et al. 1994). This transient appearance of the membrane-bound form of TNF is significant, since cell-surface TNF has been shown to be biologically active (PEREZ et al. 1990; DECOSTER et al. 1995).

Subsequently, TACE was purified independently by two groups who utilized their hydroxamate inhibitors in the purification scheme. Black and colleagues isolated THP-1 plasma membranes and used the inhibitor to profile

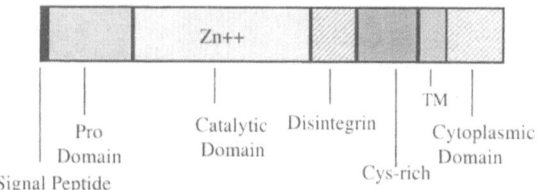

Fig. 2. Schematic representation of the domain structure of tumor necrosis factor-α converting enzyme (TACE) with the location of the individual domains identified

their activity during purification, whereas Moss used a biotinylated form of the inhibitor in an affinity-chromatography step to obtain purified porcine TACE (Black et al. 1997b; Moss et al. 1997b). The cloned enzyme soon followed and was shown to be a member of the *A Disintegrin And Metalloprotease* (ADAM)/reprolysin family of metalloproteases, which is distinct from the MMP family (see below). The general domain structure of TACE is shown in Fig. 2. Recombinant TACE expressed in insect cells has similar kinetic constants as the purified native enzyme (Moss et al. 1997a). Furthermore, the strongest validation that TACE was indeed responsible for the secretion of TNF came from the inactivation of TACE in mouse cells. TNF secretion from these cells was decreased 80–90% relative to wild-type cells (Black et al. 1997b).

Another ADAM family member, ADAM-10, has capabilities as a TACE. ADAM-10, purified from THP-1 membrane extracts, processes TNF-α at the correct site. The recombinant ADAM-10 demonstrates a preference for the TNF-α cleavage site sequence relative to peptide sequences from other shed membrane (Rosendahl et al. 1997). Likewise, ADAM-10 has been proposed to cleave proTNF in 293EBNA cells when co-transfected with rADAM-10 (Lunn et al. 1997). Yet, there still remains some mystery surrounding whether this data is physiologically relevant, since Peschon et al. (1998) demonstrated that TNF secretion was clearly dependent on TACE. An ADAM-10 knockout mouse should address the relevance of ADAM-10 to TNF processing.

III. Structural Features of TACE

The TACE sequence contains the HExxHxxGxxH zinc-binding motif characteristic of the MMP, ADAM/reprolysin, astacin and serratia families of metalloproteases (Hooper 1994). Considering the full-length TACE sequence, the most closely related sequences are a *Caenorhabditis elegans* gene of unknown function (GenBank accession number U70844) and the mammalian, *Drosophila* and *C. elegans* orthologs of kuzbanian (ADAM-10), each with 25–30% sequence identity. Other ADAM/reprolysin family members show significant but lower sequence identities. By contrast, sequences from the

MMP, astacin and serratia families show very little similarity aside from the zinc-binding motif. This classifies TACE as a member of the ADAM/reprolysin family, which currently includes ADAMs (BLACK and WHITE 1998) and snake-venom metalloproteases (Fox and Bjarnason 1996). ADAMs are membrane-anchored proteins with A Disintegrin And Metalloprotease domain. Snake-venom metalloproteases may have a disintegrin domain but are not membrane bound.

The structure of the catalytic domain of TACE has been determined by X-ray crystallography (MASKOS et al. 1998). The overall structure, with α-helices packed above and below a central β-sheet, is very similar to adamalysin II (GOMIS-RUTH et al. 1994), a snake-venom metalloprotease, and somewhat similar to collagenase and stromelysin, which are MMP family members. The β-sheet, major α-helices and the catalytic sites of adamalysin, collagenase and stromelysin can be superimposed onto the corresponding features in TACE. In adamalysin, 175 of these residues fall into structurally equivalent positions, with an average deviation of 1.3 Å. By contrast, only about 120 of the MMP residues fall into structurally equivalent positions, with the deviation increasing to 1.6 Å (MASKOS et al. 1998). The three-dimensional structure is much more highly conserved than the amino acid sequence, but the structural similarities parallel the sequence similarities, with TACE showing greater structural homology to adamalysin than to MMPs. The structures are very similar around the catalytic site, where TACE, adamalysin and the MMPs have a shallow unprimed (left-hand) side cleft, deeper primed side cleft, and all bind zinc with the same geometry. Furthermore, TACE, collagenase and stromelysin all bind an inhibitor hydroxamate group in the same position and orientation between the zinc and the catalytic glutamate residue (SPURLINO et al. 1994; MASKOS et al. 1998). This may explain why TACE is sensitive to many of the hydroxamate inhibitors designed for MMPs. TACE is apparently less sensitive to thiol and carboxylate MMP inhibitors, but the enzyme structural differences that would explain this selectivity for MMP inhibition by these compound classes are not obvious. There are substantial structural differences farther from the zinc site. For example, TACE has a deeper S3' pocket that merges with the S1' pocket. It should be possible to design TACE-selective inhibitors by filling this non-conserved pocket.

We have used the TACE and adamalysin X-ray structures, together with the NMR structure of flavoridin, a snake venom disintegrin domain (SENN and KLAUS 1993), to align the whole ADAM family (Fig. 3). This structure-based sequence alignment was carried out using the molecular viewing program (MVP) (LAMBERT 1997), using the structural information as a constraint. The typical ADAM sequence contains a signal peptide, pro-domain, catalytic domain, disintegrin domain, cysteine-rich domain, transmembrane domain and cytosolic tail as depicted in Fig. 1 and reviewed elsewhere (BLACK and WHITE 1998). The alignment shows that TACE has all of these domains, with relatively good sequence conservation to the other ADAM family members in the catalytic and disintegrin domains.

J.D. Becherer

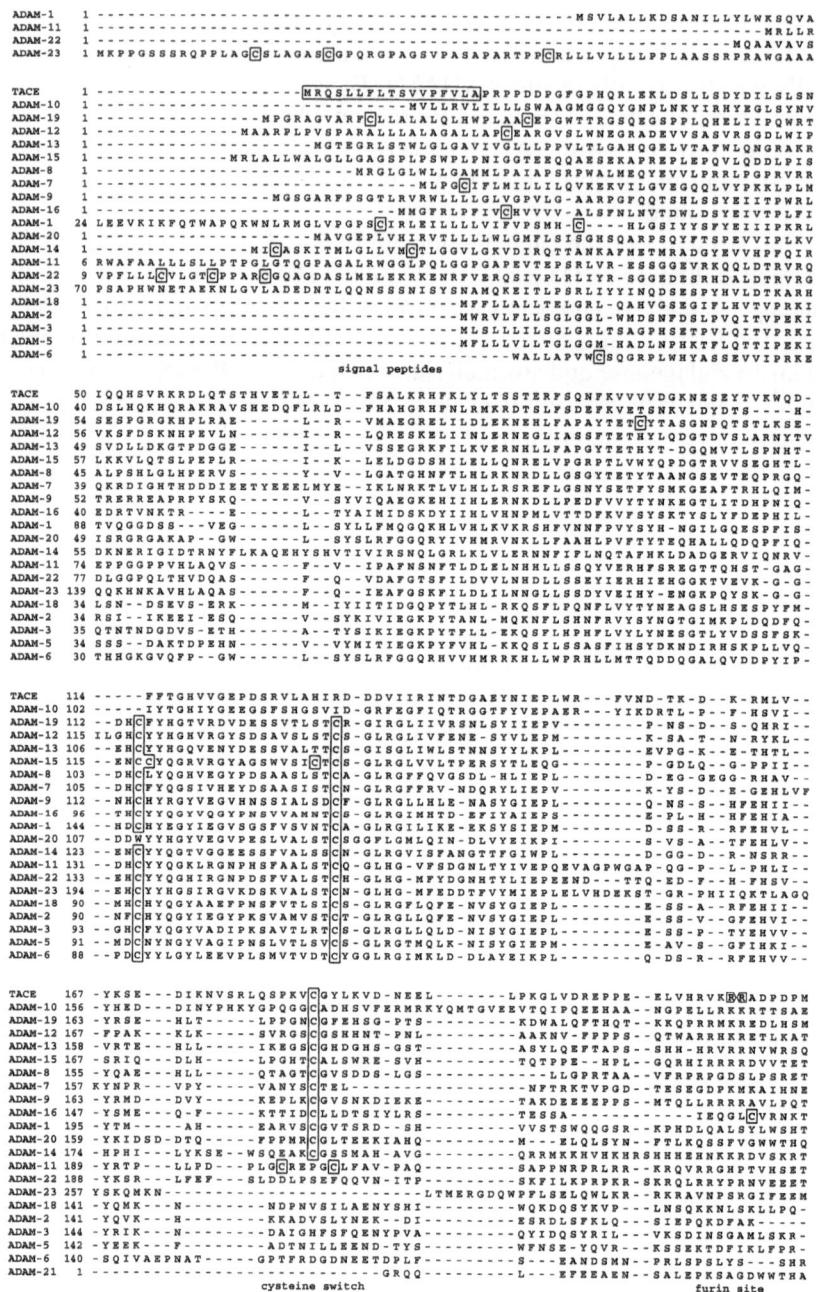

Fig. 3.

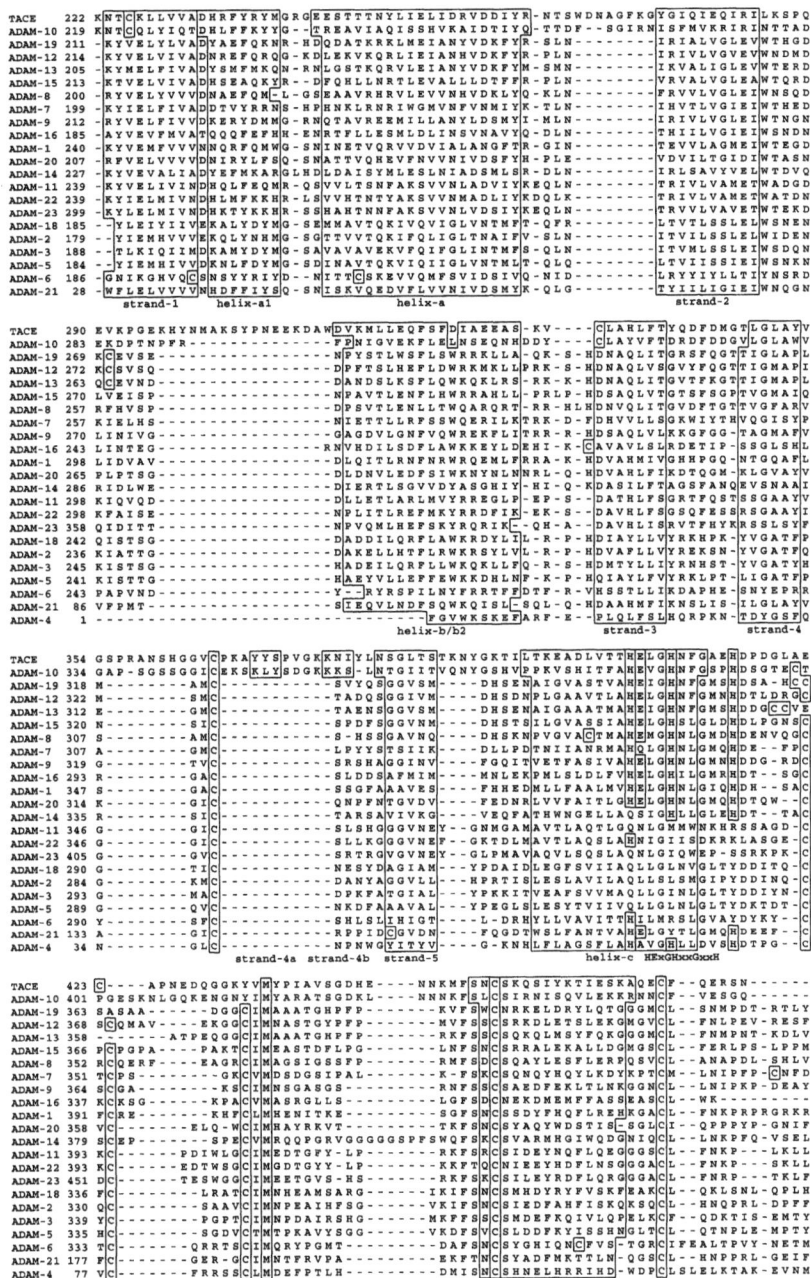

Fig. 3. *Continued*

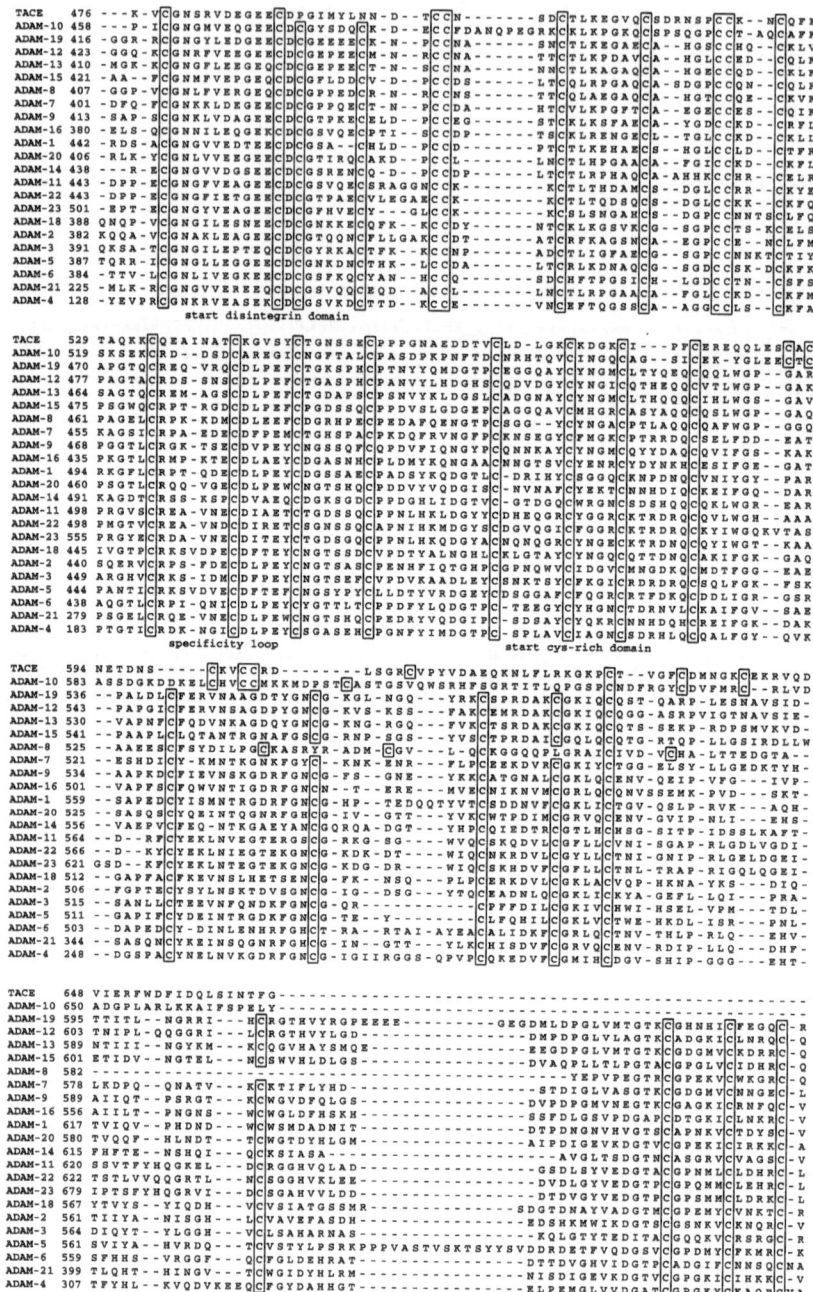

Fig. 3. *Continued*

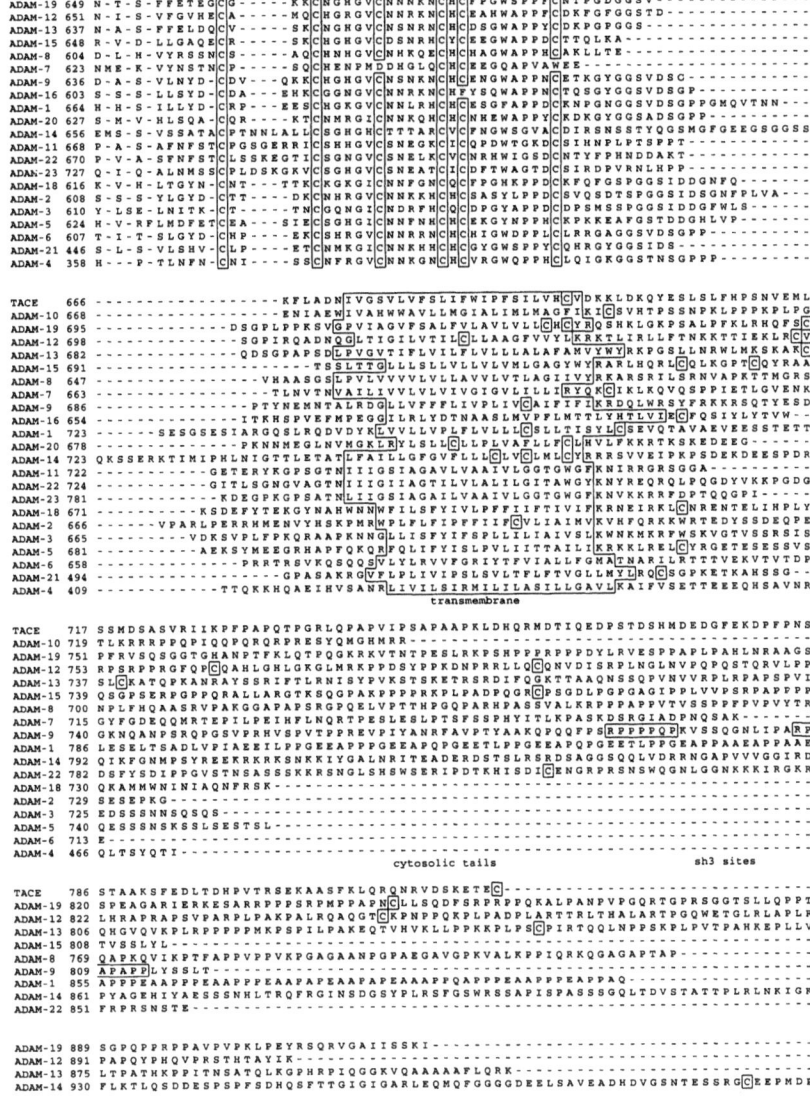

Fig. 3. Structure-based alignment of tumor necrosis factor-α converting enzyme (TACE) and *A Disintegrin and Metalloprotease* (ADAM) family sequences. Signal peptide, predicted transmembrane segments, cysteine residues and key residues in the catalytic domain are *boxed*. Secondary structure (α-helices and β-strands) in the catalytic domain was identified from the crystal structures of TACE and adamalysin II, and indicated here by *additional boxes*. The common name, organism and GenBank accession numbers for the sequences are as follows: monkey ADAM-1 (fertilin-α), X79808; monkey ADAM-2 (fertilin-β), X77653; monkey ADAM-3 (cyritestin), X76637; mouse ADAM-4, U22058; monkey ADAM-5, X77619; monkey ADAM-6, X87206; monkey ADAM-7 (epididymal apical protein), X66139; human ADAM-8 (MS2), D26579; human ADAM-9 (meltrin-γ, MDC9), U41766; human ADAM-10 (kuz gene product), AF009615; human ADAM-11, AB009675; human ADAM-12 (meltrin-α), AF023476; xenopus ADAM-13, U66003; *C. elegans* ADAM-14, U68185; human ADAM-15 (metargidin), U41767; xenopus ADAM-16, U78185; human TACE, U86755; monkey ADAM-18, Y08617; mouse ADAM-19 (meltrin-β), AF019887; human ADAM-20, AF029899; human ADAM-21, AF029900; human ADAM-22, AB009671; human ADAM-23, AB009672. All of the monkey sequences were obtained from *Macaca fascicularis*

The HExGHxxGxxH zinc-binding motif is present in TACE and about half of the ADAMs, but is completely mutated in ADAM-2, -3, -5, -6, -11, -18, -22 and –23. These proteins are presumably not active proteases, acting instead through their disintegrin domains as membrane-anchored adhesion molecules (Wolfsberg and White 1996). In ADAM-14, the HE is mutated to QS and, in ADAM-21, the middle histidine is mutated to tyrosine. It is not clear whether zinc could still bind to these mutated sites. ADAM-4 and ADAM-7 have the three histidines, but not the catalytic glutamate residue. These mutated catalytic domains might also act as adhesion or recognition molecules, binding to (but not cleaving) specific protein sequences.

TACE shows good sequence conservation to the other ADAMs in the disintegrin domain, although two of the conserved cysteines are mutated. Comparison with the structure and disulfide bonding in flavoridin (Senn and Klaus 1993) suggests that these two cysteines are disulfide bonded to other cysteines in most ADAMs. However, it should be possible to swap the disulfide partners in TACE without disrupting the overall disintegrin structure. Some snake-venom disintegrins contain an RGD or RGDC sequence that can bind to the platelet integrin $\alpha_{IIb}\beta_3$ (gpIIb/IIIa), thereby blocking fibrinogen binding and platelet aggregation (Wolfsberg and White 1996). The NMR structures show that the RGD sequence lies in a protruding loop (Senn and Klaus 1993). While the loop is somewhat conserved in snake venom disintegrins, the ADAMs show a wide range of different sequences at this position, and only ADAM-15 has an RGDC sequence. ADAM-15 binds the integrin $\alpha_v\beta_3$ (Zhang et al. 1998), and mouse ADAM-2, which has a QDEC sequence, binds to the integrin $\alpha_6\beta_1$ (Wolfsberg and White 1996). The possibility that other ADAMs might bind other integrins or extracellular proteins led us to call this a "specificity loop" in Fig. 3. However, there is currently no evidence that the NATC sequence in TACE makes specific interactions with other proteins.

MMPs and snake-venom metalloproteases are normally synthesized in an inactive form, where a "cysteine-switch" sequence in the pro-domain binds to the catalytic zinc. Cleavage between the cysteine switch and the catalytic domain releases the pro-domain to give the mature, active enzyme (Van Wart and Birkedal-Hansen 1990; Grams et al. 1993). The sequence alignment shows that TACE and most of the ADAMs contain a cysteine that can be aligned with the snake-venom cysteine switch (not shown). Synthetic peptides from this region of TACE inhibit the activity with a median inhibitory concentration (IC_{50}) of $\sim 40\,\mu M$ (Roghani et al. 1999). The ADAMs that lack this cysteine, ADAM-2, -3, -5, -6, -11, -18, -22 and -23, all have a totally mutated HExGHxxGxxH motif, suggesting that the cysteine switch is not needed in these inactive ADAMs. Interestingly, ADAM-7 and -14, which have single and double mutations in the zinc motif, appear to have intact cysteine switch sequences. The ADAM-4 and -21 sequences are incomplete, and do not reach the pro-domain.

The remaining domains are less well conserved. Within the cysteine-rich domain, TACE and ADAM-10 are similar to each other, but show little similarity to the other ADAM family members. While TACE and ADAM-10 clearly have a cysteine-rich domain, it is smaller and may be structurally different from the cysteine-rich domains of other ADAM family members. This feature, and the higher overall sequence identity between TACE and ADAM-10, emphasizes the similarity of these two proteins within the ADAM family. This similarity is consistent with their ability to cleave full-length TNF.

The cytoplasmic tail, which ranges from 11 to 273 residues in length, is generally rich in charged residues and prolines, but otherwise shows very little conservation. Sequence analysis (Moss et al. 1997b) identified a possible binding site for the src-homology 3 domain (SH3) and a possible tyrosine phosphorylation site within the TACE cytoplasmic tail. ADAM-9 has been shown to bind a SH3 domain (Weskamp et al. 1996), and several other ADAMs also have sites that could bind SH3 domains (Wolfsberg and White 1996). This opens the possibility that the cytoplasmic tail is involved in signaling or subcellular localization through interactions with cytoplasmic proteins.

C. Inhibitors of TACE and TNF-α Secretion

I. MMP Inhibitors and TACE

A growing collection of small molecules have been reported that carry a peptidomimetic, metalloprotease inhibitor-like structure (Beckett and Whittaker 1998) and inhibit cell-free TACE and/or inhibit the release of TNF-α from an appropriately stimulated cell line. A representative set of such compounds is listed in Table 1. The reader will note substantial citation of the patent literature, which indicates both the infancy of the TACE inhibitor field and the expectation of more complete disclosures of TACE inhibitor pharmacology in the open literature. TACE inhibitors reported in the literature to date are of the "RHS (right hand side) design" previously described for MMP inhibitors (Beckett et al. 1996) and are of either the succinyl hydroxamate or the RHS peptidomimetic/sulfonamide hydroxamate variety (Beckett and Whittaker 1998). The inhibitor TAPI (entry 1, Table 1) is among the most potent inhibitors of cell-free TACE. The hydroxamate BB 2516 (marimistat) is also potent versus TACE and MMP enzymes (Beckett and Whittaker 1998). BB 2516 (entry 10, Table 1) shows surprisingly good pharmacokinetics upon oral dosing in man (Beckett et al. 1996; Millar et al. 1998; Wojtowicz-Praga et al. 1998), being one of the first orally absorbed MMP-TACE inhibitors. The compound is being tested in phase-III clinical trials for cancer (Levy and Ezrin 1997).

Further entries in Table 1 beyond the succinyl hydroxamate type resemble CGS 27023 (entry 19), a low molecular weight sulfonamide peptidomimetic which is potent versus cell-free TACE. CGS 27023 is orally bioavailable in rabbits (Parker et al. 1994). Study of the putative binding mode of CGS 27023 to stromelysin-1 (Gonella et al. 1997) has led to second-

Table 1. Representative tumor necrosis factor-α converting enzyme (TACE) inhibitors

Inhibitor [entry]	Identifier	Cell TNFα inhibition, IC$_{50}$ (nM)	TACE, IC$_{50}$ or K_i (nM)
[1]	TAPI	100–200 (BLACK et al. 1997a)	8.8 (ROGHANI et al. 1999)
[2]	BB 94 batimistat	–	11 (ROGHANI et al. 1999)
[3]	GW 9471	180 (McGEEHAN et al. 1994)	4 (Moss et al. 1997a)
[4]	BB 2116	230 (GEARING et al. 1994)	–
[5]	Ro 31–9790	2900 (BARBERIA et al. 1996)	230 (BARBERIA et al. 1996)
[6]	BB 1101	100 (DiMARTINO et al. 1997)	–

Table 1. *Continued*

Inhibitor [entry]	Identifier	Cell TNFα inhibition, IC$_{50}$ (nM)	TACE, IC$_{50}$ or K_i (nM)
[7]	BB 16	1800 (XUE et al. 1998)	–
[8]	CT 572	1400 (BARBERIA et al. 1996)	130 (BARBERIA et al. 1996)
[9]	SE 205	1200 (XUE et al. 1998)	–
[10]	BB 2516 marimistat	3800 (BECKETT and WHITTAKER 1998); 4000 (GLASER et al. 1999)	22 (ROGHANI et al. 1999)
[11]	SC 903	6500 (XUE et al. 1998)	–
[12]	BB 1433	2000 (DiMARTINO et al. 1997)	–

Table 1. *Continued*

Inhibitor [entry]	Identifier	Cell TNFα inhibition, IC$_{50}$ (nM)	TACE, IC$_{50}$ or K_i (nM)
[13]	None	10500 (DeCicco et al. 1996)	–
[14]	GM 6001	2600 (Solorzano et al. 1997)	–
[15]	None	–	0.8 (Bird et al. 1997)
[16]	GW 3333	100–500 (Andrews et al. 1998)	<100 (Andrews et al. 1998)
[17]	None	365 (Broadhurst et al. 1999)	–
[18]	KB-R7785	~4000 (Hattori et al. 1997)	–

Table 1. *Continued*

Inhibitor [entry]	Identifier	Cell TNFα inhibition, IC$_{50}$ (nM)	TACE, IC$_{50}$ or K_i (nM)
[19]	CGS 27023	–	54 (ROGHANI et al. 1999)
[20]	None	700 (NANTERMET et al. 1998)	–
[21]	None	–	36.5 (VENKATESAN et al. 1998b)
[22]	None	–	9.7 (VENKATESAN et al. 1998a)
[23]	None	–	1612 (LEVIN et al. 1998b)

Table 1. *Continued*

Inhibitor [entry]	Identifier	Cell TNFα inhibition, IC$_{50}$ (nM)	TACE, IC$_{50}$ or K_i (nM)
[24]	None	2500 (PARKER 1998)	–
[25]	None	–	294 (LEVIN and FRANCES 1998)
[26]	None	–	26 (LEVIN et al. 1998a)

generation molecules based on the 27023 motif, including entries 20 and 24 in Table 1. Many exhibit potencies in the low nanomolar range for MMP enzymes (BECKETT and WHITTAKER 1998). Among the members of this family for which there is TACE inhibition data are entries 21, 22, 23, 25 and 26 (Table 1), with the compound of entry 22 reported to have a K_i of 9.7 nM versus cell-free TACE.

II. In Vivo Studies with TACE Inhibitors

Many of the compounds in Table 1 are potent inhibitors of the MMP enzymes and TACE. The pharmacological consequences of selective TACE inhibition have not been openly disclosed, and the reported small-molecule TACE-inhibitor pharmacology is complicated in that the compounds under study carry some degree of activity against MMP enzymes. TAPI (entry 1, Table 1), which has been shown to function as an inhibitor of TACE and to inhibit TNF-α release from stimulated THP-1 cells, prolongs the survival of mice injected with LPS. TAPI administered subcutaneously was shown to reduce the serum levels of TNF-α, and survival to 3 days is observed in mice treated subcutaneously with 1.5 mg of TAPI (Mohler et al. 1994).

Liver injury is observed in endotoxemia, where TNF-α is a key mediator. GW 9471 (entry 3, Table 1) given intraperitoneally at 40 mg/kg has been shown to reduce plasma TNF-α levels in D-galactosamine (GalN)/LPS-treated mice (Murakami et al. 1998). Further, biochemical markers of hepatic injury are reduced in the GW 9471-treated mice. Histopathology indicated that GW 9471 reduces the hepatic necrosis induced by D-GalN/LPS treatment. In contrast to the previous study, GM 6001 (entry 14, Table 1) was shown to not protect against liver injury in D-GalN/LPS-treated mice (Solorzano et al. 1997). When concanavalin A is employed to induce hepatitis, GM 6001 actually exacerbates the hepatic necrosis despite reduction of plasma TNF-α levels. The authors propose that cell-surface TNF-α, the level of which is believed to be increased upon treatment with TACE inhibitors, is a prime mediator of liver injury in the D-GalN/LPS or ConA model.

TNF-α and MMPs have been implicated in the demyelination associated with multiple sclerosis. BB 1101 (entry 6, Table 1) administered intraperitoneally twice daily in the rat experimental autoimmune encephalomyelitis (EAE) model is shown to preserve weight and improve clinical score (Clements et al. 1997). BB 1101 is also a potent MMP inhibitor, and MMP inhibition could play a major role in its efficacy in the rat EAE model.

TNF-α and MMPs are also implicated in the joint swelling, inflammation and tissue damage in RA (Chu et al. 1991). BB 1101 (entry 6, Table 1) and BB 1433 (entry 12, Table 1), when orally administered twice daily to adjuvant arthritic rats, significantly reduce hind-paw volume and preserve tibiotarsal bone mineral density (DiMartino et al. 1997). BB 1101 and BB 1433 are also potent inhibitors of the MMPs. The relative contribution of TACE and MMP inhibition to efficacy in this model is not known.

TNF-α is implicated in the insulin resistance characteristic of type-II diabetes (Hotamisligil and Spiegelman 1994) and TNF-α levels are increased in the adipose tissues of non-insulin-dependent diabetes mellitus (NIDDM) models such as the fa/fa rat, db/db rat and KKA^y mouse (Hotamisligil et al. 1993). KB-R7785 (entry 18, Table 1) and orally administered pioglitazone, when given to insulin-resistant KKA^y mice, reduce plasma glucose levels. KB-R7785 also reduced plasma insulin levels (Morimoto et al. 1997). Additionally, KB-R7785 administered intraperitoneally is effective in an acute mouse model of graft-versus-host disease (Hattori et al. 1997).

D. TACE and Membrane Protein Secretases

I. TACE-Mediated Shedding Events

A variety of growth factors, cell-surface receptors, and adhesion molecules are shed from the cell surface in a metalloprotease-mediated event (Hooper et al.

1997). The TACE knockout mice allowed Peschon et al. (1998) to evaluate the role of TACE in the shedding of other cell-surface proteins. The phenotype of these mutant mice resembled that reported for the TGF-α knockout. This included open eyelids at birth and perturbed hair coats due to disorganized hair follicles. Indeed, TACE$^{\Delta Zn/\Delta Zn}$ mice failed to process TGF-α relative to wild-type mice. These TACE$^{\Delta Zn/\Delta Zn}$ cells were also deficient in L-selectin and TNFR1 shedding. Another report using the same cells demonstrates that TACE may also act as the α-secretase for β-amyloid precursor protein (βAPP) (Buxbaum et al. 1998). Thus TACE appears to have a much broader substrate preference than originally thought.

Finally, TACE or a TACE-like enzyme may also be important for the shedding of Tumor-Necrosis-Factor-Related Activation-Induced Cytokine (TRANCE), another member of the TNF-α superfamily involved in osteoclastogenesis and dendritic cell survival (Lum et al. 1999). The MMP inhibitor BB-94 (entry 2, Table 1) was shown to block TRANCE release from cells at a concentration similar to that for TNF. Furthermore, TACE cleaves TRANCE in vitro and yields an amino terminus consistent with that observed for TRANCE in media from phorbol myristate acetate (PMA)-stimulated cells. Finally, TACE cleaves a synthetic peptide mimicking the TRANCE cleavage site correctly, although this is roughly 1000-fold less efficient than TACE cleaves the TNF peptide substrate. This is consistent with the relative efficiency demonstrated by TACE cleavage of L-selectin peptide substrates. One hypothesis is that, in the context of a membrane environment, TACE will cleave the native substrates more efficiently than that observed in vitro.

II. Other Putative Sheddases

The list of integral membrane proteins shed by metalloprotease activity is quite lengthy (Hooper et al. 1997; Werb and Yan 1998). While efficient transforming growth factor (TGF)-α, L-selectin and TNFR1 shedding appears linked to TACE activity, the CD23-solubilizing "sheddase" activity may be different. Faller and co-workers have studied compounds in the succinyl hydroxamate family for which there is substantial potency for inhibition of CD23 proteolysis and less potency for collagenase-1 inhibition (Bailey et al. 1998). The P3' carboxamide (Fig. 4) has an IC$_{50}$ of 160 nM for CD23 proteolysis inhi-

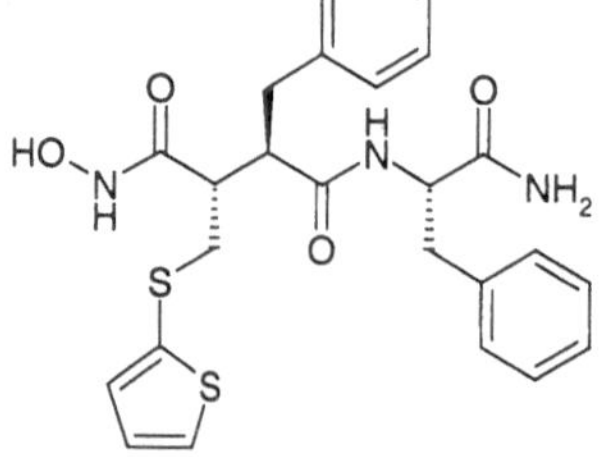

Fig. 4. CD23 proteolysis inhibitor (Bailey et al. 1998)

bition (cell assay) compared with 930nM for collagenase-1 (enzyme assay). TACE and/or cell TNF-α inhibition for these compounds has not been reported, to our knowledge. The further discovery and evaluation of inhibitors selective for TACE and/or other "sheddases" will allow less equivocal study of cell surface proteolytic processes and the relationship of such to TACE-mediated processes.

References

Andrews RC, Andersen MW, Stanford JB, Bubacz DG, Chan JH, Cowan DJ, Gaul MD, McDougald DL, Musso DL, Rabinowitz, MH, Wiethe, RW (1998) Preparation of peptidyl reverse hydroxamate derivatives as metalloprotease inhibitors. PCT Int Appl WO 9838179.

Bailey S, Bolognese B, Buckle DR, Faller A, Jackson S, Louis-Flamberg P, McCord M, Mayer RJ, Marshall LA, Smith DG (1998). Selective inhibition of low affinity IgE receptor (CD23) processing. Bioorg Med Chem Lett 8:29–34.

Barberia JT, Sweeney FJ, Carty TJ (1996) Development of a cell – based assay for TNF-α release which avoids transcription and translation steps. Poster Presentation, Fifth International Conference, Inflammation Research Association, October 1996. Hershey, PA, USA. Abstract P1.

Beckett RP, Whittaker M (1998) Matrix metalloproteinase inhibitors. Expert Opin Ther Pat 8:259–282.

Beckett RP, Davidson AH, Drummond AH, Huxley P, Whittaker M (1996) Recent advances in matrix metalloproteinase inhibitor research. Drug Discovery Today 1:16–26.

Beutler B, Cerami A (1988) Tumor necrosis, cachexia, shock, and inflammation: a common mediator. Ann Rev Biochem 57:505.

Bird TGC, Barlaam BC, Lambert CMP (1997) Preparation of sulfur-containing aminoacyl hydroxamic acid derivatives as tumor necrosis factor and matrix metalloproteinase inhibitors. PCT Int Appl WO 9742168.

Black RA, Fitzner JN, Sleath PR (1997a) Preparation of peptide derivatives as inhibitors of TNF-α secretion. US 5594106.

Black RA, Rauch CT, Kozlosky CJ, Peschon JJ, Slack JL, Wolfson MF, Castner BJ, Stocking KL, Reddy P, Srinivasan S, Nelson N, Boiani N, Schooley KA, Gerhart M, Davis R, Fitzner JN, Johnson RS, Paxton RJ, March CJ, Cerretti DP (1997b) A metalloproteinase disintegrin that releases tumour-necrosis factor-α from cells. Nature (London) 385:729–733.

Black RA, White JM (1998) ADAMs: focus on the protease domain. Curr Op Cell Biol 10:654–659

Bluethmann H (1998) Physiological, immunological, and pathological functions of tumor necrosis factor (TNF) revealed by TNF receptor-deficient mice. In: Durum SK, Muegge K (eds) Cytokine Knockouts. Humana Press, New Jersey, p 69.

Brennen FM, Chantry D, Jackson A, Maini R, Feldmann M (1989) Inhibitory effect of TNF-α antibodies on synovial cell interleukin-1 production in rheumatoid arthritis. Lancet 2:244–247.

Broadhurst MJ, Johnson WH, Walter DS (1999) Preparation of hydroxycarbamoyl-alkylcarboxylic acid hydrazides as inhibitors of tumor necrosis factor and transforming growth factor release. Deutsches Offenlegungschift DE 19829229.

Buxbaum JD, Liu K, Luo Y, Slack JL, Stocking KL, Peschon JJ, Johnson RS, Castner BJ, Cerretti DP, Black RA (1998) Evidence that tumor necrosis factor-α converting enzyme is involved in regulated α-secretase cleavage of the Alzheimer amyloid protein precursor. J Biol Chem 273:27765–27767.

Chu CQ, Field M, Feldman M, Maini R (1991) Localization of tumor necrosis factor α in synovial tissues and at cartilage-pannus junction in patients with rheumatoid arthritis. Arth Rheum 34:1125–1132.

Chung IY, Benveniste EN (1990) Tumor necrosis factor-α production by astrocytes. Induction by lipopolysaccharide, IFN-γ, and IL-1β. J. Immunol. 144:2999.

Clements JM, Cossins JA, Wells GMA, Corkill DJ, Helfrich K, Wood LM, Pigott R, Stabler G, Ward GA, Gearing AJH, Miller KM (1997) Matrix metalloproteinase expression during experimental autoimmune encephalomyelitis and effects of a combined matrix metalloproteinase and tumour necrosis factor-α inhibitor. J Neurol 74:85–94.

Cseh K, Beutler B (1989) Alternative cleavage of the cachectin/tumor necrosis factor propeptide results in a larger, inactive form of secreted protein. J Biol Chem 264:16256.

DeCicco CP, Jaffee BD, Copeland R, Jones B, DiMeo T, Gardner T, Collins R, Czerniak P, Nelson D, Magolda R (1996) TNF-c inhibitors in models of inflammation. Eur. Cytokine Netw 7:290.

Decoster E, Vanhaesebroeck B, Vandenabeele P, Grooten J, Fiers W, (1995) Generation and biological characterization of membrane-bound, uncleavable murine tumor necrosis factor. J Biol Chem 270:18473–18478.

DiMartino M, Wolff C, High W, Stroup G, Hoffman S, Laydon J, Lee JC, Bertolini D, Galloway WA, Crimmin MJ, Davis M, Davies S (1997) Anti-arthritic activity of hydroxamic acid-based pseudopeptide inhibitors of matrix metalloproteinases and TNF-α processing. Inflammation Res 46:211–215.

Djeu JY, Serbousek D, Blanchard DK (1990) Release of tumor necrosis factor by human polymorphonuclear leukocytes. Blood 76:1405.

Eugster H-P, Muller M, Le Hir M, Ryffel B (1998) Immunodeficiency of tumor necrosis factor and lymphotoxin-α double-deficient mice. In: Durum SK, Muegge K (eds) Cytokine Knockouts. Humana Press, New Jersey, p 103.

Fox JW, Bjarnason JB (1996) The reprolysins: a family of metalloproteases defined by snake venom and mammalian metalloproteases. In: Hooper NM (ed) Zinc Metalloproteases in Health and Disease. p 47. Taylor and Francis, London.

Gearing AJH, Beckett P, Christodoulou M,Churchill M, Clements J, Davidson AH, Drummond AH,Galloway WA, Gilbert R, Gordon J, Leber T, Mangan M, Miller K, Nayee P, Owen K, Patel S, Thomas W, Wells G, Wood L, Wooley K (1994) Processing of tumor necrosis factor-α precursor by metalloproteinases. Nature (London) 370:555–558.

Glaser KB, Pease L, Li J, Morgan DW (1999) Enhancement of the surface expression of tumor necrosis factor alpha (TNF-alpha) but not the p55 TNF-alpha receptor in the THP-1 monocytic cell line by matrix metalloprotease inhibitors. Biochem Pharmacol 57:291–302.

Gomis-Ruth FX, Kress LF, Kellermann J, Mayr I, Lee X, Huber R, Bode W (1994) Refined 2.0 Å X-ray crystal structure of the snake venom zinc endopeptidase adamalysin II. Primary and tertiary structure determination, refinement, molecular structure and comparison with astacin, collagenase and thermolysin. J Mol Biol 239:513–544.

Gonnella NC, Li Y-C, Zhang X, Paris CG (1997) Bioactive conformation of a potent stromelysin inhibitor determined by X-nucleus filtered and multidimensional NMR spectroscopy. Bioorg Med Chem 5:2193–2201.

Gordon JR, Galli SJ (1990) Mast cells as a source of both preformed and immunologically inducible TNF-α/cachectin. Nature (London) 346:274.

Grams F, Huber R, Kress LF, Moroder L, Bode W (1993) Activation of snake venom metalloproteases by a cysteine switch-like mechanism. FEBS Lett 335:76–80.

Hattori K, Hirano T, Ushiyama C, Miyajima H, Yamakawa N, Ebata T, Wada Y, Ikeda S, Yoshino K, Tateno M, Oshimi K, Kayagaki N, Yagita H, Okumura K (1997) A metalloproteinase inhibitor prevents lethal acute graft-versus-host disease in mice. Blood 90:542–548.

Hooper NM (1994) Families of zinc metalloproteases. FEBS Lett 354:1–6.

Hooper NM, Karran EH, Turner AJ (1997) Membrane protein secretases. Biochem J 321:265–279.

Hotamisligil GS, Shargill NS, Spiegelman BM (1993) Adipose expression of tumor necrosis factor-alpha; direct role in obesity-linked insulin resistance. Science 259:87–91.

Hotamisligil GS, Spiegelman BM (1994) Tumour necrosis factor α: a key component of the obesity – diabetes link. Diabetes 43:1271–1277.

Kim KU, Kwon OJ, Jue D-M (1993) Pro-tumor necrosis factor cleavage enzyme in macrophage membrane/particulate. Immunology 80:134.

Kriegler M, Perez C, DeFay AI, Lu SD (1988) A novel form of TNF/cachectin is a cell surface cytotoxic transmembrane protein: ramifications for the complex physiology of TNF. Cell 53:45.

Lambert MH (1997) Docking Conformationally Flexible Molecules into Protein Binding Sites. in Practical Application of Computer-Aided Drug Design, Charifson PS, editor, pp 243–303. Marcel-Dekker, New York.

Levin JI, Du Mila T, Venkatesan AM, Nelson FC, Zask A, Gu Y (1998a) The preparation and use of ortho-sulfonamido aryl hydroxamic acids as matrix metalloproteinase and TACE inhibitors. PCT Int Appl WO 9816503.

Levin JI, Zask A, Gu Y (1998b) Preparation of α-sulfonamido hydroxamic acids as matrix metalloproteinase and TACE inhibitors. PCT Int Appl WO 9816506.

Levin JI, Frances CN (1998) Preparation and use of ortho-sulfonamido heteroaryl hydroxamic acids as matrix metalloproteinase and TACE inhibitors. PCT Int Appl WO 9816520.

Levy DE, Ezrin AM. (1997) Matrix metalloproteinase inhibitor drugs. Emerging Drugs 2:205–230.

Lum, L, Wong, BR, Josien R, Becherer JD, Erdjument-Bromage H, Schloendorf J, Tempst P, Choi Y, Blobel CP (1999) Evidence for a role of a TNF-α converting enzyme-like protease in the shedding of TRANCE, a TNF family member involved in osteoclastogenesis and dendritic cell survival. J Biol Chem 274:13613–13618.

Lunn CA, Fan X, Dalie B, Miller K, Zavodny PJ, Narula SK, Lundell D (1997) Purification of ADAM 10 from bovine spleen as a TNF-α convertase. FEBS Lett 400:333–335.

MacPherson LJ, Bayburt EK, Capparelli MP, Carroll BJ, Goldstein R, Justice MR, Zhu L, Hu IS, Melton RA, Fryer L, Goldberg R., Doughty JR, Spirito S, Blancuzzi V, Wilson D, O'Byrne EM, Ganu V, Parker DT (1997) Discovery of CGS 27023 A, a non-peptidic, potent, and orally active stromelysin inhibitor that blocks cartilage degradation in rabbits. J Med Chem 40:2525–2532.

Maini RN, Breedveld FC, Kalden JR, Smolen JS, Davis D, Macfarlane JD, Antoni C, Leeb B, Elliott MJ, Woody JN, Schaible TF, Feldmann M (1998) Therapeutic efficacy of multiple intravenous infusions of anti-tumor necrosis factor α monoclonal antibody combined with low-dose weekly methotrexate in rheumatoid arthritis. Arthritis Rheum 41:1552–1563.

Maskos K, Fernandez-Catalan C, Huber R, Bourenkov GP, Bartunik H, Ellestad GA, Reddy P, Wolfson MF, Rauch CG, Castner BJ, Davis R, Clarke HRG, Petersen M, Fitzner JN, Cerretti DP, March CJ, Paxton RJ, Black RA, Bode W (1998) Crystal structure of the catalytic domain of human tumor necrosis factor-α-converting enzyme. Proc Natl Acad Sci USA 95:3408–3412.

McGeehan GM, Becherer JD, Bast RC Jr, Boyer CM, Champion B, Connolly KM, Conway JG, Furdon P, Karp S, Kidao S, McElroy A, Nichols J, Pryzwansky K, Schoenen F, Sekut L, Truesdale A, Verghese M, Warner J, Ways J (1994) Regulation of tumor necrosis factor-α processing by a metalloproteinase inhibitor. Nature (London) 370:558–61.

Mohler KM, Sleath PR, Fitzner JN, Cerretti DP, Alderson M, Kerwar SS, Torrance DS, Otten-Evans C, Greenstreet T, Black RA (1994) Protection against a lethal dose of endotoxin by an inhibitor of tumor necrosis factor processing. Nature (London) 370:218–221.

Moreland LW, Baumgartner SW, Schiff MH, Tindall EA, Fleischmann RM, Weaver AL, Ettlinger RE, Cohen S, Koopman WJ, Mohler K et al. (1997) Treatment of rheumatoid arthritis with a recombinant human tumor necrosis factor receptor (p75)-Fc fusion protein. New Engl J Med 337:141–147.

Morimoto Y, Nishikawa K, Ohash M (1997). KB-R7785, a novel matrix metalloproteinase inhibitor, exerts its antidiabetic effect by inhibiting tumor necrosis factor-α production. Life Sci 61:795–803.

Moss ML, Jin S-LC, Becherer JD, Bickett DM, Burkhart W, Chen WJ, Hassler D, Leesnitzer MT, McGeehan G, Milla M, Moyer M, Rocque W, Seaton T, Schoenen F, Warner J, Willard D (1997a) Structural features and biochemical properties of TNF-α converting enzyme (TACE). J Neuroimmunol 72:127–129.

Moss ML, Jin S-LC, Milla, ME, Bickett DM, Burkhart W, Carter HL, Chen W-J, Clay WC, Didsbury, JR, Hassler D, Hoffman CR, Kost TA, Lambert MH, Leesnitzer MA, McCauley P, McGeehan G, Mitchell J, Moyer M, Pahel G, Rocque W, Overton LK, Schoenen F, Seaton T, Su J-L, Warner J, Willard D, Becherer JD (1997b) Cloning of a disintegrin metalloproteinase that processes precursor tumour-necrosis factor-α. Nature (London) 385:733–736.

Muller R, Marmenout A, Fiers W (1986) Synthesis and maturation of recombinant human tumor necrosis factor in eukaryotic systems. FEBS Lett. 197:99.

Murakami K, Kobayashi F, Ikegawa R, Koyama M, Shintani N, Yoshida T, Nakamura N, Kondo T (1998) Metalloproteinase inhibitor prevents hepatic injury in endotoxemic mice. Eur J Pharmacol 341:105–110.

Nantermet PG, Parker DT, Macpherson LJ (1998) Certain alpha-azacycloalkyl substituted arylsulfonamido acetohydroxamic acids, useful as inhibitors of matrix-degrading metalloproteinases and TNF-α converting enzyme. US 5817822.

Niehoerster M, Tiegs G, Schade UF, Wendel A (1990) In vivo evidence for protease-catalyzed mechanism providing bioactive tumor necrosis factor α. Biochem Pharmacol 40:1601.

Old LJ (1985) Tumor necrosis factor. Science 230:630.

Parker DT (1998) Preparation of α-substituted arylsulfonamido acetohydroxamic acids as tumor necrosis factor inhibitors. US 5770624.

Parker DT, MacPherson LJ, Goldstein R, Justice MR, Zhu LJ, Caparelli M, Whaley LW, Boehm C, O'Byrne EM, Goldberg RL (1994) The development of 27023 A: a novel, potent, and orally active matrix metalloprotease inhibitor. Poster Presentation, Fourth International Conference, Inflammation Research Association, October 1994. White Haven, PA, USA Abstract P73.

Perez C, Albert I, DeFay K, Zachariades N, Gooding L, Kriegler M (1990) A nonsecretable cell surface mutant of tumor necrosis factor (TNF) kills by cell-to-cell contact. Cell 63:251–258.

Peschon JJ, Slack JL, Reddy P, Stocking KL, Sunnarborg SW, Lee DC, Russell WE, Castner BJ, Johnson RS, Fitzner JN et al (1998) An essential role for ectodomain shedding in mammalian development. Science 282:1281–1284.

Peterson PK, Gekker G, Chao CC, Hu S, Edelman C, Balfour HH Jr., Verhoef J (1992) Human cytomegalovirus-stimulated peripheral blood mononuclear cells induce HIV-1 replication via a tumor necrosis factor-α-mediated mechanism. J Clin Invest 89:574.

Piguet PF, Grau GE, Allet B, Vassalli P (1987) Tumor necrosis factor/cachectin is an effector of skin and gut lesions of the acute phase of graft-vs.-host disease. J. Exp. Med. 166:1280.

Rasmussen HS, Chiodo CA, Hawkins MJ (1998) Phase I trial of marimastat, a novel matrix metalloproteinase inhibitor, administered orally to patients with advanced lung cancer (1998) J Clin Oncol 16:2150–2156.

Roghani M, Becherer JD, Moss ML, Atherton RE, Erdjument-Bromage H, Arribas J, Blackburn KR, Weskamp G, Tempst P, Blobel CP (1999) Metalloprotease-disintegrin MDC9: intracellular maturation and catalytic activity. J Biol Chem 274:3531–3540.

Rosendahl M, Ko SC, Long DL, Brewer MT, Rosenzweig B, Hedl E, Anderson L, Pyle SM, Moreland J, Meyers MA, Kohno T, Lyons D, Lichenstein HS, (1997) Identification and characterization of a pro-tumor necrosis factor-α-processing enzyme from the ADAM family of zinc metalloproteases. J Biol Chem 272: 24588–24593.

Sawada M, Kondo N, Suzumura A, Marunouchi T (1989) Production of tumor necrosis factor-alpha by microglia and astrocytes in culture. Brain Res 491:394.

Scuderi P (1989) Suppression of human leukocyte tumor necrosis factor secretion by the serine protease inhibitor p-toluenesulfonyl-L-arginine methyl ester (TAME). J Immunol 143:168.

Senn H, Klaus W (1993) The nuclear magnetic resonance solution structure of flavoridin, an antagonist of the platelet GP IIb-IIIa receptor. J Mol Biol 232:907–925.

Shinbrot E, Moore M (1998) Cooperation between the TNF receptors demonstrated by TNF receptor knockout mice. In: Durum SK, Muegge K (eds) Cytokine Knockouts. Humana Press, New Jersey, p 89.

Shire MG, Muller GW (1998) TNF-α inhibitors and rheumatoid arthritis. Expert Opin Ther Pat 8:531–544.

Solorzano CC, Ksontini R, Pruitt JH, Hess PJ, Edwards PD, Kaibara A, Abouhamze A, Auffenberg T, Galardy RE, Vauthey JN, Copeland EM, Edwards CK, Lauwers GY, Clare-Salzler M, MacKay SLD, Moldawer LL, Lazarus DD (1997) Involvement of 26-kDa cell-associated TNF-α in experimental hepatitis and exacerbation of liver injury with a matrix metalloproteinase inhibitor. J Immunol 158:414–419.

Spurlino JC, Smallwood AM, Carlton DD, Banks TM, Vavra KJ, Johnson JS, Cook ER, Falvo J, Wahl RC, Pulvino TA, Wendoloski JJ, Smith DL (1994) 1.56Å Structure of mature truncated human fibroblast collagenase. Proteins 19:98–109.

Suffys P, Beyaert R, Van Roy F, Fiers W (1988) Involvement of a serine protease in tumor necrosis factor-mediated cytotoxicity. Eur J Biochem 178:257.

Sung S, Bjorndahl J, Wang C, Kao H, Fu S (1988) Production of tumor necrosis factor/cachectin by human T cell lines and peripheral blood T lymphocytes stimulated by phorbol myristate acetate and anti-CD3 antibody. J Exp Med 167:937.

Tracy KJ, Beutler B, Lowry SF, Merryweather J, Wolpe S, Milsark IW, Hariri RJ, Fahey TJ III, Zentella A, Albert JD, Shires GT, Cerami A (1986) Shock and tissue injury induced by recombinant human cachectin. Science 234:470.

Utsumi T, Levitan A, Hung M, Klostergaard J (1993) Effects of truncation of human pro-tumor necrosis factor transmembrane domain on cellular targeting. J Biol Chem 268:9511.

Van Wart HE, Birkedal-Hansen B (1990) The cysteine switch: A principle of regulation of metalloprotease activity with potential applicability to the entire matrix metalloprotease gene family. Proc Natl Acad Sci USA 87:5578–5581.

Vassalli P (1992) The pathophysiology of tumor necrosis factors. Annu Rev Immunol 10:411.

Venkatesan AM, Grosu GT, Davis JM, Baker JL (1998a) N-Hydroxy-2-(alkyl, aryl or heteroaryl sulfanyl, sulfinyl or sulfonyl)-3-substituted alkyl, aryl or heteroaryl amides as matrix metalloproteinase inhibitors. PCT Int Appl WO 9837877.

Venkatesan MA, Grosu GT, Davis JM, Hu B, O'Dell MJ, Cole DC, Baker JL, Jacobson MP (1998b) N-Hydroxy-2-(alkyl, aryl or heteroaryl sulfanyl, sulfinyl or sulfonyl)-3-substituted alkyl, aryl or heteroaryl amides as matrix metalloproteinase inhibitors. PCT Int Appl WO 9838163.

Weinblatt ME, Kremer JM, Bankhurst AD, Bulpitt KJ, Fleischmann RM, Fox RI, Jackson CG, Lange M, Burge DJ (1999) A trial of etanercept, a recombinant tumor necrosis factor receptor: Fc fusion protein, in patients with rheumatoid arthritis receiving methotrexate. New Engl J Med 340:253–259.

Werb Z, Yan Y (1998) A cellular striptease act. Science 282:1279–1280.

Weskamp G, Kratzschmar J, Reid MS, Blogel CP (1996) MDC9, a widely expressed cellular distintegrin containing cytoplasmic SH3 ligand domains. J Cell Biol 132:717–726.

Wojtowicz-Praga S, Torri J, Johnson M, Steen S, Marshall J, Ness E, Dickson R; Sale M; Wolfsberg TG, White JM (1996) ADAMs in fertilization and development. Dev Biol 180:389–401.

Williams RO, Feldman M, Main R (1992) Anti-tumor necrosis factor ameliorates joint disease in murine collagen-induced arthritis. Proc Natl Acad Sci USA 89:9784.

Xue C-B, He X, Roderick J, Degrado WF, Cherney RJ, Hardman KD, Nelson DJ, Copeland RA, Jaffee BD, Decicco CP (1998) Design and synthesis of cyclic inhibitors of matrix metalloproteinases and TNF-a production. J Med Chem 41:1745–1748.

Zhang X-P, Kamata T, Yokoyama K, Puzon-McLaughlin W, Takada Y (1998) Specific interaction of the recombinant disintegrin-like domain of MDC-15 (metargidin, ADAM-15) with integrin $\alpha_V\beta_3$. J Biol Chem 273:7345–7350.

CHAPTER 14

Serine Elastases in Inflammatory and Vascular Diseases

J.C. CHERONIS and M. RABINOVITCH

A. Introduction

Elastase is a general term that has traditionally described a group of proteinases that have the ability to degrade elastin, the primary extracellular matrix (ECM) protein that confers elastic qualities to a variety of tissues, including the lungs, skin and blood vessels. Different proteinases from the serine, cysteine and metallo classes have been shown to degrade elastin with varying degrees of activity. Of these different classes of proteinases, other than the metalloproteinase metalloelastase [matrix metalloprotein 12 (MMP-12) (SHAPIRO 1994; GRONSKI et al. 1997; MECHAM et al. 1997)], the serine elastases – pancreatic elastase (PE), neutrophil elastase (NE), proteinase-3 (PR-3) – have been the subject of the most intensive investigations thus far. These three enzymes have been characterized extensively, and their crystal structures have been elucidated and published. Two additional enzymes [endogenous vascular elastase (EVE) and endothelial cell elastase (ECE)] are, as yet, less well characterized and have only been identified as having "elastase-like" activities based on substrate and inhibitor activity profiles, although EVE has been shown to degrade insoluble elastin (BUSSOLINO et al. 1994; ZHU et al. 1994). Finally, at least three other serine elastases, as defined by their substrate/inhibition profiles, have been described; these were derived from transformed rat-liver epithelial or Schwann cells, human carcinoma cell lines (CAPPELLUTI and HARRIS 1994), human skin fibroblasts (CROUTE et al. 1991) and human lymphocytes (PACKARD et al. 1995). The relationships of these enzymes to PE, NE, PR-3, EVE and ECE are not known. This review will focus on NE, PR-3, EVE and, to a lesser extent, ECE, with respect to their potential roles in inflammatory and vascular diseases.

In addition to elastin, the serine elastases have been shown to degrade or process other proteins, but their relative activities against these proteins can differ markedly. The serine elastases share a preference for the cleavage of polypeptides and proteins adjacent to aliphatic amino-acid residues, primarily alanine, valine and methionine. These enzymes will also cleave, to a variable extent, polypeptides and proteins adjacent to leucine and isoleucine (NAKAJIMA et al. 1979; DEL MAR et al. 1980; McRAE et al. 1980; RAO et al. 1991). Listed in Table 1 is a selection of proteins that have been reported to be either degraded or processed by NE, the serine elastase that has been most

Table 1. Proteolytic targets of human neutrophil elastase

Soluble or membrane associated proteins	Structural proteins
Surfactant apoproteins	Collagen
Coagulation system factors	Types II, III, IV, IX
FV, FVII, FVIIIC, FIX, FX, FXII, FX111	Elastin
Complement factors	Fibrin
Fibrinogen	Fibronectin
Immunoglobulins	Laminin
Kininogen (high molecular weight)	Proteoglycans
Proenzymes	
Cathepsin B	
Matrix metalloproteinases	
Proteinase inhibitors	
TIMP-1, 2, 3	
α_2-Macroglobulin	
α_2-Plasmin inhibitor	
Antithrombin III	
Complement C1 inactivator	
Cystatin C	
Plasminogen activator inhibitor 1	
Receptors	
Platelet IIb/IIIa fibrinogen receptor	
PMNL C3b receptor (CR1)	

extensively studied. This list of reported substrates, although incomplete, demonstrates that the activity of this enzyme (and, by implication, the other serine elastases) is not restricted to elastin or ECM proteins in general. In order to assess the potential of these enzymes as targets for therapeutic intervention, it is critical that the roles they play in biological processes, other than ECM degradation or remodeling, be understood.

Just as the elastases can have subtle differences in their abilities to degrade different substrates, so too can elastase inhibitors differ with respect to their ability to inhibit different elastases even though they may be highly specific and selective with respect to other proteinases. As a result, both the spectrum of elastases inhibited by a specific drug candidate as well as the spectrum of activities these enzymes have with respect to proteins other than elastin should be taken into consideration when evaluating the potential clinical indications being targeted.

I. Neutrophil Elastase

With the exception of PE, which is generally considered to be solely a digestive enzyme, NE was the first of the serine elastases to be described (JANOFF and SCHERER 1968; JANOFF 1973; BAUGH and TRAVIS 1976; SANDHAUS and WERB 1981; WERB et al. 1982; STARKEY 1997). NE is found primarily in neutrophils but has also been described in a variety of other leukocytes, including certain subpopulations of monocytes (CAMPBELL et al. 1989). This enzyme is believed to be the primary enzyme responsible for neutrophil/leukocyte migration across vascular tissues and is responsible for the proteolytic destruction of a host of connective tissue and other proteins (Table 1) (McDONALD and KELLEY 1980; CAMPBELL et al. 1982; WEISS and REGIANI 1984; KAO et al. 1988; KLENIEWSKI and DONALDSON 1988; OKADA et al. 1988; VISSERS and WINTER-BOURN 1988; JORDAN et al. 1989; PISON et al. 1989; HECK et al. 1990; McGOWAN 1990; TOSI et al. 1990; DALET-FUMERON et al. 1993; KLEBANOFF et al. 1993; LEWIS et al. 1993; ITOH and NAGASE 1995; OWEN and CAMPBELL 1995; RICE and BANDA 1995; DELCLAUX et al. 1996; LIAU et al. 1996; EDELSTEIN et al. 1997; FERRY et al. 1997; SI-TAHAR et al. 1997).

II. Proteinase 3

PR-3 is found, along with NE, within the azurophilic granules of the neu-trophil, and is released in conjunction with NE upon stimulation (RAO et al. 1991; KAO et al. 1988; BERGENFELDT et al. 1992). In vitro experiments have demonstrated that PR-3 can degrade insoluble elastin and hydrolyze a variety of peptide-based substrates that can also be hydrolyzed by NE and/or PE. However, our data suggest that PR-3, unlike NE, is important in the process-ing of inflammatory cytokines into their active biological forms but is rela-tively unimportant with respect to the degradation of ECM by activated human neutrophils. Therefore, in contrast to NE, PR-3 appears to play an important role in the amplification of the inflammatory response rather than in the direct degradation of structural proteins. Experiments supporting this hypothesis are described below.

PR-3 was independently identified as a marker of myeloid cell precursor differentiation and of certain myeloid leukemias (BORIES et al. 1989). In addi-tion, PR-3 appears to be the autoantigen recognized by anti-cytoplasmic anti-bodies associated with Wegener's granulomatosis (CAMPANELLI et al. 1990; CSERNOK et al. 1990; CSERNOK et al. 1991; JENNETTE et al. 1990; HENSHAW et al. 1994). The importance of PR-3 in these latter two contexts will not be dis-cussed in this review.

III. Endogenous Vascular Elastase

EVE is a 23-kDa serine elastase that was first described as an enzyme associ-ated with vascular-smooth-muscle growth, proliferation and maturation (ZHU et al. 1994). In a number of different in vitro and in vivo models, EVE activ-ity is associated with vascular hypertrophy and fibrosis (MARUYAMA et al. 1991; YE and RABINOVITCH 1991; TODOROVICH-HUNTER et al. 1992; COWAN et al. 1996). In these models, inhibition of elastase activity by what appears to be EVE's primary endogenous protein antiproteinase (elafin) or other synthetic elastase inhibitors prevents and/or reverses the pathology observed (YE and RABINOVITCH 1991; COWAN et al. 1996; LEE et al. 1997; O'BLENES et al. in press).

Unlike NE, however, the correlation of EVE activity with specific human diseases has not yet been established despite the fact that the pathologic processes seen in the animal models in which EVE has been shown to play a role have been well matched to biopsy and histology specimens from humans with related conditions.

IV. Endothelial Cell Elastase

ECE appears to be a 34-kDa proteinase that is expressed by human vascular endothelial cells when stimulated by the pro-inflammatory cytokines, interleukin (IL)-1β or tumor necrosis factor α (TNFα). ECE activity appears to be responsible for the induction of 1-O-alkyl-2-lyso-glycerolphosphocholine acetyl-coenzyme A:acetyltransferase, a key enzyme in the platelet-activating factor (PAF) pathway and late-stage PAF production by vascular endothelium (Bussolino et al. 1994). Interestingly, exogenously applied NE will mimic the activity of endogenously expressed ECE in vitro. Other stimuli capable of inducing ECE expression and/or activities associated or dependent upon ECE expression are unknown. However, as will be discussed below, given the locus of ECE expression and the ability of NE to mimic ECE, its inhibition in the context of inflammatory or degenerative vascular diseases may be of therapeutic benefit.

B. Serine Elastases and Inflammation

One of the questions that we have been investigating is the relative roles of PR-3 and NE in inflammatory processes. Given their apparent redundant activities, it was unclear as to why both enzymes would be expressed in approximately equal proportions and packaged in the same intracellular granule so that they are released under identical conditions. Recently, however, reports have shown that certain pro-inflammatory cytokines, most notably IL-8 and TNFα, can be processed in vitro by PR-3 but not by NE (Padrines et al. 1994; Robache-Gallea et al. 1995). These observations led us to ask the question, "Can these enzymes serve different but synergistic roles?" Details of the experiments conducted to answer this question have been published elsewhere and are described below (Coeshott et al. 1999).

Two of the more important cytokines mediating inflammatory responses are TNFα and IL-1β, both of which require conversion to soluble mature forms through the action of specific converting enzymes, TNFα-converting enzyme (TACE) and IL-1β-converting enzyme (ICE), respectively. The importance of TACE and ICE in the production of circulating levels of active cytokines in response to systemic challenges has been demonstrated via the use of specific inhibitors of these converting enzymes.

Many inflammatory responses, however, are not systemic but are instead highly localized. In these situations, release and/or activation of cytokines may be considerably different from that seen in response to a systemically administered stimulus, particularly since intimate association of various cell populations in these foci allows for the exposure of cytokines and pro-cytokines to the potent proteolytic enzymes produced by activated neutrophils, namely NE, PR-3 and cathepsin G (Cat G). In order to investigate the possibility of alternative processing of TNFα and/or IL-1β by neutrophil-derived serine proteinases, we measured TNFα and IL-1β release from LPS-stimulated THP-1 cells (a human monocytic cell line) in the presence of activated human neutrophils from individual donors. Under these conditions, TNFα release was augmented two- to fivefold, and this was further increased in the presence of a TACE-specific inhibitor (Fig. 1). IL-1β release was also enhanced (Fig. 2). Interestingly, in the presence of a specific inhibitor of NE and PR-3 [CE-2072, a compound with both PR-3 (inhibition constant $[K_i]$ = 2.0 nM) and NE (K_i = 0.02 nM) activity], enhanced release of both cytokines was largely abolished. However, when the experiment was repeated in the presence of secretory leucocyte proteinase inhibitor (SLPI), which inhibits NE but not PR-3, reduction of the enhanced release was negligible. This profile strongly suggested that the augmented release was attributable to PR-3 but not NE or Cat G. Use of purified enzymes in concert with TACE- and ICE-specific inhibitors confirmed this conclusion and revealed exaggerated release of both TNFα and IL-1β in the presence of PR-3.

To further investigate the differential activities of PR-3 and NE, we also assessed the ability of different proteinase inhibitors to influence neutrophil-mediated degradation of ECM. In these studies, SLPI (a selective NE inhibitor) inhibits matrix degradation by activated human neutrophils to a level equal to that of CE-2072 (combined NE and PR-3 inhibition; Fig. 3).

In summary, NE appears to be primarily responsible for degradation of ECM proteins and possibly other important substrate proteins (immunoglobulins, surfactant apoproteins, etc.), whereas PR-3 appears to be particularly well-suited to the processing of pro-cytokines into their active biological forms. Our data indicate that at least two of the more important pro-inflammatory cytokines, TNF-α and IL-1β, are differentially sensitive to PR-3 relative to NE and that soluble mediator concentrations can be substantially enhanced by the action of either purified PR-3 or activated human neutrophils. It has also been shown that PR-3, but not NE, can process mature IL-8(77) into IL-8(70), which has approximately tenfold greater biological activity than the 77-amino-acid form of the cytokine (PADRINES et al. 1994). Furthermore, it is now clear from ICE-knock-out-mouse studies that IL-1β dependent inflammatory responses in vivo can occur in the absence of ICE (FANTUZZI et al. 1997). These data clearly suggest that alternative processing of this critical inflammatory cytokine can and does occur. Our data indicate that the elastases, and PR-3 in particular, may be the critical enzymes responsible for this activity.

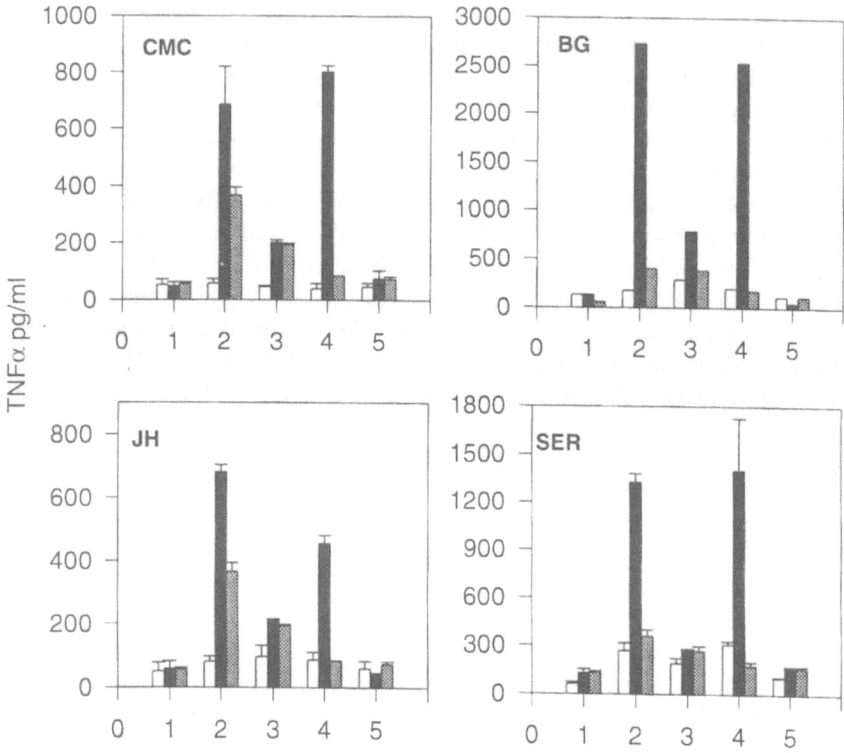

Fig. 1. Enhancement of tumor-necrosis-factor-α (TNFα) release from co-incubation of stimulated THP-1 cells and human neutrophils. Representative data are shown from three separate experiments using two neutrophil donors per day. THP-1 cells (n = 1 × 10⁶; *grey bars*) and 4 × 10⁶ neutrophils isolated from individual donors (*open bars*) were added to tissue-culture wells either separately or as a mixture of both populations (*black bars*). Cells were incubated for 4 h at 37°C either with no further additives (*position 1*), with 1 µg/ml lipopolysaccharide and 10⁻⁵ M formyl-norleucyl-leucyl-phenylalanine and no further additives (*position 2*), with 10 µM CE-2072 (*position 3*); plus 10 µM TNFα-converting enzyme (TACE) inhibitor (TACE-I; *position 4*), or with 10 µM CE-2072 and 10 µM TACE-I (*position 5*). Supernatants were recovered and assayed for TNFα by enzyme-linked immunosorbent assay (ELISA). Data represent mean TNFα levels ± standard deviation from two tissue culture wells with each well run in duplicate in ELISA

C. Serine Elastases and Vascular Diseases

Observations relating to elastase activity and the pathogenesis of vascular disease came from the study of lung-biopsy tissues in patients with pulmonary hypertension. These observations showed fragmentation of elastin in pulmonary arteries of children with a congenital heart defect and pulmonary hypertension as a very early feature, i.e. prior to the development of medial hypertrophy or neointimal formation (Rabinovitch et al. 1986). These obser-

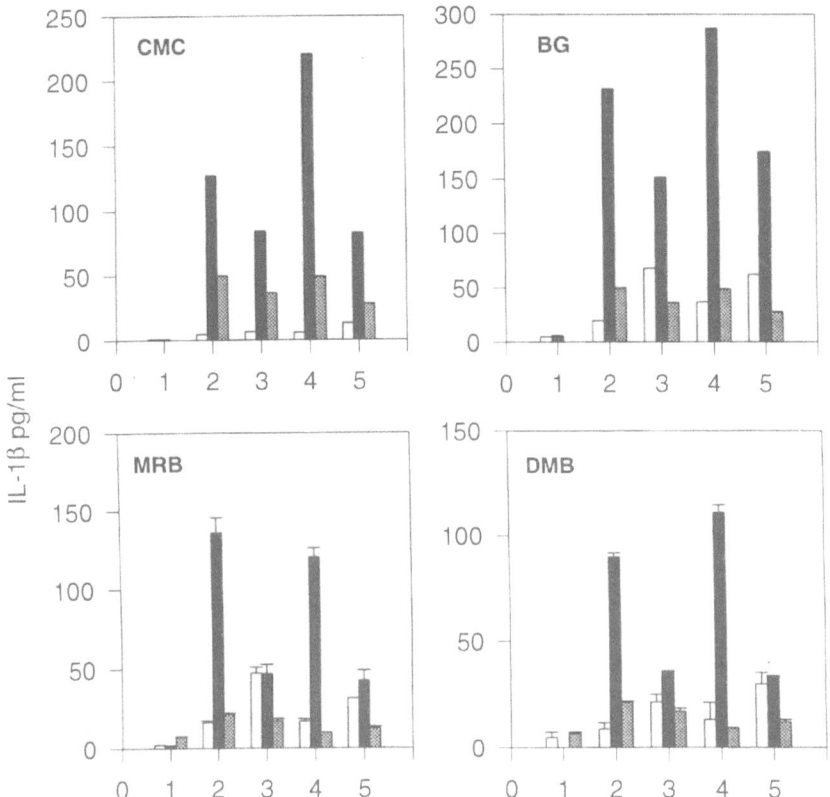

Fig. 2. Enhancement of interleukin (IL)-1β release from co-incubation of stimulated THP-1 cells and human neutrophils. Data shown are from three separate experiments using neutrophils isolated from four different donors. Experiments were set up as described for Fig. 1, and supernatants were assayed for the presence of IL-1β by enzyme-linked immunosorbent assay

vations were pursued in animal models where increased elastolytic activity in association with the progression of pulmonary hypertension was documented both by evidence of elastin breakdown and by high elastin turnover in the vessel walls (MARUYAMA et al. 1991). This was pursued further and measurement of increased activity of an EVE was shown very early after exposing infant rats to chronic hypobaric hypoxia (MARUYAMA et al. 1991) or injection of the toxin monocrotaline (within 2 days; TODOROVICH-HUNTER et al. 1992; YE and RABINOVITCH 1991). With the hypoxia model of pulmonary hypertension, the progressive development of medial hypertrophy was not associated with a further or sustained increase in elastase activity whereas, in the monocrotaline model, the increase in elastase activity paralleled the malignant course of the vascular disease (YE and RABINOVITCH 1991). Interestingly, hypoxia-induced pulmonary hypertension has the potential for regression and, in fact,

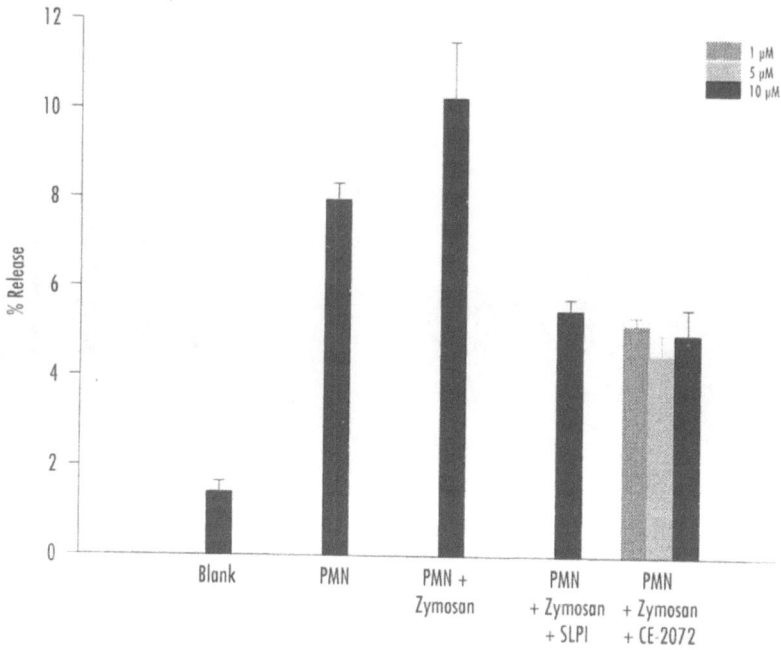

Fig. 3. Inhibition of extracellular-matrix (ECM) degradation and cytokine processing by activated human neutrophils. Neutrophils ($n = 1 \times 10^6$) isolated from an individual donor were incubated on biosynthetically radiolabeled ECM with or without zymosan in the absence or presence of either secretory leukoprotease inhibitor or CE-2072. Supernatants were sampled at 4 h, and solubilized radioactivity was measured. Total matrix radiolabel incorporation was assessed by solubilizing the remaining matrix with 10 N NaOH and adding these counts to those obtained from the supernatant. Results are expressed as the percent of total counts incorporated into the matrix solubilized under the conditions specified

in the infant model of monocrotaline-induced pulmonary hypertension (which also spontaneously regresses), the second increase in elastase activity did not occur (ZHU et al. 1994). Thus, while the initial increase in elastase activity could be expected to be associated with the induction of pathological changes, the second sustained increase in elastase activity appeared to suggest a more malignant and irreversible course of the disease. This was substantiated by studies in both the hypoxic and monocrotaline-injected rats, in which inhibition of elastase activity reduced or largely prevented the development of pulmonary hypertension and structural changes (MARUYAMA et al. 1991; YE and RABINOVITCH 1991). Moreover, in monocrotaline models of pulmonary hypertension, delayed treatment with elastase inhibitors was also highly effective in preventing the progression of pulmonary vascular disease (YE and RABINOVITCH 1991). Subsequently, a 23-kDa serine elastase, which we called EVE, was identified in pulmonary-artery tissue (ZHU et al. 1994). This elastase was purified by affinity chromatography with an antibody which recognizes the serine proteinase adipsin and, hence, may be immunologically related to adipsin.

Further studies indicated that the pathological mechanism whereby elastase is induced may require serum factors which could "leak" into the subendothelium when there is loss of barrier function in response to an injury (KOBAYASHI et al. 1994). Among the serum factors which induce smooth-muscle cell elastase activity is apolipoprotein A1 (THOMPSON et al. 1998). Once activated, elastases can liberate growth factors, such as fibroblast growth factor (FGF), from the ECM (THOMPSON and RABINOVITCH 1996), which stimulates smooth-muscle cell proliferation. Elastases can also proteolyze collagen, which induces synthesis of tenascin, a glycoprotein which amplifies the proliferative response (JONES et al. 1997a) and is upregulated in clinical and experimental pulmonary vascular disease (JONES et al. 1997b; JONES and RABINOVITCH 1996). In addition, elastin peptides can upregulate fibronectin, which switches smooth-muscle cells from a contractile to a migratory phenotype, ultimately resulting in neointimal formation (BOUDREAU et al. 1991).

Serine elastases have also been implicated in the pathobiology of coronary-artery disease after heart transplant. Specifically, there is increased elastase activity in the coronary arteries following experimental heterotopic heart transplant in piglets (OHO and RABINOVITCH 1994). When this elastase activity is inhibited by intravenous administration of the serine-elastase inhibitor elafin, both the coronary artery neointimal formation and the myocardial damage associated with rejection are remarkably reduced in animals after cardiac transplant (Fig. 4, s. appendix, page 400/401; COWAN et al. 1996). This suggests that increased elastase activity, either endogenously produced or derived from invading inflammatory cells, contributes to transplant arteriopathy and myocardial damage and that inhibition of elastase can attenuate myocyte necrosis and neointimal formation in coronary arteries.

The beneficial impact of serine-elastase inhibition on myocardial dysfunction after cardiac transplantation was unexpected and deserving of further exploration (LEE et al. 1997). A mouse acute viral myocarditis model was selected and, in this model, the concomitant administration of a biologically available oral elastase inhibitor markedly reduced the severity of myocarditis as judged by morphological evaluation of the heart, biochemical determination of collagen deposition and assessment of ventricular function. In addition, more recent studies have shown that serine-elastase inhibitors prevent neointimal formation in rabbit veins when they are interposed in an arterial position, as occurs with cardiac-bypass surgery (O'BLENES et al. in press). Since elastases have been shown to upregulate MMPs, it is likely that they also play a role in amplifying the responses associated with MMP expression (ITOH and NAGASE 1995) in atherosclerosis and restenosis, which include plaque instability and aneurysm formation (GALIS et al. 1997).

D. Potential Clinical Targets for Serine-Elastase Inhibition

Numerous publications have focused on the potential role of elastase inhibition, most notably by NE, in a variety of pulmonary disorders such as cystic fibrosis, acute respiratory-distress syndrome and chronic obstructive pulmonary diseases (emphysema and chronic bronchitis). This has recently been reviewed by VENDER (1996) and will not be discussed further in this chapter. This discussion will focus on the potential for elastase (NE, PR-3, EVE and/or ECE) inhibition in diseases that heretofore have not been considered as appropriate targets for this type of therapy. Underlying this discussion is the concept, illustrated in Fig. 5, of synergistic and/or redundant activities between serine elastases derived from a number of different cell types that can contribute to a common pathology.

Many vascular diseases have an inflammatory component and, as a result, elastases derived from the neutrophil and/or macrophage/monocyte populations can interact with or augment the elastases (EVE and ECE) derived from the vascular tissues themselves. This allows multiple biochemical pathways to effect a common endstage pathology and, as a result, may allow interventions

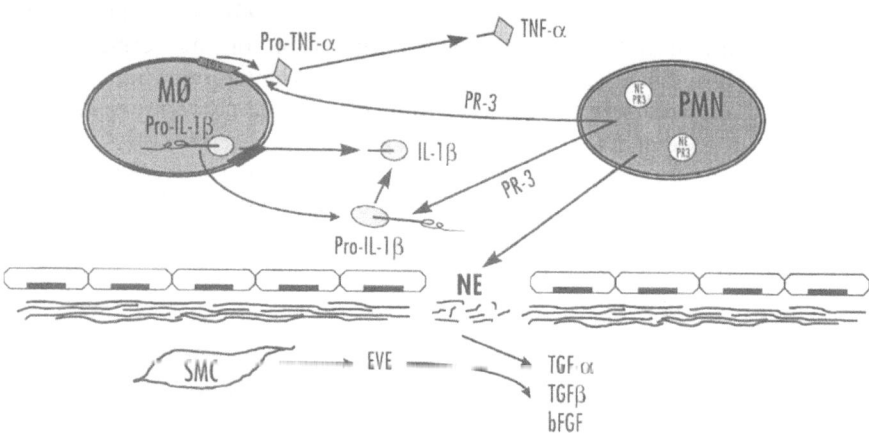

Fig. 5. Serine elastases in inflammatory and vascular diseases. Diagramatic illustration of our current understanding of the interactions between the serine elastases derived from multiple cell types in the context of inflammatory vascular conditions. Under these conditions, neutrophil elastase (NE) and endogenous vascular elastase can degrade vascular extracellular matrix, releasing latent or matrix-bound growth factors that can then go on to interact with vascular cells such as smooth muscle cells (SMC) and endothelial cells (EC). NE and endothelial cell elastase can interact with EC to stimulate release of platelet activating factor, which can both attract and stimulate inflammatory cells such as neutrophils (polymorphonuclear leukocytes) and macrophages/monocytes. Proteinase-3, then, can augment the release of monocyte derived cytokines such as tumor necrosis factor α and interleukin-1β, which can further amplify the inflammatory process

based on elastase inhibition, with appropriately designed specificities, to address a number of different diseases.

I. Restenosis, Atherosclerosis and Transplant Vasculopathy

Current estimates place the incidence of percutaneous transluminal coronary angioplasty (PTCA) at approximately 700,000 cases per year in the United States. While initial results obtained from this procedure are often satisfactory, between 30–50% of the PTCA sites will narrow again over a period of months due to restenosis. It is currently difficult to predict those patients who will have problems with restenosis. As a result, a safe and effective therapy for this disorder would likely be used in many patients undergoing PTCA.

Restenosis has been shown to be refractory to a variety of interventions, including stent placement and modification of coagulation or platelet function, interventions that appear to be more effective in preventing urgent revascularization procedures. This is borne out by a recent publication by SCHWARTZ et al. (1997), in which the rate of restenosis in arteries in which stents had been placed at the time of the original angioplasty (defined as at least 50% diameter stenosis within the stented segment) was 44% within 200 days of the first procedure. Also, there are no approved therapies for restenosis (as opposed to rethrombosis). Numerous interventions for the treatment of restenosis have been proposed, including inhibition of the factor-Xa/tissue-factor pathway, gene therapy directed at cell-cycle genes governing smooth-muscle proliferation, cell-adhesion-molecule-directed therapies and therapies directed at disrupting ECM production and deposition. While some of these therapies have shown promising results in animals, none have been advanced to the stage of clinical trials. Inhibition of the serine elastases may be particularly effective given the apparent role these enzymes play in vascular pathology.

When angioplasty with or without stenting is performed, the vascular endothelium is severely injured, thereby establishing both an inflammatory response characterized by early and robust neutrophil accumulation as well as a stimulus for vascular remodeling, including smooth-muscle cell proliferation and matrix deposition. In this regard, the elastases (NE and/or EVE) have been shown to release a variety of growth factors, including basic fibroblast growth factor (bFGF) and transforming growth factors alpha and beta (TGFα and TGFβ) from both vascular matrix and vascular smooth-muscle cells (THOMPSON and RABINOVITCH 1996; MUELLER et al. 1990). Similarly, a variety of cytokines, including IL-1β and TNFα, which can be either produced or released via proteolytic processing by PR-3, have been shown to be both potent stimulators of smooth-muscle cell fibronectin synthesis and deposition. Furthermore, elastase-specific elastin-derived peptides have been shown to be potent stimulators of smooth-muscle cell proliferation and matrix deposition, as measured by fibronectin synthesis. Finally, both elafin and α-1-proteinase inhibitor (α1PI) have been shown to reduce IL-1β-stimulated fibronectin synthesis by smooth-muscle cells. Together, these data clearly point to a pivotal role for the elastases in the response to vascular injury, leading to smooth-muscle proliferation and vascular hypertrophy that manifests anatomically and clinically as restenosis. Inhibition of this activity with a recombinant version of a naturally occurring elastase inhibitor, elafin, or with a synthetic inhibitor, SC-37698, prevents this response in two different models of vascular injury in two different species (rat and rabbit; Ye and Rabinovitch 1991; Cowan et al. 1996). In the rabbit, elafin was shown to ameliorate both the incidence and the severity of the accelerated vasculopathy associated with cardiac transplantation.

II. Myocardial Infarction

Acute myocardial infarction and myocardial ischemia is currently the leading cause of death and disability in Western societies. This problem will continue to grow with the demographic changes affecting all of the industrialized societies, most notably North America and western Europe. While significant advances have been made with respect to the management of acute myocardial infarction and both stable and unstable angina, these interventions do not address completely the underlying pathophysiology or the response to the interventions (ischemia/reperfusion injury and restenosis).

Tissue injury secondary to myocardial infarction initiates an acute inflammatory response characterized by a neutrophil-predominant cellular infiltrate. This response is exacerbated when the infarction is accompanied by ischemia/reperfusion injury that is the result of successful thrombolytic therapy. This inflammatory response can further complicate the clinical picture by increasing myocardial irritability and further damaging the tissues still at risk. As a result, altering neutrophil recruitment and activation in the context of this type of clinical setting may provide a benefit. Two different groups have shown that elastase inhibitors are capable of ameliorating the effects of myocardial ischemia/reperfusion injury in both dogs and rodents (Mehta et al. 1994; Murohara et al. 1995). Elastase inhibition, therefore, offers an opportunity to both ameliorate ischemia/reperfusion injury associated with an acute myocardial infarction and potentially improve the clinical outcome for those patients in whom angioplasty is the intervention of choice.

III. Stroke

The same arguments that apply to the treatment of myocardial infarction may also be applied to the treatment of ischemic stroke. Treatment of stroke is beginning to parallel that of myocardial infarction with the institution of thrombolytic therapy for the acute event, although this approach is controversial, and the use of carotid angioplasty to treat the underlying vascular lesions. Also, the demographics that are driving the increased incidence of myocardial infarction and myocardial ischemia are the same as those driving stroke and transient ischemic attacks of the central nervous system (CNS). As a result, elastase inhibition that addresses the inflammatory and vascular aspects of cardiovascular diseases, such as myocardial infarction and restenosis, would also be expected to have a beneficial effect in comparable diseases of other organ systems, such as the CNS.

IV. Bronchopulmonary Dysplasia

In addition to experimental studies in pulmonary hypertension and coronary-artery disease described in this chapter, elastase inhibitors might also be effective in a variety of other cardiopulmonary disorders. Of particular interest is a study in rats that led to a successful clinical trial in patients at risk for bronchopulmonary dysplasia. We had shown, in experimental studies in infant rats, that the cardiopulmonary sequelae of hyperoxic exposure could be mitigated by treatment with the elastase inhibitor Prolastin (α-1 antitrypsin, Bayer; KOPPEL et al. 1994). This agent was effective in improving lung compliance as well as associated structural changes in the pulmonary arteries associated with pulmonary hypertension and right-ventricular hypertrophy. A successful clinical trial has recently been completed and published in which we have shown a beneficial effect of Prolastin on the requirement for oxygen in premature infants at 28 and 36 weeks gestation and a significant reduction in pulmonary hemorrhage in the treated group (STISKAL et al. 1998). Other conditions in which elastase inhibitors may be promising, in addition to myocarditis (as described above), include Kawasaki disease, where a small clinical trial appears to show an impact of anti-elastase therapy on the development of coronary aneurysms (NAKANO et al. 1995).

E. Summary/Conclusion

Serine elastases are a family of related serine proteinases expressed by a wide variety of tissues and cell types that have the capability of degrading or processing a host of both soluble and matrix proteins. These enzymes are associated with a spectrum of acute and chronic diseases primarily involving tissues, such as the lung, skin and the vasculature, which require high elastin content. Serine-elastase inhibitors, whether they be highly selective for one enzyme or broadly targeted to include activity against multiple proteinases, offer significant potential opportunities for the treatment of a number of important human conditions.

References

Baugh RJ, Travis J (1976) Human leukocyte granule elastase: rapid isolation and characterization. Biochemistry 15:836–841
Bergenfeldt M, Axelsson L, Ohlsson K (1992) Release of neutrophil proteinase 4(3) and leukocyte elastase during phagocytosis and their interaction with proteinase inhibitors. Scand J Clin Lab Invest 52:823–829

Bories D, Raynal MC, Solomon DH, Darzynkiewicz Z, Cayre YE (1989) Down-regulation of a serine protease, myeloblastin, causes growth arrest and differentiation of promyelocytic leukemia cells. Cell 59:959–968

Boudreau N, Turley E, Rabinovitch M (1991) Fibronectin, hyaluronan, and a hyaluronan binding protein contribute to increased ductus arteriosus smooth muscle cell migration. Dev Biol 143:235–247

Bussolino F, Arese M, Silvestro L, Soldi R, Benfenati E, Sanavio F, Aglietta M, Bosia A, Camussi G (1994) Involvement of a serine protease in the synthesis of platelet-activating factor by endothelial cells stimulated by tumor necrosis factor-α or interleukin-1α. Eur J Immunol 24:3131–3139

Campanelli D, Melchior M, Fu Y, Nakata M, Shuman H, Nathan C, Gabay JE (1990) Cloning of cDNA for proteinase 3: a serine protease, antibiotic, and autoantigen from human neutrophils. J Exp Med 172:1709–1715

Campbell EJ, Senior RM, McDonald JA, Cox DL (1982) Proteolysis by neutrophils. Relative importance of cell-substrate contact and oxidative inactivation of proteinase inhibitors in vitro. J Clin Invest 70:845–852

Campbell EJ, Silverman EK, Campbell MA (1989) Elastase and cathepsin G of human monocytes. J Immunol 143:2961–2968

Cappelluti E, Harris RB (1994) Proteolytic maturation of transforming growth factor-α. Perspect Drug Discov Design 2:353–361

Coeshott C, et al. (1999) Converting enzyme-independent release of TNFα and IL-1β from a stimulated human monocytic cell line in the presence of activated neutrophils or purified proteinase 3. Proc Natl Acad Sci USA (in press)

Cowan B, Baron O, Crack J, Coulber C, Wilson GJ, Rabinovitch M (1996) Elafin, a serine elastase inhibitor, attenuates post-cardiac transplant coronary arteriopathy and reduces myocardial necrosis in rabbits following heterotopic cardiac transplantation. J Clin Invest 97:2452–2468

Croute F, Delaporte E, Bonnefoy JY, Fertin C, Thivolet J, Nicolas JF (1991) Interleukin-1β stimulates fibroblast elastase activity. Br J Dermatol 124:538–541

Csernok E, Ludemann J, Gross WL, Bainton DF (1990) Ultrastructural localization of proteinase 3, the target antigen of anti-cytoplasmic antibodies circulating in Wegener's granulomatosis. Am J Pathol 137:1113–1120

Csernok E, et al. (1991) Translocation of proteinase 3 on the cell surface of neutrophils: association with disease activity in Wegener's granulomatosis (abstract). Arthritis Rheum 34[Suppl 9]:571

Dalet-Fumeron V, Guinec N, Pagano M (1993) In vitro activation of pro-cathepsin B by three serine proteinases: leucocyte elastase, cathepsin G, and the urokinase-type plasminogen activator. FEBS Lett 332:251–254

Del Mar EG, Largman C, Brodrick JW, Fassett M, Geokas MC (1980) Substrate specificity of human pancreatic elastase 2. Biochemistry 19:468–472

Delclaux C, Delacourt C, D'Ortho MP, Boyer V, Lafuma C, Harf A (1996) Role of gelatinase B and elastase in human polymorphonuclear neutrophil migration across basement membrane. Am J Respir Cell Mol Biol 14:288–295

Edelstein C, Italia JA, Scanu AM (1997) Polymorphonuclear cells isolated from human peripheral blood cleave lipoprotein(a) and apolipoprotein(a) at multiple interkringle sites via the enzyme elastase. J Biol Chem 272:11079–11087

Fantuzzi G, Ku G, Harding MW, Livingston DJ, Sipe JD, Kuida K, Flavell RA, Dinarello CA (1997) Response to local inflammation of IL-1β-converting enzyme-deficient mice. J Immunol 158:1818–1824

Ferry G, Lonchampt M, Pennel L, de Nanteuil G, Canet E, Tucker GC (1997) Activation MMP-9 by neutrophil elastase in an in vivo model of acute lung injury. FEBS Lett 402:111–115

Galis ZS, Kranzhofer R, Fenton JW II, Libby P (1997) Thrombin promotes activation of matrix metalloproteinase-2 produced by cultured vascular smooth muscle cells. Arterioscler Thromb Vasc Biol 17:483–489

Gronski TJ Jr, Martin RL, Kobayashi DK, Walsh BC, Holman MC, Huber M, Van Wart HE, Shapiro SD (1997) Hydrolysis of a broad spectrum of extracellular matrix proteins by human macrophage elastase. J Biol Chem 272:12189–12194

Heck LW, Blackburn WD, Irwin MH, Abrahamson DR (1990) Degradation of basement membrane laminin by human neutrophil elastase and cathepsin G. Am J Pathol 136:1267–1274

Henshaw TJ, Malone CC, Gabay JE, Williams RC Jr (1994) Elevations of neutrophil proteinase 3 in serum of patients with Wegener's granulomatosis and polyarteritis nodosa. Arthritis Rheum 37:104–112

Itoh Y, Nagase H (1995) Preferential inactivation of tissue inhibitor of metalloproteinases-1 that is bound to the precursor of matrix metalloproteinase 9 (progelatinase B) by human neutrophil elastase. J Biol Chem 270:16518–16521

Janoff A (1973) Purification of human granulocyte elastase by affinity chromatography. Lab Invest 29:458–464

Janoff A, Scherer J (1968) Mediators of inflammation in leukocyte lysosomes. IX. Elastinolytic activity in granules of human polymorphonuclear leukocytes. J Exp Med 128:1137–1155

Jennette JC, Hoidal JR, Falk RJ (1990) Correspondence: specificity of anti-neutrophil cytoplasmic autoantibodies for proteinase 3. Blood 75:2263–2264

Jones PL, Rabinovitch M (1996) Tenascin-C is induced with progressive pulmonary vascular disease in rats is functionally related to increased smooth muscle cell proliferation. Circ Res 79:1131–1142

Jones PL, Cowan KN, Rabinovitch M (1997a)Tenascin-C, proliferation and subendothelial fibronectin in progressive pulmonary vascular disease. Am J Pathol 150:1349–1360

Jones PL, Crack J, Rabinovitch M (1997b) Regulation of Tenascin-C, a vascular smooth muscle cell survival factor that interacts with the alpha v beta 3 integrin to promote epidermal growth factor receptor phosphorylation and growth. J Cell Biol 139:279–293

Jordan RE, Nelson RM, Kilpatrick J, Newgren JO, Esmon PC, Fournel MA (1989) Inactivation of human antithrombin by neutrophil elastase. J Biol Chem 264:10493–10500

Kao RC, Wehner NG, Skubitz KM, Gray BH, Hoidal JR (1988) Proteinase 3: a distinct human polymorphonuclear leukocyte proteinase that produces emphysema in hamsters. J Clin Invest 82:1963–1973

Klebanoff SJ, Kinsella MG, Wight TN (1993) Degradation of endothelial cell matrix heparan sulfate proteoglycan by elastase and the myeloperoxidase-H_2O_2-chloride system. Am J Pathol 143:907–917

Kleniewski J, Donaldson V (1988) Granulocyte elastase cleaves human high molecular weight kininogen and destroys its clot-promoting activity. J Exp Med 167:1895–1907

Kobayashi J, Wigle D, Childs T, Zhu L, Keeley FW, Rabinovitch M (1994) Serum-induced vascular smooth muscle cell elastolytic activity through tyrosine kinase intracellular signalling. J Cell Physiol 160:121–131

Koppel R, Han RN, Cox D, Tanswell AK, Rabinovitch M (1994) a1-Antitrypsin protects neonatal rats from pulmonary vascular and parenchymal effects of oxygen toxicity. Pediatr Res 36:763–770

Lee J-K, et al. (1997) Elastase inhibitor reduces sequelae of experimental murine myocarditis. Circulation 96:738

Lewis RW, Harwood JL, Tetley TD, Harris E, Richards RJ (1993) Degradation of human and rat surfactant apoprotein by neutrophil elastase and cathepsin G. Biochem Soc Trans 21:206S

Liau DF, Yin NX, Huang J, Ryan SF (1996) Effects of human polymorphonuclear leukocyte elastase upon surfactant proteins in vitro. Biochim Biophys Acta 1302:117–128

Maruyama K, Ye CL, Woo M, Venkatacharya H, Lines LD, Silver MM, Rabinovitch M (1991) Chronic hypoxic pulmonary hypertension in rats and increased elastolytic activity. Am J Physiol 261:H1716–H1726

McDonald JA, Kelley DG (1980) Degradation of fibronectin by human leukocyte elastase. J Biol Chem 255:8848–8858

McGowan SE (1990) Mechanisms of extracellular matrix proteoglycan degradation by human neutrophils. Am J Respir Cell Mol Biol 2:271–279

McRae B, Nakajima K, Travis J, Powers JC (1980) Studies on reactivity of human leukocyte elastase, cathepsin G, and porcine pancreatic elastase toward peptides including sequences related to the reactive site of α_1-protease inhibitor (α_1-antitrypsin). Biochemistry 19:3973–3978

Mecham RP, Broekelmann TJ, Fliszar CJ, Shapiro SD, Welgus HG, Senior RM (1997) Elastin degradation by matrix metalloproteinases. J Biol Chem 272:18071–18076

Mehta JL, Nichols WW, Nicolini FA, Hendricks J, Donnelly WH, Saldeen TG (1994) Neutrophil elastase inhibitor ICI 200,880 protects against attenuation of coronary flow reserve and myocardial dysfunction following temporary coronary artery occlusion in the dog. Cardiovasc Res 28:947–956

Mueller SG, Paterson AJ, Kudlow JE (1990) Transforming growth factor α in arterioles: cell surface processing of its precursor by elastases. Mol Cell Biol 10:4596–4602

Murohara T, Guo JP, Lefer AM (1995) Cardioprotection by a novel recombinant serine protease inhibitor in myocardial ischemia and reperfusion injury. J Pharmacol Exp Ther 274:1246–1253

Nakajima K, Powers JC, Ashe BM, Zimmerman M (1979) Mapping the extended substrate binding site of cathepsin G and human leukocyte elastase. J Biol Chem 254:4027–4032

Nakano M, et al. (1995) Preventive effects of ulinastatin for coronary artery aneurysm formation in Kawasaki disease. Elsevier Science BV, Amsterdam, pp 364–371

O'Blenes S, et al. Gene transfer with elafin reduces neointimal formation and promotes adventitial remodeling in vein grafts. Circulation (in press)

Oho S, Rabinovitch M (1994) Post-cardiac transplant arteriopathy in piglets is associated with fragmentation of elastin and increased activity of a serine elastase. Am J Pathol 145:202–210

Okada Y, Watanabe S, Nakanishi I, Kishi J, Hayakawa T, Watorek W, Travis J, Nagase H (1988) Inactivation of tissue inhibitor of metalloproteinases by neutrophil elastase and other serine proteinases. FEBS Lett 229:157–160

Owen CA, Campbell EJ (1995) Neutrophil proteinases and matrix degradation. The cell biology of pericellular proteolysis. J Cell Biol 6:367–376

Packard BZ, Mostowski HS, Komoriya A (1995) Mitogenic stimulation of human lymphocytes mediated by a cell surface elastase. Biochim Biophys Acta 1269:51–56

Padrines M, Wolf M, Walz A, Baggiolini M (1994) Interleukin-8 processing by neutrophil elastase, cathepsin G and proteinase-3. FEBS Lett 352:231–235

Pison U, Tam EK, Caughey GH, Hawgood S (1989) Proteolytic inactivation of dog lung surfactant-associated proteins by neutrophil elastase. Biochim Biophys Acta 992:251–257

Rabinovitch M, Bothwell T, Hayakawa BN, Williams WG, Trusler GA, Rowe RD, Olley PM, Cutz E (1986) Pulmonary artery endothelial abnormalities in patients with congenital heart defects and pulmonary hypertension: a correlation of light with scanning electron microscopy and transmission electron microscopy. Lab Invest 55:632–653

Rao NV, Wehner NG, Marshall BC, Gray WR, Gray BH, Hoidal JR (1991) Characterization of proteinase-3 (PR-3), a neutrophil serine proteinase. J Biol Chem 266:9540–9548

Rice A, Banda MJ (1995) Neutrophil elastase processing gelatinase A is mediated by extracellular matrix. Biochemistry 34:9249–9256

Robache-Gallea S,Morand V, Bruneau JM, Schoot B, Tagat E, Realo E, Chouaib S, Roman-Roman S (1995) In vitro processing of human tumor necrosis factor-α. J Biol Chem 270:23688–23692

Sandhaus RA, Werb Z (1981) Characterization of the major form of elastase in neutrophil granules. Fed Proc 40:1004

Schwartz L, Blew B, Bui S (1997) Intracoronary-stent placement for coronary artery disease. Lancet 350:113–114

Shapiro SD (1994) Elastolytic metalloproteinases produced by human mononuclear phagocytes. Am J Respir Crit Care Med 150:S160–S164

Si-Tahar M, Pidard D, Balloy V, Moniatte M, Kieffer N, Van Dorsselaer A, Chignard M (1997) Human neutrophil elastase proteolytically activates the platelet integrin $\alpha_{IIb}\beta_3$ through cleavage of the carboxyl terminus of the α_{IIb} subunit heavy chain. J Biol Chem 272:11636–11647

Starkey PM (1997) Elastase and cathepsin G; the serine proteinases of human neutrophil leucocytes and spleen, proteinases in mammalian cells and tissues. North-Holland, Amsterdam, pp 57–58

Stiskal JA, Dunn MS, Shennan AT, O'Brien KKE, Kelly EN, Koppel RI, Cox DW, Ito S, Chappel SL, Rabinovitch M (1998) Alpha1-proteinase inhibitor therapy for the prevention of bronchopulmonary dysplasia in premature infants: a randomized controlled trial. Pediatrics 101:89–94

Thompson K, Rabinovitch M (1996) Exogenous leukocyte and endogenous elastases can mediate mitogenic activity in pulmonary artery smooth muscle cells by release of extracellular matrix-bound basic fibroblast growth factor. J Cell Physiol 166:495–505

Thompson K, Kobayashi J, Childs T, Wigle D, Rabinovitch M (1998) Endothelial and serum factors which include apolipoprotein A1 tether elastin to smooth muscle cells inducing serine elastase activity via tyrosine kinase-mediated transcription and translation. J Cell Physiol 174:78–89

Todorovich-Hunter L, Dodo H, Ye C, McCready L, Keeley FW, Rabinovitch M (1992) Increased pulmonary artery elastolytic activity and monocrotaline-induced progressive hypertensive pulmonary vascular disease in adult rats compared to infant rats with non-progressive disease. Am Rev Respir Dis 146:213–223

Tosi MF, Zakem H, Berger M (1990) Neutrophil elastase cleaves C3bi on opsonized pseudomonas as well as CR1 on neutrophils to create a functionally important opsonin receptor mismatch. J Clin Invest 86:300–308

Vender RL (1996) Therapeutic potential of neutrophil-elastase inhibition in pulmonary disease. J Investig Med 44:531–539

Vissers MCM, Winterbourn CC (1988) Activation of human neutrophil gelatinase by endogenous serine proteinases. Biochem J 249:327–331

Weiss SJ, Regiani S (1984) Neutrophils degrade subendothelial matrices in the presence of alpha-1-proteinase inhibitor. J Clin Invest 73:1297–1303

Werb Z, Banda MJ, McKerrow JH, Sandhaus RA (1982) Elastases and elastase degradation. J Invest Dermatol 79(Suppl 1):154s–159s

Ye C, Rabinovitch M (1991) Inhibition of elastolysis by SC-37698 reduces development and progression of monocrotaline pulmonary hypertension. Am J Physiol 261:H1255–H1267

Zhu L, Wigle D, Hinek A, Kobayashi J, Ye C, Zuker M, Dodo H, Keeley FW, Rabinovitch M (1994) The endogenous vascular elastase which governs development and progression of monocrotaline-induced pulmonary hypertension in rats is a novel enzyme related to the serine proteinase adipsin. J Clin Invest 94:1163–1171

CHAPTER 15
Inhibitors of Thrombin and Factor Xa

A.H. SCHMAIER

A. Introduction

In medicine today, much effort is being made to develop new protease inhibitors for use as anticoagulants in multiple disease states. Deep venous thrombosis and pulmonary embolism, in particular, are major medical problems in the developed world. Similarly, myocardial infarction, stroke, and peripheral vascular disease are arterial thromboses for which the addition of anticoagulant/antiplatelet agents have resulted in reduction of adverse events upon presentation and improved treatment outcomes. Therapies for both venous and arterial thrombosis represent major growth areas in the pharmaceutical industry. The basis for many of these anticoagulant agents has arisen from naturally occurring protease inhibitors in man and in other creatures. This interest arises from a need in clinical medicine to improve anticoagulant therapeutics for both naturally occurring disease processes and new medical therapeutic interventions. Characterization of nature's inhibitors can provide for the development of important therapeutics. Development of novel anticoagulants needs to consider the major clot-forming enzymes in the hemostatic system, thrombin and factor Xa. Thus, protease inhibitors to these two enzymes have the potential to serve as anticoagulants, i.e., to inhibit hemostatic clot formation. In this manuscript, the term "anticoagulants" will be used to describe agents that interfere with proteins that participate in the plasma coagulation system which has been traditionally termed the coagulation cascade. The term "antiplatelet agents" will be used to describe entities that specifically interfere with platelet activation only. Both anticoagulants and antiplatelet agents prevent thrombosis in blood vessels. The term "antithrombotics" will be used in this manuscript to indicate combined anticoagulant and antiplatelet activity of agents.

The hemostatic system consists of three components: the fibrin clot promoting system, the cellular system which mostly consists of platelets, and the clot lysing system. Physiologic hemostasis consists of a balance among these three systems. For the purposes of this chapter, the fibrinolytic system will not be dealt with further. Formation of the hemostatic plug is the sum of the clot-forming system and the activation of platelets (Fig. 1). The clot-forming system consists of a series of proteolytic reactions which result in the formation of factor Xa and thrombin. If nature had separated the fibrin clot-forming

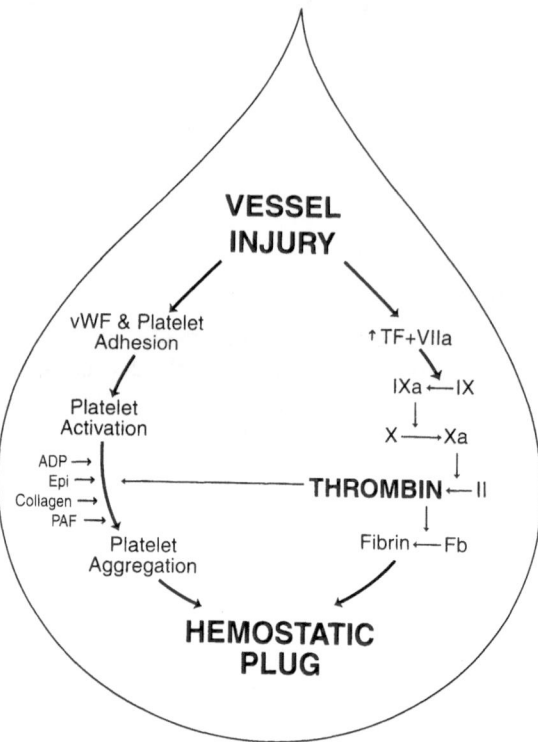

Fig. 1. Model of hemostasis. After vessel injury, there are two pathways that contribute to the hemostatic plug. On the *right* is a group of proteins which, when activated by increased expression of tissue factor which complexes with factor VIIa, initiate a series of reactions that culminates in thrombin formation. Thrombin proteolyzes fibrinogen to form fibrin and contributes to the hemostatic plug. On the *left* is the platelet contribution to the hemostatic plug. Thrombin, the main clotting enzyme, is also a potent platelet agonist. Thus interference with thrombin results in inhibition of fibrin formation and platelet activation

pathway from the platelet activation pathway, anticoagulant and antiplatelet agents would be totally separate entities (Fig. 1). However, evolution has provided us with a system where the main clot-forming enzyme, thrombin, is also a platelet agonist (Fig. 1). The full significance of thrombin in physiologic platelet activation has yet to be appreciated since no inhibitor selective to thrombin activation of platelets but not interfering with fibrin clot formation has been developed to examine this question.

Current antithrombins interfere with both the clot-promoting pathway and platelet activation. This results in increased surgical bleeding. It has been shown that anticoagulants that work at the level of thrombin are similar to antiplatelet agents in the degree of surgical bleeding present (HARKER et al.

Table 1. Classification of thrombin and factor Xa inhibitors

Direct thrombin inhibitors:	
Active site:	PPACK, argatroban
Active site & exosite I:	Heparin cofactor II, hirudin, hirulog
Exosite I:	Hirugen, DNA aptamers
Active site and exosite II:	Antithrombin and heparin
Exosite II:	Antibody to exosite II
Factor Xa inhibitors:	
Naturally occurring in man:	Antithrombin
	Tissue factor pathway inhibitor-1
	Protease nexin2/amyloid β-protein precursor
	Tissue factor pathway inhibitor-2
Other inhibitors	Low-molecular-weight heparins
	Danaparoid sodium
	Tick anticoagulant peptide
	Antistasin
	Ecostatin
	Nematode anticoagulant peptide

1995). Alternatively, using agents that inhibit at or above factor Xa are associated with less surgical bleeding, but are effective anticoagulants to prevent fibrin clots (HARKER et al. 1995). However, if one wants to prevent thromboplastin-induced venous thrombosis in a rabbit model, elimination of factor Xa or thrombin protects against thrombosis, whereas deficiency of factors IX or VII does not (ZIVELIN et al. 1993). These two in vivo experiences suggest that thrombin and factor Xa are important targets for the development of anticoagulants to prevent fibrin clots.

Inhibitors of both thrombin and factor Xa provide for adequate anticoagulation for venous thrombosis and pulmonary embolism. However, thrombin inhibition also leads to inhibition of platelet function since thrombin is a potent platelet agonist. Therapeutic inhibition of platelet function is important to prevent myocardial infarction, stroke, and graft closure. The use of potent thrombin inhibitors that interfere with all of thrombin's functions in patients with acute coronary syndromes has been associated with too much bleeding, including intracerebral hemorrhage (see below). Thus those who develop medical pharmaceuticals are being challenged to create potent and selective anticoagulants for medical conditions and interventional procedures. Much progress has been made in this field over a short time, but further progress will require deeper understanding of the actions of the targeted proteases and the consequences of their inhibition. The purpose of the present manuscript is not to provide an exhaustive review of all known thrombin and factor Xa inhibitors. Rather, it aims to present a mechanistic framework by which thrombin and factor Xa inhibitors can be evaluated as to their targets of action and presumed biologic effects (Table 1).

B. Thrombin Inhibitors

Thrombin, a 34-kDa serine protease, is the activated form of prothrombin, a 72-kDa zymogen. Prothrombin is activated to thrombin by factor Xa in a multiprotein assembly with factor V on cell membranes of platelets or endothelial cells called "prothrombinase". The critical control points in the hemostatic system are the formation of factor Xa, also in a multiprotein assembly termed "tenase", and the formation of thrombin. The catalytic efficiency ratio of factor Xa and thrombin formation in tenase and prothrombinase are 1.2×10^9 and 4.0×10^5, respectively (Jenny and Mann 1998). Thus, anticoagulant activity can be achieved by directly inhibiting thrombin or by inhibiting thrombin generation by being directed to factor Xa.

Thrombin, in addition to being the major clot-promoting enzyme by proteolysing of fibrinogen, has multiple other substrates including factor V, factor VIII, factor XIII, factor XI, protein S, protein C, carboxypeptidase U [thrombin-activable fibrinolysis inhibitor (TAFI)], and protease-activated receptors 1,3, and 4. In fact, thrombin's interaction with any of its receptors has to be considered an enzyme/substrate interaction rather than a ligand/receptor interaction (Hayes et al. 1994). Further, the affinity of thrombin towards its substrates varies which, in turn, may order its physiologic activities. Thrombin's affinity is highest for protein C in the presence of its endothelial cell receptor, thrombomodulin. This activity makes thrombin initially an anticoagulant protein, whose activity is limited by the finite number of thrombomodulin receptors in the intravascular compartment. Similarly, thrombin's activation of TAFI is potentiated by the presence of thrombomodulin (Bajzar et al. 1995, 1996). Alternatively, the inactivation of protein S by thrombin is a slow reaction, reflecting the procoagulant nature of this reaction occurring in the presence of large thrombin concentrations. Since thrombin has many substrates, its influence goes beyond fibrin clot formation. As a result of the presence of protease-activated receptors 1, 3, and 4, thrombin stimulates growth and proliferation in a number of cells (fibroblasts, smooth muscle cells, astrocytes) that express this receptor. Further, proteolysis of protease-activated receptor 1 may have a role in cancer cell metastasis (Nierodzik et al. 1998). Thus, creating inhibitors to thrombin itself will create agents that may interfere with all activities related to thrombin.

Characterization of the various mechanisms of inhibitors to thrombin requires that some fundamental structural information is appreciated about the thrombin molecule itself. Human α-thrombin has been crystallized in two laboratories (Bode et al. 1989, 1992; Rydel et al. 1990; Vijayalakshmi et al. 1994). The deep and narrow active-site cleft of thrombin consists of an S1 pocket with the presence of an asparagine 189, serine 195, and histidine 57, making thrombin a trypsin-like serine protease. In contrast to other serine proteases, thrombin's specificity toward its substrates is not determined by subsites surrounding the active site residues alone. Thrombin is a highly charged protein with two regions with clusters of positively charged residues. One is close to its β-cleavage site, forming a cleft for protein binding for fibrinogen and other substrates such as thrombomodulin, fibrin monomers, fibrin E-domain, and protease activated receptor 1. It is termed anion binding exosite I. The other is close to the carboxy-terminal B-chain helix forming a heparin-binding site. It is termed anion binding exosite II.

These regions participate in thrombin's interaction with its substrates either directly, in the case of fibrinogen, or, indirectly, as in the case of thrombomodulin-binding protein C or TAFI. Thus, naturally occurring as well as synthetic thrombin inhibitors interact with one or more of these critical structures on thrombin: its active site, anion-binding exosite I or anion-binding exosite II. Inhibitors binding to these regions on thrombin will be relatively nonselective in their activity since they should interfere with all activities of thombin. This fact may be the reason that there are too many bleeding complications with the potent direct thrombin inhibitors. In addition to these so-called nonselective inhibitors of thrombin, selective inhibitors to thrombin have begun to be created by targeting their activity to certain substrates of thrombin. Specificity of these latter thrombin inhibitors are determined by their selectivity to certain thrombin substrates. The following is a characterization of naturally occurring and artificial direct and indirect thrombin inhibitors.

I. Direct Thrombin Inhibitors

1. Naturally Occurring Thrombin Inhibitors in Humans

In humans, there are only a few endogenous inhibitors of α-thrombin. Alpha$_2$-macroglobulin and the serpins antithrombin (antithrombin III), heparin cofactor II, protease nexin I, and plasminogen activator inhibitor 1 are naturally occurring inhibitors of thrombin (ROSENBERG and DAMUS 1973; TOLLEFSEN et al. 1982; CUNNINGHAM and FARRELL 1986, NASKI et al. 1993). There are no known Kunitz-type protease inhibitors of thrombin. Alpha$_2$-macroglobulin is a general protease inhibitor that is not highly specific for thrombin. The interaction of thrombin with serpins is considerably enhanced by the acidic glycosaminoglycan heparin (LI et al. 1976; BJORK and LINDAHL 1982; OLSON and SHORE 1982).

In an elegant series of experiments, Olson and his colleagues have characterized the kinetics of thrombin inhibition by antithrombin in the presence of heparin (OLSON and SHORE 1982, 1986; OLSON 1988; OLSON et al. 1991; OLSON and BJORK 1991; BJORK et al. 1992). Since heparin binds to anion-binding exosite II, antithrombin must be considered an active site and exosite II an inhibitor. Selective mutations in the exosite-II regions of thrombin alter the affinity of the thrombin for heparin-agarose, which correlates with the rate of inhibition by antithrombin with heparin (SHEEHAN and SADLER 1994). However, occupation of thrombin's anion-binding exosite I with hirugen, an exosite-I-binding protein (see below), results in a situation where antithrombin can inhibit thrombin although at a decreased rate of interaction (BOCK et al. 1997). Interference with thrombin's exosite II may provide a novel form of anticoagulation. COLWELL and his colleagues (1997, 1998) have reported a unique naturally occurring antibody arising in a patient with myeloma which is directed to thrombin's exosite I. This antibody produced a severe hemorrhagic disorder. In addition to heparin, the thrombin–antithrombin complex associates with vitronectin in plasma and this association influences the rate of clearance of these complexes from plasma (ILL and RUOSLAHTI 1985; DE BOER et al. 1993). Alternatively, vitronectin promotes the inactivation of thrombin by plasminogen activator inhibitor 1 (NASKI et al. 1993). Unlike antithrombin, the ability of heparin cofactor II to inhibit thrombin requires that the glycosaminoglycan interact with exosite I (SHEEHAN et al. 1993, 1994). Thrombin is also inhibited by a membrane-associated serpin on fibroblasts, platelets, and other cells, named protease nexin I, whose activity is potentiated by heparin (GRONKE et al. 1987).

Other novel naturally occurring inhibitors of thrombin have been described. COUGHLIN et al. (1993) have described a serine protease from human placenta or K562 cells with a $M_r = 38\,\text{kDa}$. Thrombin can also be proteolytically inactivated by secretory granule chymase (PEJLER and KARLSTROM 1993). This inactivating enzyme is potentiated by the presence of optimal concentrations of heparin. Although heparin functions as an anticoagulant by potentiating antithrombin and other serpins' inhibition of thrombin and other proteases, heparin is limited as to its effectiveness. Interest in developing novel antithrombins was spurred by the recognition that clot-bound thrombin is resistant to inactivation by antithrombin and heparin (WEITZ et al. 1990). This finding was an impetus to discover potent thrombin inhibitors to prevent thrombolysis failures due to uninhibited clot-bound thrombin (Fig. 2). Appreciation of the relative weaknesses of heparin as an antithrombin has produced a wide range of potent thrombin inhibitors that have been applied to clinical medicine.

2. Naturally Occurring or Synthetic Thrombin Inhibitors Applied to Man

a) Active-Site Inhibitors

A number of compounds designed to interact with the active site of thrombin have been produced. The best characterized, synthetic peptide inhibitor is D-phenylalanine-L-prolyl-L-arginyl chloromethylketone (PPACK, FPRMeCl, D-Phe-Pro-Arg-CH$_2$Cl), which is a tight binding inhibitor of thrombin with a K_i of $3.7 \times 10^{-8}\,\text{M}$ (COLLEN et al. 1982). When PPACK was co-crystallized with thrombin, the inhibitor interacted with thrombin's active site by forming a hydrophobic cage formed by Ile[174], Trp[215], Leu[99], His[57], Tyr[60A], and Trp[60D] (BODE et al. 1989). The canyon-like active-site cleft of thrombin is characterized by two insertion loops around Trp[60D] and Trp[148]. With PPACK in the active site, the first loop around Trp[60D] is relatively rigid, whereas the opposite loop around Trp[148] attains different conformations depending on the complex state and crystalline environment (BODE et al. 1992). When PPACK complexes with

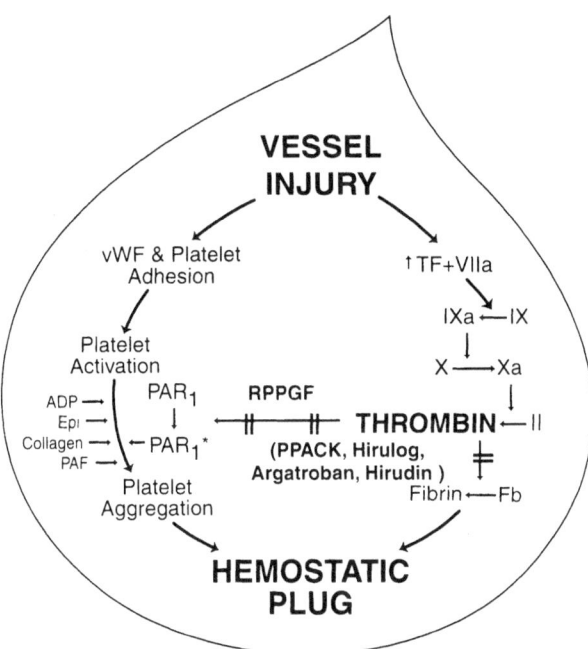

Fig. 2. Anti-thrombins. Direct antithrombins (PPACK, Hirulog, Argatroban, Hirudin) inhibit thrombin's ability to proteolyze fibrinogen to form a fibrin clot and interfere with thrombin's ability to activate platelets. The peptide RPPGF also interferes with thrombin's ability to proteolyze protease-activated receptor-1 (*PAR$_1$*)

thrombin, the side chain of Asp[189] and the segment Arg[221A]–Gly[223] move to provide space for the inhibitor (VIJAYALAKSHMI et al. 1994). Further, PPACK expels eight water molecules from the active site, but the inhibitor complex is resolved with five other water molecules.

A derivative of a naturally occurring product, nazumamide A, inhibits thrombin and on crystallographic studies was found to be a retro-binding, i.e., amino-terminal end in the active site, inhibitor of thrombin (NIENABER and AMPARO 1996). A Kazal-type serine protease inhibitor with specificity to thrombin was isolated from *Rhodnius prolixus* (FRIEDRICH et al. 1993). This active site inhibitor of thrombin inhibited it with a K_i of 2×10^{-13} M. More recently, using the three-dimensional structures of thrombin and the leech-derived tryptase inhibitor (LDTI), three variant forms of the inhibitor were produced (MORENWEISER et al. 1997). Potent forms of this inhibitor were produced by trimming the inhibitor reactive site loop to fit thrombin's narrow active site cleft (MORENWEISER et al. 1997). This designed thrombin blocker is highly potent with inhibitory activity in the picomolar range. Similarly, a macrocyclic peptide cyclotheonamide A has been isolated from the marine sponge, *Theonella* sp. (Maryanoff et al. 1993). Although substantially weaker than argatroban (see below), these latter compounds have a unique structure and represent a novel class of naturally occurring serine protease inhibitors of thrombin.

A number of synthetic active-site inhibitor compounds have been created. The best known and most advanced towards clinical development is argatroban. Argatroban, (2R,4R)-4-methyl-1-[N^2-(3(RS)-methyl-1,2,3,4,-tetrahydro-8-quinolinesulfonyl)-L-arginyl]piperidine-2-carboxylic acid monohydrate) is a potent reversible inhibitor at the active site of thrombin with K_i values of 19–39 nM (Okamoto et al. 1981; Kikumoto et al. 1984). Argatroban is able to inhibit clot-bound thrombin advantageously over heparin or hirudin (Berry et al. 1994). It has also been shown to be synergistic with $\alpha_{IIb}\beta_3$ integrin inhibitors in preventing platelet-rich clots in animal models (Imura et al. 1992). Argatroban has been used in animals and man with safe and effective anticoagulation. A series of synthetic compounds containing α-aminoboronic acids also were produced that inhibit thrombin (Kettner et al. 1990). After intravenous injection, these compounds were effective inhibitors of venous thrombosis in rabbits (Knabb et al. 1992). Another synthetic direct thrombin inhibitor, BMS-183507, was also found to be a retro-binding peptide on crystallography when complexed with α-thrombin (Tabernero et al. 1995). This compound which has a K_i for α-thrombin of about 17 nM has an alkyl-guanidino moiety which fits into the specificity pocket of the active site of thrombin.

b) Active-Site and Exosite-1 Inhibitors

Hirudin from the European medicinal leech, *Hirudo medicinalis*, is the most potent, naturally occurring inhibitor of human α-thrombin. Hirudin is a polypeptide of 65 residues which consists of a compact amino-terminal head with three internal Cys–Cys disulfide bridges and a long polypeptide carboxy-terminal tail (Rydel et al. 1990). Residues Ile[1] to Try[3] of the amino terminal end of hirudin interact with Ser[214] to Glu[217] of thrombin with the nitrogen of Ile[1] making a hydrogen bond with Ser[195] of the catalytic site, but not by occupying the specificity pocket of the active site of thrombin (Rydel et al. 1990). Simultaneously, the carboxy-terminal segment of hirudin makes numerous electrostatic interactions with the anion-binding exosite I of thrombin (Rydel et al. 1990). Peptide fragments of the amino-terminal head region of hirudin bind to thrombin, but do not inhibit the enzyme's ability to cleave small synthetic substrates (Stone and Marganore 1993). This information indicates that these peptides and hirudin itself do not occupy the active site of thrombin.

This information has led to the development of hirudin-based compounds which improve upon their ability to interact with the active site of thrombin. Modification of the Tyr[3] with Trp or Phe increases hirudin's affinity for thrombin three- to sixfold and a Thr substitution results in a 450-fold increase in the

K_i (LAZAR et al. 1991). Another group of hirudin-based peptides have been prepared which contain an amino-terminal extension capable of binding to the catalytic site joined by a linker segment to structural determinants for binding to the anion-binding exosite I (STONE and MARGANORE 1993). The bivalent compounds called "hirulogs" occupy the active site S1 pocket and the anion-binding exosite to increase inhibition of thrombin and, presumably, their anti-coagulant effect.

A group of combined active-site and exosite-I inhibitors have been created using this technology to combine PPACK as the active site inhibitor with the anion-binding exosite-I region of hirudin with the linker technology. The inhibitory constants of these compounds vary with the modifications on the active-site or exosite-binding segment modifications (STONE and MARGANORE 1993). Another modification in the hirudin structure performed to possibly increase its anticoagulant features was the replacement of a Ser-Asp-Gly-Glu sequence with an Arg-Gly-Asp-Ser (RGDS) sequence to create "hirudisins" (KNAPP et al. 1992). These agents would presumably associate with integrins on the platelet or endothelial membrane to localize thrombin inhibition. Since hirudin is such a tight binding irreversible inhibitor of thrombin, it has been used as a molecular probe to determine whether there is a basal level of thrombin present in vivo (ZOLDHELYI et al. 1993). In vivo, there is a basal pool of hirudin-accessible thrombin in the intravascular space which, if free, would be sufficient to sustain intravascular coagulation. Hirudin and related compounds have been shown to be potent thrombin inhibitors in vitro, inhibiting clot-bound thrombin. They have been efficacious in preventing both venous and arterial thrombosis (KELLY et al. 1991, 1992). Hirudin is more efficacious than heparin in inhibiting thrombin; heparin is more efficacious in preventing thrombin generation (RAO et al. 1996). However, potency and high specificity may not be what is necessary to create a safe and effective anticoagulant for arterial thrombosis. Three multicenter phase-III clinical trials with hirudin or related compounds in the management of acute coronary syndromes were terminated due to excessive hemorrhage, some of which was intracranial (ANTMAN 1994; GUSTO IIA INVESTIGATORS 1994; NEUHAUS et al. 1994). It appears that the dose of hirudin efficacious to manage acute coronary syndromes is too close to the concentration of this agent that is associated with unacceptably high incidence of serious hemorrhage.

c) Exosite-I Inhibitors

Using modifications of hirudin, additional unique thrombin inhibitors have been developed. Hirugen is a synthetic N-acetylated carboxy-terminal dode-capeptide (Ac-Asn-Gly-Asp-Phe-Glu-Glu-Ile-Pro-Glu-Glu-Tyr(SO$_3$)-Leu) of hirudin (NASKI et al. 1990). This peptide is a competitive inhibitor of human α-thrombin, cleaving fibrinogen to liberate fibrinopeptide A ($K_i = 0.54\,\mu$M). It does not interfere with the active site of thrombin at all. In order to appreciate how various mechanisms for thrombin inhibition can be useful, these agents have been examined in animals. HANSON and HARKER (1988) published an important study which showed how an active site inhibitor of thrombin, D-phenylalanyl-L-prolyl-L-arginylchloromethyl ketone (PPACK) was able to prevent platelet-dependent thrombosis as well as fibrin formation. The relative potencies of each inhibitor to thrombin's functional domains were compared (KELLY et al. 1992). The antithrombotic abilities of the active-site inhibitor PPACK and an exosite-I inhibitor which consists of the C-terminal tyrosine-sulfated dodecapeptide of hirudin (hirugen) were compared with a 20-mer peptide combining the catalytic-site antithrombin PPACK with an exosite antithrombin peptide (hirugen) conjoined with a polyglycyl linker (Hirulog-1). Inhibiting both thrombin's catalytic and exosite domains increased the antithrombotic potency by several orders of magnitude over the inhibition of either domain alone.

Other anion-binding exosite-I inhibitors of thrombin have also been discovered. Aptamers of single-stranded DNA have been shown to bind to a thrombin affinity column with affinities in the range of 25–200 nM (BOCK et al. 1992). The 15-mer nucleotide sequence GGTTGGTGTGGTTGG on crystallography interacts with the two positively charged regions on the thrombin molecule, its anion-binding exosite I (fibrinogen-binding site) and exosite II (heparin-binding site) (PADMANABHAN et al. 1993). The DNA aptamer which binds and inhibits thrombin adopts a novel motif of a highly compact and symmetrical structure which consists of two tetrads of guanosine base pairs and three loops (WANG et al. 1993). Thrombin's B chain Lys^{21} and Lys^{65} are closely associated with the DNA aptamer-binding site of α-thrombin and Arg^{70} is a key determinant of the interaction (WU et al. 1992; PABORSKY et al. 1993). The DNA aptamer binds to exosite I and inhibits the activities of thrombin by competing with exosite-binding substrates such as fibrinogen or protease-activated receptor 1 (PABORSKY et al. 1993). In the clinical situation, DNA aptamers to thrombin were shown to be potent anticoagulants in extracorporeal circulation circuits. Its onset of anticoagulation is rapid and the short in vivo half-life leads to its rapid reversal (GRIFFIN et al. 1993). Another novel exosite-I inhibitor has been created by a single amino acid substitution which dissociates the activities of thrombin from fibrinogen clotting to thrombomodulin binding (WU et al. 1991). A peptide of the thrombomodulin binding sequence (TWTANVGKGQPS), which binds to exosite I, blocks the procoagulant activity of thrombin to prevent fibrinogen clotting, factor-V activation, and platelet activation (SUZUKI and NISHIOKA 1991). This peptide shifts thrombin from being a procoagulant protein to one which is more anticoagulant since activation of protein C was not interfered with.

d) Active-Site and Exosite-II Inhibitors and Exosite-II Inhibitors Alone

As already mentioned above, the antithrombin–thrombin complex in the presence of heparin is the classic example of an active-site and exosite-II inhibitor of thrombin. No specific exosite-II inhibitors have been produced, but recently Colwell and colleagues (1997, 1998) have described a naturally occurring antibody to thrombin's exosite II, which presented as a potent anticoagulant.

Other novel approaches to direct thrombin inhibition have been explored. Interference with the activity of thrombin can arise from interaction with one or more of its key functional sites. Changing a single amino acid of thrombin (E229A) converts the protein from a procoagulant to an anticoagulant by virtue of its ability to activate protein C (Gibbs et al. 1995). Additionally, changes in the ambient sodium ion concentration have been associated with a change in thrombin's avidity for fibrinogen or protein C (Di Cera et al. 1995). A nonpeptide mimetic designated LY254603 also enhances thrombin-catalyzed generation of anticoagulant factor activated protein C, yet inhibits thrombin-dependent fibrinogen clotting (Berg et al. 1996). LY254603 mediates a change in enzymatic substrate specificity by an alteration in the S3 substrate recognition site by thrombin. This effect of LY254603 is independent of allosteric changes induced by ionic strength or thrombomodullin (Berg et al. 1996). Thus there are many opportunities to modify the activity of thrombin by creating inhibitors to various regions of the protein or by altering the enzyme's ability to interact with its various substrates.

II. Indirect Thrombin Inhibitors

To date, most attempts to develop inhibitors to thrombin have been directed to the thrombin molecule itself. As related above, the most potent direct thrombin inhibitors have been associated with too much bleeding. Thus, other kinds of thrombin inhibition need to be developed. There would be selective advantages to develop inhibitors to certain substrates of thrombin. For example, as mentioned in the previous paragraph, inhibitors to thrombin that alter its substrate specificity would be useful to convert thrombin's procoagulant activities to one which mainly activates protein C and, thus, makes thrombin an anticoagulant. Similarly, if one wants to block the mitogenic effects of thrombin, inhibitors to thrombin-mediated cellular effects could be employed without influencing the enzyme's activities on intravascular hemostasis and thrombosis. This kind of inhibitor would be important to block the inflammatory effects of thrombin on fibroblast and smooth muscle growth and proliferation (Pages et al. 1993; Shankar et al. 1994; Sower et al. 1995; Haralbopoulos et al. 1997). For example, thrombin's effects of platelets can be selectively inhibited by okadaic acid treatment (Lerea 1991).

In the anticoagulation field, there is great interest to selectively inhibit thrombin's ability to activate platelets without interfering with its ability to proteolyze fibrinogen or activate protein C. Platelet-selective thrombin inhibitors may provide a new mechanism to prevent platelet activation in arterial thrombosis that could stand alone or be additive agents to current antiplatelet agents used to manage arterial thrombotic events such as myocardial infarction or stroke. Since most inhibitors directed to thrombin interfere with all of thrombin's actions, efforts have been made to create selective inhibitors to thrombin's activation of platelets (Fig. 1).

Presently, it is believed that there are two thrombin-sensitive receptors on human platelets, protease-activated receptor 1 (PAR1) and 4 (PAR4) (Vu et al. 1991; Xu et al. 1998). Both of these structures are classic seven-transmembrane G-coupled proteins in the rhodopsin family of receptors. Although G-coupled protein receptors, activation of these structures takes place by a unique mechanism. Thrombin interacts with PAR1 and PAR4, not as a ligand–receptor interaction, but as an enzyme–substrate interaction. It is believed that PAR1 is tenfold more sensitive to thrombin activation than PAR4. When thrombin activates PAR1, it cleaves off the amino-terminus of the mature protein to expose a new amino-terminus starting with Ser[42]-Phe-Leu-Leu-Arg. This new amino-terminus then interacts with another region on the receptor to induced stimulus–response coupling and signal transduction. Thus, there are multiple means by which platelet or cell-specific thrombin inhibitors can be designed. First, thrombin can be inhibited from binding to the receptor(s). Second, thrombin could be prevented from cleaving the receptor by presenting inhibitors to cleavage or a pseudosubstrate. Third, antagonists could be developed to prevent the newly exposed amino-terminus on the proteolyzed receptor from interacting with its other regions to activate signal transduction.

Two possible inhibitors of thrombin activation of platelets have been derived from the proteins' high- and low-molecular weight kininogens. Both of these proteins inhibit thrombin-induced platelet aggregation (Meloni and Schmaier 1991; Puri et al. 1991). Both kininogens block thrombin from binding to platelets (Meloni and Schmaier 1991; Puri et al. 1991). One amidated peptide sequence from domain 3 of the kininogens (L[271]NAENNA[277]) may block thrombin from binding to the platelet glycoprotein Ib-IX-V complex (Bradford et al. 1997). Another sequence of the kininogens from its domain 4 actually inhibits α-thrombin from cleaving the thrombin substrate recognition sequence on PAR1 at its activation site, Arg[41] (Hasan et al. 1996) (Fig. 2). The minimal peptide of this sequence, Arg-Pro-Pro-Gly-Phe, which is the angiotensin converting enzyme breakdown product of bradykinin, is a bifunctional thrombin inhibitor. It interacts 175-fold less tightly with the active site of thrombin than it interacts with the thrombin-substrate cleavage sequence on PAR1, NATLDPRSFLLR (Hasan et al. 1998). This agent, when infused into coronary arteries of dogs, delays the closure of the vessels after electrolytic injury equivalent to the antiplatelet effect of aspirin (Hasan et al. 1999).

A similar approach to thrombin inhibition was also performed with an antibody to PAR1 (Cook et al. 1995). An IgG, which inhibited thrombin-induced platelet activation but not other agonists, effectively blocked cyclic flow reductions in the carotid artery of African green monkeys without statistically prolonging the bleeding time or the activated partial thromboplastin time. Another similar approach used uncleavable peptides that compete with thrombin to activate PAR1 (Hung et al. 1992). Finally, antagonist peptides have been developed to block the agonist peptide, SFLLRN, derived from the thrombin receptor. By creating a more potent PAR1 agonist peptide to activate platelets (Ser-*p*-fluoroPhe-*p*-guanidinoPhe-Leu-Arg-NH$_2$), a substitution strategy was devised to create an antagonist peptide to this agonist peptide (BERNATOWICZ et al. 1996; SEILER 1997). The efficacy of this latter approach has been questioned since current antagonists are weakly potent and appear to be partial agonists at the high concentrations required for activity (SEILER 1997).

C. Factor-Xa Inhibitors

Factor X is a 59-kDa glycoprotein which is activated by factors IXa and VIIIa or factor VIIa/tissue factor complex to factor Xa. Formation of factor Xa is a critical control step in the hemostatic process which, when it occurs in the presence of factors IXa and VIIIa and a cell membrane, its catalytic efficiency is increased 1.2×10^9 fold. In physiologic hemostasis, the factor VIIa/tissue factor complex initiates hemostasis by activation of factor IX to factor IXa. Factors IXa and VIIIa then activate factor X to Xa (BROZE 1998). The factor VIIa and tissue factor complex do not usually result in direct factor Xa formation because a quaternary complex of factor VIIa, tissue factor, factor X, and tissue factor pathway inhibitor-1 (TFPI-1) will inhibit the factor VIIa/tissue factor complex. Further, formation of factor Xa also results in the formation of the quaternary complex of factor Xa with TFPI-1, factor VIIa, and tissue factor that dampens down factor VIIa/tissue factor-initiated hemostasis and also inhibits factor Xa itself. Only sustained and amplified factor Xa formation results in catalytically efficient thrombin formation and fibrinogen clotting.

Factor Xa is highly regulated. The first control is when factor X is activated. Efficient factor-X activation requires the presence of factors IXa and VIIIa; the absence of these factors results in severe bleeding states (hemophilia A and B). In addition to these proteins, factor-X activation on cells occurs on a specific receptor. The effector cell protease receptor-1, although originally described as a receptor for activities associated with factor Xa (ALTIERI 1994, 1995; NICHOLSON et al. 1996; CIRINO et al. 1997), has been recognized to be an essential cofactor for factor-Xa binding and activity in the prothrombinase complex (BOUCHARD et al. 1997). Thus, interference with any of these proteins blocks with factor Xa function. Like thrombin, factor Xa has a very limited number of naturally occurring inhibitors. Only four known serine protease inhibitors regulate factor Xa: antithrombin, TFPI-1, protease nexin 2/amyloid β-protein precursor, or TFPI-2 (placenta protein-5).

I. Naturally Occurring Factor-Xa Inhibitors in Humans

Antithrombin is the major plasma protease inhibitor of factor Xa. In the presence of heparin, its ability to inhibit factor Xa is potentiated. Unlike thrombin, heparin does not form a trimolecular complex with antithrombin and factor Xa (Bjork et al. 1982; Craig et al. 1989; Olson et al. 1992). Rather, heparin binds to antithrombin and alters it such that antithrombin is a better inhibitor of factor Xa without heparin binding to factor Xa itself. This fact is the reason for the efficacy of low-dose heparin and for the creation of low-molecular-weight heparins (see below). TFPI-1 is the most potent inhibitor of factor Xa in intrinsic tenase. TFPI-1 is a 32-kDa Kunitz-type inhibitor with three Kunitz domains: the first Kunitz domain mediates factor-VIIa/tissue-factor binding and inhibition; the second Kunitz domain mediates factor-Xa binding and inhibition; and the third Kunitz domain may be involved in heparin binding. TFPI-1 directly inhibits factor Xa at or near its active site with a 1:1 stoichiometry (Broze et al. 1988). Both calcium ions and heparin enhance the inhibition of factor Xa by TFPI-1.

The Kunitz protease inhibitor, protease nexin-2/amyloid β-protein precursor (PN-2/AβPP), has also been shown to be an inhibitor of factor Xa (Mahdi et al. 1995). PN-2/AβPP is an active-site-directed inhibitor which forms a complex with 1:1 stoichiometry with factor Xa. Its K_i is 1.9×10^{-8} M and its second-order rate constant of inhibition of factor Xa (1.8×10^6 M^{-1}min^{-1}) is similar to that seen with antithrombin and heparin (3×10^6 M^{-1}min^{-1}) (Mahdi et al. 1995). A recombinant form of the 57-kDa Kunitz-type protease inhibitory domain of PN-2/AβPP, which was designed to be 40-fold more avid to factor Xa than wild-type protein, has been successfully used as an anticoagulant in a rabbit model of extracorporeal circulation with a more stable pharmacokinetics and pharmacodynamics than standard heparin (Annich et al. 1998). This agent, in addition to being a potent factor-Xa inhibitor, also has the ability to inhibit plasmin and kallikrein. This inhibitory spectrum combines the anticoagulant-like activity of factor Xa inhibition with aprotinin-like features. TFPI-2 (placental protein 5) is a newly recognized Kunitz protease inhibitor in man (Sprecher et al. 1994). It has much greater specificity for factor Xa than factor VIIa/tissue factor. Heparin will potentiate its ability to inhibit factor VIIa/tissue factor. It also inhibits trypsin, plasmin, factor XIa, chymotyrpsin, and plasma kallikrein (Sprecher et al. 1994).

II. Naturally Occurring or Synthetic Factor-Xa Inhibitors Applied to Man

Since factor Xa has such an important role in hemostasis, much effort has been applied to develop factor-Xa inhibitors that can be used as anticoagulants to limit thrombin generation (Fig. 3). It is a well-established concept that low-dose standard heparin potentiates antithrombin inhibition of factor Xa and thus "shifts-the-balance" to decrease thrombin formation. This notion has been proven in multiple clinical trials examining the efficacy of low-dose heparin for prophylaxis for venous thrombosis and pulmonary embolism (Kakkar et al. 1975). Thus, efforts have been made to prepare an improved heparin. Low-molecular-weight heparins (4–6kDa) are more highly purified and standardized smaller forms of heparin which potentiate antithrombin inhi-

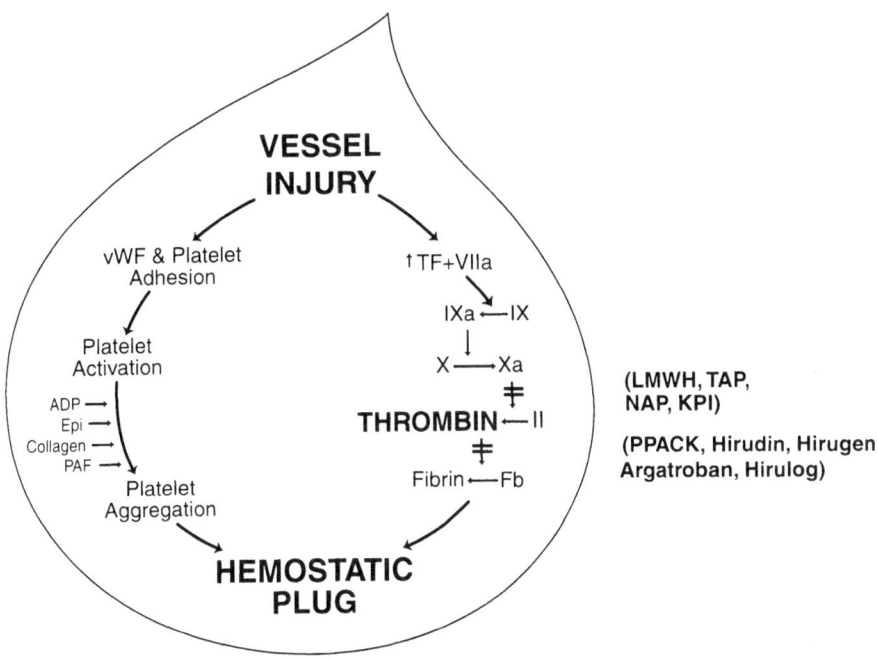

Fig. 3. Anti-fibrin generating agents. Both inhibitors to factor Xa [low-molecular-weight heparin (*LMWH*), tick anticoagulant peptide (*TAP*), nematode anticoagulant peptide (*NAP*), or the Kunitz protease inhibitory domain of the amyloid β-protein precursor (*KPI*)] and direct thrombin inhibitors (PPACK, Hirudin, Hirugen, Argatroban, and Hirulog) prevent the generation of fibrin from fibrinogen

bition of factor Xa but not thrombin (WEITZ 1997). They have proven efficacy in the prophylaxis and management of deep venous thrombosis and pulmonary embolism. They have a longer half-life than standard heparin with more stable pharmacokinetics and pharmacodynamics. Variations of low-molecular-weight heparins have been produced with differing abilities to inhibit factor Xa and thrombin. Similarly, various forms of glycosaminoglycans have been developed as potentiators of antithrombin inhibition of factor Xa. Danaparoid sodium is a factor Xa inhibitor which does not consist of heparin sulfate, but mostly consists of dermatin sulfate (85%) (CHONG et al. 1989). Since low-molecular-weight heparins contain at least two disaccharide units, they still preserve their anti-inflammatory activity. Low-molecular-weight heparins may also blunt the elevation of von Willebrand factor seen in patients presenting with acute coronary syndromes (ANTMAN and HANDIN 1998). Other pentasaccharides may prove to be effective anticoagulants as well.

A number of naturally occurring factor-Xa inhibitors from lower animals and insects have been developed as potential anticoagulants. These agents will be important to address the question whether highly specific and potent agents will be more useful clinically than less potent compounds with wider specificity. Tick anticoagulant peptide (TAP) was recognized as a novel factor-Xa inhibitor (Waxman et al. 1990; Dunwiddie et al. 1993). This 60-residue peptide from the soft tick, *Ornithodoros moubata*, is a slow, tight-binding inhibitor of factor Xa with a $K_i = 0.588\,\text{nM}$. The structure of the inhibitor has limited homology to a Kunitz-type protease inhibitor. Antistasin is a 15-kDa polypeptide from the salivary glands of the Mexican leech, *Haamenteria officinalis*, which inhibits factor Xa with a K_i between 0.31 nM and 0.62 nM (Dunwiddie et al. 1989). Ecotin, a serine protease inhibitor found in the periplasm of *Escherichia coli*, is a potent factor-Xa inhibitor with a $K_i = 54\,\text{pM}$ (Seymour et al. 1994). It does not inhibit thrombin, factor VIIa, factor XIa, activated protein C, plasmin, or tissue plasminogen activator. It is a potent inhibitor of factor XIIa, human leukocyte elastase and plasma kallikrein. Hookworm-derived inhibitor of 8.7 kDa from *Ancylostoma caninum* is a highly specific and potent factor-Xa inhibitor with a $K_i = 323\,\text{pM}$ (Cappello et al. 1995). A class of anticoagulants has been derived from this family of anti-coagulants. These agents can be delivered subcutaneously and have half-lives up to 48 h since there are no naturally occurring clearance mechanism for them in man.

D. Conclusions

There are a wide array of antithrombin and antifactor Xa agents currently available and in development. These compounds vary with regard to the specific mechanism of action and potency. Much has been learned from the development of these protease inhibitors of thrombin and factor Xa about the function of these enzymes and how therapeutic tools can be developed to better manage anticoagulation. Table 2 represents situations in hemostasis and thrombosis where specific thrombin and factor-Xa inhibitors have been shown to be useful therapeutic tools. There is no question that thrombin inhibitors can provide adequate anticoagulation for venous thrombosis and pulmonary embolus. They also serve as acceptable substitutes for heparin and related compounds for all conditions in patients with the rare condition of heparin-induced thrombocytopenia and thrombosis syndrome. Heparin, which poten-

Table 2. Proven anticoagulant activity for thrombin and factor Xa inhibitors

Inhibitor	Clinical use
Thrombin inhibitors	Venous thrombosis
	Pulmonary embolus
Factor Xa inhibitors	Venous thrombosis
	Pulmonary embolus
	Acute coronary syndromes
	Coronary bypass and extracorporeal circulation

tiates antithrombin inhibition of thrombin along with factor Xa and the other proteases of the hemostatic system, is an effective anticoagulant which serves as the first level of anticoagulation when a patient arrives with an acute coronary syndrome (cresendo angina pectoris or myocardial infarction). However, the potent direct thrombin inhibitors (hirudin, hirulog, and hirugen) have been tried in the management of acute coronary syndromes and as adjuvants in percutaneous transluminal coronary angioplasty. In most cases, they were as good as heparin, but were associated with higher than acceptable risks for bleeding. DNA aptamers have been used successfully for cardiopulmonary bypass, suggesting that a short-acting agent like this could have this therapeutic indication (GRIFFIN et al. 1993).

More studies are needed to determine how direct thrombin inhibitors can be used in cardiopulmonary bypass and extracorporeal circulation. The fact that direct thrombin inhibitors block all of thrombin's action suggests that these agents may be too potent for safe therapeutic anticoagulation. Selective thrombin inhibitors are needed to ascertain the role of thrombin in the management of acute coronary syndromes. If one can selectively inhibit thrombin's ability to activate platelets, it remains to be proven whether this agent will be additive to cyclooxygenase inhibitors and thioenpyridines in reducing adverse events associated with coronary ischemia and thrombosis associated with coronary stent placement.

However, factor-Xa inhibitors which inhibit the thrombin generation have been proven to be safe and effective anticoagulants for venous thrombosis, pulmonary embolus, acute coronary syndromes, and cardiopulmonary bypass/extracorporeal circulation (COHEN et al. 1997; GITLIN et al. 1998). The fact that these agents are usually not associated with platelet inhibition may make them safer compounds to be used in these various therapeutic arenas. Developing selective agents that can be conjoined with other agents may prove to be the best approach to anticoagulation therapy for specific medical situations. Using the paradigm of combination chemotherapy developed in medical oncology, combination anticoagulant therapy may be appropriate in certain medical situations. Further developments in the factor-Xa inhibitory field should be in the direction of creating agents that have a usable pharmacokinetic and pharmacodynamic profile for cardiopulmonary bypass and extracorporeal circulation. In developing new anticoagulants, one must recall that a single potent agent will never be adequate therapy in all situations.

References

Altieri DC (1994) Molecular cloning of effector cell protease receptor-1, a novel cell surface receptor for protease factor Xa. J Biol Chem 269:3139–3142

Altieri DC (1995) Xa receptor EPR-1. FASEB J 9:860–865

Annich G, White T, Damm D, Zhao Y, Mahdi F, Meinhardt J, Rebello S, Lucchesi BR, Bartlett RH, Schmaier AH (1998) Recombinant Kunitz protease inhibitor domain of the amyloid β-protein precursor as an anticoagulant for extracorporeal circulation in rabbits. Circulation 98[Suppl 1]:I-728

Antman EM (1994) Hirudin in acute myocardial infarction. Safety report from the thrombolysis and thrombin inhibition in myocardial infarction (TIMI) 9A trial. Circulation 90:1624–1630

Antman EM, Handin R (1998) Low-molecular weight heparins. An intriguing new twist with profound implications. Circulation 98:287–289

Bajzar L, Manuel R, Nesheim ME (1995) Purification and characterization of TAFI, a thrombin-activable fibrinolysis inhibitor. J Biol Chem 270:14477–14484

Bajzar L, Morser J, Nesheim ME (1996) TAFI, or plasma procarboxypeptidase B, couples in the coagulation and fibrinolytic cascades through the thrombin-thrombomodulin complex. J Biol Chem 271:16603–16608

Berg DT, Wiley MR, Grinnell BW (1996) Enhanced protein C activation and inhibition of fibrinogen cleavage by a thrombin modulator. Science 273:1389–1391

Bernatowicz MS, Klimas CE, Hartl KS, Peluso M, Allegretto NJ, Seiler SM (1996) Development of potent thrombin receptor antagonist peptides. J Med Chem 39:4879–4887

Berry C, Girardot C, Lecoffre C, Lunven C (1994) Effects of the synthetic thrombin inhibitor argatroban on fibrin- or clot-incorporated thrombin: comparison with heparin and recombinant hirudin. Thromb Haemost 72:381–386

Bjork I, Lindahl V (1982) Mechanism of the anticoagulant action of heparin. Mol Cell Biochem 48:161–182

Bjork I, Jackson C, Jornvall H, Lavine KK, Nordling K, Salsgiver WJ (1982) The active site of antithrombin. Release of the same proteolytically cleaved form of the inhibitor from complexes with factor IXa, factor Xa, and thrombin. J Biol Chem 257:2406–2411

Bjork I, Ylinenjarvi K, Olson ST, Bock PE (1992) Conversion of antithrombin from an inhibitor of thrombin to a substrate with reduced heparin affinity and enhanced conformation stability by binding of a tetradecapeptide corresponding to the P_1 to P_{14} region of the putative reactive bond loop of the inhibitor. J Biol Chem 267:1976–1892

Bock LC, Griffin LC, Latham JA, Vermaas EH, Toole JJ (1992) Selection of single-stranded DNA molecules that bind and inhibit human thrombin. Nature 355:564–566

Bock PE, Olson ST, Bjork I (1997) Inactivation of thrombin by antithrombin is accompanied by inactivation of regulatory exosite I. J Biol Chem 272:19837–19845

Bode W, Mayr I, Baumann U, Huber R, Stone SR, Hofsteenge J (1989) The refined 1.9A crystal structure of human α-thrombin: interaction with D-Phe-Pro-Arg chloromethylketone and significance of the Tyr-Pro-Trp insertion segment. EMBO J 8:3467–3475

Bode W, Turk D, Karshikov AJ (1992) The refined 1.9 A X ray crystal structure of D Phe-Pro-Arg chloromethylketone-inhibited human α-thrombin: structure analysis, overall structure, electrostatic properties, detailed active-site geometry, and structure-function relationships. Protein Sci 1:426–471

Bradford HN, Dela Cadena RA, Kunapuli SP, Dong J-F, Lopez JA, Colman RW (1997) Human kininogens regulate thrombin binding to platelets through the glycoprotein Ib-IX-V complex. Blood 90:1508–1515

Broze GJ (1998) The tissue factor pathway of coagulation. In: Loscalzo J, Schafer AI (eds) Thrombosis and hemorrhage, 2nd edn. Williams & Wilkins, Baltimore, pp 77–104

Broze GJ, Warren LA, Novotny WF (1988) The lipoprotein-associated coagulation factor inhibitor that inhibits the factor VII-tissue factor complex also inhibits factor Xa: insight into its possible mechanism of action. Blood 71:335–343

Bouchard BA, Catcher CS, Thrash BR, Adida C, Tracy PB. (1997) Effector cell protease receptor-1, a platelet activation dependent membrane protein, regulates prothrombinase-catalyzed thrombin generation. J Biol Chem 272:9244–9251

Cappello M, Vlasuk GP, Bergum PW, Huang S, Hotez PJ (1995) Ancylostoma canninum anticoagulant peptide: a hookworm-derived inhibitor of human coagulation factor Xa. Proc Natl Acad Sci 92:6152–6156

Chong BH, Ismail F, Cade J, Gallus AS, Gordon S, Chesterman CN (1989) Heparin-induced thrombocytopenia: studies with a new low molecular weight heparinoid, Org 10172. Blood 73:1592–1596

Cirino G, Cicala C, Bucci M, Sorrentino L, Ambrosini G, DeDominicis G, Altieri D (1997) Factor xa as an interface between coagulation and inflammation. Molecular mimicry of factor Xa association with effector cell protease receptor-1 induces acute inflammation in vivo. J Clin Invest 99:2446–2451

Cohen M, Demers C, Gurfinkel EP, Turpie AGG, Fromell GJ, Goodman S, Langer A, Califf RM, Fox KAA, Premmereur J, Bigonzi F (1997) A comparison of low-molecular-weight heparin with unfractionated heparin for unstable coronary artery disease. N Engl J Med 337:447–452

Collen D, Matsuo O, Stassen JM, Kettner C, Shaw E (1982) In vivo studies of a synthetic inhibitor of thrombin. J Lab Clin Med 99:76–83

Colwell NS, Tollefson DM, Blinder MA (1997) Identification of a monoclonal thrombin inhibitor associated with multiple myeloma and a severe bleeding disorder. Br J Haematol 97:219–226

Colwell NS, Blinder MA, Bock PE, Tollefsen DM (1998) Allosteric effects of a monoclonal antibody against thrombin exosite II. Circulation 98[Supple I]:I-519

Cook JJ, Sitko GR, Bednar B, Condra C, Mellott MJ, Feng D-M, Nutt RF, Shafer JA, Gould RJ, Connolly TM (1995) An antibody against the exosite of the cloned thrombin receptor inhibits experimental arterial thrombosis in the African Green Monkey. Circulation 91:2961–2971

Coughlin PB, Tetaz T, Salem HH (1993) Identification and purification of a novel serine proteinase inhibitor. J Biol Chem 268:9541–9547

Craig PA, Olson ST, Shore JD (1989) Transient kinetics of heparin-catalyzed protease inactivation by antithrombin III. J Biol Chem 264:5452–5461

Cunningham DD, Farrell DH (1986) Thrombin interactions with cultured fibroblasts: relationship to mitogenic stimulation. Ann N Y Acad Sci 485:240–248

De Boer HC, de Groot PG, Bouma BN, Preissner KT (1993) Ternary vitronectin-thrombin-antithrombin III complexes in human plasma. J Biol Chem 268:1279–1283

Di Cera E, Guinto ER, Vindigni A, Dang QD, Ayala YM, Wuyil M, Tulinsky A (1995) The Na+ binding site of thrombin. J Biol Chem 270:22089–22092

Dunwiddie C, Thornberry NA, Bull HG (1989) Antistasin, a leech-derived inhibitor of factor Xa. Kinetic analysis of enzyme inhibition and identification of the reactive site. J Biol Chem 264:16694–16699

Dunwiddie CT, Waxman L, Vlasuk GP, Friedman PA (1993) Purification and characterization of inhibitors of blood coagulation factor Xa from hematophagous organisms. Methods Enzymol 223:291–312

Friedrich T, Kroger B, Bialojan S, Lemaire HG, Hoffken HW, Reuschenbach P, Otte M, Dodt J (1993) A Kazal-type inhibitor with thrombin specificity from Rhodnius prolixus. J Biol Chem 268:16216–16222

Gibbs CS, Coutre SE, Tsiang M, Li W-X, Jain AK, Dunn KE, Law VS, Mao CT, Matsumura SY, Mejza SJ, Paborsky LR, Leung LLK (1995) Conversion of thrombin into an anticoagulant by protein engineering. Nature 378:413–416

Gitlin SD, Deeb GM, Yann C, Schmaier AH (1998) Intraoperative monitoring of danaparoid sodium (Organan) anticoagulation during cardiovascular surgery. J Vasc Surg 27:568–575

Griffin LC, Tidmarsh GF, Bock LC, Toole JJ, Leung LLK (1993) In vivo anticoagulation properties of a novel nucleotide-based thrombin inhibitor and demonstration of region anticoagulation in extracorporeal circuits. Blood 81:3271–3276

Gronke RS, Bergman BL, Baker JB (1987) Thrombin interaction with platelets. Influence of a platelet protease nexin. J Biol Chem 262:3030–3036

Gusto IIa Investigators (1994) Randomized trial of intravenous heparin versus recombinant hirudin for acute coronary syndromes. Circulation 90:1631–1637

Hanson SR, Harker LA (1988) Interruption of acute platelet-dependent thrombosis by the synthetic antithrombin D-phenylalanyl-L-prolyl-L-arginyl chloromethyl ketone. Proc Natl Acad Sci USA 85:3184–3188

Haralbopoulos GC, Grant DS, Kleinman HK, Maragoudakis ME (1997) Thrombin promotes endothelial cell alignment in Matrigel in vitro and angiogenesis in vivo. Am J Physiol 273:C239–C245

Harker L, Hanson SR, Kelly AB (1995) Antithrombotic benefits and hemorrhagic risks of direct thrombin inhibitors. Thromb Haemost 74:464–472

Hasan AAK, Amenta S, Schmaier AH (1996) Bradykinin and its metabolite, ARG-PRO-PRO-GLY-PHE, are selective inhibitors of α-thrombin-induced platelet activation. Circulation 94:517–528

Hasan AAK, Krishnan R, Tulinsky A, Schmaier AH (1998) The mechanism of thrombostatin's inhibition of thrombin-induced platelet activation. Circulation 98[Suppl 1]:I-800

Hasan AAK, Rebello SS, Smith E, Srikanth S, Werns S, Driscoll E, Faul J, Brenner D, Normolle D, Lucchesi BR, Schmaier AH (1999) Thrombostatin inhibits induced canine coronary thrombosis. Thromb Haemost 82:in press

Hayes KL, Leong L, Henriksen RA, Bouchard BA, Ouellette L, Church WR, Tracy PB (1994) α-Thrombin-induced human platelet activation results solely from formation of a specific enzyme-substrate complex. J Biol Chem 269:28606–28612

Hung DT, Vu T-KH, Wheaton VI, Charo IF, Nelken NA, Esmon N, Esmon CT, Coughlin SR (1992) "Mirror image" antagonists of thrombin-induced platelet activation based on thrombin receptor structure. J Clin Invest 89:444–450

Ill CR, Ruoslahti E (1985) Association of thrombin-antithrombin III complex with vitronectin in serum. J Biol Chem 260:15610–15615

Imura Y, Stassen J-M, Vreys I, Lesaffre E, Gold HK, Collen D (1992) Synergistic antithrombotic properties of G4120, a RGD-containing synthetic peptide, and argatroban, a synthetic thrombin inhibitor, in a hamster femoral vein platelet-rich thrombosis model. Thromb Haemost 68:336–340

Jenny NS, Mann KG (1998) Coagulation cascade: an overview. In: Loscalzo J, Schafer AI (eds) Thrombosis and hemorrhage, 2nd edn. Williams & Wilkins, Philadelphia, p 11

Kakkar VV, Corrigan TP, Fossard DP, Sutherland I, Shelton MG, Thirlwall J (1975) Prevention of fatal postoperative pulmonary embolism by low doses of heparin. Lancet II:45–51

Kelly AB, Marzec UM, Krupski W, Bass A, Cadroy Y, Hanson SR, Harker LA (1991) Hirudin interruption of heparin-resistant arterial thrombus formation in baboons. Blood 77:1006–1012

Kelly AB, Maraganore JM, Bourdon P, Hanson SR, Harker LA (1992) Antithrombotic effects of synthetic targeting various functional domains of thrombin. Proc Natl Acad Sci 89:6040–6044

Kettner C, Mersinger L, Knabb R (1990) The selective inhibition of thrombin by peptides of boroarginine. J Biol Chem 265:18289–18297

Kikumoto R, Tamao Y, Tezuka T, Tonomura S, Hara H, Ninomiya K, Hijikata A, Okamoto S (1984) Selective inhibition of thrombin by (2R,4R)-4-methyl-1-[N²-(3(RS)-methyl-1,2,3,4,-tetrahydro-8-quinolinyl)sulfonyl]-L-arginyl]-2-piperidinecarboxylic acid. Biochemistry 23:85–90

Knabb RM, Kettner CA, Timmermans PBMWM, Reilly TM (1992) In vivo characterization of a new synthetic thrombin inhibitor. Thromb Haemost 67:56–59

Knapp A, Degenhardt T, Dodt J (1992) Hirudisins. Hirudin-derived thrombin inhibitors with disintegrin activity. J Biol Chem 267:24230–24234

Lazar JB, Winant RC, Johnson PH (1991) Hirudin: amino-terminal residues play a major role in the interaction with thrombin. J Biol Chem 266:685–688

Lerea KM (1991) Thrombin-induced effects are selectively inhibited following treatment of intact human platelets with okadaic acid. Biochemistry 30:6819–6824

Li EHH, Fenton JW II, Feinman RD (1976) The role of heparin in the thrombin-antithrombin III reaction. Arch Biochem Biophys 175:153–159

Mahdi F, Van Nostrand WE, Schmaier AH (1995) Protease nexin-2/amyloid β-protein precursor inhibits factor Xa in the prothrombinase complex. J Biol Chem 270:23468–23474

Maryanoff BE, Qui X, Padmanabhan KP, Tulinsky A, Almond HR, Andrade-Gordon P, Greco MN, Kauffman JA, Nicolaou KC, Liu A, Brungs PH, Fusetani N (1993) Molecular basis for the inhibition of human α-thrombin by macrocyclic peptide cyclotheonamide A. Proc Natl Acad Sci 90:8048–8052

Meloni FJ, Schmaier AH (1991) Low molecular weight kininogen binds to platelets to modulate thrombin-induced platelet activation. J Biol Chem 266:6786–6794

Morenweiser R, Auerswald EA, de Locht A, Fritz H, Sturzebaecher J, Stubbs MT (1997) Structure-based design of a potent chimeric thrombin inhibitor. J Biol Chem 272:19938–19942

Naski MC, Fenton JW III, Maraganore JM, Olson ST, Shafer JA (1990) The COOH-terminal domain of hirudin. An exosite-directed competitive inhibitor of the action of α-thrombin on fibrinogen. J Biol Chem 265:13484–13489

Naski MC, Lawrence DA, Mosher DF, Rodor TJ, Ginsburg D (1993) Kinetics of inactivation of α-thrombin by plasminogen activator. J Biol Chem 268:12367–12373

Nienaber VL, Amparo EC (1996) A noncleavable retro-binding peptide that spans the substrate binding cleft of serine proteases. Atomic structure of nazumamide A: human thrombin. J Am Chem Soc 118:6807–6810

Nierodzik ML, Chen K, Takeshita K, Li J-J, Huang Y-Q, Feng X-S, D'Andrea MR, Andrade-Gordon P, Karpatkin S (1998) Protease-activated receptor 1 (PAR1) is required and rate-limiting for thrombin-enhanced experimental pulmonary metastasis. Blood 92:3694–3700

Neuhaus K-L, Essen Rv, Tebbe U, Jessel A, Heinrichs H, Maurer W, Doring W, Harmjanz D, Kotter V, Kalhammer E, Simon H, Horacek T (1994) Safety observations from the pilot phase of the randomized r-hirudin for improvement of thrombolysis (HIT-III) study. Circulation 90:1638–1642

Nicholson AC, Nachman RL, Altieri DC, Summers BD, Ruf W, Edgington TS, Hajjar DP (1996) Effector cell protease receptor-1 is a vascular receptor for coagulation factor Xa. J Biol Chem 271:28407–28413

Okamoto S, Hijikata A, Kikumoto R, Tonomara S, Hara H, Ninomiya K, Maruyama A, Sugano M, Tamao Y (1981) Potent inhibition of thrombin by the newly synthesized arginine derivative no. 805. The importance of sterostructure of its hydrophobic caroxamide portion. Biochem Biophys Res Commun 101:440–446

Olson ST (1988) Transient kinetics of heparin-catalyzed protease inactivation by antithrombin III. Linkage of protease-inhibitor-heparin interactions in the reaction with thrombin. J Biol Chem 263:1698–1708

Olson ST, Bjork I (1991) Predominant contribution of surface approximation to the mechanism of heparin acceleration of the antithrombin-thrombin reaction. Elucidation from salt concentration effects. J Biol Chem 266:6353–6364

Olson ST, Shore JD (1982) Demonstration of a two-step reaction mechanism for inhibition of α-thrombin by antithrombin III and identification of the step affected by heparin. J Biol Chem 257:14891–14895

Olson ST, Shore JD (1986) Transient kinetics of heparin-catalyzed protease inactivation by antithrombin III. The reaction step limiting heparin turnover in thrombin neutralization. J Biol Chem 261:13151–13159

Olson ST, Halvorson HR, Bjork I (1991) Quantitative characterization of the thrombin-heparin interaction. Discrimination between specific and nonspecific models. J Biol Chem 266:6342–6352

Olson ST, Bjork I, Sheffer R, Craig PA, Shore JD, Choay J (1992) Role of the antithrombin-binding pentasaccharide in heparin acceleration of antithrombin-proteinase reactions. J Biol Chem 267:12528–12538

Paborsky LR, McCurdy SN, Griffin LC, Toole JJ, Leung LLK (1993) The single-stranded DNA aptamer-binding site of human thrombin. J Biol Chem 268: 20808–20811

Padmanabhan K, Padmanabhan KP, Ferrara JD, Sadler JE, Tulinsky A (1993) The structure of α-thrombin inhibited by a 15-mer single-stranded DNA aptamer. J Biol Chem 268:17651–17654

Pages G, Lenormand P, L'Allemain G, Chambard J-C, Meloche S, Pouyssegur J (1993) Mitogen-activated protein kinases p42^{mapk} and p44^{mapk} are required for fibroblast proliferation. Proc Natl Acad Sci USA 90:8319–8323

Pejler G, Karlstrom A (1993) Thrombin is inactivated by mast cell secretory granule chymase. J Biol Chem 268:11817–11822

Puri RN, Zhou F, Hu C-J, Colman RF, Colman RW (1991) High molecular weight kininogen inhibits thrombin-induced platelet aggregation and cleavage of aggregin by inhibiting binding of thrombin to platelets. Blood 77:500–507

Rao AK, Sun L, Chesebro JH, Fuster V, Harrington RA, Schwartz D, Gallo P, Matos D, Topol EJ (1996) Distinct effects of recombinant desulfatohirudin (Revasc) and heparin on plasma levels of fibrinopeptide A and protrhombin fragment F1.2 in unstable angina. A multicenter trial. Circulation 94:2389–2395

Rosenberg RD, Damus PS (1973) The purification and mechanism of action of human antithrombin-heparin cofactor. J Biol Chem 248:6490–6505

Rydel TJ, Ravichandran KG, Tulinsky A, Bode W, Huber R, Roitsch C, Fenton JW II (1990) The structure of a complex of recombinant hirudin and human α-thrombin. Science 249:277–280

Seiler SM (1997) Thrombin receptor antagonists. Semin Thromb Hemost 22:223–232

Seymour J, Lindquist RN, Dennis MS, Moffat B, Yansura D, Reilly D, Wessinger ME, Lazarus RA (1994) Ecotin is a potent anticoagulant and reversible tight-binding inhibitor of factor Xa. Biochemistry 33:3949–3958

Shankar R, de la Motte C, Poptic EJ, DiCorleto PE (1994) Thrombin-receptor-activating peptides differentially stimulate platelet-derived growth factor production, monocyte cell adhesion, and E-selectin expression inhuman umbilical vein endothelial cells. J Biol Chem 269:13936–13941

Sheehan JP, Sadler JE (1994) Molecular mapping of the heparin-binding exosite of thrombin. Proc Natl Acad Sci USA 91:5518–5522

Sheehan JP, Wu Q, Tollefsen DM, Sadler JE (1993) Mutagenesis of thrombin selectively modulates inhibition by serpins heparin cofactor II and antithrombin III. J Biol Chem 268:3639–3645

Sheehan JP, Tollefsen DM, Sadler JE (1994) Heparin cofactor II is regulated allosterically and not primarily by template effects. J Biol Chem 269:32747–32751

Sower LE, Froelich CJ, Carney DH, Fenton II JW, Klimpel GR (1995) Thrombin induced IL-6 production in fibroblasts and epithelial cells. J Immunol 155:895–901

Sprecher CA, Kisiel W, Amathewes S, Foster D (1994) Molecular cloning, expression, and partial characterization of a second human tissue-factor-pathway-inhibitor. Proc Natl Acad Sci USA 91:3353–3357

Stone SR, Maraganore JM (1993) Hirudin and hirudin-based peptides. Methods Enzymol 223:312–336

Suzuki K, Nishioka J (1991) A thrombin-based peptide corresponding to the sequence of the thrombomodulin-binding site blocks the procoagulant activities of thrombin. J Biol Chem 266:18498–18501

Tollefsen DM, Majerus DW, Blank MK (1982) Heparin cofactor II. Purification and properties of a heparin-dependent inhibitor of thrombin in human plasma. J Biol Chem 257:2162–2169

Tabernero L, Chang CYY, Ohringer SL, Lau WF, Iwanowicz EJ, Han W-C, Wang TC, Seiler SM, Roberts DGM, Sack JS (1995) Structure of a retro-binding peptide inhibitor complexed with human α-thrombin. J Mol Biol 246:14–20

Vijayalakshmi J, Padmanabhan KP, Mann KG, Tulinsky A (1994) The isomorphous structures of prethrombin2, hirugen-, and PPACK-thrombin: changes accompanying activation and exosite binding to thrombin. Protein Sci 3:2254–2271

Vu T-K, Hung DT, Wheaton VI, Coughlin SR (1991) Molecular cloning of a functional thrombin receptor reveals a novel proteolytic mechanism of receptor activation. Cell 64:1057–1068

Wang KY, McCurdy S, Shea RG, Swaminathan S, Bolton PH (1993) A DNA aptamer which binds to and inhibits thrombin exhibits a new structural motif for DNA. Biochemistry 32:1899–1904

Waxman L, Smith DE, Arcuri KE, Vlasuk GP (1990) Tick anticoagulant peptide (TAP) is a novel inhibitor of blood coagulation factor Xa. Science 248:593–596

Weitz JI (1997) Low-molecular weight heparins. N Engl J Med 337:688–698

Weitz JI, Hudoba M, Massel D, Marganore J, Hirsh J (1990) Clot-bound thrombin is protected from inhibition by heparin-antithrombin III but is susceptible to inactivation by antithrombin III-independent inhibitors. J Clin Invest 86:385–391

Wu Q, Sheehan JP, Tsiang M, Lentz SR, Birktoft JJ, Sadler JE (1991) Proc Natl Acad Sci USA 88:6775–6779

Wu Q, Tsiang M, Sadler JE (1992) Localization of the single-stranded DNA binding site in the thrombin anion-binding exosite. J Biol Chem 267:24408–24412

Xu W-F, Andersen H, Whitmore TE, Presnell SR, Yee DP, Ching A, Gilbert T, Davie EW, Foster DC (1998) Cloning and characterization of human protease-activated receptor 4. Proc Natl Acad Sci 95:6642–6646

Zivelin A, Rao LVM, Rapaport SI (1993) Mechanism of anticoagulant effect of warfarin as evaluated in rabbits by selective depression of individual procoagulant vitamin K-dependent clotting factors. J Clin Invest 92:2131–2140

Zoldhelyi P, Chesebro JH, Owen WG (1993) Hirudin as a molecular probe for thrombin in vitro and during systemic coagulation in the pig. Proc Natl Acad Sci USA 90:1819–1823

CHAPTER 16

Inhibitors of Papain-Like Cysteine Peptidases in Cancer

R. Shridhar, B.F. Sloane, and D. Keppler

A. Introduction

Since the discovery of a polypeptide in chicken egg white (cew cystatin) that is able to inhibit the plant cysteine peptidase papain, investigators have identified numerous mammalian cysteine peptidases and cloned more than 40 cystatin genes. The cloning frenzy will probably continue for a few more years, raising classification and nomenclature problems in the papain- and cystatin super families. We now know that cysteine peptidases are involved in many biological processes, such as growth, differentiation, immunity, virulence, and death. Deregulation of these processes leads to tumor formation and metastasis. The present challenge is to define when and where each gene product is expressed and against what target(s) it is directed. The diversity of cancers and their intrinsic heterogeneity renders this task very difficult. In this review, we summarize what is presently known about cystatins, with particular emphasis on their function as inhibitors of lysosomal/endosomal cysteine peptidases. However, there is also evidence for other functions of cystatins. Is there a relationship between inhibitor function and these other functions? Why are there cytosolic and secreted inhibitors, when the target peptidases are vesicular? These are only a few questions that may be answered during the next 10 years.

B. General Overview

I. Cysteine Peptidases

Cysteine peptidases have been isolated from a wide variety of biological sources including viruses, bacteria, fungi, plants, and mammals. They form one of the major groups of peptidases and are divided into clans and families[1] based on their amino acid sequence homologies and active-site configurations (Barrett et al. 1998). In mammals, the main families and subfamilies comprise the calpains (C2), caspases (C14) and papain-like cysteine peptidases (C1A). Additional families and subfamilies are the bleomycin hydrolases (C1B),

---

[1] For on-line access to an updated classification of peptidases, go to the MEROPS Database, at the following URL: http://www.bi.bbsrc.ac.uk/Merops/

ubiquitin C-terminal hydrolases (UCHs, C12), ubiquitin-specific peptidases (UBPs, C19) and legumains (Asn-endopeptidases, C13). Papain-like cysteine peptidases are mostly present in the lysosomes of cells and are well adapted to function under the reducing conditions of this cellular compartment. Lysosomal cysteine peptidases that have been isolated and well characterized include cathepsins B, C/J, H, L, S and legumain (KIRSCHKE and WIEDERANDERS 1994; DOLENC et al. 1995; CHEN et al. 1997). Several further lysosomal cysteine peptidases have been cloned and partially characterized after expression in various systems such as bacteria, yeast or insect cells: cathepsins F, O, O2/K, L2/U/V, Q, W/lymphopain, Y, and P/X/Z (INAOKA et al.1995; LINNEVERS et al. 1997; SANTAMARIA et al. 1998). Characterization after purification from the tissues in which these latter cathepsins are most highly expressed still needs to be done. The alphabet being already exhausted, it is not clear how additional members should be named in the future.

Lysosomal cysteine peptidases are synthesized as preproenzymes and acquire N-linked carbohydrates co-translationally. Their maturation involves limited proteolytic processing and modification of the oligosaccharide side chains by the addition of mannose 6-phosphate residues in the Golgi. In many cell types, this allows trafficking to the lysosome via mannose-6-phosphate receptors (GEUZE et al. 1985). Lysosomal cysteine peptidases are involved in important cellular functions, such as intracellular protein turnover, degradation of extracellular matrix components, activation of precursor proteins, proteolysis of the invariant chain during maturation of the major histocompatability complex (MHC) class-II complex, antigen processing, and osteoclastic bone resorption. They have been implicated in numerous pathological conditions including rheumatoid arthritis, Alzheimer's disease, pulmonary emphysema and cancer progression; for a compilation of references, see KIRSCHKE et al. (1995).

II. Cystatin Super-family

The cystatin super-family contains over 40 members; many members possess inhibitory activity against the presently characterized papain-like cysteine peptidases. The super-family is divided into six, maybe seven, families on the basis of molecular structure (Table 1).

1. Family 1 (Stefins)

Family-1 members include human stefins A and B, which have homologues in other species such as cystatins α and β in rat (TSUKAHARA et al. 1987). The lack of a signal sequence and internal disulfide bonds makes family-1 members quite distinct from all other members of the cystatin super-family. The name "stefin" underlines these differences and avoids confusion with family-2 cell-secreted cystatins (see below) (TURK and BODE 1991). Thus, stefin C and cystatin C will not be mistaken as the same inhibitor.

Table 1. Characteristics of human members of the cystatin superfamily

Families	Name or abbreviation	Molecular mass (kDa)	Chromosome and locus
1-Stefins	A	11–12	3cen-q21
	B	11–12	21q22.3
2-Cystatins	C	15	20p11.2
	D	15–17	"
	E/M	15–20	11q13
	F	15–20, 34	n.a.
	S	15	20p11.2
	SA	15	"
	SN	15	"
3-Kininogens	HK	112–120	3q26-qter
	LK	64–68	"
4-Fetuins	Fetuin (α_2-HS*)	50–55	3q27–q29
5-CSTRPs (?)	CSTRP-1	20, 22	(20)
	CSTRP-2	20, 22	(20)
	CSTRP-3, etc.	n.a.	(20)
6-HRGs	HRG	80	3q28–29
7-CRES	CRES	15–20	n.a.

*, α_2-Heremans-Schmid plasma glycoprotein.
n.a., not available.

The genes for human stefins A and B have been mapped to chromosomes 3cen-q21 and 21q22.3, respectively (HSIEH et al. 1991; PENNACCHIO et al. 1996). Two mutations, a 3'-splice site mutation and a stop-codon mutation, were identified in the gene encoding human stefin B in Finnish and American patients with progressive myoclonus epilepsy of the Unverricht-Lundborg type (EPM1). These mutations were not present in unaffected individuals (PENNACCHIO et al. 1996) or in Mediterranean patients with EPM1 (LABAUGE et al. 1997). The biochemical pathways that are affected by the decreased levels of stefin B in this autosomal recessive disease are not known. Unlike in other progressive myoclonus epilepsies, inclusion bodies or storage material are not observed with EPM1 (PENNACCHIO et al. 1996). Two independent studies show that, in mice, the stefin A gene locus may represent a multi-gene family composed of three to four closely related genes (*Stf1–4*) (HAWLEY-NELSON et al. 1988; TSUI et al. 1993). Additional members include bovine thymus stefin C (TURK et al. 1993) and porcine thymus stefins D1 and D2 (LENARCIC et al. 1996).

Stefins consist of 98–103 amino acids and have a molecular mass of 11–12 kDa. They are cytosolic inhibitors initially believed to protect cytoskeletal proteins from degradation by cysteine peptidases that might leak out of lysosomes during stress conditions. However, given the recent identification of numerous novel cytosolic papain-like cysteine peptidases, such as calpains (SORIMACHI et al. 1997), bleomycin hydrolases (SEBTI et al. 1989), ubiquitin C-terminal hydrolases (Johnston et al. 1997) or viral polyprotein processing enzymes (Chaps. 6 and 7), stefins might have surprising novel regulatory functions. Stefin A is found in most epithelial cells, as well as in liver, spleen, placenta, uterus, and in squamous epithelial cells of the mouth, esophagus and vagina (Davies and Barrett 1984). In addition, it was found in polymorphonuclear leukocytes, suggesting a protective role against cysteine peptidases produced by pathogens invading the body (Brzin et al. 1983; Davies and Barrett 1984). Stefin B is uniformly distributed among various cell types, including lymphocytes and monocytes. Stefin B levels are generally higher than those of stefin A, although stefin A levels are higher in polymorphonuclear leukocytes (Barrett et al. 1986; Jarvinen et al. 1987). Stefin B has a cysteine at position 3 that might act like a redox-sensitive switch turning the inhibitor on or off. Accordingly, oxidizing conditions are thought to lead to the formation of a mixed disulfide that mediates inactivation of the inhibitor (Tsukahara et al. 1987).

2. Family 2 (Cystatins)

Family-2 cystatins contain seven human members: cystatins C, D, E/M, S, SA, SN, and F/leukocystatin. These polypeptides consist of 120–126 amino acids and have a molecular mass ranging from 14 kDa to 20 kDa. They contain two very well conserved internal disulfide bonds and some members, such as rat cystatin C (Esnard et al. 1992) or human cystatins E/M (Ni et al. 1997; Sotiropoulou et al. 1997), and F/leukocystatin (Halfon et al. 1998; Ni et al. 1998) are glycosylated. Whether the additional two cysteine residues in cystatin F form an inter- (Halfon et al. 1998) or intramolecular disulfide bond (Ni et al. 1998) or mixed disulfides as in stefin B (Tsukahara et al. 1987) awaits elucidation of the three-dimensional structure. Cystatins are secreted from the cell and are believed to inhibit extracellular cysteine peptidases. Very few reports have actually identified enzyme-inhibitor complexes in extracellular fluids (Assfalg-Machleidt et al. 1988; Lah et al. 1992b; Luthgens et al. 1993), although the equilibrium constants for dissociation (K_i) of some enzyme complexes are in the subpicomolar range (for review, see Abrahamson 1994). The genes for cystatins C, D, S, SA and SN (*Cst1 5*) as well as two pseudogenes (*Cstp1–2*) are found on a 1.2-megabase fragment of human chromosome 20p11.2 (Thiesse et al. 1994). The gene for cystatin E/M has been localized to 11q13 (Table 1) by means of fluorescent in situ hybridization (Stenman et al. 1997).

Human cystatin C was initially detected in cerebrospinal fluid (Clausen 1961) and, later, in serum and was called γ-trace because of its electrophoretic mobility and low levels in serum (Lofberg and Grubb 1979). The complete amino acid sequence of cystatin C (15 kDa) was determined in 1982 (Grubb and Lofberg 1982). The biological function was identified in 1984 when the sequence was shown to have high homology to that of a cysteine peptidase inhibitor isolated from patients suffering from autoimmune disease (Barrett et al. 1984). Besides cerebrospinal fluid and serum, cystatin C is found in saliva, seminal fluid, tears, milk, and synovial fluid (ABRAHAMSON et al. 1986). The wide extracellular distribution would suggest a general protective role against host or exogenous cysteine peptidases. A point mutation (L68Q) in the coding region of the human cystatin C gene (GHISO et al. 1986) is tightly linked to an Icelandic form of amyloid angiopathy with cerebral hemorrhage (ABRAHAMSON et al. 1992). The recombinant mutant protein appears to be partially misfolded and undergoes dimerization in conditions where the recombinant wild-type protein remains unaffected (EKIEL et al. 1996). Transfection studies have recently shown that the mutant protein accumulates within the endoplasmic reticulum of NIH/3T3 mouse fibroblasts (BJARNADOTTIR et al. 1998).

Cystatins S, SA, and SN were isolated from saliva and shown to have a 54% amino acid homology to cystatin C (ISEMURA et al. 1991). Cystatin S consists of three isoforms (non-phosphorylated, mono-phosphorylated, and diphosphorylated) (ISEMURA et al. 1991) and is found in whole saliva, submandibular saliva, and parotid saliva (AL-HASHIMI et al. 1988; HENSKENS et al. 1996). Cystatin SN is a neutral, non-phosphorylated protein and cystatin SA is an acidic, non-phosphorylated protein (ISEMURA et al. 1991). Cystatin SN is found in submandibular saliva and parotid saliva. Cystatin SA is found in only trace amounts in submandibular saliva and parotid saliva (HENSKENS et al. 1996). Cystatin D was discovered when a genomic library was screened with a cystatin C probe. Cystatin D is expressed in parotid gland tissue and in tears (FREIJE et al. 1993). Two independent groups found cystatin E/M. SOTIROPOULOU and co-workers (1997) found cystatin M by differential display of mRNA from a matched primary tumor and metastatic lesion. NI and colleagues (1997) found cystatin E by analyzing expressed sequence tags (ESTs) from fetal-skin epithelial and amniotic cDNA libraries. On a sodium dodecyl sulfate (SDS)-polyacrylamide gel, cystatin E/M migrates as two bands: a nonglycosylated form at 14–15kDa and a glycosylated form at 16–20kDa (SOTIROPOULOU et al. 1997; NI et al. 1997). Based on Northern-blot analysis, cystatin E/M is distributed primarily in uterus, liver, placenta, pancreas, brain, heart, spleen, and small intestine. Lower levels are detected in the kidney and testes. The fact that cystatin E/M expression is upregulated in fetal epithelial cells might suggest a protective role for this cystatin during fetal development (NI et al. 1997).

Using a similar approach to screen an EST database established from multiple cDNA libraries, two independent groups recently identified a novel cystatin-related sequence which they named cystatin F or leukocystatin, respectively. The leukocystatin cDNA was isolated from human CD34[+] hematopoietic progenitor cells. The genomic organization of this cystatin is different from other cystatins or stefins in that it contains three introns instead of two (HALFON et al. 1998). The cystatin F cDNA, however, was isolated from CD34-depleted human cord blood cells (NI et al. 1998). Northern analyses showed that the protein is primarily expressed in peripheral blood leukocytes and spleen. B-cell lines in culture express barely detectable levels of cystatin F/leukocystatin, whereas several T-cell lines, monocytes, dendritic cells and natural killer cells show moderate expression at the protein and mRNA levels (Halfon et al. 1998; Ni et al. 1998). Since hematopoietic cells selectively produce cystatin F/leukocystatin, this inhibitor may play a specific role in immune regulation through interaction with a unique target in the hematopoietic system.

3. Family 3 (Kininogens)

The kininogens are major plasma glycoproteins, synthesized primarily by the liver and kidney (Hermann et al. 1996). The human kininogens map to 3q26-qter (Fong et al. 1991). A single gene, composed of eleven exons, codes for two kininogens: high-molecular-mass kininogen (HK) is the full-length gene product and low-molecular-mass kininogen (LK) is a smaller splice variant (Kitamura et al. 1985). Both are of much larger molecular size than the members of families 1 and 2 (Ohkubo et al. 1984). Using SDS-polyacrylamide gel electrophoresis, HK and LK were shown to have molecular masses of 112–120 kDa and 64–68 kDa, respectively. Kininogens contain nine internal disulfide bonds and three domains (D1, D2, and D3) with significant homology to family 2 cystatins (Muller-Esterl et al. 1986). However, only cystatin domains D2 and D3 were found to have inhibitory activity when assayed against the limited set of human papain-like cysteine peptidases that were known or available at that time (cathepsins B, H, L and μ- and m-calpains) (Barrett et al. 1986; Ishiguro et al. 1987). HK and LK are biosynthetic precursors of potent vasoactive kinins, such as bradykinin. Part of the cystatin domain D3 was shown to interact specifically with a platelet–endothelial cell surface receptor (Jiang et al. 1992). This interaction is believed to greatly favor the release of bradykinin from kininogens and to allow immediate binding of the vasoactive kinin to specific receptors (Herwald et al. 1995). In addition, HK is involved in contact activation of the blood-coagulation cascade. Deficiencies in HK, a co-factor of blood coagulation, are related to bleeding disorders (Muller-Esterl 1989). There are two additional LK species in rat, T-kininogens 1 and 2 (TK-1 and -2) (Okamoto and Greenbaum 1986; Enjyoji and Kato 1992). TKs are potent acute-phase reactants (Sierra et al. 1989), but only TK-1 appears to increase in rat plasma (Enjyoji and Kato 1992). The dramatic increase in TK serum and tissue levels 4 months before death of rats raises many questions as to their biological role in aging (Sierra et al. 1995).

4. Families 4, 5, 6 (Fetuins, Cystatin-Related Proteins, Histidine-Rich Glycoproteins)

Families 4, 5, and 6 represent the fetuins, cystatin-related proteins (CSTRPs), and histidine-rich glycoproteins (HRGs), respectively. These proteins have homology to family-2 cystatins and presumably are derived from an ancestral gene by a single duplication event at some point in evolution. The human HRG gene has been localized to chromosome 3q28-q29, i.e., within a region that already contains the loci 3q27-q29 for the fetuin, 3q26-qter for the kininogen and 3cen-q21 for the stefin A genes (Table 1) (for review, see BROWN and DZIEGIELEWSKA 1997). The fetuins and HRGs are major plasma glycoproteins, like the kininogens, but contain only two well-conserved cystatin domains with a total of 6 and 7 disulfide bonds, respectively. In fetal calf serum, which is widely used in cell culture, fetuin represents about 40% of the serum proteins (DEMETRIOU et al. 1996). As for kininogen domain D1, the two cystatin domains in fetuins were not found to be inhibitory against a limited set of human papain-like cysteine peptidases (TURK and BODE 1991). The CSTRP genes are located on rat chromosome 3, which has been proposed to have homology to human chromosome 20.

From Southern analyses, the existence of approximately six different rat CSTRP genes has been suggested (DEVOS et al. 1995). Only two CSTRP genes, *Cstrp-1* and *Cstrp-2*, have been cloned. They have a partially duplicated exon 2 and contain a total of four exons (DEVOS et al. 1993). CSTRP-1 and 2 are tissue-specific secretory glycoproteins, synthesized in the ventral prostate and lacrimal glands of the male rat (WINDERICKX et al. 1990). Their expression is transcriptionally regulated by androgens (VERCAEREN et al. 1996). It has been stated that CSTRP-1 and 2 have no inhibitory activity against cysteine peptidases, but the enzymes that were tested were not mentioned (DEVOS et al. 1995). A unique testis-regulated gene related to the cystatin family has been identified and shown to be highly restricted in its expression to the proximal region of the mouse epididymis (CORNWALL et al. 1992). This cystatin-related epididymal-specific protein (CRES) appears to be involved in spermatogenesis and caput-epididymal sperm maturation (CORNWALL and HANN 1995). A full-length 910-bp cDNA sequence for the human homologue of mouse CRES has been cloned (accession AF059244). The deduced amino acid sequence clearly corresponds to a family-2 cystatin with quite well-conserved motifs for inhibition of papain-like cysteine peptidases. However, the presence of a 5'-splice variant of the 910-bp human CRES transcript (GenBank accession AA436982) indicates that the CRES gene may have more than three exons. For this reason, CRES might (like CSTRPs) form a novel, evolutionary distinct family of secreted cystatins.

III. Mechanism of Inhibition

There are three motifs within the primary sequence of cystatins (Table 2) that are responsible for their inhibitory activity and selectivity against papain-like cysteine peptidases (BODE et al. 1988). The first motif is a stretch of three amino acids at the N-terminus of stefins, cystatins or cystatin domains. This motif is assumed to bind in the S_3 to S_1 substrate-binding pockets of the target enzymes, based on direct interaction studies between human and chicken cystatin C and papain (ABRAHAMSON et al. 1987), docking experiments of chicken cystatin C with papain (BODE et al. 1988), and site-directed mutagenesis studies on human cystatin C (LINDAHL et al. 1994). The motif contains a very well conserved Gly in the third position, which is important for orientation toward the active site of the target cysteine peptidase. This has been confirmed by engineering truncated forms of human cystatin C missing the first eleven amino acids including the three-amino acid motif, -Leu-Val-Gly$_{11}$ (Table 2, LVG). When compared with the wild-type inhibitor, truncated cystatin C displayed decreased affinity for all tested enzymes. The K_i increased more than 1000-fold for human cathepsin B, more than 100-fold for human cathepsin L and bovine cathepsin S, and about 50-fold for human cathepsin H (HALL et al. 1998). From a physiological point of view it is noteworthy that polymorphonuclear leukocyte elastase has been found to cleave human cystatin C within this N-terminal motif, i.e., between Val$_{10}$ and Gly$_{11}$. This truncation reduces the inhibitory potential of the cystatin 700-fold for cathepsin B, more than 400-fold for cathepsin L and only 6-fold for cathepsin H (ABRAHAMSON et al. 1991). The substrate-like binding of cystatins to their target enzymes is underlined by the finding that human cystatin C is cleaved at Gly$_{11}$ upon complex formation with papain and that the modified inhibitor is at least a 1000-fold weaker inhibitor of papain (ABRAHAMSON et al. 1987). Similar truncated forms of human cystatin C have been identified in cancerous ascites fluids (LAH et al. 1992b). Interestingly, the replacement of Val$_{10}$ by Gly (V10G) in human cystatin C also decreased the affinity of the inhibitor for cathepsin B by more than 1000-fold and for cathepsin H by 100-fold, but had no effect on inhibition of cathepsins L and S (HALL et al. 1995). These results provide support for the inability of human cystatin D and porcine stefins D1 and 2 that have an N-terminal LAG motif (Table 2) to inhibit human cathepsin B (LENARCIC et al. 1996; HALL et al. 1998). However, replacement of the 11-amino acid N-terminal segment of cystatin D by that of cystatin C did not improve the inhibitory potential of the hybrid cystatin towards cathepsin B, indicating that other structural elements are required for tight binding (HALL et al. 1998).

The second and third motifs represent the first and second β-hairpin loops of the β-sheet that forms the main body of the well-conserved cystatin domain (BODE et al. 1988). The first hairpin loop is a highly conserved sequence of five amino acids, Gln$_{53}$-Ile-Val-Ala-Gly$_{57}$ in human cystatin C, and Gln-X-Val-X-Gly as a general consensus sequence (Table 2). Cystatin F (leukocystatin) is quite unique in that it has a Lys in the fourth position of this motif (Table 2, QIVKG) and is extremely tissue specific (HALFON et al. 1998; NI et al. 1998). Together with its poor inhibition of cathepsins B and L when compared with cystatin C (HALFON et al. 1998), these results might indicate that cystatin F may target an equally tissue-specific enzyme, e.g., cathepsin S or W (LINNEVERS et al. 1997). The importance of the second motif was investigated in human stefin A by producing recombinant mutants displaying the sequence KVVAG or QVTAG instead of the wild-type sequence QVVAG. The mutants were found to have the same inhibitory potential against papain and cathepsins B, H and L as wild-type stefin (NIKAWA et al. 1989). In contrast, amino acid substitutions in the second motif of chicken egg-white cystatin C reduced the inhibitory activity against cathepsin B by 1000-fold (AUERSWALD et al. 1992). The third motif, or second β-hairpin loop, represents a three-amino acid

Table 2. Comparative list of the three motifs composing the active site of members of the cystatin superfamily and inhibitory potential

Families	Name or abbrev.	Animal species	Motif-1 N-terminal segment	Motif-2 1st β-hairpin loop	Motif-3 2nd β-hairpin loop	Inhibitory potential
1-Stefins	A	human	IPG	QVVAG	PGQ	yes
	α	rat	IVG	QVVAG	SGD	yes
	A1	mouse	IPG	QAVAG	TGK	yes
	A2	"	IKG	QVVAG	SGE	n.d.
	A3	"	IIG	QVVQG	SSE	n.d.
	A4	"	SLG	QVVAG	TGK	n.d.
	B	human	MCG	QVVAG	PHE	yes
	β	rat	MCG	QVVAG	PHE	yes
	C	bovine	NLG	QVVAG	PHE	yes
	D1	porcine	LAG	QVVAG	PHQ	yes
	D2	"	LAG	n.d.	n.d.	yes
2-Cystatins	C	human	LVG	QIVAG	PWQ	yes
	"	rat	LLG	QLVAG	PWK	yes
	"	mouse	MLG	QLVAG	PWK	yes
	D	human	LAG	QIVGG	PWE	yes
	E/M	human	MVG	QLVAG	PWQ	yes
	F	human	KPG	QIVKG	PWL	yes
	"	mouse	KPG	QVVKG	PWL	n.d.
	S	human	IPG	QTFGG	PWE	yes
	SA	"	IEG	QIVGG	PWE	yes
	SN	"	IPG	QTVGG	PWE	yes
	S?	rat	FLG	QVVAG	PWE	yes
3-Kininogens	D1	human	QES	TVGSD	SST	no
	D1	rat	QEE	KDGAE	EN–	no
	TK-1/D1	"	QEE	KDGAE	EN–	no
	TK-2/D1	"	QEE	KDGAE	RNN	no
	D2	human	CLG	QVVAG	IQL	yes
	D2	rat	CVG	QVVAG	IHN	yes
	TK-1/D2	"	CVG	QVVAG	IHN	yes
	TK-2/D2	"	CVG	QVVAG	IHK	yes
	D3	human	CVG	QVVAG	PWE	yes
	D3	rat	CFG	QVVAG	PWE	yes
	TK-1/D3	"	CRG	QVVAG	PWE	yes
	TK-2/D3	"	CPG	QVVAG	PWE	yes
4-Fetuins	D1	human	PGL	QQPSG	KLD	no
	D1	rat	AGL	RRPFG	KQD	no
	D1	mouse	TGL	RRPFG	KQD	no
	D2	human	CQD	VPLPP	KLG	no
	D2	rat	CPR	VPFPV	RLG	no
	D2	mouse	CPR	VPLPV	NLG	no
5-CSTRPs	CSTRP-1	rat	PME	32-aa loop §	YIE	no
	CSTRP-2	"	GME	32-aa loop §	YIE	no
6-HRG	D1	human	n.d.	VENTT	HES	no
	D2	"	SPV	GGEGT	LDL	no
7-CRES	CRES	human	ETG	QVTNL	PWN	n.d.
	"	mouse	n.d.	QITDR	PWN	n.d.

n.d., not determined; §, inserted 32-amino acids loop due to duplicated exon-2 rat cystatin S? This cystatin does not seem to be the rat homologue of human cystatin S.

sequence, Pro-Trp_{106}-Gln, in human cystatin C (Table 2, PWQ). The importance of this motif was established by direct chemical modification of the Trp_{104} residue in chicken cystatin C. The modified inhibitor displayed a 10^5-fold lower affinity for papain than the intact cystatin (Nycander and Bjork 1990). Similarly, site-directed mutagenesis studies showed that the human cystatin C mutant W106G has a more than 1000-fold decreased affinity for cathepsins B and H, more than 10-fold for cathepsin L, but has the same affinity for cathepsin S (Hall et al. 1995). In the case of chicken cystatin C, the second and third motifs have been reported to fit into the wide S_1'-subsite pocket of papain (Bode et al. 1988).

Altogether, the above studies indicate that the specificity of the interaction between cystatins and their respective target enzymes is governed by at least three non-continuous sequences in the inhibitor molecule. These sequences involve about 10 amino acids from a total of 120. The three motifs align in the three-dimensional fold of cystatins, as exemplified in the crystal structure of chicken cystatin C, where they form a contiguous structure known as the wedge-shaped "edge" (Bode et al. 1988). This feature should aid in the design of selective inhibitors as well as quenched fluorescent substrates for each enzyme.

C. Cystatins in Cancer

Metastasis is a multi-step process in which tumor cells escape from the primary tumor and establish colonies at distant sites. Invading tumor cells must penetrate basement membrane surrounding stromal tissues and microvasculature. Thus, basement-membrane invasion has been hypothesized to be a critical component of the metastatic potential (Woodhouse et al. 1997). Many different proteolytic enzymes have been implicated in tumor-cell invasion and metastasis. Metallo-, serine-, aspartyl-, and cysteine peptidases are able to cleave and degrade basement-membrane components (for reviews, see Hewitt and Dano 1996; Rochefort et al. 1996; Chambers and Matrisian 1997; Lah and Kos 1998). Initially, it was thought that peptidases were involved primarily in the later stages of cancer progression including extravasation from the circulation and formation of micrometastases. Several recent studies have implicated peptidases in the early stages of tumor progression including tumor growth (for review, see Chambers and Matrisian 1997). The role of cysteine peptidases and their inhibitors in the earliest stages of tumor progression has not been elucidated.

I. Inhibitory Activity

There have been a number of reviews on cathepsin B in cancer (Keppler and Sloane 1996; Yan et al. 1998). Novel cysteine peptidases have been cloned from different tumor cells or tissues, but their roles in the progression of the disease have not yet been studied. There are many levels of regulation of cysteine peptidases, but the final level of regulation is through inhibition of enzyme activity. Lysosomal cysteine peptidases, such as cathepsins B and L, are regulated extracellularly by secretory cystatins, such as cystatin C, and intracellularly by stefins A or B. Therefore, it has been suggested that an imbalance in the expression of cathepsins and stefins and/or cystatins could contribute to the malignant progression of tumors, as has been observed in other proteolytic systems. Analysis of metastatic variants of the murine B16 melanoma revealed an increase in the cathepsin B activity that is associated with low papain inhibitory activity (PIA) (SLOANE et al. 1990). Human lung tumors show an imbalance between cathepsin B and PIA (EBERT et al. 1994; HEIDTMANN et al. 1997). This could be a plausible explanation for the observed inverse correlation between cathepsin B expression and basement-membrane staining in lung adenocarcinomas (INOUE et al. 1994; SUKOH et al. 1994).

When compared with normal human breast tissue, two-thirds of breast carcinomas have a decreased PIA. Those samples that have a decreased PIA often have increased activity of cathepsins B and L. The decreased PIA can, in part, be attributed to a decreased expression of stefin A at both the RNA and protein levels (see below). Low PIA is also observed in poorly differentiated carcinomas of the breast (LAH et al. 1992a,c). However, other studies in breast carcinomas showed that increased activities of cathepsins B, H, and L are not always accompanied by a decrease in PIA (GABRIJELCIC et al. 1992a,b). In human colorectal tumors, PIA is unchanged between matched pairs of tumors and normal mucosa (SHEAHAN et al. 1989). These differences may reflect different tissue types or protocols for tissue collection and protein extraction. PIA does not account for the individual contribution of each and every cystatin present in a given tissue. It is rather the sum of inhibitory activities from different freely available cystatins at the time of the assay (KEPPLER et al. 1997). In order to pinpoint the specific contribution of a cystatin biochemical, immunochemical and molecular techniques have to be employed.

II. Stefins A and B

Several groups have analyzed whether decreased expression or activity of stefins A and/or B accompanies overexpression of cathepsins B, H and/or L. The rationale for this is not obvious, since cathepsins are vesicular and/or secreted and stefins are cytosolic. Nevertheless, both stefins A and B have been detected in normal and pathologic extracellular fluids (ABRAHAMSON et al. 1986; LAH et al. 1992b; LETO et al. 1997). These inhibitors could therefore also potentially be involved in the extracellular regulation of cathepsin activities.

Purified stefin A from human sarcomas has reduced inhibitory activity against cathepsin B and only negligible amounts can be recovered from ovarian carcinomas. In contrast, stefin B retains its inhibitory capacity in both tumor types (LAH et al. 1989). The effect of purified human stefins A and B on adhesion and motility of melanoma cells on collagen type-IV-coated filters was investigated. Stefin A is more inhibitory than stefin B on migration of W256 rat carcinosarcoma cells, whereas stefin B is more efficient than stefin A on motility of A2058 human melanoma cells. Stefins appear also to affect adhesion of the cells to collagen (Boike et al. 1992). Since stefin A is of epithelial origin, and most tumors are of epithelial origin, many studies have compared stefin A levels in normal and tumor tissues or cell lines. An immuno-histochemical study conducted on human prostate revealed that stefin A expression decreases from normal prostate to benign prostate hyperplasia and is undetectable in prostatic carcinoma (Soderstrom et al. 1995). Differential display analysis of mRNAs from a primary carcinoma of the esophagus and local lymph-node metastases recently established that human stefin B is down-regulated upon progression towards invasive and metastatic tumors (Shiraishi et al. 1998). In a squamous cell carcinoma of the esophagus, no immunostaining for stefins A and B is observed in malignant keratinocytes. Stefins A and B are present in the most clearly differentiated cells of the carcinoma, but not in the undifferentiated basal-like cells (Jarvinen et al. 1987). Similar results were obtained in squamous cell carcinomas of the skin (Pernu et al. 1990), uterus (Eide et al. 1992), and lung (Jarvinen et al. 1987).

At the molecular level, stefin A mRNA levels are reduced in mouse carcinomas when compared with skin papillomas. However, in the same study, it was also observed that stefin A mRNA is localized to the less differentiated basal and lower spinous layers of the newborn mouse epidermis by means of in situ hybridization. In addition, Northern-blot analysis of mouse stefin A mRNA revealed lower expression in Ca^{2+}-induced differentiating cells than in cells grown in low Ca^{2+}. These data suggest that stefin A may be important in the control of normal keratinocyte proliferation and differentiation (Hawley-Nelson et al. 1988). However, as we have mentioned before, mice may express more than one isoform of stefin A (Hawley-Nelson et al. 1988; Tsui et al. 1993). Therefore, probes that can distinguish between these isoforms are needed to assess the precise contribution of each stefin A gene product. A study conducted on human breast cancer found that stefin A-positive cancers confer a poorer prognosis than stefin A-negative cancers (Kuopio et al. 1998). Using both different breast samples and immunoprobes, another group showed the opposite, i.e., that stefin A-negative cancers conferred a poorer prognosis (Lah et al. 1992c, 1998). More studies of this sort need to be conducted before any definitive conclusion can be drawn. The exchange of specific probes, samples and standardization of techniques would also greatly reduce discrepancies between independent studies.

III. Cystatins C and E/M

Many normal cells and cancer cell lines have been shown to secrete cystatin C in vitro (Warfel et al. 1987; Chapman et al. 1990; Solem et al. 1990; Barka et al. 1992; Esnard et al. 1992; Keppler et al. 1994b; Bjarnadottir et al. 1995; Lerner et al. 1997; Sexton and Cox 1997). Cystatin C has also been detected in vivo in ascites fluid from patients with ovarian cancer and its activity against cathepsin B is unaltered (LAH et al. 1992b). In an experimental intraperitoneal carcinosis model in nude mice, human colon carcinoma cells were shown to release cystatin C in vivo (KEPPLER et al. 1994a). The use of cystatin C as a marker for gastro-entero-pancreatic tumors has been investigated using a commercial polyclonal antibody preparation, but this antigen is not a reliable marker, as expression is detected in normal and tumor samples (LIGNELID and JACOBSSON 1992). Using the same immunoprobe, human cystatin C was not found to be a reliable marker for renal cell carcinomas (JACOBSSON et al. 1995) or any specific brain tumor type. However, in neoplastic tissues of the brain and pituitary, the cellular production and secretion of cystatin C does change with malignant progression (LIGNELID et al. 1997). Using both a monospecific rabbit polyclonal antibody (LOFBERG and GRUBB 1979) and a mouse mono-clonal antibody (HCC3) (OLAFSSON et al. 1988), we have observed that expression and localization of cystatin C changed during early progression of human breast cancers ($n = 30$) (Fig. 1). In normal breast tissue, cystatin C immunos-taining was mainly observed in the lumen of ducts with some staining also of adipocytes (Fig. 1A,B). In metaplastic and moderate hyperplastic lesions, cys-tatin C localization changed from extracellular to cytoplasmic (Fig. 1C,D). In biopsies with a clear invasive component, cystatin C immunostaining could not be observed (Fig. 1E). On a frozen serial section stained with a monoclonal antibody directed against human platelet-endothelial cell adhesion molecule-1 (PECAM-1, CD-31), strong immunostaining of the activated vasculature was seen (Fig. 1F). Based on these preliminary results on a few human breast tissue samples, it appeared that there might be an inverse relationship between cys-tatin C- and PECAM-1 immunostainings. Downregulation of cystatin C might be part of the angiogenic switch during progression of human breast cancer. Cystatin E/M mRNA expression is also downregulated in a metastatic breast lesion when compared with its matched primary tumor. Based on this finding, cystatin E/M has been suggested to be a novel metastasis suppressor gene (SOTIROPOULOU et al. 1997). Cystatin E/M may be important during differen-tiation of the breast epithelium because its mRNA can be induced by retinoids (SOTIROPOULOU et al. 1997). The respective roles of cystatins C and E/M in the breast epithelium have yet to be determined in order to understand their downregulation during invasion and metastasis.

Compared with the large amount of biochemical data on cystatins, there are only a few studies that have addressed their biological function, and most of them were carried out on cystatin C. Overexpression of mouse cystatin C in an invasive and motile mouse melanoma cell line results in inhibition of motility and invasion in vitro (SEXTON and COX 1997). In murine B lymphoma A20 cells, cystatin C inhibits invariant (Ii) chain proteolysis within an endo-somal compartment (PIERRE and MELLMAN 1998). Ii chain cleavage by cathep-sin S plays a key role in many MHC class-II antigen-presenting cells (RIESE et al. 1996). Other potential biological functions that have been ascribed to cystatins include promotion of cell growth (TAVERA et al. 1992), increase in

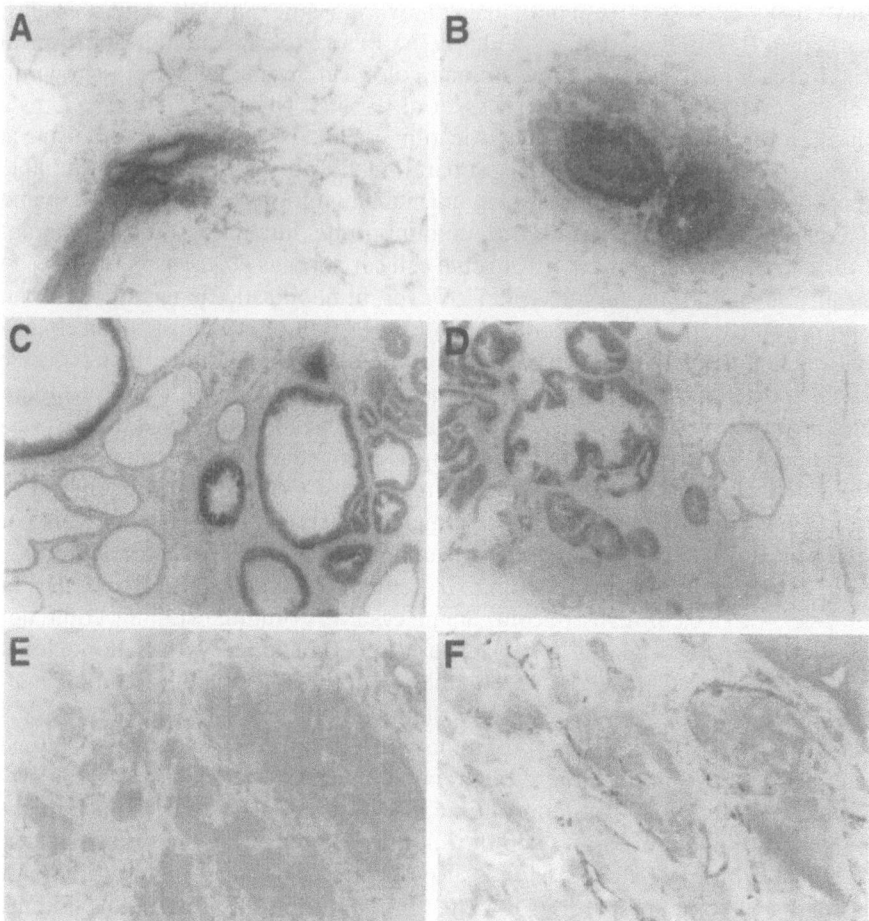

Fig. 1A–F. Immunoperoxidase staining for cystatin C in early human proliferative breast disease. **A, B** Frozen sections of normal breast tissue: cystatin C immunostaining of the lumen of ducts using a rabbit antiserum at 1:400 dilution (**A** 100×; **B** 200×). **C, D** Frozen sections of metaplastic/moderate hyperplastic breast tissue: cytoplasmic immunostaining for cystatin C in dysplastic, but not in normal epithelium using mouse monoclonal antibody HCC3 at 2.5 μg/ml (100×). **E, F** Frozen serial sections of an invasive breast adenocarcinoma (100×): **E** Little or no immunostaining for cystatin C using monoclonal HCC3 at 2.5 μg/ml. **F** Platelet–endothelial cell adhesion molecule (PECAM)-1 immunostaining of the activated tumor vasculature using monoclonal BBA-7 (R&D Systems) at 1.0 μg/ml. Hematoxylin and Eosin counterstaining

nitric oxide production (Verdot et al. 1996), and inhibition of viral replication (see Chaps. 7 and 8).

IV. Synthetic Inhibitors

Synthetic cysteine peptidase inhibitors have become an area of intense research. Indeed, various cysteine peptidases have been implicated in tumor progression, inflammation, and viral diseases. There are several classes of synthetic inhibitors, which include irreversible inhibitors such as halomethyl- and diazomethyl-ketones, peptidyl sulfonium salts, peptidyl vinyl sulfones, epoxysuccinyl-, and aziridine derivatives, as well as reversible inhibitors such as peptide aldehydes, α-keto aldehydes, and peptidyl nitriles (for review, see SHAW et al. 1986; SHAW 1988; KIRSCHKE et al. 1995; BROMME et al. 1996). The epoxysuccinyl peptides and aziridines are specific for cysteine peptidases. Diazomethanes can also inactivate subtilisin-like serine peptidases, and peptide aldehydes inactivate both serine and cysteine peptidases (KIRSCHKE et al. 1995). Peptidyl nitriles may represent more selective reversible inhibitors of cysteine peptidases than peptide aldehydes. However, the nitrile derivatives are much less potent than the aldehydes. The α-keto aldehydes inhibit cathepsin B three times and chymotrypsin ten times more efficiently than the aldehydes but do not offer selectivity towards cysteine peptidases. E-64 is an epoxide isolated from an extract of a solid culture of Aspergillus japonicus. This antibiotic rapidly inactivates lysosomal cysteine peptidases (BARRETT et al. 1982). One exception, however, is the legumains that are inhibited by human cystatin C, but not by E-64 (CHEN et al. 1997). E-64 inhibits papain-like cysteine peptidases by alkylation of the active-site thiolate group. An E-64 derivative, CA-074, is highly selective for cathepsin B in vitro as well as in vivo (MURATA et al. 1991; TOWATARI et al. 1991). Among the peptidyl diazomethanes, Z-Phe-Tyr(OBu^t)-CHN$_2$ is extremely potent and selective for cathepsin L (KIRSCHKE et al. 1988). None of the new cysteine peptidases with cathepsin L-like substrate specificities have been compared with cathepsin L for their ability to be inhibited by this agent.

The development of selective inhibitors against individual cysteine peptidases has received renewed attention with the development of various combinatorial approaches. Inhibition of individual lysosomal cysteine peptidases by their respective propeptides also shows much promise, as these have been shown by crystallographic studies to completely cover the active-site cleft of their respective enzymes (COULOMBE et al. 1996). In addition, the 60- to 110-amino acid prosegments that have been naturally designed to keep enzymes latent show much higher selectivity than cystatins. Thus, at pH 6.0, recombinant cathepsin B is inhibited by its 62-amino acid propeptide with a K_i of 0.4 nM, whereas the interaction with papain is about 5000-fold weaker giving a K_i in the micromolar range (Fox et al. 1992). Similarly, inhibition by the cathepsin L prosegment of 96 amino acids is selective for cathepsin L over other cathepsin L-like peptidases, such as papain and cathepsin S (COULOMBE et al. 1996). There are no structural homologies between cystatins and the proregions of cathepsins so that the high selectivity of the latter offers additional structural information on the configuration of the active site cleft of each enzyme. Limitations in the use of synthetic propeptide sequences as inhibitors, however, are of two types: (1) at pH < 6.0, the propeptides are good substrates for their respective enzymes (Fox et al. 1992) and, in general, may be good substrates for other peptidases. Agents designed according to propeptide sequences could therefore be used mainly to target extracellularly released enzymes; (2) recombinant propeptides or shorter propeptide sequences (Chagas et al. 1996) may be immunogenic in vivo and limit applications. Interestingly, differential screening of a subtracted cDNA library from resting and activated mouse T lymphocytes permitted isolation of two novel genes, CTL-2α and CTL-2β, specifically expressed in activated T cells. The products of these genes show 90% homology to each other and 42% homology to the propeptide region of mouse cathepsin L (Denizot et al. 1989). Indeed, CTL-2β was shown to inhibit cathepsins H and L with a K_i in the nanomolar range but did not inhibit cathepsin B (Delaria et al. 1994).

To date, the most widely used synthetic cysteine peptidase inhibitor is E-64. Many groups have studied whether E-64 could block invasion through Matrigel (a reconstituted basement membrane made of solubilized components, without intramolecular cross-links) or block experimental metastasis in mice. E-64 does partially block in vitro migration of invasive human bladder carcinoma EJ cells and inhibits their experimental metastasis to the lungs of nude mice (Redwood et al. 1992). E-64 also reduces the number of spontaneous metastases in mice bearing ovarian sarcomas, but not in mice bearing mammary carcinomas or leukemias (Leto et al. 1994). Invasion of ovarian cancer cells through Matrigel is blocked by E-64 due to interference with the proteolytic activation of the zymogen of urinary-type plasminogen activator (Goretzki et al. 1992; Kobayashi et al. 1992), a serine peptidase also implicated in tumor invasion (Hewitt and Dano 1996). This study suggests that cysteine peptidases may contribute indirectly to invasion by proteolytically activating latent zymogens.

D. Potential Transcriptional Regulation

One of the explanations for the lower cystatin activity found in tumors is decreased expression at the level of transcription. To study transcriptional regulation, one must have knowledge of the DNA sequence upstream of the gene of interest. The upstream sequences of human stefins A, B and cystatin C are known and have been deposited in the GenBank.[2] We retrieved the sequences and performed a computer analysis to find potential transcription-factor-binding sites.[3] Many sites were found in the promoters of these three genes. The search is based on a statistical analysis of the likelihood that a certain transcription factor will bind to the specific DNA sequences being analyzed.

---

[2] For on-line access to gene sequences, go to the Genome database, at the following URL: http://www.gdb.org/ (the accession numbers for the human stefin A, stefin B and cystatin C genes are D88439, U46692 and X52255, respectively).

[3] For access to the on-line service of the Baylor College of Medicine, go to the BCM Search Launcher, at the following URL: http://dot.imgen.bcm.tmc.edu:9331/. The TESS String-Based Search is hyperlinked to TransFac v3.2, the European transcription factor database.

All putative binding sites to be mentioned herein are either a perfect match or had only a one base-pair mismatch.

I. Stefins A and B

The complete promoter region of stefin A (1800 bp), the partial promoter region of stefin B (300 bp) and the putative transcription-factor-binding sites are illustrated in Fig. 2A. An analysis of the promoter region of stefin B (Fig. 2A) reveals several characteristics of housekeeping-type genes, including multiple potential Sp1 sites, high GC content, a large number of CpG dinucleotides, and lack of TATA- and CAAT-boxes. However, analysis of the stefin A promoter (Fig. 2A) shows a very low GC content, a low number of CpG dinucleotides, very few Sp1-binding sites, and two TATA-boxes. In addition, two potential transcriptional start sites are found just downstream of the TATA-boxes indicating possible transcript variants of stefin A or use of different promoters. These characteristics would explain the tissue distribution of stefin A which appears to be quite restricted to epithelial cells and polymorphonuclear leukocytes. Regional hypermethylation is an efficient mechanism for the silencing of genes at the level of transcription. This enzymatic process has been implicated in carcinogenesis (for a review, see MOSTOSLAVSKY and BERGMAN 1997). The potential of this mechanism to inactivate the stefin B gene exists due to the large CpG island, although this needs now to be addressed more directly. In contrast, stefin A has very few CpG dinucleotides and, therefore, methylation as an explanation for the downregulation in cancer seems highly unlikely.

We cited studies in Sect. C.II. indicating that stefins A and B are localized to the more differentiated cells of a tumor. This could be explained by the presence of certain transcription-factor-binding sites within their promoters. In stefin A, there are putative binding sites for C/EBPα (CCAAT-enhancer binding protein-α) and PPARs (peroxisome proliferator-activated receptors). These transcription factors are involved in differentiation of keratinocytes, hepatocytes, monocytes/macrophages, and adipocytes (SPIEGELMAN 1998). In addition, stefin A contains eight putative GATA-1 sites. GATA transcription factors are required for differentiation of stem cells, erythroid cells and cardiomyocytes (SHIVDASANI 1997). The role of stefin A in differentiation has yet to be elucidated, but the presence of multiple transcription-factor-binding sites that are involved in differentiation would suggest that stefin A does play an important role in this process. In contrast, stefin B lacks all of the aforementioned binding sites in its promoter. This would suggest that stefin B is probably not important for differentiation even though it is localized to more differentiated cells in breast cancers and melanomas. Interestingly, stefin B does contain a putative binding site for Wilms' tumor factor (WT1), a transcription factor that has been classified as a tumor-suppressor gene (MENKE et al. 1998).

R. Shridhar et al.

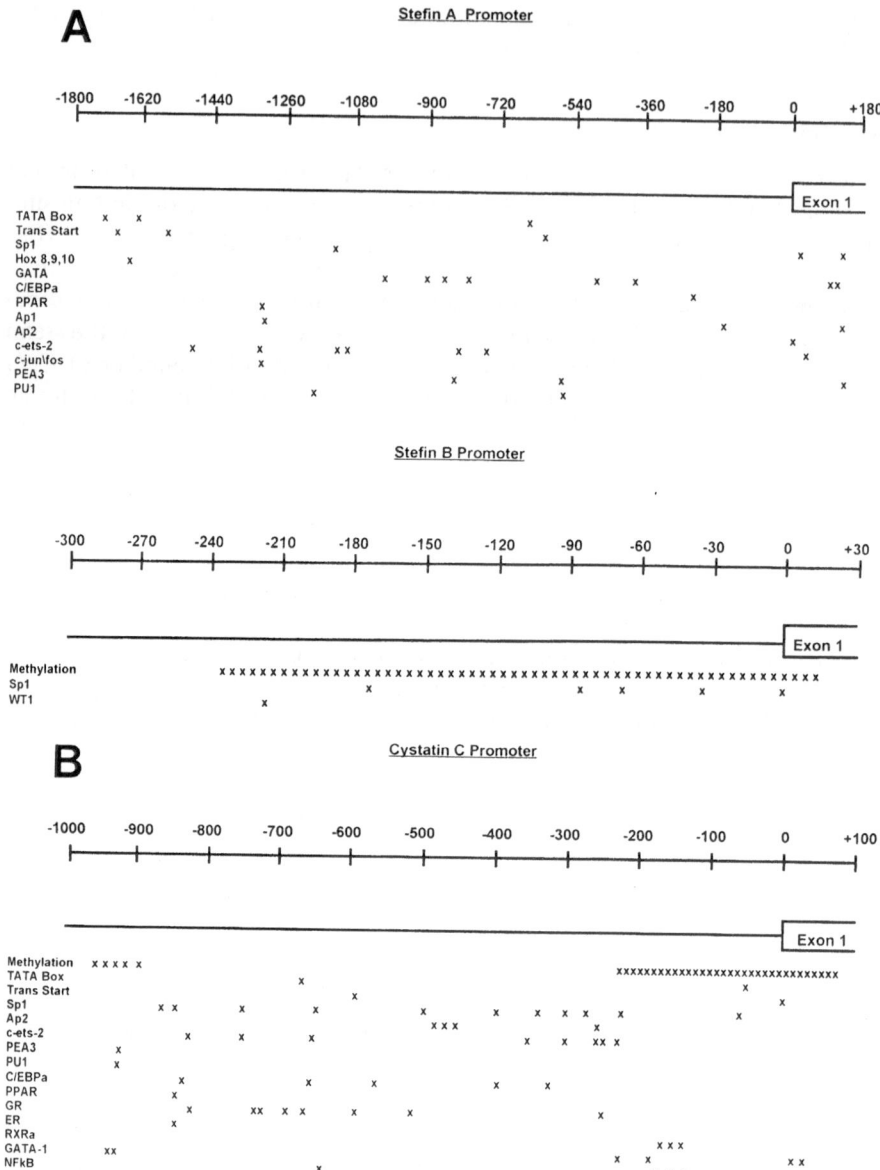

Fig. 2A,B. Analysis of the 5′-flanking regions of the genes for human stefins A, B (**A**) and cystatin C (**B**). *X* indicates a putative binding site for a transcription factor (listed on the *left*). Abbreviations are taken from the TRANSFAC database v3.2. TATA-boxes, transcription start sites and methylation of CpG islands are also marked

II. Cystatin C

The complete promoter region and potential transcription-factor-binding sites for human cystatin C are illustrated in Fig. 2B. Cystatin C may be a house-keeping-type gene, because it contains numerous Sp1 binding sites, has a high GC content, and contains a large number of CpG dinucleotides. The large number of CpG dinucleotides would suggest that regional hypermethylation could reduce or silence the expression of cystatin C mRNA and ultimately result in the increased cysteine peptidase activity that is seen in invasive can-cers. Interestingly, only one TATA-box was identified in 1990 (ABRAHAMSON et al. 1990), whereas this computer analysis of the cystatin C promoter reveals a second putative TATA-box. Others have found a CAAT-box upstream of the first TATA-box, but, in agreement with ABRAHAMSON et al. (1990), no CAAT-boxes were identified in our search. Two transcriptional start sites are detected upstream of exon 1. In addition, our analysis reveals four potential transcription start sites downstream of exon 1 (not shown in Fig. 2B). The use of the downstream transcription start sites would not result in expression of a functional inhibitor though. Multiple transcription start sites are characteris-tic for housekeeping-type genes and have also been observed in the cathep-sin B and D genes (BERQUIN et al. 1995).

Nevertheless, analysis of the 5'-flanking region of the cystatin C gene reveals several interesting binding sites. There are a number of transcription-factor-binding sites that are involved in differentiation, including C/EBPα, PPAR, RXRα (retinoic acid receptor that preferentially binds 9-cis retinol), GATA-1 and/or inflammatory responses such as NF-κB (nuclear factor-κB). The presence of RXRs would suggest that the therapeutic effectiveness of retinoids might be, in part, mediated via enhanced expression of cystatin C, although this needs to be investigated directly. Another interesting set of factors that were identified are two other members of the sterol-binding family: the glucocorticoid receptor (GR) and the estrogen receptor (ER). The GR response elements appear to be functional, since a synthetic glucocorti-coid dexamethasone induces increased expression and secretion of cystatin C in human cervical adenocarcinoma HeLa cells (BJARNADOTTIR et al. 1995). Perhaps, cystatin C levels could be used as a marker for people with abnor-mal adrenal functions, since the adrenal cortex is the primary site of action and metabolism of glucocorticoids. ER-negative breast cancers are more inva-sive than ER-positive cancers (THOMPSON et al. 1993). One possible explana-tion is that estrogen induces expression of peptidase inhibitors, such as cystatin C, resulting in lower proteolytic activity and lower invasive potential.

III. Links to Cancer Progression

In both the stefin A and the cystatin C promoter regions, there are many binding sites for transcription factors that belong to the erythroblast-transforming sequence (ETS) family. Some of the family members are c-ets-

1, c-ets-2, PEA3, and PU.1 (for review, see Dittmer and Nordheim 1998). Binding sites for AP1, AP2, c-jun, and c-fos are included in Fig. 2A,B (if present), because members of the ETS family are known to co-operate in transcription with these factors. C-ets-1 and 2 as well as PEA3 form complexes with AP1, a member of the fos and jun proto-oncogene family. ETS family members have been implicated in oncogenic transformation of cells. Since there are many ETS-binding sites in cystatin C and stefin A, these two inhibitors may be involved in transformation of cells. The increased expression of stefin A during breast cancer progression (Kuopio et al. 1998) could be due to the presence of oncogenic members of the ETS family. Cystatin C could help mediate early oncogenic transformation by ETS family members as cystatin C can act as a growth factor in rat glomerular mesangial cells (Tavera et al. 1992). At later stages of tumor progression, cystatin C gene expression would then be silenced by hypermethylation of the CpG island or some other mechanism. An understanding of transcriptional regulation can provide powerful insights into mechanisms of action of certain agents through the up- or downregulation of certain genes. For instance, retinoids induce differentiation of several different tumor types, and the demethylating agent, 5-aza-2-deoxycytidine, although not selective, has been clinically tested in the treatment of certain leukemias. How these agents affect expression of cystatins in different cancers should help clarify why cystatins are overexpressed in some tumor types or stages but downregulated in others.

E. Perspectives: Therapeutic Implications

The potential use of E-64 or its derivatives, such as CA-074, in the therapy of cancer still remains an open question. Further animal studies and, possibly, phase-I clinical trials are needed to determine the effectiveness of these types of agents. A combination therapy using a cocktail of peptidase inhibitors, such as E-64 (cysteine peptidase), batimastat (metallo peptidase), and PAI-1 (serine peptidase), would seem promising considering that several classes of peptidases are implicated in tumor progression. Since E-64 irreversibly inhibits a number of cysteine peptidases, it could well present a problem of host toxicity. E-64 is reported to have powerful teratogenic effects on rat embryogenesis (Tachikura 1990). Perhaps a more selective inhibitor such as CA-074 could be utilized. However, it would seem that cysteine peptidase inhibitors in general would disrupt normal fetal development as evidenced by the observation that the double knockout for cathepsins B and L was lethal (J. Deussing and C. Peters, personal communication).

An alternative to synthetic inhibitors might be the administration of either a recombinant cystatin or a propeptide. These are selective, reversible, probably less toxic and would not induce acquired drug resistance because they are endogenous inhibitors. As an example, recombinant mutants of cystatin C show selectivity towards cathepsin L (Mason et al. 1998). In theory, recombinant mutant cystatins that are selective for individual cysteine peptidases could be used to treat tumors that have increased activity of a given cysteine peptidase. However, recombinant polypeptides may be misfolded, degraded or immunogenic and, therefore, may require further refinements. Finally, gene therapy could be used to deliver the genes for certain cystatins or stefins to a tissue of interest. The design of newer and more effective therapies will evolve as we have a better understanding of the role of proteases and their inhibitors at each stage of tumor progression.

Acknowledgements. We thank Dr. Magnus Abrahamson for the kind gifts of a rabbit polyclonal antibody and a mouse monoclonal antibody to human cystatin C; Pamela Tabaczki and Dr. Daniel W. Visscher, Department of Pathology, for performing the immunoperoxidase stainings and helpful discussions regarding histopathology, respectively. Ravi Shridhar is a first year PhD/MD student in the Department of Pharmacology. This work was supported by a Virtual Discovery Grant from the Barbara Ann Karmanos Cancer Institute and a US Public Health Service Center Grant CA36481 from the National Cancer Institute.

References

Abrahamson M (1994) Cystatins. Methods Enzymol 244:685–700

Abrahamson M, Barrett AJ, Salvesen G, Grubb A (1986) Isolation of six cysteine proteinase inhibitors from human urine. Their physicochemical and enzyme kinetic properties and concentrations in biological fluids. J Biol Chem 261:11282–11289

Abrahamson M, Ritonja A, Brown MA, Grubb A, Machleidt W, Barrett AJ (1987) Identification of the probable inhibitory reactive sites of the cysteine proteinase inhibitors human cystatin C and chicken cystatin. J Biol Chem 262:9688–9694

Abrahamson M, Olafsson I, Palsdottir A, Ulvsback M, Lundwall A, Jensson O, Grubb A (1990) Structure and expression of the human cystatin C gene. Biochem J 268:287–294

Abrahamson M, Mason RW, Hansson H, Buttle DJ, Grubb A, Ohlsson K (1991) Human cystatin C. Role of the N-terminal segment in the inhibition of human cysteine proteinases and in its inactivation by leucocyte elastase. Biochem J 273:621–626

Abrahamson M, Jonsdottir S, Olafsson I, Jensson O, Grubb A (1992) Hereditary cystatin C amyloid angiopathy: identification of the disease-causing mutation and specific diagnosis by polymerase chain reaction based analysis. Hum Genet 89:377–380

Al-Hashimi I, Dickinson DP, Levine MJ (1988) Purification, molecular cloning, and sequencing of salivary cystatin SA-I. J Biol Chem 263:9381–9387

Assfalg-Machleidt I, Jochum M, Klaubert W, Inthorn D, Machleidt W (1988) Enzymatically active cathepsin B dissociating from its inhibitor complexes is elevated in blood plasma of patients with septic shock and some malignant tumors. Biol Chem Hoppe Seyler 369[Suppl]:263–269

Auerswald EA, Genenger G, Assfalg-Machleidt I, Machleidt W, Engh RA, Fritz H (1992) Recombinant chicken egg white cystatin variants of the QLVSG region. Eur J Biochem 209:837–845

Barka T, van der Noen H, Patil S (1992) Cysteine proteinase inhibitor in cultured human medullary thyroid carcinoma cells. Lab Invest 66:691–700

Barrett AJ, Kembhavi AA, Brown MA, Kirschke H, Knight CG, Tamai M, Hanada K (1982) L-trans-Epoxysuccinyl-leucylamido(4-guanidino)butane (E-64) and its analogues as inhibitors of cysteine proteinases including cathepsins B, H and L. Biochem J 201:189–198

Barrett AJ, Davies ME, Grubb A (1984) The place of human gamma-trace (cystatin C) amongst the cysteine proteinase inhibitors. Biochem Biophys Res Commun 120:631–636

Barrett AJ, Rawlings ND, Davies ME, Machleidt W, Salvesen G, Turk V (1986) Cysteine proteinase inhibitors of the cystatin superfamily. In: Barrett AJ, Salvesen G (eds) Proteinase inhibitors. Elsevier, Amsterdam, pp 515–569

Barrett AJ, Rawlings ND, Woessner JF (1998) Handbook of proteolytic enzymes. Academic Press, London

Berquin IM, Cao L, Fong D, Sloane BF (1995) Identification of two new exons and multiple transcription start points in the 5′-untranslated region of the human cathepsin-B-encoding gene. Gene 159:143–149

Bjarnadottir M, Grubb A, Olafsson I (1995) Promoter-mediated, dexamethasone-induced increase in cystatin C production by HeLa cells. Scand J Clin Lab Invest 55:617–623

Bjarnadottir M, Wulff B, Sameni M, Sloane BF, Keppler D, Grubb A, Abrahamson M (1998) Intracellular accumulation of the amyloidogenic L68Q variant of human cystatin C in NIH/3T3 cells. Mol Pathol 51:316–323

Bode W, Engh R, Musil D, Thiele U, Huber R, Karshikov A, Brzin J, Kos J, Turk V (1988) The 2.0A X-ray crystal structure of chicken egg white cystatin and its possible mode of interaction with cysteine proteinases. EMBO J 7:2593–2599

Boike G, Lah T, Sloane BF, Rozhin J, Honn K, Guirguis R, Stracke ML, Liotta LA, Schiffmann E (1992) A possible role for cysteine proteinase and its inhibitors in motility of malignant melanoma and other tumour cells. Melanoma Res 1:333–340

Bromme D, Klaus JL, Okamoto K, Rasnick D, Palmer JT (1996) Peptidyl vinyl sulphones: a new class of potent and selective cysteine protease inhibitors: S2P2 specificity of human cathepsin O2 in comparison with cathepsins S and L. Biochem J 315:85–89

Brown WM, Dziegielewska KM (1997) Friends and relations of the cystatin superfamily – new members and their evolution. Protein Sci 6:5–12

Brzin J, Kopitar M, Turk V, Machleidt W (1983) Protein inhibitors of cysteine proteinases. I. Isolation and characterization of stefin, a cytosolic protein inhibitor of cysteine proteinases from human polymorphonuclear granulocytes. Hoppe Seylers Z Physiol Chem 364:1475–1480

Chagas JR, Ferrer-Di Martino M, Gauthier F, Lalmanach G (1996) Inhibition of cathepsin B by its propeptide: use of overlapping peptides to identify a critical segment. FEBS Lett 392:233–236

Chambers AF, Matrisian LM (1997) Changing views of the role of matrix metalloproteinases in metastasis. J Natl Cancer Inst 89:1260–1270

Chapman HA J, Reilly JJ J, Yee R, Grubb A (1990) Identification of cystatin C, a cysteine proteinase inhibitor, as a major secretory product of human alveolar macrophages in vitro. Am Rev Respir Dis 141:698–705

Chen JM, Dando PM, Rawlings ND, Brown MA, Young NE, Stevens RA, Hewitt E, Watts C, Barrett AJ (1997) Cloning, isolation, and characterization of mammalian legumain, an asparaginyl endopeptidase. J Biol Chem 272:8090–8098

Clausen J (1961) Proteins in normal cerebrospinal fluid not found in serum. Proc Soc Exp Biol Med 107:170–172

Cornwall GA, Hann SR (1995) Transient appearance of CRES protein during spermatogenesis and caput epididymal sperm maturation. Mol Reprod Dev 41:37–46

Cornwall GA, Orgebin-Crist MC, Hann SR (1992) The CRES gene: a unique testis-regulated gene related to the cystatin family is highly restricted in its expression to the proximal region of the mouse epididymis. Mol Endocrinol 6:1653–1664

Coulombe R, Grochulski P, Sivaraman J, Menard R, Mort JS, Cygler M (1996) Structure of human procathepsin L reveals the molecular basis of inhibition by the prosegment. EMBO J 15:5492–5503

Davies ME, Barrett AJ (1984) Immunolocalization of human cystatins in neutrophils and lymphocytes. Histochemistry 80:373–377

Delaria K, Fiorentino L, Wallace L, Tamburini P, Brownell E, Muller D (1994) Inhibition of cathepsin L-like cysteine proteases by cytotoxic T-lymphocyte antigen-2 beta. J Biol Chem 269:25172–25177

Demetriou M, Binkert C, Sukhu B, Tenenbaum HC, Dennis JW (1996) Fetuin/alpha2-HS glycoprotein is a transforming growth factor-beta type II receptor mimic and cytokine antagonist. J Biol Chem 271:12755–12761

Denizot F, Brunet JF, Roustan P, Harper K, Suzan M, Luciani MF, Mattei MG, Golstein P (1989) Novel structures CTLA-2 alpha and CTLA-2 beta expressed in mouse activated T cells and mast cells and homologous to cysteine proteinase proregions. Eur J Immunol 19:631–635

Devos A, De Clercq N, Vercaeren I, Heyns W, Rombauts W, Peeters B (1993) Structure of rat genes encoding androgen-regulated cystatin-related proteins (CRPs): a new member of the cystatin superfamily. Gene 125:159–167

Devos A, Zhang J, Riviere M, Vercaeren I, Heyns W, Cassiman JJ, Rombauts W, Marynen P, Szpirer J, Szpirer C (1995) The genes coding for rat cystatin-related prostate protein (Cstrp) map to chromosome 3q41. Cytogenet Cell Genet 68:239–242

Dittmer J, Nordheim A (1998) Ets transcription factors and human disease. Biochim Biophys Acta 1377:F1–F11

Dolenc I, Turk B, Pungercic G, Ritonja A, Turk V (1995) Oligomeric structure and substrate induced inhibition of human cathepsin C. J Biol Chem 270:21626–21631

Ebert W, Knoch H, Werle B, Trefz G, Muley T, Spiess E (1994) Prognostic value of increased lung tumor tissue cathepsin B. Anticancer Res 14:895–899

Eide TJ, Jarvinen M, Hopsu-Havu VK, Maltau J, Rinne A (1992) Immunolocalization of cystatin A in neoplastic, virus and inflammatory lesions of the uterine cervix. Acta Histochem 93:241–248

Ekiel I, Abrahamson M (1996) Folding-related dimerization of human cystatin C. J Biol Chem 271:1314–1321

Enjyoji K, Kato H (1992) Purification and characterization of two isoforms of T-kininogens from rat liver microsomes. J Biochem (Tokyo) 111:670–675

Esnard A, Esnard F, Guillou F, Gauthier F (1992) Production of the cysteine proteinase inhibitor cystatin C by rat Sertoli cells. FEBS Lett 300:131–135

Fong D, Chan MM, Hsieh WT (1991) Gene mapping of human cathepsins and cystatins. Biomed Biochim Acta 50:595–598

Fox T, de Miguel E, Mort JS, Storer AC (1992) Potent slow-binding inhibition of cathepsin B by its propeptide. Biochemistry 31:12571–12576

Freije JP, Balbin M, Abrahamson M, Velasco G, Dalboge H, Grubb A, Lopez-Otin C (1993) Human cystatin D. cDNA cloning, characterization of the Escherichia coli expressed inhibitor, and identification of the native protein in saliva. J Biol Chem 268:15737–15744

Gabrijelcic D, Svetic B, Spaic D, Skrk J, Budihna J, Turk V (1992a) Determination of cathepsins B, H, L and kininogen in breast cancer patients. Agents Actions Suppl 38:350–357

Gabrijelcic D, Svetic B, Spaic D, Skrk J, Budihna M, Dolenc I, Popovic T, Cotic V, Turk V (1992b) Cathepsins B, H and L in human breast carcinoma. Eur J Clin Chem Clin Biochem 30:69–74

Geuze HJ, Slot JW, Strous GJ, Hasilik A, von Figura K (1985) Possible pathways for lysosomal enzyme delivery. J Cell Biol 101:2253–2262

Ghiso J, Jensson O, Frangione B (1986) Amyloid fibrils in hereditary cerebral hemorrhage with amyloidosis of Icelandic type is a variant of gamma-trace basic protein (cystatin C). Proc Natl Acad Sci USA 83:2974–2978

Goretzki L, Schmitt M, Mann K, Calvete J, Chucholowski N, Kramer M, Gunzler WA, Janicke F, Graeff H (1992) Effective activation of the proenzyme form of the urokinase-type plasminogen activator (pro-uPA) by the cysteine protease cathepsin L. FEBS Lett 297:112–118

Grubb A, Lofberg H (1982) Human gamma-trace, a basic microprotein: amino acid sequence and presence in the adenohypophysis. Proc Natl Acad Sci USA 79:3024–3027

Halfon S, Ford J, Foster J, Dowling L, Lucian L, Sterling M, Xu Y, Weiss M, Ikeda M, Liggett D, Helms A, Caux C, Lebecque S, Hannum C, Menon S, McClanahan T, Gorman D, Zurawski G (1998) Leukocystatin, a new class II cystatin expressed selectively by hematopoietic cells. J Biol Chem 273:16400–16408

Hall A, Hakansson K, Mason RW, Grubb A, Abrahamson M (1995) Structural basis for the biological specificity of cystatin C. Identification of leucine 9 in the N-terminal binding region as a selectivity-conferring residue in the inhibition of mammalian cysteine peptidases. J Biol Chem 270:5115–5121

Hall A, Ekiel I, Mason RW, Kasprzykowski F, Grubb A, Abrahamson M (1998) Structural basis for different inhibitory specificities of human cystatins C and D. Biochemistry 37:4071–4079

Hawley-Nelson P, Roop DR, Cheng CK, Krieg TM, Yuspa SH (1988) Molecular cloning of mouse epidermal cystatin A and detection of regulated expression in differentiation and tumorigenesis. Mol Carcinog 1:202–211

Heidtmann HH, Salge U, Abrahamson M, Bencina M, Kastelic L, Kopitar-Jerala N, Turk V, Lah TT (1997) Cathepsin B and cysteine proteinase inhibitors in human lung cancer cell lines. Clin Exp Metastasis 15:368–381

Henskens YMC, Veerman ECI, Nieuw Amerongen AV (1996) Cystatins in health and disease. Biol Chem Hoppe-Seyler 377:71–86

Hermann A, Braun A, Figueroa CD, Muller-Esterl W, Fritz H, Rehbock J (1996) Expression and cellular localization of kininogens in the human kidney. Kidney Int 50:79–84

Herwald H, Hasan AA, Godovac-Zimmermann J, Schmaier AH, Muller-Esterl W (1995) Identification of an endothelial cell binding site on kininogen domain D3. J Biol Chem 270:14634–14642

Hewitt R, Dano K (1996) Stromal cell expression of components of matrix-degrading protease systems in human cancer. Enzyme Protein 49:163–173

Hsieh WT, Fong D, Sloane BF, Golembieski W, Smith DI (1991) Mapping of the gene for human cysteine proteinase inhibitor stefin A, STF1, to chromosome 3cen-q21. Genomics 9:207–209

Inaoka T, Bilbe G, Ishibashi O, Tezuka K, Kumegawa M, Kokubo T (1995) Molecular cloning of human cDNA for cathepsin K: novel cysteine proteinase predominantly expressed in bone. Biochem Biophys Res Commun 206:89–96

Inoue T, Ishida T, Sugio K, Sugimachi K (1994) Cathepsin B expression and laminin degradation as factors influencing prognosis of surgically treated patients with lung adenocarcinoma. Cancer Res 54:6133–6136

Isemura S, Saitoh E, Sanada K, Minakata K (1991) Identification of full-sized forms of salivary (S-type) cystatins (cystatin SN, cystatin SA, cystatin S, and two phosphorylated forms of cystatin S) in human whole saliva and determination of phosphorylation sites of cystatin S. J Biochem (Tokyo) 110:648–654

Ishiguro H, Higashiyama S, Namikawa C, Kunimatsu M, Takano E, Tanaka K, Ohkubo I, Murachi T, Sasaki M (1987) Interaction of human calpains I and II with high molecular weight and low molecular weight kininogens and their heavy chain: mechanism of interaction and the role of divalent cations. Biochemistry 26:2863–2870

Jacobsson B, Lignelid H, Bergerheim US (1995) Transthyretin and cystatin C are catabolized in proximal tubular epithelial cells and the proteins are not useful as markers for renal cell carcinomas. Histopathology 26:559–564

Jarvinen M, Rinne A, Hopsu-Havu VK (1987) Human cystatins in normal and diseased tissues–a review. Acta Histochem 82:5–18

Jiang YP, Muller-Esterl W, Schmaier AH (1992) Domain 3 of kininogens contains a cell-binding site and a site that modifies thrombin activation of platelets. J Biol Chem 267:3712–3717

Johnston SC, Larsen CN, Cook WJ, Wilkinson KD, Hill CP (1997) Crystal structure of a deubiquitinating enzyme (human UCH-L3) at 1.8A resolution. EMBO J 16:3787–3796

Keppler D, Sloane BF (1996) Cathepsin B: multiple enzyme forms from a single gene and their relation to cancer. Enzyme Protein 49:94–105

Keppler D, Abrahamson M, Sordat B (1994a) Secretion of cathepsin B and tumour invasion. Biochem Soc Trans 22:43–49

Keppler D, Waridel P, Abrahamson M, Bachmann D, Berdoz J, Sordat B (1994b) Latency of cathepsin B secreted by human colon carcinoma cells is not linked to secretion of cystatin C and is relieved by neutrophil elastase. Biochim Biophys Acta 1226:117–125

Keppler D, Sordat B, Sierra F (1997) T-kininogen present in the liver of old rats is biologically active and readily forms complexes with endogenous cysteine proteinases. Mech Ageing Dev 98:151–165

Kirschke H, Wiederanders B (1994) Cathepsin S and related lysosomal endopeptidases. Methods Enzymol 244:500–511

Kirschke H, Wikstrom P, Shaw E (1988) Active center differences between cathepsins L and B: the S1 binding region. FEBS Lett 228:128–130

Kirschke H, Barrett AJ, Rawlings ND (1995) Lysosomal cysteine proteinases. Protein Profile 2:1613–1619

Kitamura N, Kitagawa H, Fukushima D, Takagaki Y, Miyata T, Nakanishi S (1985) Structural organization of the human kininogen gene and a model for its evolution. J Biol Chem 260:8610–8617

Kobayashi H, Ohi H, Sugimura M, Shinohara H, Fujii T, Terao T (1992) Inhibition of in vitro ovarian cancer cell invasion by modulation of urokinase-type plasminogen activator and cathepsin B. Cancer Res 52:3610–3614

Kuopio T, Kankaanranta A, Jalava P, Kronqvist P, Kotkansalo T, Weber E, Collan Y (1998) Cysteine proteinase inhibitor cystatin A in breast cancer. Cancer Res 58:432–436

Labauge P, Ouazzani R, MqRabet A, Grid D, Genton P, Dravet C, Chkili T, Beck C, Buresi C, Baldy-Moulinier M, Malafosse A (1997) Allelic heterogeneity of Mediterranean myoclonus and the cystatin B gene. Ann Neurol 41:686–689

Lah TT, Kos J (1998) Cysteine proteinases in cancer progression and their clinical relevance for prognosis. Biol Chem 379:125–130

Lah TT, Clifford JL, Helmer KM, Day NA, Moin K, Honn KV, Crissman JD, Sloane BF (1989) Inhibitory properties of low molecular mass cysteine proteinase inhibitors from human sarcoma. Biochim Biophys Acta 993:63–73

Lah TT, Kokalj-Kunovar M, Drobnic-Kosorok M, Babnik J, Golouh R, Vrhovec I, Turk V (1992a) Cystatins and cathepsins in breast carcinoma. Biol Chem Hoppe Seyler 373:595–604

Lah TT, Kokalj-Kunovar M, Kastelic L, Babnik J, Stolfa A, Rainer S, Turk V (1992b) Cystatins and stefins in ascites fluid from ovarian carcinoma. Cancer Lett 61:243–253

Lah TT, Kokalj-Kunovar M, Strukelj B, Pungercar J, Barlic-Maganja D, Drobnic-Kosorok M, Kastelic L, Babnik J, Golouh R, Turk V (1992c) Stefins and lysosomal cathepsins B, L and D in human breast carcinoma. Int J Cancer 50:36–44

Lenarcic B, Krizaj I, Zunec P, Turk V (1996) Differences in specificity for the interactions of stefins A, B and D with cysteine proteinases. FEBS Lett 395:113–118

Lerner UH, Johansson L, Ranjso M, Rosenquist JB, Reinholt FP, Grubb A (1997) Cystatin C, and inhibitor of bone resorption produced by osteoblasts. Acta Physiol Scand 161:81–92

Leto G, Pizzolanti G, Tumminello FM, Gebbia N (1994) Effects of E-64 (cysteine-proteinase inhibitor) and pepstatin (aspartyl-proteinase inhibitor) on metastasis formation in mice with mammary and ovarian tumors. In Vivo 8:231–236

Leto G, Tumminello FM, Pizzolanti G, Montalto G, Soresi M, Gebbia N (1997) Lysosomal cathepsins B and L and Stefin A blood levels in patients with hepatocellular carcinoma and/or liver cirrhosis: potential clinical implications. Oncology 54:79–83

Lignelid H, Jacobsson B (1992) Cystatin C in the human pancreas and gut: an immuno-histochemical study of normal and neoplastic tissues. Virchows Arch A Pathol Anat Histopathol 421:491–495

Lignelid H, Collins VP, Jacobsson B (1997) Cystatin C and transthyretin expression in normal and neoplastic tissues of the human brain and pituitary. Acta Neuropathol (Berl) 93:494–500

Lindahl P, Ripoll D, Abrahamson M, Mort JS, Storer AC (1994) Evidence for the inter-action of valine-10 in cystatin C with the S2 subsite of cathepsin B. Biochemistry 33:4384–4392

Linnevers C, Smeekens SP, Bromme D (1997) Human cathepsin W, a putative cysteine protease predominantly expressed in CD8+ T-lymphocytes. FEBS Lett 405:253–259

Lofberg H, Grubb AO (1979) Quantitation of gamma-trace in human biological fluids: indications for production in the central nervous system. Scand J Clin Lab Invest 39:619–626

Luthgens K, Ebert W, Trefz G, Gabrijelcic D, Turk V, Lah T (1993) Cathepsin B and cysteine proteinase inhibitors in bronchoalveolar lavage fluid of lung cancer patients. Cancer Detect Prev 17:387–397

Mason RW, Sol-Church K, Abrahamson M (1998) Amino acid substitutions in the N-terminal segment of cystatin C create selective protein inhibitors of lysosomal cys-teine proteinases. Biochem J 330:833–838

Menke A, McInnes L, Hastie ND, Schedl A (1998) The Wilms' tumor suppressor WT1: approaches to gene function. Kidney Int 53:1512–1518

Mostoslavsky R, Bergman Y (1997) DNA methylation: regulation of gene expression and role in the immune system. Biochim Biophys Acta 1333:F29–F50

Muller-Esterl W (1989) Kininogens, kinins and kinships. Thromb Haemost 61:2–6

Muller-Esterl W, Iwanaga S, Nakanishi S (1986) Kininogens revisited. Trends Biochem Sci 11:336–339

Murata M, Miyashita S, Yokoo C, Tamai M, Hanada K, Hatayama K, Towatari T, Nikawa T, Katunuma N (1991) Novel epoxysuccinyl peptides. Selective inhibitors of cathepsin B, in vitro. FEBS Lett 280:307–310

Ni J, Abrahamson M, Zhang M, Fernandez MA, Grubb A, Su J, Yu GL, Li Y, Parmelee D, Xing L, Coleman TA, Gentz S, Thotakura R, Nguyen N, Hesselberg M, Gentz R (1997) Cystatin E is a novel human cysteine proteinase inhibitor with structural resemblance to family 2 cystatins. J Biol Chem 272:10853–10858

Ni J, Alvarez Fernandez M, Danielsson L, Chillakuru RA, Zhang J, Grubb A, Su J, Gentz R, Abrahamson M (1998) Cystatin F is a glycosylated human low molecu-lar weight cysteine proteinase inhibitor. J Biol Chem 273:24797–24804

Nikawa T, Towatari T, Ike Y, Katunuma N (1989) Studies on the reactive site of the cys-tatin superfamily using recombinant cystatin A mutants. Evidence that the QVVAG region is not essential for cysteine proteinase inhibitory activities. FEBS Lett 255:309–314

Nycander M, Bjork I (1990) Evidence by chemical modification that tryptophan-104 of the cysteine-proteinase inhibitor chicken cystatin is located in or near the proteinase-binding site. Biochem J 271:281–284

Ohkubo I, Kurachi K, Takasawa T, Shiokawa H, Sasaki M (1984) Isolation of a human cDNA for alpha 2-thiol proteinase inhibitor and its identity with low molecular weight kininogen. Biochemistry 23:5691–5697

Okamoto H, Greenbaum LM (1986) Isolation and properties of two rat plasma T-kininogens. Adv Exp Med Biol 198:69–75

Olafsson I, Lofberg H, Abrahamson M, Grubb A (1988) Production, characterization and use of monoclonal antibodies against the major extracellular human cysteine proteinase inhibitors cystatin C and kininogen. Scand J Clin Lab Invest 48:573–582

Pennacchio LA, Lehesjoki AE, Stone NE, Willour VL, Virtaneva K, Miao J, DqAmato E, Ramirez L, Faham M, Koskiniemi M, Warrington JA, Norio R, de la Chapelle A, Cox DR, Myers RM (1996) Mutations in the gene encoding cystatin B in progressive myoclonus epilepsy (EPM1). Science 271:1731–1734

Pernu H, Rasanen O, Salo T, Rinne A, Herva R, Jarvinen M (1990) Cystatin A and B in the development of human squamous epithelia. Acta Histochem 88:53–57

Pierre P, Mellman I (1998) Developmental regulation of invariant chain proteolysis controls MHC class II trafficking in mouse dendritic cells. Cell 93:1135–1145

Redwood SM, Liu BC, Weiss RE, Hodge DE, Droller MJ (1992) Abrogation of the invasion of human bladder tumor cells by using protease inhibitor(s). Cancer 69:1212–1219

Riese RJ, Wolf PR, Bromme D, Natkin LR, Villadangos JA, Ploegh HL, Chapman HA (1996) Essential role for cathepsin S in MHC class II-associated invariant chain processing and peptide loading. Immunity 4:357–366

Rochefort H, Liaudet E, Garcia M (1996) Alterations and role of human cathepsin D in cancer metastasis. Enzyme Protein 49:106–116

Santamaria I, Velasco G, Pendas AM, Fueyo A, Lopez-Otin C (1998) Cathepsin Z, a novel human cysteine proteinase with a short propeptide domain and a unique chromosomal location. J Biol Chem 273:16816–16823

Sebti SM, Mignano JE, Jani JP, Srimatkandada S, Lazo JS (1989) Bleomycin hydrolase: molecular cloning, sequencing, and biochemical studies reveal membership in the cysteine proteinase family. Biochemistry 28:6544–6548

Sexton PS, Cox JL (1997) Inhibition of motility and invasion of B16 melanoma by the overexpression of cystatin C. Melanoma Res 7:97–101

Shaw E (1988) Peptidyl sulfonium salts. A new class of protease inhibitors. J Biol Chem 263:2768–2772

Shaw E, Angliker H, Rauber P, Walker B, Wikstrom P (1986) Peptidyl fluoromethyl ketones as thiol protease inhibitors. Biomed Biochim Acta 45:1397–1403

Sheahan K, Shuja S, Murnane MJ (1989) Cysteine protease activities and tumor development in human colorectal carcinoma. Cancer Res 49:3809–3814

Shiraishi T, Mori M, Tanaka S, Sugimachi K, Akiyoshi T (1998) Identification of cystatin B in human esophageal carcinoma, using differential displays in which the gene expression is related to lymph- node metastasis. Int J Cancer 79:175–178

Shivdasani RA (1997) Stem cell transcription factors. Hematol Oncol Clin North Am 11:1199–1206

Sierra F, Fey GH, Guigoz Y (1989) T-kininogen gene expression is induced during aging. Mol Cell Biol 9:5610–5616

Sierra F, Walter R, Vautravers P, Guigoz Y (1995) Identification of several isoforms of T-kininogen expressed in the liver of aging rats. Arch Biochem Biophys 322:333–338

Sloane BF, Rozhin J, Robinson D, Honn KV (1990) Role for cathepsin B and cystatins in tumor growth and progression. Biol Chem Hoppe Seyler 371[Suppl]:193–198

Soderstrom KO, Laato M, Wu P, Hopsu-Havu VK, Nurmi M, Rinne A (1995) Expression of acid cysteine proteinase inhibitor (ACPI) in the normal human prostate, benign prostatic hyperplasia and adenocarcinoma. Int J Cancer 62:1–4

Solem M, Rawson C, Lindburg K, Barnes D (1990) Transforming growth factor beta regulates cystatin C in serum-free mouse embryo (SFME) cells. Biochem Biophys Res Commun 172:945–951

Sorimachi H, Ishiura S, Suzuki K (1997) Structure and physiological function of calpains. Biochem J 328:721–732

Sotiropoulou G, Anisowicz A, Sager R (1997) Identification, cloning, and characterization of cystatin M, a novel cysteine proteinase inhibitor, down-regulated in breast cancer. J Biol Chem 272:903–910

Spiegelman BM (1998) PPARgamma in monocytes: less pain, any gain? Cell 93:153–155

Stenman G, Astrom AK, Roijer E, Sotiropoulou G, Zhang M, Sager R (1997) Assignment of a novel cysteine proteinase inhibitor (CST6) to 11q13 by fluorescence in situ hybridization. Cytogenet Cell Genet 76:45–46

Sukoh N, Abe S, Nakajima I, Ogura S, Isobe H, Inoue K, Kawakami Y (1994) Immunohistochemical distributions of cathepsin B and basement membrane antigens in human lung adenocarcinoma: association with invasion and metastasis. Virchows Arch 424:33–38

Tachikura T (1990) The teratogenic effects of E-64 on rat embryogenesis. Acta Paediatr Jpn 32:495–501

Tavera C, Leung-Tack J, Prevot D, Gensac MC, Martinez J, Fulcrand P, Colle A (1992) Cystatin C secretion by rat glomerular mesangial cells: autocrine loop for in vitro growth-promoting activity. Biochem Biophys Res Commun 182:1082–1088

Thiesse M, Millar SJ, Dickinson DP (1994) The human type 2 cystatin gene family consists of eight to nine members, with at least seven genes clustered at a single locus on human chromosome 20. DNA Cell Biol 13:97–116

Thompson EW, Brunner N, Torri J, Johnson MD, Boulay V, Wright A, Lippman ME, Steeg PS, Clarke R (1993) The invasive and metastatic properties of hormone-independent but hormone-responsive variants of MCF-7 human breast cancer cells. Clin Exp Metastasis 11:15–26

Towatari T, Nikawa T, Murata M, Yokoo C, Tamai M, Hanada K, Katunuma N (1991) Novel epoxysuccinyl peptides. A selective inhibitor of cathepsin B, in vivo. FEBS Lett 280:311–315

Tsui FW, Tsui HW, Mok S, Mlinaric I, Copeland NG, Gilbert DJ, Jenkins NA, Siminovitch KA (1993) Molecular characterization and mapping of murine genes encoding three members of the stefin family of cysteine proteinase inhibitors. Genomics 15:507–514

Tsukahara T, Kominami E, Katunuma N (1987) Formation of mixed disulfide of cystatin-beta in cultured macrophages treated with various oxidants. J Biochem (Tokyo) 101:1447–1456

Turk V, Bode W (1991) The cystatins: protein inhibitors of cysteine proteinases. FEBS Lett 285:213–219

Turk B, Krizaj I, Kralj B, Dolenc I, Popovic T, Bieth JG, Turk V (1993) Bovine stefin C, a new member of the stefin family. J Biol Chem 268:7323–7329

Vercaeren I, Vanaken H, Devos A, Peeters B, Verhoeven G, Heyns W (1996) Androgens transcriptionally regulate the expression of cystatin-related protein and the C3 component of prostatic binding protein in rat ventral prostate and lacrimal gland. Endocrinology 137:4713–4720

Verdot L, Lalmanach G, Vercruysse V, Hartmann S, Lucius R, Hoebeke J, Gauthier F, Vray B (1996) Cystatins up-regulate nitric oxide release from interferon-gamma-activated mouse peritoneal macrophages. J Biol Chem 271:28077–28081

Warfel AH, Zucker-Franklin D, Frangione B, Ghiso J (1987) Constitutive secretion of cystatin C (gamma-trace) by monocytes and macrophages and its downregulation after stimulation. J Exp Med 166:1912–1917

Winderickx J, Hemschoote K, De Clercq N, Van Dijck P, Peeters B, Rombauts W, Verhoeven G, Heyns W (1990) Tissue-specific expression and androgen regulation of different genes encoding rat prostatic 22-kilodalton glycoproteins homologous to human and rat cystatin. Mol Endocrinol 4:657–667

Woodhouse EC, Chuaqui RF, Liotta LA (1997) General mechanisms of metastasis. Cancer 80:1529–1537

Yan S, Sameni M, Sloane BF (1998) Cathepsin B and human tumor progression. Biol Chem 379:113–123

Caspases and Their Natural Inhibitors as Therapeutic Targets for Regulating Apoptosis

Q.L. DEVERAUX, J.C. REED, and G.S. SALVESEN

A. Apoptosis

Apoptosis, or programmed cell death, is a physiological cell suicide program that occurs in all animal species (STELLER 1995). Apoptosis ensures that the genesis of new cells via division is appropriately controlled and offset by cell loss. Cell death is a natural accompaniment of the physiology of fully differentiated cells in the skin, intestine, immune system, mammary gland and uterus. Developmental organization requires removal of many cells for achieving the final desired structures and ensures proper cyto-architectures of most organs such as the kidney, heart and brain. Moreover, elimination of cells that have been compromised by viral infection, oxidation, hypoxia and DNA damage is important for maintaining healthy tissues. Thus, it can be appreciated that dysfunctional programmed cell death contributes to several human diseases (DUKE et al. 1998). An illustration of this concept is that apoptotic defects appear to be the primary lesion in some types of cancer and leukemia, allowing malignant cells to survive longer than their intended life span and endowing these cells with a selective survival advantage relative to their normal counterparts. Therefore, cell death can contribute to neoplastic expansion in the absence of increased cellular division (REED 1998). Multiple examples exist where excessive apoptosis has been implicated with human disease including acquired immunodeficiency syndrome (AIDS), Alzheimer's disease, myocardial infarction and stroke (DUKE et al. 1998).

B. Apoptosis Is Mediated by Caspases

Though perhaps not directly involved in cell death, the interleukin converting enzyme (ICE) was the first member of a family of cysteine proteases – now termed caspases – to be identified (Fig. 1) (SALVESEN and DIXIT 1997). The "ICE branch" of the human caspase family (caspases 1, 4, 5 and mouse caspase-11) is involved in the processing of pro-inflammatory cytokines, for example interleukin (IL)-1 and IL-18 (KUIDA et al. 1995; GHAYUR et al. 1997; GU et al. 1997). Nevertheless, most if not all apoptotic programs employ caspases as mediators of cell death. Located in the cell as latent precursors (zymogens), these enzymes can become rapidly activated and are essential to the execution phase of all apoptotic programs known thus far. In mammalian cells,

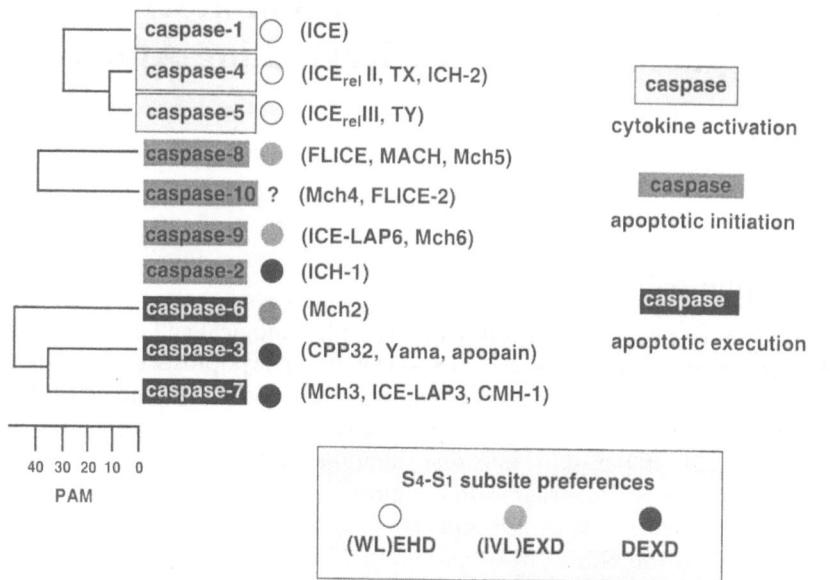

Fig. 1. The human caspase family. Currently ten caspases are known in humans. They can be subdivided based on either of two criteria: (1) they fall into groupings based on their known or suspected role in activating pro-inflammatory cytokines, initiating apoptosis, or executing apoptosis; (2) they can be categorized according to their related enzyme specificitys from experiments using tetrapeptide substrates (Thornberry et al. 1997). Interestingly, there is some overlap in the two criteria, and the familial relationships observed in this partial dendrogram, where *PAM* is point accepted mutations

activation of the caspase zymogens is achieved through at least two independent mechanisms (Fig. 2) initiated by distinct upstream members of the caspase family, but resulting in activation of common executioner caspases (Salvesen and Dixit 1997). In this regard, many apoptotic programs rely on a common set of caspases that includes various combinations of caspases 3, 6 and 7 – the executioners. These enzymes appear to be directly or indirectly involved in events that characterize the apoptotic phenotype including DNA fragmentation, chromosome condensation, membrane blebbing, and cell shape alterations. Understanding the mechanisms of how these caspases are regulated is critical to their therapeutic exploitation.

C. Lessons Learned from Natural Caspase Inhibitors

Viruses have evolved mechanisms that target common points in apoptotic programs for their benefit. Examples are the cowpox viral serpin CrmA, a caspase inhibitor that exhibits specificity for caspase-1 and -8, with inhibitory constants (K_is) of 0.01 nM and 0.1 nM, respectively (Komiyama et al. 1994; Zhou and Salvesen 1997), and the baculovirus protein p35 (K_i ~1 nM for all caspases

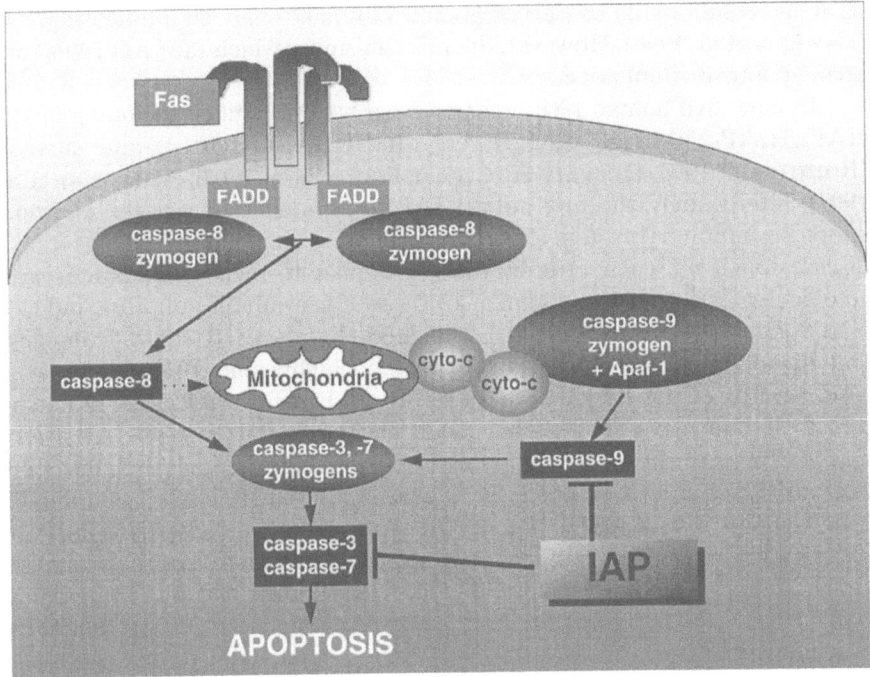

Fig. 2. Apoptotic pathways are mediated by caspases. Several members of the tumor necrosis family (TNF) family of death receptors (exemplified here by Fas) recruit caspase-8 to their cytosolic domains upon binding their respective ligands, resulting in proteolytic activation of this proximal caspase (WALLACH et al. 1997). Once activated, caspase-8 can induce, either directly or indirectly, the activation of a number of distal caspases such as caspase-3 and 7 (MUZIO et al. 1997). Another, although not mutually exclusive, pathway for caspase activation involves cytochrome-c (*cyto-c*) which, in mammalian cells, is often released from the mitochondria into the cytosol as an early event in apoptosis (LIU et al. 1996; KHARBANDA et al. 1997; KLUCK et al. 1997; YANG et al. 1997; BOSSY-WETZEL et al. 1998). Upon entering the cytosol, cyto-c induces the ATP or dATP-dependent formation of a complex of proteins that results in the proteolytic activation of the executioner caspases (LIU et al. 1996). Among the members of this complex are the CED-4 homolog Apaf-1, and caspase-9 (LIU et al. 1996; LI et al. 1997; ZOU et al. 1997)

tested). The viral caspase inhibitors are important for suppressing the host cell-death response, thereby allowing the virus to propagate (CLEM and MILLER 1994; HAY et al. 1994; XUE and HORVITZ 1995; BERTIN et al. 1996). Baculoviruses contain another family of cell-death inhibitors termed the *i*nhibitor of *ap*optosis proteins or IAPs (CLEM and MILLER 1994). Like p35, the baculoviral IAPs are found to suppress the host cell-death response (CROOK et al. 1993; CLEM and MILLER 1994). Moreover, ectopic expression of some of the viral IAPs protects mammalian cells from apoptosis induced by overexpression of caspases – observations that are consistent with the idea that the IAPs block apoptosis at an evolutionarily conserved point common to many apoptotic programs (HAWKINS et al. 1996). However, the mechanism by which the viral IAPs suppress apoptosis is unknown.

To date, five human IAP relatives have been identified including NAIP, cIAP1/HIAP-2/hMIHB, cIAP2/HAIP-1/hMIHC, XIAP/hILP and survivin (ROTHE et al. 1995; DUCKETT et al. 1996; LISTON et al. 1996; AMBROSINI et al. 1997). Interestingly, the first human IAP to be identified was the *n*euronal *a*poptosis *i*nhibitory *p*rotein (NAIP), based on its linkage to the degenerative disease *s*pinal *m*uscular *a*trophy (SMA). Similar to their viral counterparts, ectopic expression of these human IAP genes can inhibit apoptosis induced by a variety of stimuli (DUCKETT et al. 1996; LISTON et al. 1996). Consistent with the idea that IAPs function by blocking a highly conserved step in apoptosis, several of the human IAPs (XIAP, cIAP1 and cIAP2) are found to directly inhibit caspases (DEVERAUX et al. 1997; ROY et al. 1997). XIAP, cIAP1 and cIAP2 were shown to bind and potently inhibit caspases 3, 7 and 9 but not caspases 1, 6, 8 or 10 or CED3 (DEVERAUX et al. 1997, 1998, and unpublished data; ROY et al. 1997). The K_is for XIAP, cIAP1 and cIAP2 against caspases 3 and 7 range from ~0.2 nM to 10 nM. Similar results have been obtained for NAIP and survivin (D. Nicholson and A. McKenzie, personal communication). Thus, IAPs represent the first family of endogenous caspase inhibitors in mammals.

D. Structural Characteristics of the IAPs

Structurally, the IAPs contain a conserved sequence of ~70 amino acids referred to as the baculoviral inhibitory repeat (BIR) domain (CLEM and DUCKETT 1998). The BIR motif is present from one to three copies depending on the particular IAP. Among the known human IAPs, most contain three BIR domains – the exception being survivin which has only one BIR. Although particular to the IAPs, the BIR domain may represent a novel type of zinc binding fold characterized by unique spacing of cysteines and histidines. The BIR regions (BIR 1, 2 and 3) from human XIAP, cIAP1 and cIAP2 were found to be essential for caspase inhibitory activity (DEVERAUX et al. 1997; ROY et al. 1997) consistent with previous observations that these domains are required for anti-apoptotic activity of the IAP's (HAY et al. 1995; CLEM and DUCKETT 1998). Recently, the caspase inhibitory activity of XIAP was localized specifically to the second of its three tandem BIR domains (BIR2). Thus, despite the identity in primary amino acid sequence between BIR2 and BIR1 (~42%) or BIR3 (~32%), all BIR's may not be created equal, at least with respect to caspase inhibition and anti-apoptotic function (TAKAHASHI et al. 1998).

Given the notion that at least some IAPs regulate apoptosis at a highly conserved step – caspases – it is perhaps not surprising to find sequences with high similarity to BIR2 of human XIAP in numerous species including yeast:

Fig. 3. Alignment of sequences exhibiting similarity to the caspase inhibitory region (BIR2) of XIAP. Sequences were identified by a PSI search of the non-redundant database at NCBI (ALTSCHUL et al. 1997) using the XIAP BIR2 region as a query sequence. For proteins with multiple BIR domains, only the BIR region with highest similarity to XIAP BIR2 is pictured. Sequences were aligned using the MegAlign program (DNASTAR) employing the Clustal method and designated by species and the residue number at the beginning of each BIR domain. No functional relation is implied. The following is a list of the species designation used for the above alignment and primary accession number or database entry number for each sequence. Human IAPs: hXIAP; P98170, hIAP1; Q13489, hIAP2; Q13490, hNAIP; 1737213, hSurvivin; 2315863. Pig: pIAP; 2957175. Chick: Q90660. Mouse: mXIAP; Q60989, mNAIP; 2352685, mIAP1; Q13489, mIAP2; Q62210, mSurvivin; (TIAP); d1029206. Drosophilia: dIAP1; Q24306, dIAP2; Q24307. Chilo iridescent virus: CiIAP; 2738454. Orgyia psuedotsugta nuclear polyhedrosis virus: OpIAP; P41437. Cydia pomonella granulosis virus: CpIAP; P41436. Autographa California nuclear polyhedrosis virus: AcIAP; D36828. African swine fever virus: AsIAP; 011452. *Caenorhabditis elegans*: CeIAP1; e249029, CeIAP2; e348121. *Schizosaccharomyces pombe*: SpIAP; e339290, *Sacromyces cerevisiae*: SciAP, P47134

Caenorhabditis elegans which contains several apoptosis genes similar to those found in humans (Fig. 3). It is interesting to note, however, the identification of similar sequences in yeast, in which the occurrence of apoptosis is still controversial and appears not to involve caspases (SHAHAM et al. 1998). Since yeast apparently lacks caspases and may not undergo apoptosis, not all BIR-containing molecules should be viewed as inhibitors of apoptosis. Moreover, it is unclear as to whether the anti-apoptotic activity of IAPs is always mediated through caspase inhibition. Further studies should resolve these questions and future structural studies should elucidate the exact mechanism by which human XIAP, cIAP1 and cIAP2 mediate caspase inhibition, revealing the precise contacts important for their interactions. These structural data may well facilitate the development of synthetic inhibitors that block the IAP-caspase interaction, thus freeing the caspases to induce apoptosis in clinical scenarios where this would be desirable.

E. Biology of the Human IAPs

Human IAPs were found to inhibit both Bax- and Fas-induced apoptotic events; however, the caspases blocked within each pathway by the IAPs were distinct (Deveraux et al. 1998). Bax is a pro-apoptotic molecule that has been implicated in altering the function of mitochondria and more recently has been shown to promote the release of cytochrome c from these organelles (Wolter et al. 1997; Bossy-Wetzel et al. 1998; Jurgensmeier et al. 1998; Mahajan et al. 1998). Cytochrome c, in the presence of dATP and the *a*poptosis *p*rotease *a*ctivating *f*actor (APAF1), binds to the zymogen form of caspase-9 resulting in its activation and the subsequent activation of caspase-3 – events that lead to nuclear and cellular apoptotic destruction (Liu et al. 1996; Li et al. 1997; Reed 1997; Zou et al. 1997). The IAPs block cytochrome-c-induced apoptosis by inhibiting the activation of caspase-9; thus all apoptotic events mediated by caspase-9 are suppressed, including the activation of caspase-3 (Deveraux et al. 1998). These data are consistent with in vivo observations that the human IAPs (at least XIAP, cIAP1 and cIAP2) can function downstream of cytochrome-c to block apoptosis (Orth and Dixit 1997; Roy et al. 1997; Deveraux et al. 1997; Duckett et al. 1998).

Fas (CD95), a *t*umor *n*ecrosis *f*amily (TNF) member, appears to initiate apoptosis through the activation of caspase-8, which either directly or indirectly activates caspase-3 (Muzio et al. 1996; Medema et al. 1997; Games et al. 1998; Martin et al. 1998). Following Fas (CD95) stimulation or caspase-8 activation, IAPs were found to bind and inhibit the executioner caspase-3, but not the initiator caspase-8 (Deveraux et al. 1998). Complete maturation and activity of caspase-3 was blocked as was all further caspase-3-mediated apoptotic events. Thus, the IAPs can inhibit different apoptotic programs by targeting distinct caspases (Fig. 2).

F. IAPs as Therapeutic Targets

As previously mentioned, IAPs may play an important role in proper neuromuscular development. Deletions of NAIP have been linked to severe forms of SMA. Possibly, this is due to increased susceptibility to apoptotic signals in cells destined to become motorneurons or in the maintenance of viable spinal muscle cells. Recent observations have implicated cIAP1 as a mediator of the transforming oncogene v-Rel (You et al. 1997). These studies suggest that cIAP1 is induced during the v-Rel-mediated transformation process and functions as a suppressor of apoptosis in v-Rel-transformed cells. Thus, at least in some cases, IAPs appear to contribute to tumorigenesis. Experiments employing an inhibitor of nuclear factor (NF)-κB demonstrated suppression of the expression of porcine aortic endothelial cell pIAP and induced apoptosis in response to TNF-α, suggesting that pIAP is one of the NF-κB-regulated genes that operates to prevent endothelial cell apoptosis during inflammation (Stehlik et al. 1998). Therefore, dysfunctional regulation of the IAPs might be detrimental in sepsis and in some inflammatory conditions, inducing endothelial cell death and precipitating disseminated intravascular inflammation (DIC). Other studies indicate that IAPs may be involved in the suppression of granulosa cell apoptosis by gonadotropin in antral follicles and, therefore, may play an important role in determining the fate of the these cells, and thus, the eventual follicular destiny (LI et al. 1998).

Survivin is a molecule that contains a single BIR domain and, interestingly, exhibits the highest similarity to what appears to be the most archaic IAPs – those predicted from yeast genome sequences (Fig. 2). Many human carcinomas appear to over express survivin, whereas survivin appears to be virtually absent from normal adult tissues; thus, survivin has potential as a diagnostic marker for cancer (YOU et al. 1997; LU et al. 1998). Possibly elevated levels of survivin allow tumor cells to evade cell death. In this regard, recent experiments employing antisense strategies to reduce survivin expression in some tumor cell lines revealed increased apoptosis and loss of cell viability, implying that survivin expression is important for survival of at least some carcinomas. (AMBROSINI et al. 1998). Based on these data and the evidence that expression of survivin is highly restricted to malignant cells, this IAP appears to be an attractive target for therapeutic intervention.

G. Potential for Caspase Inhibitor Therapy

The demonstration that natural caspase inhibitors can have a profound influence on cell life and death decisions gives sound rational for therapeutic intervention by pharmaceuticals that target specific caspases. A significant barrier in the development of anti-degenerative drugs based on caspase inhibition is the potential side effects that might be caused by inhibition of normal programmed cell death which occurs in humans, accounting for over 50×10^9 deaths per day in rapidly turning-over cell populations in adults. Thus, the use of broad-spectrum caspase inhibitors is likely to be appropriate only for acute conditions, such as the rescue of cells destined to die by apoptosis following stroke or myocardial ischemia. Development of caspase inhibitors targeting individual caspases or delivered to specific cells may be required for chronic disorders such as arthritis or neurodegenerative disorders such as amoyotrophic lateral sclerosis, Alzheimer's, Parkinson's or Huntington's diseases (BERGERON and YUAN 1998). In this context, observations from caspase-1 knockout mice are encouraging since these animals have a complete absence of IL-1 and IL-18 activity (the cytokines whose processing is dependent on caspase-1) but develop normally (KUIDA et al. 1995; GHAYUR et al. 1997; GU et al. 1997). Further studies will be necessary to determine whether these mice are less susceptible to inflammatory disease, but the initial observations bode well for the use of therapeutics that selectively target caspase-1.

Viral infection is combated with the aid of caspases and cell death (TALANIAN et al. 1997). T lymphocytes known as cytotoxic or killer T cells attack virally infected cells – activating directly the caspases and resulting in targeted cell death. However, the Fas-ligand system may inadvertently induce apoptosis in healthy cells, resulting in excessive cell death (STRASSER 1995; WONG et al. 1997;

Duke et al. 1998). This may be an important factor that contributes to the extensive liver damage observed following hepatitis virus infection (Irie et al. 1998). In this regard, peptidyl inhibitors of caspases have been used to prevent lethal Fas-mediated hepatitis and TNF/galactosamine- and Salmonella-mediated hepatic cell death in mice (Jaeschke et al. 1998; Suzuki 1998) An unresolved issue, however, concerns long-term exposure to broad-spectrum caspase inhibitors in viral diseases which potentially could be detrimental – some cells are simply better off dead.

In contrast to many inflammatory diseases, viral infection with human immunodeficiency virus (HIV) results in destructive alterations in T lymphocytes that increase their susceptibility to apoptosis (Cohen 1995; Hellerstein and McCune 1997; Stricker et al. 1998). Helper T cells in HIV-infected individuals are overly sensitive to apoptotic signals (Duke et al. 1998). In some experimental systems, broad-spectrum caspase inhibitors block HIV-induced apoptosis but increase virus production (Glynn et al. 1996; Chinnaiyan et al. 1997). Further studies will be required to determine whether more selective caspase inhibitors, once they are developed, could find a role in the treatment of HIV.

Possibly, the best pathways in which to target caspases are those where caspases are both initiators and effectors of the apoptotic programs, such as in the Fas and TNF pathways. In this context, TNF and Fas have been implicated as mediators of chronic heart failure and stroke (Geng 1997). In this regard, the broad-spectrum caspase inhibitor benzoxycarbonyl-Val-Ala-Asp-fluoromethyl ketone (ZVAD-fmk) reduced myocardial reperfusion injury in rats, which appeared to be, at least in part, attributed to the attenuation of cardiomyocyte apoptosis (Yaoita et al. 1998). Given that clonogenic survival can be maintained by the caspase inhibitor ZVAD-fmk when Fas, but not other stimuli such as growth-factor deprivation or chemotherapeutic reagents, is employed (Bossy-Wetzel et al. 1998), indicates that use of this caspase inhibitor may be appropriate in cases where hyperactivity of the Fas pathway is the cell-death initiator.

H. Conclusions

Current understanding of the apoptotic pathway, combined with an appreciation of the role of dysregulation of cell death in human disease, inspires potential strategies for novel therapies. Theoretically, one can either target caspases to improve the outcome of conditions that involve excessive apoptosis, such as stroke or myocardial ischemia, or promote apoptosis in the case of hyperplasias and cancers by targeting natural caspase inhibitors such as the IAPs. Continuing efforts to understand the detailed regulation and mechanisms of the natural inhibitors of caspases should facilitate these endeavors. However, much remains to be learned about apoptosis and several fundamental questions must be addressed. Can very selective caspase inhibitors be made and will they be effective in vivo? Will inhibition of apoptosis have serious side effects? Clearly these and other major hurdles must be overcome before successful therapies are derived. Nevertheless, targeting the caspases and their natural regulators is likely to be a priority for future drug development efforts.

References

Altschul SF, Madden TL, Schaffer AA, Zhang J, Zhang Z, Miller W, Lipman DJ (1997) Gapped BLAST and PSI-BLAST: a new generation of protein database search programs. Nucleic Acids Res 25:3389–3402

Ambrosini G, Adida C, Altieri D (1997) A novel anti-apoptosis gene, survivin, expressed in cancer and lymphoma. Nat Med 3:917–921

Ambrosini G, Adida C, Sirugo G, Altieri DC (1998) Induction of apoptosis and inhibition of cell proliferation by survivin gene targeting. J Biol Chem 273:11177–11182

Bergeron L, Yuan J (1998) Sealing one's fate: control of cell death in neurons. Curr Opin Neurobiol 8:55–63

Bertin J, Mendrysa SM, LaCount DJ, Gaur S, Krebs JF, Armstrong RC, Tomaselli KJ, Friesen PD (1996) Apoptotic suppression by baculovirus p35 involves cleavage by and inhibition of a virus-induced CED-3/ICE-like protease. J Virol 70:6251–6259

Bossy-Wetzel E, Newmeyer DD, Green DR (1998) Mitochondrial cytochrome c release in apoptosis occurs upstream of DEVD-specific caspase activation and independently of mitochondrial transmembrane depolarization. EMBO J 17:37–49

Chinnaiyan AM, Woffendin C, Dixit VM, Nabel GJ (1997) The inhibition of pro-apoptotic ICE-like proteases enhances HIV replication. Nat Med 3:333–337

Clem RJ, Duckett CS (1998) The IAP genes: unique arbitrators of cell death. Trends Biochem Sci 23:159–162

Clem RJ, Miller LK (1994) Control of programmed cell death by the baculovirus genes p35 and iap. Mol Cell Biol 14:5212–5222

Cohen J (1995) Researchers air alternative views on how HIV kills cells. Science 269:1044–1045

Crook NE, Clem RJ, Miller LK (1993) An apoptosis-inhibiting baculovirus gene with a zinc finger-like motif. J Virol 67:2168–2174

Deveraux Q, Takahashi R, Salvesen GS, Reed JC (1997) X-linked IAP is a direct inhibitor of cell death proteases. Nature 388:300–303

Deveraux QL, Roy N, Stennicke HR, Van Arsdale T, Zhou Q, Srinivasula M, Alnemri ES, Salvesen GS, Reed JC (1998) IAPs block apoptotic events induced by caspase-8 and cytochrome c by direct inhibition of distinct caspases. EMBO J 17:2215–2223

Duckett CS, Nava VE, Gedrich RW, Clem RJ, Van Dongen JL, Gilfillan MC, Shiels H, Hardwick JM, Thompson CB (1996) A conserved family of cellular genes related to the baculovirus iap gene and encoding apoptosis inhibitors. EMBO J 15:2685–2689

Duckett CS, Li F, Tomaselli KJ, Thompson CB, Armstrong RC (1998) Human IAP-like protein regulates programmed cell death downstream of Bcl-X$_L$ and cytochrome c. Mol Cell Biol 18:608–615

Duke RC, Ojcius DM, Youn JD (1998) Cell suicide in health and disease. Sci Amz

Games S, Anel A, Pineiro A, Naval J (1998) Caspases are the main executioners of Fas-mediated apoptosis, irrespective of the ceramide signalling pathway. Cell Death Differ 5:241–249

Geng YJ (1997) Regulation of programmed cell death or apoptosis in atherosclerosis. Heart Vessels Suppl 12:76–80

Ghayur T, Banerjee S, Hugunin M, Butler D, Herzog L, Carter A, Quintal L, Sekut L, Talanian R, Paskind M, Wong W, Kamen R, Tracey D, Allen, H (1997) Caspase-1 processes IFN-γ-inducing factor and regulates LPS-induced IFN-γ production. Nature 386:619–622

Glynn JM, McElligott DL, Mosier DE (1996) Apoptosis induced by HIV infection in H9T cells is blocked by ICE-family protease inhibition but not by a Fas(CD95) antagonist. J Immunol 157:2754–2758

Gu Y, Kuida K, Tsutsui H, Ku G, Hsiao K, Fleming MA, Hayashi N, Higashino K, Okamura H, Nakanishi K, Kurimoto M, Tanimoto T, Flavell RA, Sato V, Harding MW, Livingston DJ, Su MS (1997) Activation of interferon-gamma inducing factor mediated by interleukin-1beta converting enzyme. Science 275:206–209

Hawkins CJ, Uren AG, Hacker G, Medcalf RL, Vaux DL (1996) Inhibition of interleukin 1β-converting enzyme-mediated apoptosis of mammalian cells by baculovirus IAP. Proc Natl Acad Sci USA 93:13786–13790

Hay BA, Wolff T, Rubin GM (1994) Expression of baculovirus p35 prevents cell death in Drosophila. Development 120:2121–2129

Hay BA, Wassarman DA, Rubin GM (1995) Drosophila homologs of baculovirus inhibitor of apoptosis proteins function to block cell death. Cell 83:1253–1262

Hellerstein MK, McCune JM (1997) T cell turnover in HIV-1 disease. Immunity 7:583–589

Irie H, Koyama H, Kubo H, Fukuda A, Aita K, Koike T, Yoshimura A, Yoshida T, Shiga J, Hill T (1998) Herpes simplex virus hepatitis in macrophage-depleted mice: the role of massive, apoptotic cell death in pathogenesis. J Gen Virol 79:1225–1231

Jaeschke H, Fisher MA, Lawson JA, Simmons CA, Farhood A, Jones DA (1998) Activation of caspase 3 (CPP32)-like proteases is essential for TNF-alpha-induced hepatic parenchymal cell apoptosis and neutrophil-mediated necrosis in a murine endotoxin shock model. J Immunol 160:3480–3486

Jurgensmeier JM, Xie Z, Deveraux Q, Ellerby L, Bredesen D, Reed JC (1998) Bax directly induces release of cytochrome c from isolated mitochondria. Proc Natl Acad Sci USA 95:4997–5002

Kharbanda S, Pandey P, Schofield L, Israel S, Roncinske R, Yoshida K, Bharti A, Yuan Z-M, Saxena S, Weichselbaum R, Nalin C, Kufe D (1997) Role for Bcl-X$_L$ as an inhibitor of cytosolic cytochrome C accumulation in DNA damage-induced apoptosis. Proc Natl Acad Sci USA 94:6939–6942

Kluck RM, Bossy-Wetzel E, Green DR, Newmeyer DD (1997) The release of cytochrome c from mitochondria: a primary site for Bcl-2 regulation of apoptosis. Science 275:1132–1136

Komiyama T, Ray CA, Pickup DJ, Howard AD, Thornberry NA, Peterson EP, Salvesen G (1994) Inhibition of interleukin-1 beta converting enzyme by the cowpox virus serpin CrmA. An example of cross-class inhibition. J Biol Chem 269:19331–19337

Kuida K, Lippke JA, Ku G, Harding,MW, Livingston DJ, Su MS-S, Flavell RA (1995) Altered cytokine export and apoptosis in mice deficient in interleukin-1β converting enzyme. Science 267:2000–2003

Li P, Nijhawan D, Budihardjo I, Srinivasula S, Ahmad M, Alnemri E, Wang X (1997) Cytochrome c and dATP-dependent formation of Apaf-1/Caspase-9 complex initiates an apoptotic protease cascade. Cell 91:479–489

Li J, Kim JM, Liston P, Li M, Miyazaki T, Mackenzie AE, Korneluk RG, BK T (1998) Expression of inhibitor of apoptosis proteins (IAPs) in rat granulosa cells during ovarian follicular development and atresia. Endocrinology 139:1321–1328

Liston P, Roy N, Tamai K, Lefebvre C, Baird S, Cherton-Horvat G, Farahani R, McLean M, Ikeda J, MacKenzie A, Korneluk RG (1996) Suppression of apoptosis in mammalian cells by NAIP and a related family of IAP genes. Nature 379:349–353

Liu X, Kim CN, Yang J, Jemmerson R, Wang X (1996) Induction of apoptotic program in cell-free extracts: requirement for dATP and cytochrome c. Cell 86:147–157

Lu CD, Altieri DC, Tanigawa N (1998) Expression of a novel anti-apoptosis gene, survivin, correlated with tumor cell apoptosis and p53 accumulation in gastric carcinomas. Cancer Res 58:1808–1812

Mahajan N, Linder K, Berry G, Gordon G, Heinm R, Herman B (1998) Bcl-2 and bax interactions in individual mitochondria probed with mutant green fluorscent proteins and fluorescence resonance energy transfer. Nat Biotechnol 16:547–552

zMartin SJ, Green DR (1995) Protease activation during apoptosis: death by a thousand cuts? Cell 82:349–352

Martin D, Siegel R, Zheng L, Lenardo M (1998) Membrane oligomerization and cleavage activates the caspase-8 (FLICE/MACHα1) death signal. J Biol Chem 273: 4345–4349

Medema JP, Scaffidi C, Kischkel FC, Shevdhenko A, Mann M, Krammer PH, Peter ME (1997) FLICE is activated by association with the CD95 death-inducing signaling complex (DISC). EMBO J 16:2794–2804

Muzio M, Chinnaiyan AM, Kischkel FC, O'Rourke K, Shevchenko A, Ni J, Scaffidi C, Bretz JD, Zhang M, Gentz R, Mann M, Krammer PH, Peter ME, Dixit VM (1996) Flice, a novel FADD-homologous ICE/CED-3-like protease, is recruited to the CD95 (Fas/APO-1) death–inducing signaling complex. Cell 85:817–827

Muzio M, Salvesen GS, Dixit VM (1997) FLICE induced apoptosis in a cell-free system. J Biol Chem 272:2952–2956

Orth K, Dixit VM (1997) Bik and Bak induce apoptosis downstream of CrmA but upstream of inhibitor of apoptosis. J Biol Chem 272:8841–8844

Reed JC (1997) Cytochrome c: can't live with it; can't live without it. Cell 91:559–562

Reed J (1998) Chronic lymphocytic leukemia: a disease of disregulated programmed cell death clinical. Immunol Lett 17:125–140

Rothe M, Pan M-G, Henzel WJ, Ayres TM, Goeddel DV (1995) The TNFR2-TRAF signaling complex contains two novel proteins related to baculoviral inhibitor of apoptosis proteins. Cell 83:1243–1252

Roy N, Deveraux QL, Takahashi R, Salvesen GS, Reed JC (1997) The c-IAP-1 and c-IaP-2 proteins are direct inhibitors of specific caspases. EMBO J 16:6914–6925

Salvesen GS, Dixit VM (1997) Caspases: intracellular signaling by proteolysis. Cell 91:443–446

Shaham S, Shuman MA, Herskowitz I (1998) Death-defying yeast identify novel apoptosis genes. Cell 92:425–427

Stehlik C, de Martin R, Binder BR, Lipp J (1998) Cytokine induced expression of porcine inhibitor of apoptosis protein (iap) family member is regulated by NF-kappa B. Biochem Biophys Res Commun 243:827–832

Steller H (1995) Mechanisms and genes of cellular suicide. Science 267:1445–1449

Strasser A (1995) Death of a T cell. Nature 373:385–386

Stricker K, Knipping E, Bohler T, Benner A, Krammer P, Debatin K (1998) Anti-CD95 (APO-1/Fas) autoantibodies and T cell depletion in human immunodeficiency virus Type 1 (HIV-1)-infected children. Cell Death Differ z:222–230

Suzuki A (1998) The dominant role of CPP32 subfamily in fas-mediated hepatitis. Proc Soc Exp Biol Med 217:450–454

Takahashi R, Deveraux Q, Tamm I, Welsh K, Assa-Munt N, Salvesen G, Reed J (1998) A single BIR domain of XIAP sufficient for inhibiting caspases. J Biol Chem 273:7787–7790

Talanian RV, Yang X, Turbov J, Seth P, Ghayur T, Casiano CA, Orth K, Froelick CJ (1997) Granule-mediated killing: pathways for granzyme B-initiated apoptosis. J Exp Med 186:1323–1331

Thornberry N, Rano T, Peterson E, Rasper D, Timkey T, Garcia-Calvo M, Houtzager V, Nordstrom P, Roy S, Vaillancourt J, Chapman K, Nicholson D (1997) A combinatorial approach defines specificities of members of the caspase family and granzyme B. J Biol Chem 272:17907–17911

Wallach D, Boldin M, Varfolomeev E, Beyaert R, Vandenabeele P, Fiers W (1997) Cell death induction by receptors of the TNF family: towards a molecular understanding. FEBS Lett 410:96–106

Wolter KG, Hsu YT, Smith CL, Nechushtan A, Xi XG, Youle RJ (1997) Movement of bax from the cytosol to mitochondria during apoptosis. J Cell Biol 139:1281–1292

Wong B, Arron J, Choi Y (1997) T cell receptor signals enchance susceptibility to fas-mediated apoptosis. J Exp Med 186:1939–1944

Xue D, Horvitz HR (1995) Inhibition of the Caenorhabditis elegans cell-death protease CED-3 by a CED-3 cleavage site in baculovirus p35 protein. Nature 377: 248–251

Yang J, Liu X, Bhalla K, Kim CN, Ibrado AM, Cai J, Peng I-I, Jones DP, Wang X (1997) Prevention of apoptosis by Bcl-2: release of cytochrome c from mitochondria blocked. Science 275:1129–1132

Yaoita H, Ogawa K, Maehara K, Maruyama Y (1998) Attenuation of ischemia/reperfusion injury in rats by a caspase inhibitor. Circulation 97:276–281

You M, Ku PT, Hrdlickova R, Bose JHR (1997) ch-IAP1, a member of the inhibitor-of-apoptosis protein family, is a mediator of the antiapoptotic activity of the v-rel oncoprotein. Mol Cell Biol 17:7328–7341

Zhou Q, Salvesen GS (1997) Activation of pro-caspase-7 by serine proteases includes a non-canonical specificity. Biochem J 324:361–364

Zou H, Henzel WJ, Liu X, Lutschg A, Wang X (1997) Apaf-1, a human protein homologous to C. elegans CED-4, participates in cytochrome c-dependent activation of caspase-3. Cell 90:405–413

CHAPTER 18

Proteasome and Apoptosis

K. TANAKA and H. KAWAHARA

A. Introduction

Proteasome and ubiquitin (Ub) are the principal components of an energy-dependent major proteolytic system in eukaryotic cells. Selective destruction of intracellular proteins by this system is ensured by two distinct, concerted pathways: first, a process that selectively marks appropriate proteins with a degradation signal by covalently attaching multiple Ubs, and second a subsequent process involving proteolytic attack on the poly-ubiquitinated proteins by the proteasome. Metabolic energy is required for both steps. The tremendous progress in research on the proteasome and ubiquitination system during the past decade provides new insights on the essence of proteolysis in various biological systems.

The Ub–proteasome pathway is a mechanism for regulating divergent cellular functions in eukaryotic cells by controlling the level of intracellular proteins rapidly, irreversibly and in a timely manner. Indeed, the turnover of a wide variety of proteins, including cell-cycle regulators, transcription factors, signal transducers, tumor suppressors, oncoproteins, short-lived enzymes, viral gene products, membrane polypeptide receptors, and incompletely assembled or mutated proteins with aberrant structures, is known to be controlled by proteolysis via the Ub–proteasome pathway. Accordingly, the proteasome is involved in various biologically important processes, such as the cell cycle, cellular metabolism, signal transduction, the immune response, and protein quality control. Moreover, accumulating evidence reveals the involvement of the proteasome in apoptosis; its mechanism must be further elucidated. In this chapter, we will focus on the functional relationship between the proteasome and apoptosis and discuss the proteasome system as a possible therapeutic target. As there is a close relationship between cell growth and apoptosis, we wish to briefly address the indispensable role of the Ub–proteasome system as a control system for cell-cycle progression before describing known mechanisms of apoptosis and the proteasome machinery.

B. The Ub System

I. The Ub-Ligating Pathway

The Ub system acts by covalently attaching Ub to selected substrate proteins mediated by a cascade of three enzymes termed E1 (Ub-activating), E2 (Ub-

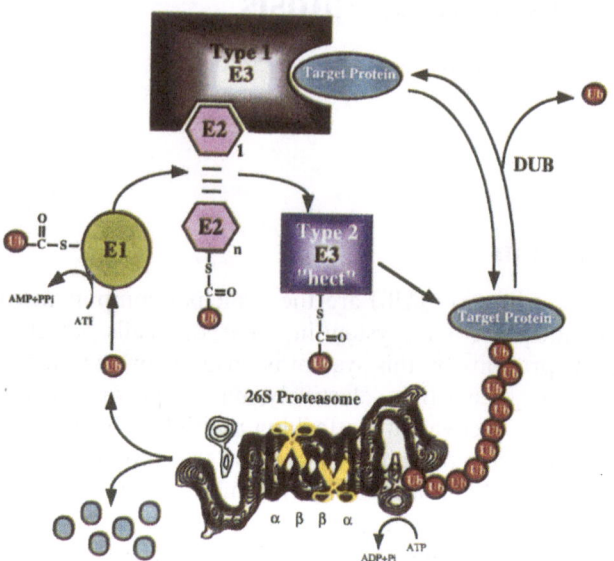

Fig. 1. The ubiquitin–proteasome system. *DUB*, deubiquitinating enzyme; *hect*, homologous to the E6-AP carboxyl terminus; *Ub*, ubiquitin; *E1*, Ub-activating enzyme; *E2* Ub-conjugating enzyme; *E3* Ub-ligating enzyme. For details, see text

conjugating), and E3 (Ub-ligating) enzymes (Hershko and Ciechanover 1992; Hass and Siepmann 1997; Varshavsky 1997). Figure 1 depicts how Ub is attached covalently to target proteins via an isopeptide linkage between the *C*-terminal Gly of Ub and the ε-NH$_2$ group of Lys residues of the acceptor substrate. Protein ubiquitination is initiated by the formation of a high-energy thioester bond between Ub and an E1 in a reaction that requires adenosine triphosphate (ATP) hydrolysis. The activated Ub is then transferred to an E2, forming a thioester bond with E2 that catalyzes the formation of the isopeptide bond between Ub and the substrate protein. In some cases, Ub is transferred directly to target proteins by E2 but, frequently, the additional participation of E3 is required. Finally, a poly-Ub chain can be formed by linking the *C*-terminus of one Ub to a Lys within another Ub. The resultant poly-Ub chain acts as a degradation signal for proteolytic attack by the proteasome (see below).

In the ubiquitination pathway, Ub-protein ligase E3 presumably plays a decisive role in the selection of proteins to be degraded, because it specifically binds to protein substrates (Hochstrasser 1996; Hershko and Ciechanover 1998). The precise role played by E3 is still not entirely clear; there seem to be multiple mechanisms for E3 and its two subtypes, type 1 and type 2, to recognize target proteins. Type-1 E3 is supposed to ligate E2 by associating the target protein and E2, and type-2 E3 is linked to Ub via a thioester bond, which acts directly to form an isopeptide bond between Ub and the substrate protein (see below). A simplified view of this Ub–proteasome pathway is depicted in Fig. 1.

II. Ubiquitination and Cell Cycle

Various cell-cycle regulators have been shown to be ubiquitinated prior to their breakdown, and two species of E3 Ub ligases have been reported for ubiquitination of various cell-cycle factors. A recent exciting finding is that the ubiquitination of these cell-cycle factors involving G1/S transition is catalyzed by a large multisubunit Ub-ligase called SCF (Skp1-Cdc53/cullins-F-box protein) complex, and the progression of mitosis is catalyzed by and cyclosome/APC (anaphase-promoting complex) (HERSHKO 1997; PATTON et al. 1998). Both of these catalysts are type-1 E3s.

It is conceivable that G1 cyclins, such as Cln1p, Cln2p, Cln3p, cyclin E and cyclin D1, and inhibitors to Cdk (cyclin-dependent protein kinase), such as Sic1p, Rum1p, Far1p, p21$^{Cip1/Waf1}$ and p27^{Kip1}, all of which regulate the G1/S transition, are degraded by the Ub pathway, in which ubiquitination is probably mediated by SCF E3 ligase. Remarkably, three SCF components, Skp1, Cdc53/Cullin, and the F-box protein, all are members of a large family of proteins, strongly indicating the existence of many species of SCF complexes in the cells (DESHAIES 1997; PATTON et al. 1998; HARPER and ELLEDGE 1999).

A major clue to the role of proteolysis in cell-cycle regulation was first obtained by the discovery that B-type cyclin is degraded by the Ub–proteasome pathway. Cyclin is a key regulator in *Saccharomyces cerevisiae* Cdc28p and *Schizosaccharomyces pombe* Cdc2p, which acts as an engine for M-phase progression of the cell cycle. The cyclosome/APC is a large protein complex consisting of 8–13 components which catalyzes the programmed breakdown of cyclin proteins and Pds1p/Cut2p (anaphase inhibitor), which is required for the transition from metaphase to anaphase. Moreover, APC is involved in degradation of additional factors, such as Ase1 and cohesin, in addition to mitotic cyclin B, which are required for the traverse of M-phase and the exit from mitosis.

The action of E3 was extensively studied (SCHEFFNER et al. 1995; HUIBREGTSE et al. 1998) in the case of ubiquitination of a tumor-suppressive gene product, p53, a labile nuclear protein with a half-life of 20–35 min. The p53 is ubiquitinated by a particular E3 called E6-AP, which is stimulated by E6, a papilloma-virus-encoded gene product (Sect. F). This E3 has a unique common region with a thiol residue capable of accepting Ub from E2, termed the "*hect* domain" (homologous to the E6-AP carboxyl terminus), which appears in many proteins, indicating that these enzymes carrying the *hect* domain are members of the family of Ub ligases. Here, we have defined these *hect*-containing enzymes as a type-2 E3 to distinguish them from other known E3s provisionally called type-1 E3 (Fig. 1). However, in normal, non-human papilloma virus (HPV) infected cells, p53 is also an unstable protein.

Interestingly, an oncogene product, Mdm-2 – known as a potent inhibitor of p53 – was found to promote also the rapid degradation of p53, and it was demonstrated to function as another type-2 Ub-ligase for the ubiquitination of p53 (HONDA et al. 1997). Intriguingly, Mdm-2 was itself found to be degraded by the Ub–proteasome pathway (CHANG et al. 1998). In addition, many other cell-cycle mediators, such as c-Mos and c-Myb, and immediate–early gene products, such as c-Fos, c-Jun, and c-Myc, are also degraded by the Ub–proteasome pathway, but the E3s required for ubiquitination of these factors have not yet been identified (CIECHANOVER and SCHWARTZ 1998; TANAKA 1998a; TANAKA and CHIBA 1998).

III. Deubiquitinating Enzymes and Cell Proliferation

It is important to note that all eukaryote cells also contain deubiquitinating enzymes (DUB), catalyzing a reaction to an intact Ub moiety from poly-Ub and Ub-fused proteins in cells. They belong to a family of cysteine proteases sub-classified into at least two gene families that are structurally unrelated: the UCH (Ub C-terminal hydrolase) family and the UBP (Ub-specific protease) family (HOCHSTRASSER 1996; WILKINSON 1997). The UBP family proteins contain a conserved Cys and His domain assumed to be responsible for catalytic activity. These DUBs constitute an unexpectedly large protein family in eukaryotes. For example, 17 genes for DUBs are present in the yeast genome (HOCHSTRASSER 1996). Alltogether, more than 60 full-length DUB sequences have been identified so far (WILKINSON 1997), although little is known about their biological roles. The total number of DUBs present in mammalian cells still remains to be elucidated. Irrespective of clarification of their exact roles, the reversibility of the ubiquitinating reaction implies that the proteasome is undoubtedly an important enzyme system responsible for determining the final fate/stability of proteins in cells (TANAKA and CHIBA 1998).

Accumulating evidence also indicates that deubiquitination of polyubiquitinated proteins plays a distinct role in a variety of physiological processes. For example, the product of the fat facets genes (*Faf*) in *Drosophila* is required for eye-facet development (HUANG et al. 1995). Very recently, Fam, a mammalian homolog of Faf, has been shown to catalyze the removal of the Ub moiety from ubiquitinated AF-6, which is known to serve as one of the peripheral components responsible for cell–cell adhesions and to function in the downstream Ras signalling pathway (TAYA et al. 1998). A neural UCH is important for long-term facilitation in aplysia (HEGDE et al. 1997). Yeast UBP3, a 110-kDa SIR4-binding protein, functions as an inhibitor of silencing of transcription (MOAZED and JOHNSON 1996). HAUSP (herpes virus-associated Ub-specific protease), a 135-kDa UBP, is dynamically associated with the PML nuclear bodies and herpes virus protein Vmw110, suggesting its involvement in the control of viral gene expression (EVERETT et al. 1997). It has recently been reported that two mouse DUBs are immediate–early gene products induced by cytokines. The mouse DUB-1 is induced by interleukin 3 (IL-3) (ZHU et al. 1996), and DUB-2 is induced by IL-2 (ZHU et al. 1997); intriguingly, DUB-1 inhibits cell growth when overexpressed. Moreover, the human *tre-2* oncogene encodes a DUB similar to Doa4, indicating a role for the Ub system in mammalian growth control (NAKAMURA et al. 1992; PAPA and HOCHSTRASSER 1993). The recently identified UBPY accumulates upon growth stimulation, and its levels decrease in response to growth arrest induced by cell–cell contact, suggesting that it correlates stringently with cell proliferation and plays a role in regulating the overall function of the Ub–proteasome pathway (NAVIGLIO et al. 1998). In contrast, a nuclear localized UCH called BAP1 is suggested to be a new tumor suppressor gene (JENSEN et al. 1998). It remains to be elucidated how these DUBs are responsible for these pheno-types, although it is known that a loss of function of these genes by deletion or mutation or an excessive high activity by their overexpression causes the induction of the abnormalities mentioned above.

C. The Proteasome: a Protein-Killing Machine

Proteasomes are unusually large multisubunit proteolytic complexes consist-ing of a central catalytic machine (equivalent to the 20S proteasome) and two terminal regulatory subcomplexes termed PA700, which are attached to both ends of the central portion in opposite orientations to form the enzymatically active proteasome (BAUMEISTER et al. 1998; DEMARTINO and SLAUGHTER 1999). In total, about 40 subunits with sizes of 20–110kDa are known to be assem-bled to form two types of the proteasomal complexes with the same catalytic core but different regulatory modules. To date, cDNAs or genes encoding almost all subunits of human and the budding yeast proteasomes have been isolated and characterized (TANAKA 1998b). The proteasomal multicomponent system appears to act as a highly organized apparatus for efficient protein degradation (COUX et al. 1996; HOCHSTRASSER 1996; BAUMEISITER et al. 1998; RECHSTEINER 1998; TANAKA 1998a; TANAKA and CHIBA 1998).

The 20S proteasome is a protease complex with a molecular mass of 700–750kDa composed of 28 subunits. It is a barrel-like particle formed by the axial stacking of four rings made up of two outer a rings and two inner β rings associated in the order $\alpha\beta\beta\alpha$ (Fig. 1). The α and β rings are each made up of seven structurally similar α and β subunits, respectively (BAUMEISTER et al. 1998). The X-ray crystal structures of archaebacterial and yeast 20S protea-somes reveal that three β-type subunits of each inner ring have catalytically active threonine residues at their N-termini, defining the proteasome as a thre-onine protease. These active sites face the interior of the cylinder and reside in a chamber formed by the centers of the abutting β rings. The highly ordered structure indicates that substrates reach the active sites only after passing through a narrow (13 Å) opening corresponding to the center of the a rings. Thus, the N-termini of the α subunits form a physical barrier for substrates on their way to the active sites.

A sort of concealment of the 20S proteasome is supported by the recent structural observation that the center space of the a ring is almost closed, pre-venting the penetration of proteins into the inner surface of the β ring on which the proteolytically active sites are located. PA700 (also known as the 19S regulatory complex) was found to be an activator of the proteasome;

(DeMartino and Slaughter 1999). By interacting with the 20S core particles, it presumably opens the proteasome channel for the entry of the protein substrate. PA700 can associate with the 20S proteasome in an ATP-dependent manner to form the 2000-kDa protein complex termed the 26S proteasome (Fig. 1). This structure is a dumbbell-shaped particle consisting of a centrally located, cylindrical 20S proteasome and two large, terminal PA700 modules attached to the 20S core particle in opposite orientations (Baumeister et al. 1998). PA700 is a 700-kDa protein complex composed of about 20 subunits with sizes of 25–110 kDa, which can be divided into two subgroups: 6 homologous ATPases and approximately 14 non-ATPase subunits that are structurally unrelated (Tanaka 1998b).

One role of these ATPases is to supply continuously energy for the degradation of target proteins. Presumably, the energy is utilized to unfold (or restructure) the proteins, enabling then to penetrate the channel of the α and β rings of the 20S proteasome (Braun et al. 1999). In addition, the metabolic energy supplied by ATP is also thought to be utilized for the assembly of the 26S proteasome from the 20S proteasome and PA700. However, it is still unknown why multiple homologous ATPases are present in the 26S proteasome complex. The PA700 regulatory complex has approximately 14 non-ATPase subunits that seem to play a pivotal role in the functions of the 26S proteasome. The functions of most of these non-ATPase subunits are largely unknown, with the exception of two subunits: the multi-Ub receptor and the DUB. They are involved in the trapping of ubiquitinated target proteins and the recycling of Ub moieties, respectively. The functions mediated by the other non-ATPase subunits await further elucidation (Baumeisiter et al. 1998; Rechsteiner 1998; Tanaka and Chiba 1998).

D. Regulatory Control of Ub and the Proteasome in Apoptosis

Apoptosis, or programmed cell death, has been shown to play a pivotal role in many important biological processes, such as differentiation and development. In multicellular organisms, apoptosis is a process inducing the suicide of a large number of cells in a predictable pattern. For example, the death of most T-cells by apoptosis in the thymus is an essential process for maintaining immunological tolerance. There are multiple pathways for apoptosis. One typical route operates in response to extracellular death signals through Fas/APO-1 and a type-1 receptor of the tumor necrosis factor α (TNFα); another route is activated upon on the deprivation of nutrients and/or growth factors. Currently, accumulating evidence indicates that multiple members of a family of caspases contribute as key players in the process of cell death (Green 1998; Chap. 15).

Ub and/or the proteasome also appears to be involved in the cell suicide pathway(s). Several lines of evidence indicate abnormal accumulations of Ub and/or Ub-conjugated proteins in a wide variety of neurodegenerative disorders that often lead to apoptotic cell death (Lowe et al. 1993; Mayer et al. 1998). Accordingly, alterations in the regulation of the Ub pool in cells may play an important role in the pathogenesis of such diseases. Alternatively, functional disorder of the proteasome or dysfunction of DUB may also cause abnormal accumulations of ubiquitinated proteins in the cells, which then leads to apoptosis.

The activation of both Ub gene expression and nuclear ubiquitination has been observed in γ-irradiated human lymphocytes undergoing apoptosis (Delic et al. 1993). Muscle cell death in the hawkmoth *Manduca sexta* seemed to be coupled to the transient upregulation of Ub mRNA (Dawson et al. 1995). Intriguingly, the inactivation of a temperature-sensitive mutated E1 (Ub-activating enzyme) induced apoptosis (Monney et al. 1998). In this case, caspase inhibitors were unable to block this type of cell death; indeed, the activation of caspase-3, i.e. the main enzyme responsible for apoptosis, was not observed, whereas the overexpression of Bcl-2 (known to be an anti-apoptotic protein) was capable of protecting cells from apoptosis induced by the defect in the E1 at high temperature. Thus, the ubiquitination pathway may play an important role – directly or indirectly – in the cell suicide pathway(s).

One attractive target for the Ub-proteasome pathway became known recently; it is β-catenin, a central component of the cadherin cell adhesion complex or transcriptional mediator of the Wnt/Wingless signalling pathway. Intriguingly, the protein adenomatous polyposis coli, which was found to be associated with β-catenin, reduces the amount of cytoplasmic β-catenin. Adenomatous polyposis coli also binds to the glycogen synthetase kinase 3β (GSK-3β) gene product, which phosphorylates β-catenin directly and induces its ubiquitination-dependent destabilization (Aberle et al. 1997). Thus, the degradation of β-catenin appears to be similar to that of I-κB, a specific inhibitor protein for NF-κB (nuclear factor kB, a heterodimeric complex of p50/p65), which is known to be degraded by the proteasome after undergoing phosphorylation by a specific I-κB kinase that is required for its ubiquitination (Maniatis 1997, 1999). Moreover, a mutation of D-APC (a *Drosophila* homolog of adenomatous polyposis coli) causes the apoptotic cell death, resulting in retinal degeneration (Ahmed et al. 1998). Interestingly, the reduction of *Drosophila* β-catenin armadillo prevents apoptosis of D-APC mutant cells. The expression of the *N*-terminal mutant of armadillo, which is degraded more slowly than wild armadillo, mimicked the inactivation of D-APC that promotes cell death.

Qualitative and quantitative changes in the 26S proteasome have been found to occur specifically in the hawkmoth *Manduca sexta*, undergoing programmed cell death (Dawson et al. 1995; Jones et al. 1995). In contrast, treatment with dexamethasone resulted in a decrease in proteasome activity, which is apparently correlated with the degree of apoptosis observed in thymocytes (Beyette et al. 1998). Beyette et al. (1998) also showed that the loss of 20S and 26S proteasome activities during apoptosis appears to be due to a down-regulation of their proteolytic activities and not to a decrease in their concentrations.

The reorganization of the actin cytoskeleton as well as the disappearance of the proteasome from the nucleus in immortalized cyclic adenosine monophosphate-stimulated rat ovarian granulosa cells suggest a possible function of the proteasome in apoptotic regulation (Pitzer et al. 1996). These findings indicate that changes in the function and subcellular distribution of proteasomes may be closely related to the process(es) undergoing apoptosis.

E. Proteasome Inhibitors Help Elucidate the Biological Roles of the Proteasome in the Apoptotic Pathway

It is still difficult to determine the individual biological roles of the mentioned intracellular proteases. However, recent use of membrane-permeable proteasome inhibitors (PI) has greatly contributed to our understanding of the in vivo functions of proteasomes (Tanaka 1998a). For example, substrate-related peptidyl aldehydes, such as N-benzyloxycarbonyl-Leu-Leu-norvalinal (Z-LLnV-H) and N-benzyloxycarbonyl-Leu-Leu-leucinal (Z-LLL-H), have been employed as potent inhibitors of proteasomes. Moreover, a new microbial metabolite, lactacystin (LC), was found to be a covalently binding PI which does not affect other proteases examined, indicating that the proteasome might be the sole or at least main physiological target of LC (Fenteany et al. 1995). Recently, *clasto*-lactacystin β-lactone, synthesized from LC spontaneously by lactonization, seems to be the actual form of the PI and is considerably more effective than LC (Dick et al. 1997). Both LC and its β-lactone derivative are very effective in living cells, inducing almost complete loss of proteasome-mediated proteolysis.

In 1995, it was reported for the first time that LC induces apoptosis in U937 myeloid leukemia cells, showing the typical morphological change of apoptotic cell death and DNA fragmentation comparable to the apoptotic activity induced by TNFα (Imaioh-Ohmi et al. 1995). There is now accumulating evidence for a relationship between PI and apoptosis in a variety of mammalian cells (reviewed in Orlowski 1999), as listed in Table 1.

The following are examples of the experimental use of PI in various cell systems that primarily seem to be unrelated to each other; they are listed for the sake of complete coverage of this topic. For example, the inhibitor Z-LLL-H was found to induce apoptosis in human leukemic MOLT-4 cells, but N-benzyloxycarbonyl-Leu-leucinal (Z-LL-H), a highly sensitive inhibitor for calpain, had no effect on the apoptotic cell death (Shinohara et al. 1996). Z-Ile-Glu(O-t-butyl)-Ala-leucinal (PSI) and N-benzyloxycarbonyl-Leu-Leu-leucinal (Z-LLnL-H) (Table 1) also induced apoptosis in human leukemic HL60 cells (Drexler 1997). Similarly, Z-LLnV-H and PSI could induce apoptosis in RVC lymphoma cells (Tanimoto et al. 1997) and L1210 leukemic cells

Table 1. Relationship between proteasome inhibition and apoptosis in various mammalian cells

Inhibitor	Cell type	Apoptosis	Comments	Reference
LC	U937	Induction		(Imajoh-Ohmi et al. 1995)
Z-LLL-H	MOLT-4	Induction	p53 (+)	(Shinohara et al. 1996)
PSI, Z-LLnL-H, DCI	HL60	Induction		(Drexler 1997)
PSI	L1210	Induction		(Wojcik et al. 1997)
Z-LLnV-H	RVC	Induction		(Tanimoto et al. 1997)
LC	CLL	Induction	NF-κB (−)	(Delic et al. 1998)
LC	Hepatocytes	Induction	NF-κB (−)	(Bellas and Gerald 1997)
MG-132	J774A.1	Induction	NF-κB (−)	(Ruckdeschel et al. 1998)
LC	T-cells	Induction		(Wang et al. 1998)
Z-LLL-H, LC	M-07e	Induction	GM-CSF (−)	(Zhang et al. 1999)
MG101, MG-132	Thymocyte	Protection	Ionizing radiation	(Grimm et al. 1996)
LC	Sympathetic neuron	Protection	NGF-deprivation	(Sadoul et al. 1996)
	U937	Protection	TNF treatment	(Fujita et al. 1996)
Z-LLL-H, Z-LLnL-H	Thymocyte	?	CPD32-Lactivity (+)	
Z-LLnV-H		Induction	Etoposide treatment	(Stefanell et al. 1998)
		Protection		
LC, MG-132	Thymocyte	Protection	Dexamethasone treatment	(Hirsch et al. 1998)
MG132	U937, 292	Induction	JNK (+)	(Merin et al. 1998)
		Protection	Hsp72 (+)	
LC	T-cells	Induction	Anti-CD3 treatment	(Cui et al. 1997)
		Protection		
PSI, MG115	PC12, Rat-1	Induction	p53 (+)	(Lopes et al. 1997)
	Rat-1	Protection	Serum deprivation	
	PC12	Induction	NGF treatment	
MG132, LC	AT3	Protection	Sindbis virus infection	(Lin et al. 1998)

The magnitude of the effects shows decrease (−) or increase (+).
DCI, 3,4-dichloroisocoumarin; LC, lactacystin; H, peptidyl aldehyde; Z, N-benzyloxycarbonyl; PSI, Z-Ile-Glu(O-t-butyl)-Ala-leucinal; Z-LLL-H (= MG132), Z-Leu-Leu-leucinal; Z-LLnL-H (= MG101), Z-Leu-Leu-norleucinal; Z-LLnV-H (= MG115), Z-Leu-Leu-norvalinal; JNK, c-Jun N-terminal kinase; NF-κB, nuclear factor κB; NGF, nerve growth factor; TNF, tumor mecrosis factor.

(Wojcik et al. 1997), respectively. Thus, PIs have been shown to cause apoptotic cell death in various hematopoietic tumor cells (Fenteany and Schreiber 1998). It is conceivable that the degradation of I-κB by the Ub–proteasome pathway resulted in marked activation of NF-κB in a variety of mammalian cells (Maniatis 1998). LC or Z-LLL-H suppresses the activation of NF-κB by blocking the degradation of I-κB, which inhibits programmed cell death, causing induction of apoptosis in a variety of cells such as CLL (Delic and Gerald 1998), hepatocytes (Bellas et al. 1997), and J774A.1 cells (Ruckdeschel et al. 1998).

However, inhibition of proteasomal activity has also elicited conflicting results for apoptosis, because PIs also play a protective role in the apoptosis of some cell types (Table 1). Actually, Z-LLL-H and acetyl-Leu-Leu-leucinal blocked thymocyte death induced by ionizing radiation, glucocorticoids, and phorbol ester, suggesting that the proteasome may degrade either regulatory protein(s) that normally inhibit the apoptotic pathway or proteolytically activate protein(s) that promote cell death (Grimm et al. 1996). Moreover, LC (Table 1) blocks the initiation of apoptosis in sympathetic neurons following deprivation of nerve growth factor (NGF) (Sadoul et al. 1996). For an explanation of these apparently contradictory observations, it is generally accepted that PIs induce the apoptosis of rapidly growing cells, whereas they greatly inhibit apoptosis (induced by extracellular signals) of non-dividing cells, such as thymocytes and sympathetic neurons.

The effect of PIs, however, does not appear to be explained by such a simple scenario, because LC induced apoptosis in a T-cell hybridoma but inhibited activation-induced cell death in anti-CD3-coated cultured cells (Cui et al. 1997). Similarly, Z-LLnV-H induced apoptosis weakly in thymocyte but prevented the cell suicide induced by etoposide, an inhibitor of topoisomerase II (Hirsch et al. 1998; Stefanell et al. 1998). As depicted in Fig. 2, Z-LLL-H induced a significant accumulation of p53, which may cause apoptosis in MOLT-4 cells (Shinohara et al. 1996). Adenovirus E1A induced apoptosis in the KB cell variant MA1 strain and greatly enhanced the Ub-mediated degradation of topoisomerase IIα, which seemed to be correlated with the inhibition of p53 degradation by E1A (Nakajima et al. 1997). Interestingly, BAG-1, found as a Bcl-2-binding protein with a Ub-like domain, could negatively modulate the chaperone activity of Hsp70/Hsc70 (Takayama et al. 1997). BAG-1 also inhibited p53-induced growth arrest in 293 cells by suppressing the actions of Siah-1A, a BAG-1-binding protein, indicating that Siah-1A is an important mediator of p53-dependent cell-cycle arrest (Matsuzawa et al. 1998). In additon, Lin et al. (1998) reported that the concentration of PI critically affected the cell fate, leading to promotion or suppression of cell death. Apparently, higher doses of PI induce apoptosis in AT3 cells, whereas treatment with lower concentrations results in inhibition of sindbis virus-induced AT3 cell death. They proposed that a high concentration of the inhibitor stabilizes pro-apoptotic proteins, such as p53, which leads to the induction of apoptosis, whereas lower concentrations do not influence p53 but incompletely affect the stability of sindbis virus-related

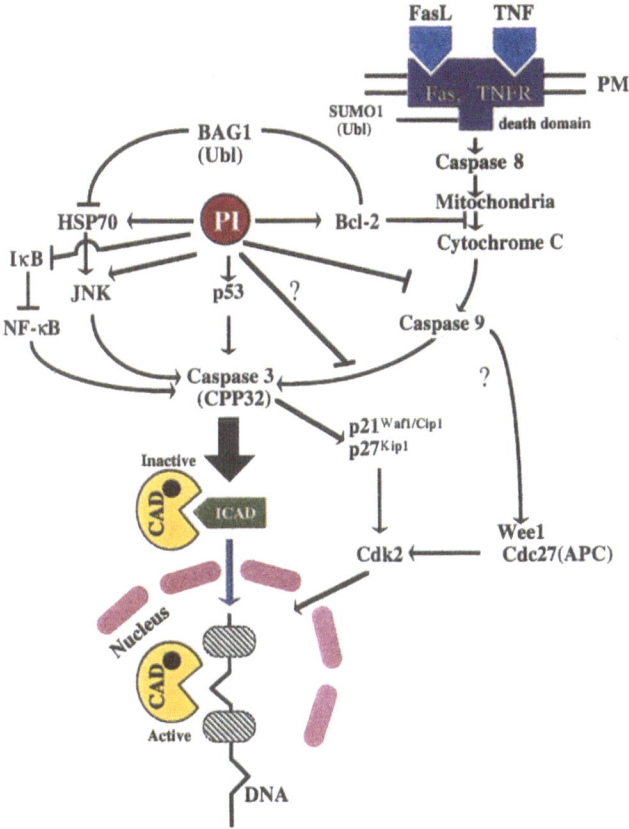

Fig. 2. Model of the role of proteasome function in regulating apoptotic cell death. See text for detailed explanations. *APC*, anaphase-promoting complex; *CAD*, caspase-activated DNase; *FasL*, Fas ligand; *HSP*, heat-shock protein; *ICAD*, inhibitor of CAD; *JNK*, c-Jun *N*-terminal kinase; *PI*, proteasome inhibitor; *TNF*, tumor necrosis factor; *TNFR*, TNF receptor; *Ubl*, ubiquitin-like protein

anti-apoptotic proteins, thus leading to partial protection from cell death. However, why the same inhibitors have opposite effects on apoptosis depending on type of cells used remains unknown.

In PC12 and Rat-1 cells, PSI and Z-LLnV-H induced apoptosis and accumulation of p53, p21$^{Cip1/Waf1}$ and Mdm-2. The induction of apoptosis is blocked by dominant negative p53 mutants, suggesting that p53 plays a key role in apoptosis induced by PIs. Indeed, overexpression of p53 induced apoptosis in Rat-1 cells. Interestingly, Bcl-2 or CrmA (an inhibitor of caspases) resisted treatment with PSI and Z-LLnV-H, indicating that the apoptosis induced by the latter PIs was regulated by Bcl-2 and mediated by IL-1β-converting enzyme (ICE) family proteases. Z-LLnV-H failed to induce apoptosis of quiescent Rat-1 cells by serum deprivation but was still able to induce the apoptosis of nonproliferating, differentiated PC12 cells induced by NGF. Thus, cell proliferation seems not to be the sole determinant of cellular sensitivity to apoptosis by PIs (Lopes et al. 1997). In U937 and 293 kidney cells, Z-LLL-H led to an increase in the activity of c-Jun *N*-terminal kinase (JNK1), which is known to initiate apoptosis in response to stress. Inhibition of the JNK signalling pathway strongly suppresses Z-LLL-H-induced apoptosis, indicating that JNK is critical for the cell death caused by PIs. An anti-apoptotic action of PIs is revealed by a short incubation of cells with Z-LLL-H, followed by its withdrawal. Under these conditions, Hsp72 accumulated and caused suppression of JNK activation during stresses. Accordingly, pretreatment with Z-LLL-H reduced JNK-dependent apoptosis caused by heat shock, but it was unable to block JNK-independent apoptosis induced by TNFα. PIs activating JNK (which initiates apoptosis) induced Hsp72 simultaneously, which suppresses JNK-dependent apoptosis. Thus, a balance between these two effects might define the fate of cells exposed to the inhibitors (Meriin et al. 1998).

LC inhibits the ICE-like activity stimulated by lipopolysaccharide and ATP in macrophages, indicating that the proteasome can regulate the activation of the proteases of the ICE/Ced-3 family (Sadoul et al. 1996). However, ICE itself does not participate in thymocyte death, since the ICE inhibitor YVAD does not inhibit apoptosis in thymocytes (Grimm et al. 1996). CPP32 (equivalent to caspase 3), which is closely related to CED-3, the apoptotic protease in *Caenorhabditis elegans*, is activated during apoptosis induced by anti-Fas and TNF in U937 cells. Surprisingly, PIs enhanced CPP32-like activity in the TNF-treated U937 cells, but did not affect the activity in the untreated condition (Fujita et al. 1996), indicating that the proteasome seems to protect cells from apoptosis by degrading the CPP32-like protease or its processing enzyme. Similarly, LC and Z-LLL-H prevented all manifestations of thymocyte apoptosis induced by dexamethasone and etoposide and simultaneously caused an increase of proteasome activity, suggesting that proteasome activation occurs at an early, premitochondrial step of thymocyte apoptosis (Hirsch et al. 1998). Moreover, it was suggested that the rapid induction of apoptosis in HL60 caused by Z-LLvL-H is due to the activation of CPP32 but occurs independent of ICE activity (Drexler 1997).

Apoptosis in HL60 cells is accompanied by an increase in the Cdk inhibitor p27[Kip1], implying that the HL60 cells undergoing apoptosis are primarily in the G1 phase of the cell cycle. Intriguingly, Fas induced the activation of Cdc2 and Cdk2 in Jurkat cells; the induction was shut down early during apoptosis before caspase-3 activation (Zhou et al. 1998). Activation of these kinases seems to result from both a rapid cleavage of Wee1 (an inhibitory kinase of Cdc2/Cdk2) and inactivation of APC/cyclosome Ub ligase, in which Cdc27, a component of APC/cyclosome, is cleaved during apoptosis (Fig. 2). In addition, apoptosis of human endothelial cells after growth-factor deprivation is associated with rapid upregulation of Cdk2 activity. In these apoptotic cells, loss of the Cdk inhibitors p21[Cip1/Waf1] and p27[Kip1] leads to dramatic induction of Cdk2 activity. Thus, Cdk2 activation through caspase-mediated cleavage of the Cdk inhibitors, may be instrumental in the execution of apoptosis following caspase activation (LEVKAU et al. 1998). However, these two Cdks are also known to be rapidly degraded by the Ub–proteasome pathway, indicating the existence of dual destabilizing mechanisms for these Cdk inhibitors. Recently, caspase-3 was shown to cleave the Ub-protein ligase Nedd4, demonstrating that an enzyme of the Ub pathway is cleaved by caspases during apoptosis (HARVEY et al. 1998).

It is noteworthy that sentrin/SUMO1, a Ub-like protein, associates with the cytopasmic tail, called the "death domain", of Fas/APO-1 or type-1 TNF receptor and seemingly prevents the transduction of these death signals to downstream molecules (OKURA et al. 1996). Moreover, FAF1, a Fas-associated protein that participates in apoptosis, was found to contain two domains with structural similarity to Ub (BECKER et al. 1997), suggesting that proteins related to ubiquitination may modulate the Fas signalling pathway. Further research will reveal further aspects and the context of this exciting topic.

F. The Ub–Proteasome System and Cancer Therapy

Abnormalities of cell-cycle progression may result in apoptotic cell death, which appears to participate somehow in the surveillance of cell-cycle defects or check-points. For cancer therapy, one conceivable strategy is to control the accumulation of anti-oncoprotein(s), such as p53 and Cdk inhibitors, in abnormally proliferating cells. Thus, pharmacological intervention that alters the half-lives of cell-cycle regulatory proteins may have obvious therapeutic potential. If we can manipulate the cell-death program universally or if we devise reagents that regulate apoptosis, we may have valuable therapeutic tools for a variety of human cancers (ROLFE et al. 1997; SPATARO et al. 1998).

Obviously, the levels of various anti-oncoproteins, such as p53 and $p27^{Kip1}$, are known to be reduced dramatically in a variety of tumor cells. As described in this article, the quantity of cellular p53 is tightly controlled by the Ub–proteasome system. The extreme instability of the p53 protein is well exemplified in the case of HPV-related cancers. The oncogenicity of HPV, which is involved in the majority of human anogenital carcinomas, is mediated by the enhancement of p53 degradation by the Ub–proteasome pathway. The E6 oncoprotein encoded by high-risk viruses (HPV 16 and 18) but not low-risk virus types binds to p53 and promotes its degradation by the proteasome (HUIBREGTSE et al. 1998). For this selective degradation, E6-AP, belonging to type-2 E3 (see above), plays an essential role. Moreover, Mdm-2 generally acts analogously to (but independent of) E3 to promote the ubiquitination of p53. Accordingly, specific inhibitors for E6-AP and Mdm-2 may help to significantly increase the cellular levels of p53, which probably inhibits cell proliferation and may block tumorigenesis of rapidly growing cells.

In addition, the low level of $p27^{Kip1}$ protein, but not its mRNA, commonly appeared in various tumors, indicating that $p27^{Kip1}$ levels could be regulated post-translationally, perhaps through the enhanced breakdown by the Ub–proteasome pathway (CATZAVELOS et al. 1997; LODA et al. 1997; PORTER et al.

1997). Recently, the Ub-ligase for the ubiquitination of p27^{Kip1} was found to be as a SCF^{Skp2} complex (Tsvetkov et al. 1999; Carrano et al. 1999; Meng et al. 1999). It, thus, is of interest to search the inhibitory compounds capable of suppressing the ubiquitination of p27^{Kip1} by blockage of the function of E3-SCF^{Skp2}.

As mentioned in Sect. D, a loss of function of several DUBs by deletion or mutation of these genes causes cell-cycle arrest, although it remains to be explained how DUBs are responsible for these phenotypic abnormalities. Irrespective of these mechanisms, various compounds leading to the inhibition of the activities of these growth-related DUBs may be effectively applied in cancer therapy.

Finally, specific inhibitors of proteasomes, such as LC, may be effective as drugs inducing, for distinct time periods, both cell-cycle arrest and apoptosis, thus preventing cell destruction, which occurs through a variety of apoptotic signals. Actually, it was recently reported that epoxomicin found as an antitumor agent in actinomycete natural products is a highly sensitive and selective inhibitor of the proteasome (Meng et al. 1999). Interestingly, epoxomicin has strong anti-angiogenic activity (Oikawa et al. 1991) like LS (Oikawa et al. 1998), because treatment of both compounds resulted in almost complete prevention of in vivo neovascularization in the developing chick embryo chorioallantoic membrane. Moreover, inhibitors of PA700, the regulatory subunit of the 26S proteasome, but not of catalytic 20S proteasome, which is essential for keeping cell viability (Tanaka and Chiba 1998), would be expected to selectively destroy cells via activation of apoptotic machinery. Considerations for using proteases as therapeutic targets are merely beginning; further analysis of the various regulatory mechanisms of the proteasome machinery are required.

G. Perspectives

There is no doubt that the Ub–proteasome system plays an indispensable role not only in the cell-cycle progression but also in the cell suicide pathway. Moreover, this proteolytic pathway also appears to be involved in the immune response, i.e. acting as a processing enzyme for endogenous antigens, in which it plays a critical role for distinguishing self from non-self at the molecular level. This seems to be of fundamental importance in immunity (Tanaka and Kasahara 1998). Abnormalities of these biological events presumably result in various pathological diseases. In considering the important contribution of the proteasome function to the regulation of these biological processes, it may be extremely useful to use those agents capable of modulating specific functions of ubiquitination and proteasomes as tools for the therapy of diseases, such as cancers, autoimmune diseases, and others.

The discovery that LC epoxomicin are selective PIs would infer the existence of microbial metabolites that may specifically suppress or enhance the functional diversity of the enzymes responsible for ubiquitination, particularly the Ub-ligase-E3 protein family. A major advantage of a drug mechanism controlling ubiquitination is that the Ub system is based on multiple enzymatic reactions consisting of E1, E2, and E3 enzymes. We would expect to find target compounds by comprehensive screening once the assay systems are established. Moreover, since the Ub–proteasome system has an unusual diversity (judging from the large numbers of E2 and E3 species and the complex functions of the proteasome), one would expect the mechanism of the drugs will be capable of interrupting a specific route without causing severe side effects.

Acknowledgments. This work was supported in part by grants from the program Grants in Aid of Scientific Research on Priority Areas (Intracellular Proteolysis) from the Ministry of Education, Science, Sports, and Culture of Japan, and the Human Frontier Science-Promotion Organization.

References

Aberle H, Bauer A, Stappert J, Kispert A, Kemler R (1997) β-Catenin is a target of the ubiquitin–proteasome pathway. EMBO J 16:3797–3804

Ahmed Y, Hayashi S, Levine A, Wieschaus E (1998) Regulation of armadillo by a *Drosophila* APC inhibits neuronal apoptosis during retinal development. Cell 93:1171–1182

Baumeister W, Walz J, Zühl F, Seemüller E (1998) The proteasome: Paradigm of a self-compartmentalizing protease. Cell 92: 367–380

Becker K, Schneider P, Hofmann K, Mattmann C, Tschopp J (1997) Interaction of Fas(apo-1/CD95) with proteins implicated in the ubiquitination pathway. FEBS Lett 412:102–106

Bellas RE, Gerald MJF (1997) Inhibition of NF-κB activity induces apoptosis in murine hepatocytes. Am J Pathol 151:891–896

Beyette J, Mason GG, Murry RZ, Cohen GM, Rivett AJ (1998) Proteasome activities decrease during dexamethasone-induced apoptosis of thymocytes. Biochem J 32: 315–320

Braun BC, Glickman M, Kraft R, Dahlmann B, Kloetzel P-M, Finley D, Schmidt M (1999) The base of the proteasome regulatory particle exhibits chaperone-like activity. Nature Cell Biol 1:221–226

Carrano AC, Eytan E, Hershoko A, Pagano M (1999) SKP2 is required for ubiquitin-mediated degradation of the CDK inhibitor p27. Nature Cell Biol 1:193–199

Catzavelos C, Bhattacharya N, Ung YC, Wilson JA, Roncari L, Sandhu C, Shaw P, Yeger H, Morava-Protzner I, Kapusta L, Franssen C, Pritchard KI, Slingerland JM (1997) Decreased levels of the cell cycle inhibitor p27^{Kip1} protein: prognostic implications in primary breast cancer. Nature Med 3:227–230

Chang Y-C, Lee YS, Tejima T, Tanaka K, Omura S, Heintz NH, Mitsui Y, Magae J (1998) Degradation of p53-responsive proteins Bax and Mdm-2 by ubiquitin/proteasome proteolytic system. Cell Growth and Differ 9:79–84

Ciechanover A, Schwartz AL (1998) The ubiquitin–proteasome pathway: the complexity and myriad functions of protein death. Proc Natl Acad Sci USA 95: 2727–2730

Clem RJ, Fechheimer M, Miller K (1991) Prevention of apoptosis by a baculovirus gene during infection of insect cells. Science 254:1388–1390

Coux O, Tanaka K, Goldberg AL (1996) Structure and functions of the 20S and 26S proteasomes. Annu Rev Biochem 65: 801–847

Cui H, Matsui K, Omura S, Schauer SL, Matulka RA, Sonenshein GE, Ju ST (1997) Proteasome regulation of activation-induced T cell death. Proc Natl Acad Sci USA 94:7515–7520

Dawson SP, Arnold JE, Mayer NJ, Reynolds SE, Billett MA, Gordon C, Colleaux L, Kloetzel PM, Tanaka K, Mayer RJ (1995) Developmental changes of the 26S proteasome in abdominal intersegmental muscles of *Manduca sexta* during programmed cell death. J Biol Chem 270:1850–1858

Delic JP, Morange M, Magdelenat H (1993) Ubiquitin pathway involvement in human lymphocyte gamma-irradiation-induced apoptosis. Mol Cell Biol 8: 4875–4883

Delic JP, Masdehors P, Omura S, Cosset JM, Dumont J, Binet JL, Magdelenat H (1998) The proteasome inhibitor lactacystin induces apoptosis and sensitizes chemo- and radioresistant human chronic lymphocytic leukemia lymphocytes to TNF-α-initiated apoptosis. Br J Cancer 77:1103–1107

DeMartino GN, Slaughter CA (1999) J Biol Chem 274:22123–22126

Deshaies RJ (1997) Phosphorylation and proteolysis: partners in the regulation of cell division in budding yeast. Curr Opin Gen Dev 7:7–16

Dick LR, Cruikshank AA, Destree AT, Grenier L, McCormack TA, Melandri FD, Nunes SL, Palombella VJ, Parent LA, Plamondon L, Stein RL (1997) Mechanistic studies on the inactivation of the proteasome by lactacystin in cultured cells. J Biol Chem 272:182–188

Drexler HCA (1997) Activation of the cell death program by inhibition of proteasome function. Proc Natl Acad Sci USA 94:855–860

Everett RD, Meredith M, Orr A, Cross A, Kathoria M, Parkinson J (1997) A novel ubiquitin-specific protease is dynamically associated with the PML nuclear domain and binds to a herpesvirus regulatory protein. EMBO J 16:1519–1530

Fenteany GF, Schreiber SL (1998) Lactacystin, proteasome function, and cell fate. J Biol Chem 273:8545–8548

Fenteany G, Standaert RF, Lane WS, Choi S, Corey EJ, Scheiber SL (1995) Inhibition of proteasome activities and subunit-specific amino-terminal threonine modification by lactacystin. Science 268:726–731

Fujita E, Mukasa T, Tsukahara T, Arahata K, Omura S, Momoi T (1996) Enhancement of CPP32-like activity in the TNF-treated U937 cells by the proteasome inhibitors. Biochem Biophys Res Comm 224:74–79

Green DR (1998) Apoptotic pathways: the road to Ruin. Cell 94:695–698

Grimm LM, Goldberg AL, Poirier GG, Schwatz LM, Osborne BA (1996) Proteasomes play an essential role in thymocyte apoptosis. EMBO J 15:3835–3844

Harper JW, Elledge SJ (1999) Skippong into the E2F1-destruction pathway. Nature Cell Biol 1:E5–E7

Harvey KF, Harvey HL, Michael JM, Parasivam G, Waterhouse N, Alnemri ES, Watters D, Kumar S (1998) Caspase-mediated cleavage of the ubiquitin-protein ligase Nedd-4 during apoptosis. J Biol Chem 273:13524–13530

Hass AL, Siepmann TJ (1997) Pathways of ubiquitin conjugation. FASEB J 11:1257–1268

Hauser HP, Badroff M, Pyrowolakis G, Jentsch S (1998) A giant ubiquitin-conjugating enzyme related to IAP apoptosis inhibitors. J Cell Biol 141:1415–1422

Hegde AN, Inokuchi K, Pei W, Casadio A, Ghirardi M, Chain DG, Martin KC, Kandel ER, Schwartz JH (1997) Ubiquitin C-terminal hydrolase is an immediate-early gene essential for long-term facilitation in aplysia. Cell 89:115–126

Hershko A (1997) Roles of ubiquitin-mediated proteolysis in cell cycle control. Curr Opin Cell Biol 9:788–799

Hershko A, Ciechanover A (1992) The ubiquitin system for protein degradation. Annu Rev Biochem 61:761–807

Hershko A, Ciechanover A (1998) The ubiquitin system. Annu Rev Biochem 67:425–479

Hirsch T, Dallaporta B, Zamzami N, Susin SA, Ravagnan L, Marzo I, Brenner C, Kroemer G (1998) Proteasome activation occurs at an early, premitochondrial step of thymocyte apoptosis. J Immunol 161:35–40

Hochstrasser M (1996) Ubiquitin-dependent protein degradation. Annu Rev Genet 30:405–439

Honda R, Tanaka H, Yasuda H (1997) Oncoprotein MDM2 is a ubiquitin ligase E3 for tumor suppressor p53. FEBS Lett 420:25–27

Huang Y, Baker RT, Fischer-Vize JA (1995) Control of cell fate by a deubiquitinating enzyme encoded by the fat facets gene. Science 270:1828–1831

Huibregtse JM, Maki CG, Howley PM (1998) Ubiquitination of the p53 tumor suppressor. In: Peters J-M, Harris JR, Finley D (eds) Ubiquitin and the biology of the cell. Plenum, New York, pp 147–189

Imajoh-Ohmi S, Kawaguchi T, Sugiyama S, Tanaka K, Omura S, Kikuchi H (1995) Lactacystin, a specific inhibitor of the proteasome, induces apoptosis in human monoblast U937 cells. Biochem Biophys Res Comm 27:1070–1077

Jensen DE, Proctor M, Marquis ST, Gardner HP, Ha SI, Chodosh LA, Ishov AM, Tommerup N, Vissing H, Sekido Y, Minna J, Borodovsky A, Schultz DC, Wilkinson KD, Maul GG, Barlev N, Berger SL, Prendergast GC, Rauscher III FJ (1998) BAP1: a novel ubiquitin hydrolase which binds to the BRCA1 RING finger and enhances BRCA1-mediated cell growth suppression. Oncogene 16:1097–1112

Jones ME, Haire MF, Kloetzel PM, Mykles DL, Schwartz LM (1995) Changes in the structure and function of the multicatalytic proteinase (proteasome) during programmed cell death in the intersegmental muscles of the hawkmoth, Manduca sexta. Dev Biol 169:436–447

Levkau B, Koyama H, Raines EW, Clurman BE, Herren B, Orth K, Roberts JM, Ross R (1998) Cleavage of p21$^{Cip1\,Waf1}$ and p27^{Kip1} mediates apoptosis in endothelial cells through activation of cdk2: Role of a caspase cascade. Mol Cell 1:553–563

Lin KI, Baraban JM, Ratan RR (1998) Inhibition versus induction of apoptosis by proteasome inhibitors depends on concentration. Cell Death Differ 5:577–583

Loda M, Cuker B, Tam SW, Lavin P, Fiorentino M, Draetta GF, Jessup JM, Pagano M (1997) Increased proteasome-dependent degradation of the cyclin-dependent kinase inhibitor p27 in aggressive colorectal carcinomas. Nature Med 3:231–234

Lopes UG, Erhardt P, Yao R, Cooper GM (1997) p53-Dependent induction of apoptosis by proteasome inhibitors. J Biol Chem 272:12893–12896

Lowe J, Mayer RJ, Landon M (1993) Ubiquitin in neurodegerative diseases. Brain Pathol 3:55–65

Maniatis T (1997) Catalysis by a multiprotein IκB kinase complex. Science 278:818–819

Maniatis T (1999) A ubiquitin ligase complex essential for the NF-κB, Wnt/Wingless, and Hedgehog signaling pathways. Genes Dev 13:505–510

Margottin F, Bour SP, Durand H, Selig L, Benichou S, Richard V, Thomas D, Strebel K, Benarous R (1998) A novel human WD protein, h-βTrCP, that interacts with HIV-1 Vpu connects CD4 to the ER degradation pathway through an F-box motif. Mol Cell 1:585–574

Matsuzawa S, Takayama S, Froesch BA, Zapata JM, Reed JC (1998) p53-Inducible human homologue of *Drosophila* seven in absentia (Siah) inhibit cell growth: suppression by BAG-1. EMBO J 17:2736–2747

Mayer RJ, Landon M, Lowe J (1998) Ubiquitin and the molecular pathology of human disease. In: Peters J-M, Harris JR, Finley D (eds) Ubiquitin and the biology of the cell. Plenum, New York, pp 147–189

Meng L, Kwok BHB, Sin N, Crews CM (1999) Eponemycin exerts its antitumor effect through the inhibition of proteasome function. Cancer Res 59:2798–2801

Meriin AB, Gabai VL, Yaglom J, Shifrin VI, Sherman MY (1998) Proteasome inhibitors activate stress kinases and induce Hsp72. Diverse effects on apoptosis. J Biol Chem 273:6373–6379

Moazed D, Johnson AD (1996) A deubiquitinating enzyme interacts with SIR4 and regulates silencing in *S. cerevisiae*. Cell 86:667–677

Monney L, Otter I, Oliver R, Ozer HL, Haas AL, Omura S, Borner C (1998) Defects in the ubiquitin pathway induce caspase-independent apoptosis blocked by bcl-2. J Biol Chem 273:6121–6131

Nakajima T, Kimura M, Kuroda K, Tanaka M, Kikuchi A, Seino H, Yamao F, Oda K (1997) Induction of ubiquitin conjugating enzyme activity for degradation of topoisomerase II alpha during adenovirus E1A-induced apoptosis. Biochem Biophys Res Comm 239:823–829

Nakajima T, Morita K, Tsunoda H, Imajoh-Ohmi S, Tanaka H, Yasuda H, Oda K (1998) Stabilization of p53 by adenovirus E1A occurs through its amino-terminal region by modification of the ubiquitin–proteasome pathway. J Biol Chem 273:20036–20045

Naviglio S, Matteucci C, Matoskova B, Nagase T, Nomura N, Fiore PP, Draetta GF (1998) UBPY: a growth-regulated human ubiquitin isopeptidase. EMBO J 17:3241–4350

Oikawa T, Hasegawa M, Shimamura M, Ashino H, Murota S, Morita I (1991) Eponemycin, a novel antibiotic, is a highly powerful angiogenesis inhibitor. Biochem Biophys Res Commun 181:1070–1076

Oikawa T, Sasaki T, Nakamura M, Shimamura M, Tanahashi N, Ōmura S, Tanaka K (1998) The proteasome is involved in angiogenesis. Biochem Biophys Res Commun 246:243–248

Okura T, Gong L, Kamitani T, Wada TM, Okura I, Wei CF, Chang HM, Yeh ETH (1996) Protection against Fas/Apo-1- and tumor necrosis factor-mediated cell death by a novel protein, sentrin. J Immunol 157:4277–4281

Orlowski RZ (1999) The role of the ubiquitin-proteasome pathway in apoptosis. Cell Death Differ 6:303–313

Papa FR, Hochstrasser M (1993) The yeast *DOA4* gene encodes a deubiquitinating enzyme related to a product of the human *tre-2* oncogene. Nature 366:313–319

Patton EE, Willems AR, Tyers M (1998) Combinatorial control in ubiquitin-dependent proteolysis: don't Skp the F-box hypothesis. Trends Genet 14:236–243

Pitzer F, Dantes A, Fuchs T, Baumeister W, Amsterdam A (1996) Removal of proteasomes from the nucleus and their accumulation in apoptotic blebs during programmed cell death. FEBS Lett 394:47–50

Porter PL, Malone KE, Heagerty PJ, Alexander GM, Gatti LA, Firpo EJ, Daling JR, Roberts JM (1997) Expression of cell-cycle regulators p27/Kip1 and cyclin E, alone or in combination, correlate with survival in young breast cancer patients. Nature Med 3:222–235

Rechsteiner M (1998) The 26S proteasome. In: Peters J-M, Harris JR, Finley D (eds) Ubiquitin and the biology of the cell. Plenum, New York, pp 147–189

Rolfe M, Chiu MI, Pagano M (1997) The ubiquitin-mediated proteolytic pathway as a therapeutic area. J Mol Med 75:5–17

Ruckdeschel K, Harb S (1998) *Yersinia enterocolitica* impairs activation of transcription factor NF-κB: involvement in the induction of programmed cell death and in the suppression of the macrophage tumor necrosis factor alpha production. J Exp Med 187:1069–1079

Sadoul R, Fernandez PA, Quiquerez AL, Martinou I, Maki M, Schroter M, Becherer JD, Irmler M, Tschopp J, Martinou JC (1996) Involvement of the proteasome in the programmed cell death of NGF-deprived sympathetic neurons. EMBO J 15:3845–3852

Saitoh H, Pu RT, Dasso M (1997) SUMO-1: wrestling with a new ubiquitin-related modifier. Trends Biochem Sci 22:374–376

Scheffner M, Nuber U, Huibregtse JM (1995) Protein ubiquitination involving an E1–E2–E3 enzyme ubiquitin thioester cascade. Nature 373:81–83

Shinohara K, Tomioka M, Nakano H, Tone S, Ito H, Kawashima S (1996) Apoptosis induction resulting from proteasome inhibition. Biochem J 317:384–388

Spataro V, Norbury C, Harris AL (1998) The ubiquitin–proteasome pathway in cancer. Br J Cancer 77:448–355

Stefanelli C, Bonavita F, Stanic I, Pignatti C, Farruggia G, Masotti L, Guarnieri C, Caldarera CM (1998) Inhibition of etoposide-induced apoptosis with peptide aldehyde inhibitors of proteasome. Biochem J 332:661–665

Sutterluty H, Chatelain E, Marti A, Wirbelauer C, Senften M, Muller U, Kred W (1999) p45^{SKP2} promotes p27^{Kip1} degradation and induces S phase in quiescent cells. Nature Cell Biol 1:207–214

Takayama S, Bimston DN, Matsuzawa S, Freeman BC, Aime-Sempe C, Xie Z, Morimoto RI, Reed JC (1997) BAG-1 modulates the chaperone activity of Hsp70/Hsc70. EMBO J 16:4887–4896

Tanaka K (1998a) Proteasomes: structure and biology. J Biochem 123:195–204

Tanaka K (1998b) Molecular biology of the proteasome. Biochem Biophys Res Commun 247:537–541

Tanaka K, Chiba T (1998) The proteasome: a protein-destroying machine. Genes Cells 3:485–498

Tanaka K, Kasahara M (1998) The MHC class-I ligand generating system:Roles of immunoproteasomes and INF-γ inducible PA28. Immunol Rev 163:161–176

Tanimoto Y, Onishi Y (1997) Peptidyl aldehyde inhibitors of proteasome induce apoptosis rapidly in mouse lymphoma RVC cells. J Biochem 121:542–549

Taya S, Yamamoto T, Kano K, Kawano Y, Iwamatsu A, Tsuchiya T, Tanaka K, Kanai M, Wood SA, Kaibuchi K (1998) The Ras target AF-6 is a physiological substrate of the Fam deubiquitinating enzyme. J Cell Biol 142:1053–1062

Tsvetkov LM, Yeh KH, Lee SJ, Sun H, Zhang H (1999) p27^{Kip1} ubiquitination and degradation is regulated by the SCF^{Skp2} complex through phosphorylated Thr187 in p27. Curr Biol 9:661–664

Varshavsky A (1997) The ubiquitin system. Trends Biochem Sci 22:383–387

Wang X, Luo H, Chen H, Duguid W, Wu J (1998) Role of proteasomes in T cell activation and proliferation. J Immunol 16:788–801

Wilkinson KD (1997) Regulation of ubiquitin-dependent processes by deubiquitinating enzymes. FASEB J 11:1245–1256

Wojcik C, Stokosa T, Giermasz A, Golab J, Zagozdzon R, Kawiak J, Wilk S, Komar A, Kaca A, Malejczyk J, Jakobisiak M (1997) Apoptosis induced in L1210 leukemia cells by an inhibitor of the chymotrypsin-like activity of the proteasome. Apoptosis 2:455–462

Zhang XM, Lin H, Chen C, Chen BD (1999) Inhibition of ubiquitin-proteasome pathway activates a caspase-3-like protease and induces Bcl-2 cleavage in human M-07e leukaemic cells. Biochem J 340:127–133

Zhou BB, Junying HL, Kirschner MW (1998) Caspase-dependent activation of cyclin-dependent kinases during Fas-induced apoptosis in Jurkat cells. Proc Natl Acad Sci USA 95:6785–6790

Zhu Y, Carroll M, Papa FZ, Hochstrasser M, D'Andrea AD (1996) Dub-1, a deubiquitinating enzyme with growth-suppressing activity. Proc Natl Acad Sci USA 93:3275–3279

Zhu Y, Lambert K, Carroll M, Copeland NG, Gilbert DJ, Jenkins NA, D'Andrea AD (1997) Dub-2 is a member of a novel family of cytokine-inducible deubiquitinating enzymes. J Biol Chem 272:51–57

CHAPTER 19

Proteolytic Processing of the Amyloid Precursor Protein of Alzheimer's Disease

S.F. Lichtenthaler, C.L. Masters, and K. Beyreuther

A. Introduction

Alzheimer's disease (AD) is a neurodegenerative disease and the most common form of progressive dementia in the elderly. It is characterized by a progressive loss of memory, declining cognitive function and, ultimately, leads to decreasing physical functions and death. The neuropathological hallmarks of AD are the senile plaques and the neurofibrillary tangles, which are both protein aggregates deposited in the brain. The neurofibrillary tangles represent intraneuronal bundles of paired helical filaments that mainly consist of the microtubule-associated protein tau in an abnormally phosphorylated form (for review on tau see Goedert et al. 1996). The extracellular amyloid plaques mainly consist of the 42-residue-long amyloid β-peptide (Aβ, βA4; Glenner and Wong 1984; Masters et al. 1985), which is proteolytically derived from the much larger amyloid precursor protein (APP, AβPP, βAPP; Kang et al. 1987). The generation and deposition of Aβ seem to be at the origin of the disease and are believed to trigger a complex pathological cascade that causes neuronal dysfunction, the appearance of the neurofibrillary tangles and, finally, the onset of the disease. This current view for the central role of Aβ in the pathogenesis of AD is supported by a wealth of data (for review, see Hardy 1997), including mutations and a polymorphism in four genes that lead to an inherited form or an early-onset form of AD. On the cellular level, these mutations either lead to an enhanced generation of the Aβ-peptide or to an increased rate of its aggregation. Consequently, approaches used to develop a causative therapy for AD try to determine in detail the proteolytic mechanisms that lead to the release of Aβ from APP and the mechanisms of Aβ aggregation and Aβ neurotoxicity (for reviews about Aβ aggregation and neurotoxicity see Harper and Lansbury 1997 and Yankner 1996).

Most AD cases are sporadic, meaning that no disease-causing genetic mutations have been found. However, about 50% of sporadic AD patients carry the allele ε4 of the apolipoprotein E, which is a risk factor for AD. On the molecular level, this allele seems to enhance the rate of aggregation of Aβ (for a review see Strittmatter and Roses 1996).

About 5% of all AD cases are familial (autosomal dominant inheritance) and are caused by mutations in three genes encoding either APP, presenilin 1 (PS1) or presenilin 2 (PS2). The common effect of these mutations is to alter the proteolytic processing of APP in such a way that more of the pathogenic, rapidly aggregating Aβ-peptide is generated (HARDY 1997; SCHEUNER et al. 1996). The presenilins, their proteolytic processing and the PS mutations are the subject of the following chapter of this volume. A detailed description of the APP mutations is given in Sect. C.

The aim of this review is to summarize the current knowledge about the different proteolytic activities cleaving APP and how they lead to the generation of the Aβ peptide. First we present an overview of the different pathways of the proteolytic processing of APP; then, we describe the individual protease activities in detail. Finally, we discuss the potential for a pharmacological modulation of the activity of these proteases for the development of potent drugs against AD.

B. Molecular Biology of AD

I. The Amyloid Precursor Protein

The Aβ-peptide is proteolytically derived from APP, as depicted in Fig. 1. APP is a ubiquitously expressed, glycosylated type-I membrane protein (KANG et al. 1987; WEIDEMANN et al. 1989). Due to alternative splicing of its gene, APP occurs in eight different membrane-bound isoforms with a length of 677–770 amino acids, with APP695 (695 residues) being the major isoform of APP expressed in neuronal cells (SANDBRINK et al. 1994). The function of APP is unknown so far, although a variety of different functions have been proposed, including an involvement of APP in copper homeostasis (MULTHAUP 1997), cell adhesion (SHIVERS et al. 1988) and receptor signaling (KANG et al. 1987).

APP has a large N-terminal ectodomain of up to 699 residues, a transmembrane domain of 24 residues (grey hatched in Fig. 1A) and a short C-terminal, cytoplasmic domain of 47 residues. The ectodomain consists of several subdomains; in the cytoplasmic domain, a signal (sequence NPTY) for receptor-mediated endocytosis of APP has been described as well as binding sites for proteins involved in signal transduction (for a review see SELKOE 1994). The Aβ domain comprises 28 residues of the ectodomain of APP and 12–14 residues of its transmembrane domain.

II. Overview of the Proteolytic Processing of APP

The proteolytic processing of APP by the so-far unidentified proteases termed α-, β-, γ- and δ-secretases leads to a variety of different soluble and membrane-bound proteins (Fig. 1). Two main pathways contribute to this proteolytic processing of APP. The *amyloidogenic pathway* leads to the generation of the Aβ-peptide (HAASS et al. 1992b; SEUBERT et al. 1992; SHOJI et al. 1992), which is deposited in the amyloid plaques of AD. In the *non-amyloidogenic pathway*,

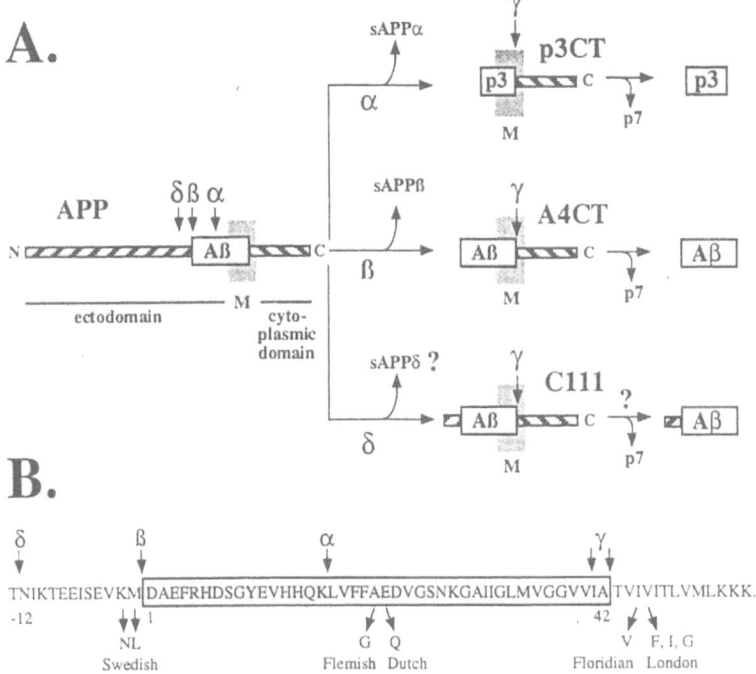

Fig. 1A,B. Proteolytic processing of the amyloid precursor protein (APP). **A** APP may be cleaved by three proteases (α-, β- and δ-secretase) within its ectodomain. The cleaved large N-terminal fragments [soluble APP (sAPPα, sAPPβ, sAPPδ, respectively) are secreted, whereas the C-terminal fragments (p3CT, A4CT, C111, respectively) are cleaved by γ-secretase. The membrane (*M*) is *shaded*. The non-amyloidogenic processing pathway involves α- and γ-secretases, whereas the amyloidogenic processing pathway involves β- and γ-secretases. Cleavage or cleavage products with a *question mark* have not yet been unequivocally identified. **B** Depiction of the Aβ sequence, comprising the secretase cleavage sites. The amino acids are given in the one-letter code. The *boxed* residues indicate the sequence of Aβ_{42}. The numbers of the residues are relative to the Aβ sequence. Familial APP mutations close to the secretase cleavage sites are indicated by the *lower arrows*

APP is processed in a way that precludes the formation of the amyloidogenic Aβ-peptide (Esch et al. 1990).

Expression of APP in cultured cells leads to its predominant processing in the non-amyloidogenic pathway, which involves the two proteases α- and γ-secretase (Fig. 1). Close to the plasma membrane, α-secretase cleaves APP within the Aβ domain and thus precludes the generation of Aβ. This yields a secreted, soluble form of APP (sAPPα), comprising most of the N-terminal ectodomain of APP. The second product generated by α-secretase is the remaining membrane-bound C-terminal fragment p3CT, which is further cleaved within its transmembrane domain by γ-secretase, leading to the generation of the p3 peptide and its subsequent release into the conditioned medium. Additionally, the p7 peptide is formed, which is the C-terminal fragment and has a mass of 7 kDa. The fate of the p7 peptide is not known, but it may be rapidly degraded.

Those APP molecules that are not cleaved by α-secretase may be reinternalized by endocytosis and then be cleaved in the amyloidogenic pathway, which involves β- and γ-secretase. In most cell lines, the amyloidogenic pathway is the minor pathway of APP processing but may be a preferred one in neuronal cells (SIMONS et al. 1996). β-secretase cleaves APP within its ectodomain at the N-terminus of the Aβ-peptide domain. Thus, the β-secretase cleavage site is 16 residues further away from the transmembrane domain than the α-secretase cleavage site (Fig. 1B). As in the non-amyloidogenic pathway, the N-terminal fragment (sAPPβ) is secreted and a C-terminal fragment (A4CT, C99) remains membrane bound. A4CT comprises the C-terminal 99 residues of APP and is the direct precursor of Aβ; it is cleaved within its transmembrane domain by γ-secretase at the C-terminus of the Aβ-peptide domain (DYRKS et al. 1993; HIGAKI et al. 1995; LICHTENTHALER et al. 1997). Aβ is generated and released from the cells. The remaining C-terminal protein fragment, p7, is the same as in the non-amyloidogenic pathway.

Besides α- and β-secretase, a third proteolytic activity, called δ-secretase, may cleave within the ectodomain of APP (Simons et al. 1996). δ-Secretase cleavage occurs 12 residues N-terminal of the β-secretase cleavage site and does not seem to be an important proteolytic processing pathway of APP.

C. Description of the Proteolytic Activities Cleaving APP

The main APP-protease activities (α-, β- and γ-secretase) have been found in all mammalian cell lines studied so-far (for a review see EVIN et al. 1994). α-Secretase, at least, is also present in yeast (ZHANG et al. 1994) and in insect Sf9 cells (ESSALMANI et al. 1996). The identity of the proteases is not yet known. Thus, α-, β-, γ- and δ-secretase are simply the names for the proteolytic activities that cleave APP at defined cleavage sites. Hence, it is not clear whether each secretase activity is represented by a single protease or by several different proteases.

Most of the current knowledge about the secretases comes from studies using cell lines expressing APP or its C-terminal fragment, A4CT. The secretase activities are monitored indirectly via the detection of Aβ and p3 in the conditioned medium. The expression of full-length APP allows the study of all secretases at the same time, whereas the use of A4CT offers the possibility to study γ-secretase exclusively or the combination of α- and γ-secretases (DYRKS et al. 1993; LICHTENTHALER et al. 1997). *In vitro* assays allowing the direct study of the enzymatic properties of the individual enzymes have not yet been established.

In this section, we describe what is known about the secretases with regard to their cellular localization, their substrate specificity and the availability of specific inhibitors. Furthermore, we will discuss candidates for the protease activities as well as the class of proteolytic enzymes to which they might belong.

I. α-Secretase

α-Secretase cleaves the peptide bond Lys687-Leu688 of APP770 (corresponding to residues 16 and 17 of Aβ; Fig. 1; ESCH et al. 1990). To a minor extent, the peptide bonds after residues 10, 11 and 15 of Aβ have been described as additional cleavage sites of α-secretase (for a review see EVIN et al. 1994). Studies with transfected cell lines expressing mutant APP proteins suggest that α-secretase has a broad substrate specificity and is capable of cleaving many different peptide bonds. The cleavage site of α-secretase seems to be determined by the distance (12 residues) from the transmembrane domain. Accordingly, APP was still cleaved by α-secretase when point mutations, deletions or insertions had been introduced into the APP sequence at or near the α-secretase cleavage site (SISODIA 1992). Howevery a proline close to the α-secretase cleavage site almost completely abolished α-cleavage, suggesting that the APP sequence around the cleavage site needs to adapt an α-helical conformation in order to be cleared (SISODIA 1992). This assumption is further supported by the mutation A692G, which is likely to destabilize the helical conformation and which partly inhibited α-secretase cleavage (HAASS et al. 1994). This mutation, which is close to the α-secretase cleavage site, has been found in a Flemish family suffering from AD (HENDRIKS et al. 1992). A second mutation close to the α-secretase cleavage site (E693Q), which is also found in patients (LEVY et al. 1990; VAN BROECKHOVEN et al. 1990), does not seem to inhibit α-secretase but rather leads to an enhanced aggregation of the released Aβ peptide (WISNIEWSKI et al. 1991). Furthermore, it was shown that APP needs to be membrane anchored in order to be cleaved by α-secretase (HAASS et al. 1992a; SISODIA 1992), indicating that α-secretase itself might also be a membrane-bound protease.

Among the different APP proteases, β- and γ-secretase seem to be constitutively active, as no regulators of their activity have been found so far. In contrast, the activity of α-secretase seems to be tightly regulated in the cell. The processing of APP in the non-amyloidogenic pathway, and hence the α-secretase activity, is stimulated by phorbol esters (BUXBAUM et al. 1990), indicating an involvement of protein kinase C (PKC). Consequently, first messengers which stimulate cell-surface receptors that can activate the phospholipase C/PKC pathway are also able to stimulate the activity of α-secretase and, thus, lead to an increased secretion of sAPPα (for a review see CHECLER 1995). The involvement of PKC suggests that α-secretase itself might be phosphorylated, whereas phosphorylation of APP was determined to be unnecessary for the increased α cleavage (HUNG and SELKOE 1994).

A variety of different experiments have been carried out to determine the cellular location of α-secretase (HAASS et al. 1992a; SAMBAMURTI et al. 1992; SISODIA 1992; DE STROOPER et al. 1993; HAASS et al. 1993). APP needs to be fully maturated, including protein sulfation, in order to be cleaved. APP, biotinylated at the cell surface, is cleaved by α-secretase. Inhibition of endo-somal and lysosomal proteases did not decrease secretion of sAPPα, thus elim-inating these intracellular compartments as sites of α-secretase cleavage. APP lacking the cytoplasmic domain, and hence its internalization signal (NPTY), is not reinternalized but is still cleaved by α-secretase. The release of sAPPα may be inhibited by the weak base methylamine. Taken together, these exper-iments suggest that α-secretase cleavage occurs in a late, acidic compartment of the default secretory pathway, either in the vesicles budding off from the trans-Golgi network or at the plasma membrane. This assumption is in agree-ment with a recent study suggesting that α-secretase cleavage occurs in cave-olae, which are plasma-membrane microdomains (Ikezu et al. 1998).

α-Secretase activity was not inhibited using a variety of specific protease inhibitors, with the exception of metalloprotease inhibitors like 1,10-phenan-throline, tumor necrosis factor α (TNF-α) protease inhibitor-2 (Arribas et al. 1996) or batimastat (Parvathy et al. 1998), indicating that α-secretase could be a metalloprotease. What is the molecular identity of α-secretase? The use of synthetic peptides encompassing the α-secretase cleavage site of APP has led to the proposal of various proteases as candidates for α-secretase (for a review see Evin et al. 1994). However, none of these proteases has unequivo-cally been shown to be α-secretase. Interestingly, α-secretase shares a lot of similarities with proteolytic activities that generate soluble isoforms of unrelated type-I membrane proteins. Examples of such proteins include angiotensin-converting enzyme and transforming growth factor-α (TGF-α) (for a review see Hooper et al. 1997 and chap. 13). For all of these proteins, the corresponding α-secretase-like protease has not yet been identified but shares many features with the α-secretase activity that cleaves APP. The cleav-age generally occurs in the ectodomain at a short distance from the trans-membrane domain and may be stimulated by phorbol esters and inhibited with metalloprotease inhibitors. Moreover, these proteolytic activities seem to have a broad substrate specificity and cleave within α-helical protein conformations. Interestingly, a mutated Chinese hamster ovary (CHO) cell line defective in the α-secretase-like cleavage of TGF-α was also defective in the α-secretase cleavage of APP (Arribas and Massague 1995), suggesting that different cell-surface proteins might be cleaved by a single α-secretase or by a family of closely related metalloproteases.

Not only type-I but also type-II membrane proteins such as TNF-α or Fas are cleaved within their ectodomains, leading to the secretion of a soluble counterpart of the corresponding membrane protein. Again, these protease activities show most of the features of α-secretase. The protease cleaving TNF-α has, meanwhile, been identified (TNF-α-converting enzyme) and belongs to the protease family of the ADAMs (A disintegrin and metalloproteases; Black et al. 1997; Moss et al. 1997; see also chap. 13). The α-secretase might turn out to be a similar protease.

II. β-Secretase

β-Secretase cleaves the peptide bond Met671-Asp672 of APP770 (corresponding to residues Met–1 and Asp1 of the Aβ sequence) and, thus, at the N-terminus of the Aβ domain (Fig. 1). Minor cleavage sites were reported around the main cleavage site (for a review see EVIN et al. 1994). As in the case of α-secretase, it remains unclear whether all cleavages are carried out by a single or by several β-secretases. APP needs to be membrane anchored in order to be cleaved by β-secretase, suggesting that β-secretase itself might also be membrane bound (CITRON et al. 1995), but experimental proof has yet to be shown.

Although α- and β-secretase have not yet been identified, it seems clear that, in neuronal as well as in non-neuronal cells, β-secretase is different from α-secretase. Intensive mutagenesis around the β-secretase cleavage site revealed that β-secretase, in contrast to α-secretase, does not cleave at a certain distance from the membrane surface and that β-secretase has a high sequence specificity (CITRON et al. 1995). The first evidence for a high sequence specificity came with the identification of a double mutation (Lys–2Asn and Met–1Leu) found in the APP gene of a Swedish family suffering from familial AD (MULLAN et al. 1992). On the cellular level, this so-called Swedish mutation leads to an increased β-secretase cleavage of APP and, as a consequence, to a six- to eightfold increase in Aβ generation (CITRON et al. 1992; CAI et al. 1993; SCHEUNER et al. 1996).

Based on studies of the cellular localization of β-secretase, it appears that the proteolytic activities cleaving wild-type APP (APP-wt) and APP with the Swedish mutation (APP-Swe) are different enzymes. APP-wt is cleaved after internalization from the cell surface, probably in the endosomes (PERAUS et al. 1997). In contrast, APP-Swe seems to be cleaved in a late compartment of the secretory pathway (HAASS et al. 1995). A variety of experiments are the basis for this assumption of two different β-secretases, among them the following. APP-wt lacking the cytoplasmic domain with the NPTY-internalization signal is hardly cleaved by β-secretase (HAASS et al. 1993; KOO and SQUAZZO 1994), demonstrating the necessity of internalization of APP-wt. However, APP-Swe, which lacks the cytoplasmic domain and thus cannot be internalized, is still processed by β-secretase, indicating that this cleavage event should occur in the secretory pathway (CITRON et al. 1995). Although β-secretase cleavage of both APP-wt and APP-Swe occurs in acidic compartments (DYRKS et al. 1993; KOO and SQUAZZO 1994; SHOJI et al. 1992), it does not occur in the lysosomes, as inhibitors of lysosomal proteases, such as leupeptin, did not inhibit the generation of sAPPβ (BUSCIGLIO et al. 1993; HAASS et al. 1993).

The only known inhibitor (with mean inhibitory concentration of $300\,\mu M$ to $1\,mM$) of β-secretase activity is the broad-spectrum serine-protease inhibitor 4-(2-aminoethyl)-benzenesulphonyl fluoride (CITRON et al. 1996a), suggesting that β-secretase might be a serine protease. Several different proteases have been proposed as candidates for β-secretase (for a review see EVIN et al. 1994). However, as in the case of candidates for α-secretase, most experiments have been carried out in vitro using synthetic peptides, and none of these candidates could unequivocally be shown to be β-secretase. One of the candidate proteases, cathepsin D, has recently been ruled out as a candidate (Saftig et al. 1996).

III. γ-Secretase

γ-Secretase cleaves the APP-derived fragments A4CT and p3CT at the C-terminus of the Aβ domain. Two requirements must be met so that γ-secretase cleavage can occur. First, APP has to be processed by α-, β- or δ-secretase before it may be cleaved by γ-secretase, indicating that γ-secretase requires substrate protein bearing a short ectodomain. Second, the presence of the protein PS1 is required. p3CT and A4CT, both having a short ectodomain of 12 or 28 residues, respectively (Fig. 1), are cleaved by γ-secretase. The same holds true for C111 (ectodomain of 40 residues), which is the C-terminal fragment of APP that arises from δ-secretase cleavage (Tienari, personal communication; Fig. 1). However, full-length APP has never been found to be cleaved by γ-secretase. The reason why γ-secretase cleavage requires substrates with a short ectodomain is not known. The second requirement for γ-secretase activity is the presence of PS1. De Strooper et al. (1998) have shown, that neuronal cells originating from a mouse embryo with a PS1 knock-out show a decrease of 80% in the γ-secretase cleavage of A4CT. The residual 20% of γ-secretase cleavage has been attributed to the fact that, in the absence of PS1, the cells still contain PS2, which is 67% homologous to PS1. As the presenilins do not seem to be proteases, it has been proposed that they need to bind to A4CT and then present it to γ-secretase as a prerequisite for γ-secretase cleavage (De Strooper et al. 1998). However, it is still under debate whether the presenilins and APP directly interact with each other. In any case, there is a striking resemblance between the necessity of PS1 for γ-secretase cleavage and the necessity of the protein sterol-regulatory-element-binding protein (SREBP)-cleavage-activating protein for the γ-secretase-like cleavage of SREBP (Rawson et al. 1997).

Among the APP-secretases, the enzymatic mechanism of γ-secretase is most enigmatic, due to the fact that γ-secretase has two cleavage sites and that both are located within the transmembrane domain of APP. However, it is not clear whether γ-secretase cleaves A4CT and p3CT while they are still inserted into the membrane or after the release of these proteins from the membrane so that the cleavage site could be accessible to a potentially soluble γ-secretase.

The γ-secretase cleavage site is identical in both p3CT and A4CT, so that both p3 and Aβ have the same C-terminus and only differ by 16 residues at the N-terminus (Fig. 1B). Even the C-terminal APP fragment C111, which arises through d-secretase cleavage, is cleaved by γ-secretase at the same site as A4CT and p3CT (Lichtenthaler, unpublished observation). The γ-

secretase cleavage occurs after residue Val40 or Ala42 of $A\beta$ (corresponding to residues 711 and 713 of APP770, respectively). The major $A\beta$ species secreted into human cerebrospinal fluid and by cultured cells expressing APP-wt ends at the C-terminus with residue 40 of $A\beta$ ($A\beta_{40}$); the minor species ends with residue 42 ($A\beta_{42}$; WANG et al. 1996). However, this minor species $A\beta_{42}$ is the pathologically relevant species. This is illustrated by different findings. First, the amyloid plaques in AD (IWATSUBO et al. 1994; GRAVINA et al. 1995) contain mainly $A\beta_{42}$ but hardly any $A\beta_{40}$. $A\beta_{42}$ has also been shown to aggregate *in vitro* much faster than $A\beta_{40}$ (JARRETT et al. 1993). Second, all familial AD mutations in the APP and PS genes lead to an enhanced generation of $A\beta_{42}$ when compared with the wild-type proteins. One of the mutations (Swedish mutation, which affects β-secretase cleavage) also increases the amount of $A\beta_{40}$ (SCHEUNER et al. 1996). Third, the peptide p3 ending with residue 42 of $A\beta$ ($p3_{42}$) but not $p3_{40}$ is a major constituent of Down's syndrome cerebellar preamyloid, which represents an early form of amyloid plaques (LALOWSKI et al. 1996). Thus, γ-cleavage after residue 42 of $A\beta$ is crucial for the pathogenesis of AD.

At present, it is not clear whether the cleavage event after residue 42 represents a second cleavage site for a single γ-secretase in addition to the main cleavage after residue 40, or whether different γ-secretases could be involved in generating the $A\beta_{40}$ and $A\beta_{42}$ species. Three studies have shown that the use of inhibitors for γ-secretase leads to a reduction of the concentration of $A\beta$ in the conditioned medium of cultured cells, with the reduction being more pronounced for $A\beta_{40}$ than $A\beta_{42}$ (CITRON et al. 1996b; KLAFKI et al. 1996; YAMAZAKI et al. 1997). However, as the inhibitors may not only affect γ-secretase cleavage (generation of $A\beta$) but also the rate of degradation of $A\beta$ in the conditioned medium of the cultured cells, it is not clear whether the differential reduction of the amount of both $A\beta$-species is indeed an indication of the existence of two unrelated γ-secretases, each specific for the generation of either $A\beta_{40}$ or $A\beta_{42}$.

As in the case of β-secretase and α-secretase, APP mutations close to the γ-secretase cleavage site have been found in families suffering from familial AD (Fig. 1). These mutations either occur three (Floridian mutation Ile45Val; ECKMAN et al. 1997) or four residues (London mutations Val46 to Phe, Ile or Gly; CHARTIER-HARLIN et al. 1991; GOATE et al. 1991; MURRELL et al. 1991) away from the γ-secretase cleavage site at $A\beta$-residue 42. On the cellular level, these mutations lead to an increased cleavage of γ-secretase after residue 42, thus generating more of the pathogenic, rapidly aggregating $A\beta_{42}$ (SUZUKI et al. 1994; ECKMAN et al. 1997; LICHTENTHALER et al. 1997). In addition, the familial AD mutations found in the proteins PS1 and PS2 have also been shown to influence the cleavage site of γ-secretase by leading to an enhanced $A\beta_{42}/A\beta_{40}$ ratio (SCHEUNER et al. 1996), but it is not clear whether the mutant presenilins exert this effect by directly interacting with γ-secretase or via an indirect effect.

There seem to be different cellular localizations of γ-secretase activity. One is localized at or close to the plasma membrane, generating the secreted

Aβ and p3 (HARTMANN et al. 1997). This localization is in good agreement with previous results showing that (a) the weak base methylamine, that raises the pH of acidic cellular compartments, had an inhibitory effect on the release of Aβ in A4CT expressing cells (DYRKS et al. 1993) and (b) A4CT, lacking the cytoplasmic domain and thus the ability to be internalized, is still processed to Aβ (LICHTENTHALER and BEYREUTHER, manuscript in preparation). Additional γ-secretase activities have mainly been found in neuronal cells (COOK et al. 1997; HARTMANN et al. 1997; TIENARI et al. 1997), but also in kidney 293 cells (WILD-BODE et al. 1997), and generate intracellular Aβ that does not seem to be secreted. These activities are mainly localized at the membrane of the endoplasmic reticulum (generation of intracellular Aβ_{42}) and in the trans-Golgi network (generation of intracellular Aβ_{40}). It is not known if the different γ-secretase activities are related enzymes or not.

Three years ago, HIGAKI et al. (1995) published a report about a peptide aldehyde that inhibited γ-secretase activity in CHO cells. Since then, related peptide aldehydes and derivatives have been found that inhibit γ-secretase activity in different cell lines at low micromolar concentrations (KLAFKI et al. 1996; ALLSOP et al. 1997; WOLFE et al. 1998). All these inhibitors are serine- and cysteine-protease inhibitors, suggesting that γ-secretase belongs to either class. Inhibitors specific for other classes of proteases have not been found to inhibit γ-secretase activity.

As γ-secretase does not only cleave at a single peptide bond but at two different peptide bonds (Val40-Ile41 and Ala42-Thr43), the substrate specificity of γ-secretase consists of two factors: the sequence specificity and the cleavage specificity. The sequence specificity (this expression is often used as a synonym for substrate specificity) means the necessity of specific amino acids within the A4CT sequence for γ-secretase activity. In contrast, the cleavage specificity of γ-secretase reflects the extent to which γ-secretase cleaves at one or the other peptide bond and may be measured by determining the product ratio Aβ_{42}/Aβ_{40}. Intensive mutagenesis around the γ-secretase cleavage site of A4CT showed (LICHTENTHALER et al. 1997; MARUYAMA et al. 1996; TISCHER and CORDELL 1996) that γ-secretase must be a protease with a broad sequence specificity. However, the cleavage specificity is determined by the amino-acid composition of the transmembrane domain of A4CT, as mutations in the transmembrane domain strongly alter the cleavage specificity. A further factor determining the cleavage specificity seems to be the length of the transmembrane domain of A4CT (LICHTENTHALER and BEYREUTHER, manuscript in preparation). These experiments also suggest that γ-secretase itself is a membrane protein.

The identity of γ-secretase remains unclear, but, interestingly, three proteins have been found to be cleaved within their transmembrane domain in a way similar to APP: Notch (SCHROETER et al. 1998), SREBP (RAWSON et al. 1997) and proteins of the inner mitochondrial membrane (LEONHARD et al. 1996). The proteases cleaving SREBP (S2P) and the inner mitochondrial proteins (m-AAA) have been identified (see following chapter). Both proteases are integral membrane proteins, with S2P being a single polypeptide with several transmembrane domains (RAWSON et al. 1997) and m-AAA being a heterooligomeric protease complex (LEONHARD et al. 1996). γ-Secretase might be similar to one or both proteases; however, it is not *identical* to either protease. S2P has recently been ruled out as being γ-secretase (Ross et al. 1998), and m-AAA is located in the inner mitochondrial membrane, where γ-secretase activity has not been found (HARTMANN et al. 1997).

IV. δ-Secretase

δ-Secretase is the least understood proteolytic activity among the APP-secretases. This activity was first described by SIMONS et al. (1996), who found three C-terminal fragments of APP in the cell lysates of primary hippocampal neurons: p3CT, A4CT and a fragment with a N-terminus which exceeded that of A4CT by 12 residues. This newly identified d-secretase activity thus cleaves the peptide bond Thr584-Asn585 of APP695 (Fig. 1B). Similar to A4CT and p3CT, the C-terminal APP fragment generated by δ-secretase seems to be further cleaved by γ-secretase, leading to the secretion of a 6-kDa peptide (LICHTENTHALER et al. 1997). No specific type of protease has yet been associated with the d-Secretase activity.

D. Therapeutic Potential of the APP Secretases

To what extent might the described APP-secretases be reasonable targets for the development of an effective treatment or prevention for AD? Potential drugs need to modulate the activity of the APP secretases in such a way that less of the pathogenic Aβ peptide is generated. Among the APP secretases, δ-secretase does not seem to be a promising target for a therapeutic approach, as a possible inhibition of δ-secretase would most likely still allow the cleavage of APP by α-, β- and γ-secretases and thus still lead to the generation of Aβ and p3.

The activity of β- and γ-secretase leads to the generation of Aβ. Thus, the inhibition of β- or γ-secretase might be a reasonable therapeutic approach. Highly specific inhibitors for both enzymes are currently lacking. An inhibition of β-secretase will reduce the amount of Aβ but at the same time enhance the generation of p3 (through α- and γ-cleavage). This might be an undesired effect; although p3 has hardly been found in the amyloid plaques of AD, it has nevertheless been shown to be a major constituent of the preamyloid aggregates that represent an early form of amyloid plaques (LALOWSKI et al. 1996). However, due to its high sequence specificity β-secretase might still be a promising drug target.

Currently, γ-secretase seems to be the most promising target for the development of a therapeutic intervention for AD. A specific inhibitor of γ-secretase would abolish the generation of both Aβ and p3. However, as γ-secretase obviously has a broad sequence specificity, it might not exclusively cleave APP but rather turn out to be a more general proteolytic mechanism for the degradation of transmembrane domains of various proteins. In this case, the inhibition of γ-secretase would interfere with the inhibition of such other proteins.

An alternative therapeutical approach involving the APP-secretases might be stimulation of α-secretase, as α-secretase cleavage precludes the formation of Aβ. However, α-secretase might be a general cleavage activity for different membrane proteins. Furthermore, its activity seems to be regulated in a PKC-dependent manner. Therefore, stimulation of α-secretase activity for therapeutic purposes might interfere with signaling pathways in the cell and lead to an enhanced secretion not only of secretory APP but also of the soluble counterparts of the other membrane proteins.

In the next few years new drugs are likely to become available that allow the specific modulation of the activity of the APP secretases. Both the identification of the secretases and drug development will greatly facilitate the evaluation of the therapeutic potential of the APP secretases.

References

Allsop D, Christie G, Gray C, et al. (1997) Studies on inhibition of β-amyloid formation in APP751-transfected IMR-32 cells, and SPA4CT-transfected SHSY5Y cells. In: Iqbal K, Winblad B, Nishimura T, Takeda M, Wisniewski HM (eds) Alzheimer's disease: biology, diagnosis and therapeutics. John Wiley & Sons Ltd, Chichester, pp 717–727

Arribas J, Massague J (1995) Transforming growth factor-alpha and beta-amyloid precursor protein share a secretory mechanism. J Cell Biol 128:433–441

Arribas J, Coodly L, Vollmer P, et al. (1996) Diverse cell surface protein ectodomains are shed by a system sensitive to metalloprotease inhibitors. J Biol Chem 271:11376–11382

Black RA, Rauch CT, Kozlosky CJ, et al. (1997) A metalloproteinase disintegrin that releases tumour-necrosis factor-alpha from cells. Nature 385:729–733

Busciglio J, Gabuzda DH, Matsudaira P, Yankner BA (1993) Generation of beta-amyloid in the secretory pathway in neuronal and nonneuronal cells. Proc Natl Acad Sci USA 90:2092–2096

Buxbaum JD, Gandy SE, Cicchetti P, et al. (1990) Processing of Alzheimer beta/A4 amyloid precursor protein: modulation by agents that regulate protein phosphorylation. Proc Natl Acad Sci USA 87:6003–6006

Cai XD, Golde TE, Younkin SG (1993) Release of excess amyloid β protein from a mutant amyloid β protein precursor. Science 259:514–516

Chartier-Harlin MC, Crawford F, Houlden H, et al. (1991) Early-onset Alzheimer's disease caused by mutations at codon 717 of the β-amyloid precursor protein gene. Nature 353:844–846

Checler F (1995) Processing of the b-amyloid precursor protein and its regulation in Alzheimer's disease. J Neurochem 65:1431–1444

Citron M, Oltersdorf T, Haass C, et al. (1992) Mutation of the β-amyloid precursor protein in familial Alzheimer's disease increases β-protein production. Nature 360:672–674

Citron M, Teplow DB, Selkoe DJ (1995) Generation of amyloid beta protein from its precursor is sequence specific. Neuron 14:661–670

Citron M, Diehl TS, Capell A, et al. (1996a) Inhibition of amyloid β-protein production in neural cells by the serine protease inhibitor AEBSF. Neuron 17:171–179

Citron M, Diehl TS, Gordon G, et al. (1996b) Evidence that the 42- and 40-amino acid forms of amyloid beta protein are generated from the beta-amyloid precursor protein by different protease activities. Proc Natl Acad Sci USA 93:13170–13175

Cook DG, Forman MS, Sung JC, et al. (1997) Alzheimer's A beta(1–42) is generated in the endoplasmic reticulum/intermediate compartment of NT2N cells. Nat Med 3:1021–1023

De Strooper B, Umans L, Van Leuven F, Van Den Berghe H (1993) Study of the synthesis and secretion of normal and artificial mutants of murine amyloid precursor protein (APP): cleavage of APP occurs in a late compartment of the default secretion pathway. J Cell Biol 121:295–304

De Strooper B, Saftig P, Craessaerts K, et al. (1998) Deficiency of presenilin-1 inhibits the normal cleavage of amyloid precursor protein. Nature 391:387–390

Dyrks T, Dyrks E, Monning U, et al. (1993) Generation of beta A4 from the amyloid protein precursor and fragments thereof. FEBS Lett 335: 89–93

Eckman CB, Mehta ND, Crook R, et al. (1997) A new pathogenic mutation in the APP gene (I716V) increases the relative proportion of A beta 42(43). Hum Mol Genet 6:2087–2089

Esch FS, Keim PS, Beattie EC, et al. (1990) Cleavage of amyloid β peptide during constitutive processing of its precursor. Science 248:1122–1124

Essalmani R, Guillaume JM, Mercken L, Octave JN (1996) Baculovirus-infected cells do not produce the amyloid peptide of Alzheimer's disease from its precursor. FEBS Lett 389:157–161

Evin G, Beyreuther K, Masters CL (1994) Alzheimer's disease amyloid precursor protein (AβPP): proteolytic processing, secretases and βA4 amyloid production. Amyloid. Int J Exp Clin Invest 1:263–280

Glenner GG, Wong CW (1984) Alzheimer's disease: initial report of the purification and characterization of a novel cerebrovascular amyloid protein. Biochem Biophys Res Commun 120:885–890

Goate A, Chartier Harlin MC, Mullan M, et al. (1991) Segregation of a missense mutation in the amyloid precursor protein gene with familial Alzheimer's disease. Nature 349:704–706

Goedert M, Spillantini MG, Hasegawa M, et al. (1996) Molecular dissection of the neurofibrillary lesions of Alzheimer's disease. Cold Spring Harb Symp Quant Biol 61:565–573

Gravina SA, Ho LB, Eckman CB, et al. (1995) Amyloid beta protein (A beta) in Alzheimer's disease brain – biochemical and immunocytochemical analysis with antibodies specific for forms ending at A beta 40 or A beta 42(43). J Biol Chem 270:7013–7016

Haass C, Koo EH, Mellon A, et al. (1992a) Targeting of cell-surface beta-amyloid precursor protein to lysosomes: alternative processing into amyloid-bearing fragments. Nature 357:500–503

Haass C, Schlossmacher MG, Hung AY, et al. (1992b) Amyloid beta-peptide is produced by cultured cells during normal metabolism. Nature 359:322–325

Haass C, Hung AY, Schlossmacher MG, et al. (1993) β-Amyloid peptide and a 3-kDa fragment are derived by distinct cellular mechanisms. J Biol Chem 268:3021–3024

Haass C, Hung AY, Selkoe DJ, Teplow DB (1994) Mutations associated with a locus for familial Alzheimer's disease result in alternative processing of amyloid beta-protein precursor. J Biol Chem 269:17741–17748

Haass C, Lemere CA, Capell A, et al. (1995) The Swedish mutation causes early-onset Alzheimer's disease by beta-secretase cleavage within the secretory pathway. Nat Med 1:1291–1296

Hardy J (1997) The Alzheimer family of diseases: many etiologies, one pathogenesis? Proc Natl Acad Sci USA 94:2095–2097

Harper JD, Lansbury PJ (1997) Models of amyloid seeding in Alzheimer's disease and scrapie: mechanistic truths and physiological consequences of the time-dependent solubility of amyloid proteins. Annu Rev Biochem 66:385–407

Hartmann T, Bieger SC, Bruhl B, et al. (1997) Distinct sites of intracellular production for Alzheimer's disease A beta40/42 amyloid peptides. Nat Med 3:1016–1020

Hendriks L, van Duijn CM, Cras P, et al. (1992) Presenile dementia and cerebral haemorrhage linked to a mutation at codon 692 of the β-amyloid precursor protein gene. Nat Genet 1:218–221

Higaki J, Quon D, Zhong Z, Cordell B (1995) Inhibition of beta-amyloid formation identifies proteolytic precursors and subcellular site of catabolism. Neuron 14: 651–659

Hooper NM, Karran EH, Turner AJ (1997) Membrane protein secretases. Biochem J 321:265–279

Hung AY, Selkoe DJ (1994) Selective ectodomain phosphorylation and regulated cleavage of beta-amyloid precursor protein. EMBO J 13:534–542

Ikezu T, Trapp BD, Song KS, et al. (1998) Caveolae, plasma membrane microdomains for α-secretase-mediated processing of the amyloid precursor protein. J Biol Chem 273:10485–10495

Iwatsubo T, Odaka A, Suzuki N, et al. (1994) Visualization of Aβ42(43) and Aβ40 in senile plaques with end-specific Aβ monoclonals: evidence that an initially deposited species is Aβ42(43). Neuron 13:45–53

Jarrett JT, Berger EP, Lansbury PT Jr (1993) The carboxy terminus of the beta amyloid protein is critical for the seeding of amyloid formation: implications for the pathogenesis of Alzheimer's disease. Biochemistry 32:4693–4697

Kang J, Lemaire HG, Unterbeck A, et al. (1987) The precursor of Alzheimer's disease amyloid A4 protein resembles a cell-surface receptor. Nature 325:733–736

Klafki H, Abramowski D, Swoboda R, et al. (1996) The carboxyl termini of beta-amyloid peptides 1–40 and 1–42 are generated by distinct gamma-secretase activities. J Biol Chem 271:28655–28659

Koo EH, Squazzo SL (1994) Evidence that production and release of amyloid beta-protein involves the endocytic pathway. J Biol Chem 269:17386–17389

Lalowski M, Golabek A, Lemere CA, et al. (1996) The "nonamyloidogenic" p3 fragment (amyloid beta17–42) is a major constituent of Down's syndrome cerebellar preamyloid. J Biol Chem 271:33623–33631

Leonhard K, Herrmann JM, Stuart RA, et al. (1996) AAA proteases with catalytic sites on opposite membrane surfaces comprise a proteolytic system for the ATP-dependent degradation of inner membrane proteins in mitochondria. EMBO J 15:4218–4229

Levy E, Carman MD, Fernandez MI, et al. (1990) Mutation of the Alzheimer's disease amyloid gene in hereditary cerebral hemorrhage, Dutch type. Science 248: 1124–1126

Lichtenthaler SF, Ida N, Multhaup G, et al. (1997) Mutations in the transmembrane domain of APP altering γ-secretase specificity. Biochemistry 36:15396–15403

Maruyama K, Tomita T, Shinozaki K, et al. (1996) Familial Alzheimer's disease-linked mutations at Val717 of amyloid precursor protein are specific for the increased secretion of A beta 42(43). Biochem Biophys Res Commun 227:730–735

Masters CL, Simms G, Weinman NA, et al. (1985) Amyloid plaque core protein in Alzheimer disease and Down syndrome. Proc Natl Acad Sci USA 82:4245–4249

Moss ML, Jin SL, Milla ME, et al. (1997) Cloning of a disintegrin metalloproteinase that processes precursor tumour-necrosis factor-alpha. Nature 385:733–736

Mullan M, Crawford F, Axelman K, et al. (1992) A pathogenic mutation for probable Alzheimer's disease in the APP gene at the N-terminus of β-amyloid. Nat Genet 1:345–347

Multhaup G (1997) Amyloid precursor protein, copper and Alzheimer's disease. Biomed Pharmacother 51:105–111

Murrell J, Farlow M, Ghetti B, Benson MD (1991) A mutation in the amyloid precursor protein associated with hereditary Alzheimer's disease. Science 254:97–99

Parvathy S, Hussain I, Karran EH, et al. (1998) Alzheimer´s amyloid precursor protein a-secretase is inhibited by hydroxamic acid-based zinc metalloprotease inhibitors: similarities to the angiotensin converting enzyme secretase. Biochemistry 37: 1680–1685

Peraus GC, Masters CL, Beyreuther K (1997) Late compartments of amyloid precursor protein transport in SY5Y cells are involved in beta-amyloid secretion. J Neurosci 17:7714–7724

Rawson RB, Zelenski NG, Nijhawan D, et al. (1997) Complementation cloning of S2P, a gene encoding a Putative metalloprotease required for intramembrane cleavage of SREBPs. Mol Cell 1:47–57

Ross SL, Martin F, Simonet L, et al. (1998) Amyloid precursor protein processing in sterol regulatory element-binding protein site 2 protease-deficient Chinese hamster ovary cells. J Biol Chem 273:15309–15312

Saftig P, Peters C, von Figure K, et al. (1996) Amyloidogenic processing of human amyloid precursor protein in hippocampal neurons devoid of cathepsin D. J Biol Chem 271:27241–27244

Sambamurti K, Shioi J, Anderson JP, et al. (1992) Evidence for intracellular cleavage of the Alzheimer's amyloid precursor in PC12 cells. J Neurosci Res 33:319–329

Sandbrink R, Masters CL, Beyreuther K (1994) Beta A4-amyloid protein precursor mRNA isoforms without exon 15 are ubiquitously expressed in rat tissues including brain, but not in neurons. J Biol Chem 269:1510–1517

Scheuner D, Eckman C, Jensen M, et al. (1996) Secreted amyloid β-protein similar to that in the senile plaques of Alzheimer's disease is increased in vivo by the presenilin 1 and 2 and APP mutations linked to familial Alzheimer's disease. Nat Med 2:864–870

Schroeter EH, Kisslinger JA, Kopan R (1998) Notch-1 signalling requires ligand-induced proteolytic release of intracellular domain. Nature 393:382–386

Selkoe DJ (1994) Cell biology of the amyloid β-protein precursor and the mechanism of Alzheimer's disease. Ann Rev Cell Dev Biol 10:373–403

Seubert P, Vigo-Pelfrey C, Esch F, et al. (1992) Isolation and quantification of soluble Alzheimer's beta-peptide from biological fluids. Nature 359:325–327

Shivers BD, Hilbich C, Multhaup G, et al. (1988) Alzheimer's disease amyloidogenic glycoprotein: expression pattern in rat brain suggests a role in cell contact. EMBO J 7:1365–1370

Shoji M, Golde TE, Ghiso J, et al. (1992) Production of the Alzheimer amyloid beta protein by normal proteolytic processing. Science 258:126–129

Simons M, Destrooper B, Multhaup G, et al. (1996) Amyloidogenic processing of the human amyloid precursor protein in primary cultures of rat hippocampal neurons. J Neurosci 16:899–908

Sisodia SS (1992) Beta-amyloid precursor protein cleavage by a membrane-bound protease. Proc Natl Acad Sci USA 89:6075–6079

Strittmatter WJ, Roses AD (1996) Apolipoprotein E and Alzheimer's disease. Annu Rev Neurosci 19:53–77

Suzuki N, Cheung TT, Cai XD, et al. (1994) An increased percentage of long amyloid β protein secreted by familial amyloid β protein precursor (βAPP717) mutants. Science 264:1336–1340

Tienari PJ, Ida N, Ikonen E, et al. (1997) Intracellular and secreted Alzheimer beta-amyloid species are generated by distinct mechanisms in cultured hippocampal neurons. Proc Natl Acad Sci USA 94:4125–4130

Tischer E, Cordell B (1996) Beta-amyloid precursor protein. Location of transmembrane domain and specificity of gamma-secretase cleavage. J Biol Chem 271: 21914–21919

Van Broeckhoven C, Haan J, Bakker E, et al. (1990) Amyloid β protein precursor gene and hereditary cerebral hemorrhage with amyloidosis (Dutch). Science 248:1120–1122

Wang R, Sweeney D, Gandy SE, Sisodia SS (1996) The profile of soluble amyloid beta protein in cultured cell media. J Biol Chem 271:31894–31902

Weidemann A, König G, Bunke D, et al. (1989) Identification, biogenesis, and localization of precursors of Alzheimer's disease A4 amyloid protein. Cell 57:115–126

Wild-Bode C, Yamazaki T, Capell A, et al. (1997) Intracellular generation and accumulation of amyloid beta-peptide terminating at amino acid 42. J Biol Chem 272:16085–16088

Wisniewski T, Ghiso J, Frangione B (1991) Peptides homologous to the amyloid protein of Alzheimer's disease containing a glutamine for glutamic acid substitution have accelerated amyloid fibril formation. Biochem Biophys Res Commun 179:1247–1254

Wolfe MS, Citron M, Diehl TS, et al. (1998) A substrate-based difluoro ketone selectively inhibits Alzheimer's γ-secretase activity. J Med Chem 41:6–9

Yamazaki T, Haass C, Saido TC, et al. (1997) Specific increase in amyloid beta-protein 42 secretion ratio by calpain inhibition. Biochemistry 36:8377–8383

Yankner BA (1996) Mechanisms of neuronal degeneration in Alzheimer's disease. Neuron 16:921–932

Zhang H, Komano H, Fuller RS, et al. (1994) Proteolytic processing and secretion of human beta-amyloid precursor protein in yeast. Evidence for a yeast secretase activity. J Biol Chem 269:27799–27802

Presenilins and β-Amyloid Precursor Protein-Proteolytically Processed Proteins Involved in the Generation of Alzheimer's Amyloid β Peptide

C. HAASS

A. Introduction

Alzheimer's disease (AD) is the most common dementia worldwide. In most of the cases, AD occurs sporadically, with an increased risk during aging. In some cases, however, mutations in three genes (see below) were found which caused early-onset familial Alzheimer's diseases (FAD; for review, see SELKOE 1996). Furthermore, a polymorphism in the apoE allele is a major risk factor, that can significantly increase the chance of late-onset AD (ROSES 1996). A pathological hallmark of AD is the invariant accumulation of numerous senile plaques in certain areas of the brain. Senile plaques are composed of the amyloid β-peptide (Aβ), a proteolytic derivative of the β-amyloid precursor protein (βAPP; HAASS and SELKOE 1993).

B. Proteolytic Generation of the Amyloid β-Peptide

The primary structure of βAPP closely resembles a cell-surface receptor with a signal sequence, a large extramembranous N-terminal region, a single trans-membrane (TM) domain, and a small cytoplasmic C-terminal tail (KANG et al. 1987). Aβ represents only a small fragment of βAPP, and proteolytic processing of the precursor is required for the formation of this peptide. The last 11–15 amino acids of Aβ are located within the TM domain (Fig. 1). This indicates that the C-terminus of Aβ is protected from proteolytic processing, since proteases do not cleave within a hydrophobic environment such as the phospholipid bilayer. Moreover, Aβ generation is inhibited, since normal secretory processing of βAPP (WEIDEMANN et al. 1989) results in a cleavage of the precursor at amino acid 16 within Aβ (ESCH et al. 1990). Based on these data, it has been widely assumed that only aberrant processing under pathological conditions could lead to the formation of Aβ and its release from the cell membrane. For description of such proteolytic processing see the preceding chap. 19, briefly: The cut at amino acid 16 is made by an enzyme designated α-secretase. α-Secretase cleavage leads to the secretion of the large soluble ectodomain of βAPP (APP$_s$) and the retention of the small 10-kDa C-terminal fragment within the membrane (Fig. 1). The α-secretase-mediated cleavage of βAPP can occur at the cell surface or intracellularly (HAASS et al.

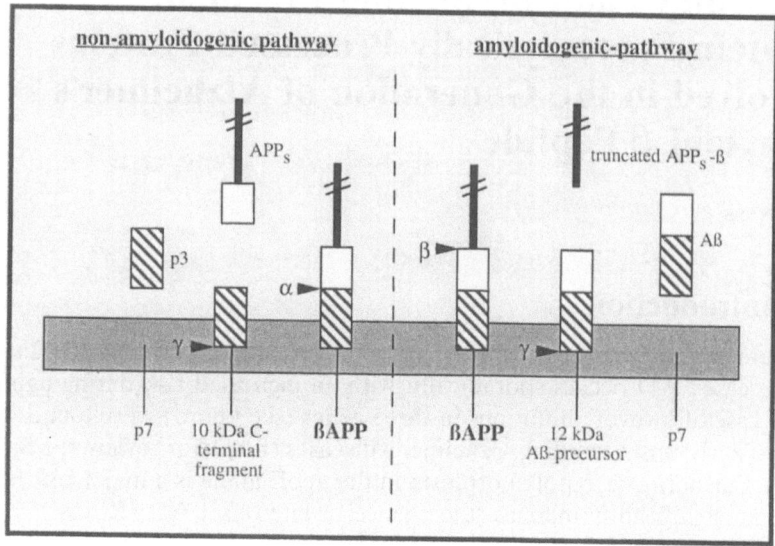

Fig. 1. Proteolytic processing of β amyloid precursor protein by α, β or γ-secretase, depending on the pathway

1993). α-Secretase cleavage is stimulated by the activation of protein kinase C (PKC) and PKA (Nitsch et al. 1992). However, βAPP is phosphorylated by casein-kinase-like enzymes on its ectodomain at amino-acid residues 198 and 206 (Hung and Selkoe 1994; Walter et al. 1997). It therefore appears that PKC/PKA could stimulate α-secretase directly (without phosphorylating βAPP). Interestingly, proteases of the a-disintegrin-and-metalloprotease family exhibit very similar characteristics, such as α-secretase cleavage. These proteases are located on the cell surface and within the trans-Golgi network, and their activity is also increased by stimulation of PKC/PKA. Recent evidence indeed indicates that α-secretase is a member of this metalloprotease family (Arribas and Massague 1995; Lammich et al. 1999).

I. Endosomal/Lysosomal Processing Generates Amyloidogenic Precursors

C-terminal fragments of βAPP were identified in cultured cells (Golde et al. 1992; Haass et al. 1992a) or brain tissue (Estus et al. 1992) that contained the complete Aβ sequence and could thus serve as potential degradative intermediates for Aβ formation. These fragments were found to be stabilized by leupeptin, ammonium chloride or chloroquine, agents known to inhibit endosomal/lysosomal proteases. These results led to the proposal that βAPP could also be processed in an endosomal/lysosomal pathway. However, it is not clear whether some of these fragments may initially be made within the Golgi or at the cell surface and then accumulate within the lysosome, where they could be subjected to further processing. Surface biotinylation and antibody-binding experiments on living cells revealed that full-length βAPP can be reinternalized from the cell surface in an coated-pit-mediated pathway (HAASS et al. 1992a). Indeed, full-length βAPP and the 10-kDa C-terminal fragment have been found within isolated clathrin-coated vesicles (NORDSTEDT et al. 1993). Furthermore, isolation of late endosomes/lysosomes from leupeptin-treated cells directly demonstrated that full-length βAPP, the 10-kDa C-terminal fragment and a range of larger C-terminal fragments of βAPP containing the intact $A\beta$ sequence accumulate within lysosomes (HAASS et al. 1992a). Taken together, these observations indicate that some full-length βAPP molecules, together with the 10-kDa and probably other C-terminal fragments, are reinternalized from the cell surface and targeted to lysosomes for final degradation. In addition to this reinternalization pathway, a pathway which targets βAPP from the trans-Golgi network directly to endosomes and lysosomes could exist.

II. $A\beta$ Is Produced by a Physiological Processing Pathway

The description of a normal processing pathway for βAPP that generates $A\beta$-bearing fragments under physiological conditions suggested that aberrant processing of βAPP might not be necessary to generate $A\beta$. This concept has now been validated by the discovery that $A\beta$ is normally secreted into the media of a wide array of cultured cells which express βAPP (HAASS et al. 1992b; SEUBERT et al. 1992; SHOJI et al. 1992; BUSCIGLIO et al. 1993). In such conditioned media, peptides of mass 4 kDa ($A\beta$) could be isolated by immunoprecipitation with $A\beta$ specific antibodies (HAASS et al. 1992b; SHOJI et al. 1992; BUSCIGLIO et al. 1993). Detailed biochemical analysis revealed that $A\beta$ secreted from cultured cells is identical to $A\beta$ isolated from AD-afflicted brain. Based on these results, two additional secretases must be postulated: β-secretase cleaving on the N-terminus and γ-secretase cleaving on the C-terminus of the $A\beta$ domain (Fig. 1) (see also preceding chap.19). In addition to the secreted $A\beta$, a 3-kDa peptide (p3) was found to be precipitated by $A\beta$ antibodies (HAASS et al. 1992b; HAASS et al. 1993). p3 is generated by the combined action of the α-secretase (which generates the 10-kDa fragment; Fig. 1) and the γ-secretase (Fig. 1). A detailed description of the three secretase activities is in the preceding chapter of this volume. Importantly, $A\beta$ has been detected not only in the supernatants of cultured cells but also in normal body fluids, e.g., in human cerebrospinal fluid and plasma (SEUBERT et al. 1992; SHOJI et al. 1992).

III. FAD-Linked Mutations in the βAPP Gene
Affect $A\beta$ Generation

In about 10–15% of cases, AD is caused by famial autosomal dominant (FAD) mutations within three genes (SELKOE 1996). A very limited set of families was

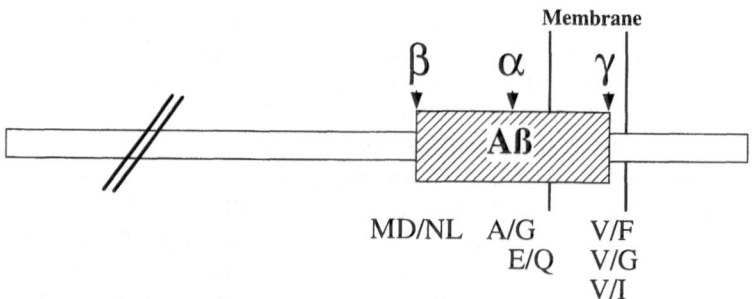

Fig. 2. Familial-Alzheimer's-disease-associated mutations are located close to the three secretase sites and affect Aβ generation

identified carrying mutations in the βAPP gene (Selkoe 1996). These mutations occur at or close to the cleavage sites of the βAPP-processing enzymes (Fig. 2), the secretases (Haass and Selkoe 1993). βAPP mutations located close to the β- and γ-secretase sites cause an enhanced production of Aβ, specifically the highly pathogenic 42-amino-acid variant, Aβ42 (Fig. 2; Selkoe 1996). The longer form of Aβ plays a central role in AD. Aβ42 is a major component of amyloid plaques (Selkoe 1996), which are the pathological hallmark of the disease (Selkoe 1996). Furthermore, Aβ42 exhibits enhanced neurotoxicity, which might be due to its increased ability to form insoluble fibers (Burdick et al. 1992; Jarret et al. 1993). Therefore, these findings strongly support a central role of Aβ for the pathogenesis of AD.

C. Role of Mutant Presenilins in Amyloid Generation

Mutations in the βAPP gene are extremely rare. The majority of mutations were linked to a gene on chromosome 14 that was recently cloned and designated presenilin (PS)-1 (Sherrington et al. 1995). A second gene (PS2), located on chromosome 1, was cloned shortly thereafter (Levy-Lahad et al. 1995; Rogaev et al. 1995) and shown to be responsible for some FAD cases as well. The integral membrane proteins encoded by PS1 and PS2 genes are highly homologous (Levy-Lahad et al. 1995; Rogaev et al. 1995). The identity of the proteins is approximately 63%; within the TM domains, the homology can reach 95% (Levy-Lahad et al. 1995; Rogaev et al. 1995). Mutations within the PS1 gene are responsible for the most aggressive form of FAD recorded and can cause typical AD pathology as early as 35 years of age (reviewed by Tanzi et al. 1996). Those investigators who support the "amyloid hypothesis" expected mutant PS genes to alter proteolytic processing of βAPP. This has indeed turned out to be the case. It was found that plasma and conditioned media of primary fibroblasts of patients with chromosome-14-associated FAD contained significantly elevated levels of Aβ42 (Scheuner et al. 1996). Similar results were obtained by measuring the Aβ42 concentrations in conditioned media of cells transfected with complementary DNAs encoding FAD-associated PS mutations (BORCHELT et al. 1996; XIA et al. 1996; CITRON et al. 1997; TOMITA et al. 1997). Therefore, all PS1 and PS2 mutations analyzed have been found to significantly increase Aβ42 levels but not Aβ40 levels. These results were also confirmed in transgenic mice overexpressing PS1 mutations (BORCHELT et al. 1996; DUFF et al. 1996; CITRON et al. 1997). These findings therefore provide very strong evidence for a direct link between altered proteolytic processing of βAPP and PS mutations that cause early-onset FAD. However, the levels of Aβ42 secreted into culture media do not correlate with the age of onset of clinical symptoms caused by individual mutations. A rather trivial explanation for this apparent discrepancy may be the presence of genetic modifiers in individual patients. Nevertheless, it is still very striking that mutations in both PS genes change βAPP metabolism in a manner similar to mutations within the βAPP gene, thus making it very likely that the increased production of Aβ42 is indeed a cause of early-onset FAD.

I. Structure and Topology of PS Proteins

The topology of PS1 (Fig. 3) was recently determined by utilizing two strategies. First, putative TM domains were used to determine whether they were able to export a protease-sensitive substrate (DOAN et al. 1996). Second, cells permeabilized with the pore-forming toxin streptolysine O were probed with epitope-specific antibodies (DOAN et al. 1996; DE STROOPER et al. 1997). These experiments suggest that PS1 probably contains six or eight TM domains, with the N'- and C'-terminal domains and the large hydrophilic loop between TM6 and TM7 oriented towards the cytoplasm (Fig. 3). LI and GREENWALD (1996) carried out very elegant experiments to determine the topology of the PS-homologous protein of *Caenorhabditis elegans* (SEL-12; LEVITAN and GREENWALD 1995) by using SEL-12 βGal fusion proteins. The membrane orientation of SEL-12 appears to be very similar to PS1 (LI and GREENWALD 1996), indicating structural as well as functional conservation of PS proteins during evolution. Although the topology of PS2 has not yet been determined, it is very likely that PS2 adopts a structure similar to that of PS1 and SEL-12 in view of its extensive sequence homologies with these proteins.

All FAD-associated mutations found so far occur at conserved positions in both human PS proteins, indicating that the mutations affect functionally or structurally important amino acids within the presenilins. Moreover, these amino acids are conserved within SEL-12 as well (LEVITAN and GREENWALD 1995). Numerous mutations have been found within the PS1 gene, but only two mutations have been localized to the PS2 gene (Fig. 3; reviewed by TANZI et al. 1996). The mutations are predominantly located within TM domains (specifically TM2) and in the large hydrophilic loop (Fig. 3). All except one are missense mutations. The one exceptional mutation causes a splicing error, that results in the elimination of exon 10 (originally called exon 9; reviewed by TANZI et al. 1996). Interestingly, no mutations have been identified that

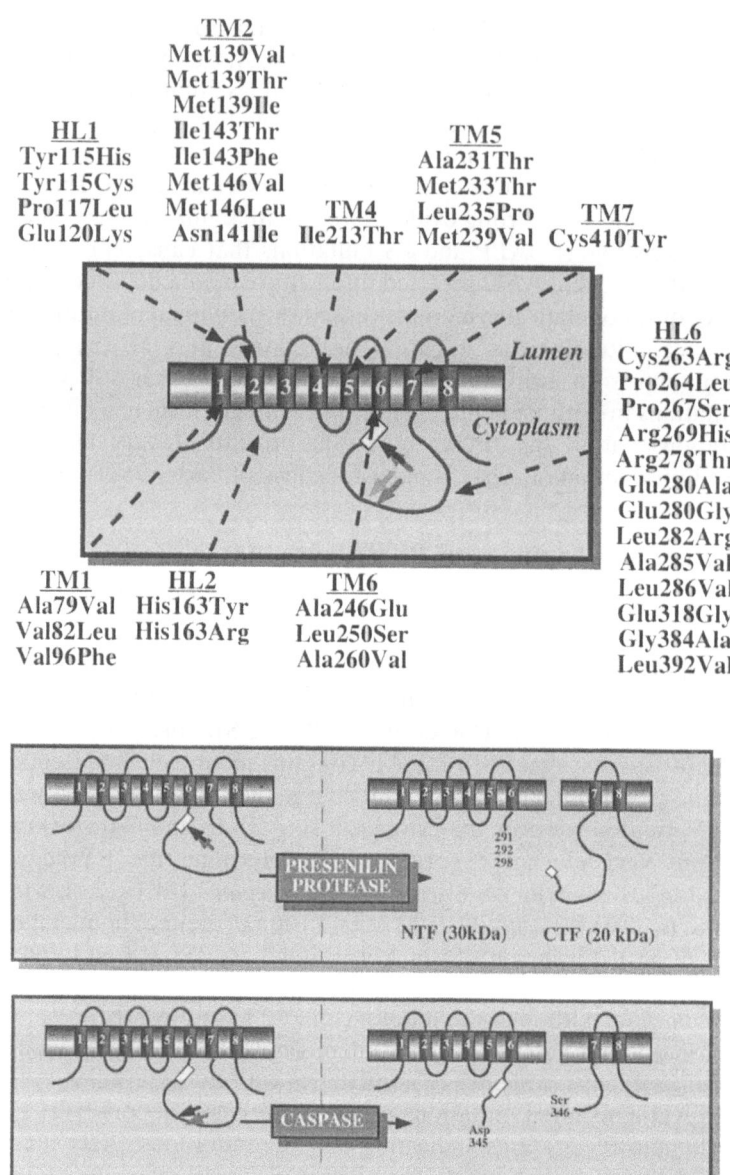

TM2
Met139Val
Met139Thr
Met139Ile
HL1   Ile143Thr   TM5
Tyr115His Ile143Phe   Ala231Thr
Tyr115Cys Met146Val   Met233Thr
Pro117Leu Met146Leu TM4   Leu235Pro TM7
Glu120Lys Asn141Ile Ile213Thr Met239Val Cys410Tyr

HL6
Cys263Arg
Pro264Leu
Pro267Ser
Arg269His
Arg278Thr
Glu280Ala
Glu280Gly
Leu282Arg
Ala285Val
Leu286Val
Glu318Gly
Gly384Ala
Leu392Val

Lumen
Cytoplasm

TM1   HL2   TM6
Ala79Val   His163Tyr Ala246Glu
Val82Leu   His163Arg Leu250Ser
Val96Phe   Ala260Val

PRESENILIN PROTEASE

NTF (30kDa)   CTF (20 kDa)

CASPASE

NTF_H   CTF_10

Fig. 3. The topology of presenilins (PSs) and location of familial-Alzheimer's-disease-associated mutations (*upper panel*). Proteolytic processing by the PS protease and by caspases (*lower panel*)

cause premature termination of the protein or a frame shift, suggesting that severe structural changes in the PS proteins cannot be tolerated.

D. Conventional Proteolytic Processing of PS Proteins

Shortly after the PS genes were cloned, a very surprising observation was made. When PS1 expression in cultured cells was analyzed, the expected protein of mass 45–50 kDa could not be found. Instead, smaller peptides of masses ~30 kDa and ~18–20 kDa were observed (THINAKARAN et al. 1996, MERCKEN et al. 1996). The 30-kDa fragment [N-terminal fragment (NTF)$_{30}$] was found to lack C-terminal epitopes, whereas the ~18–20-kDa peptide [C-terminal fragment (CTF)$_{18-20}$] exhibited no reactivity with antibodies raised to N-terminal epitopes (Fig. 3). This result led to the conclusion that PS1 undergoes proteolytic processing by a cleavage most likely within the large cytoplasmic loop. Very similar results were obtained for PS2 as well (TOMITA et al. 1997), indicating that proteolytic processing is a conserved feature of the two homologous PS proteins. A detailed analysis of PS1-protein expression in the human brain revealed an 80-fold excess of PS1 fragments over the holoprotein (MORI, pers. communication). These findings certainly raise the question of whether the fragments themselves are biologically active, since they are by far the predominant PS molecules in vivo (see below).

I. Identification of the Cleavage Site

A first idea about the location of the cleavage site came from the analysis of a naturally occurring FAD mutation (Fig. 3) which, due to a splicing error, deletes exon 9 (PEREZ-TUR et al. 1995). Analysis of this mutation in fibroblasts derived from patients as well as transfected cells and transgenic animals showed an accumulation of unclipped PS1 $\Delta 9$ protein (THINAKARAN et al. 1996; LEE et al. 1997). Since fragment formation was no longer observed (THINAKARAN et al. 1996; LEE et al. 1997), these data indicated that the domain encoded by exon 9 harbors the cleavage site. The observation that the cleavage occurs within the exon-9-encoded domain was strongly supported by the direct radiosequencing of CTF$_{18-20}$. This work revealed heterogeneous cleavages after amino acids 291, 292, and 298 of PS1 (PODLISNY et al. 1997). Very similar data were obtained recently for PS2 as well (SHIROTANI et al. 1997).

The proteolytic cleavage of PS proteins at amino acids 291,292, and 298 appears to be regulated. In primary hippocampal neurons and brain tissue, an alternative cleavage occurs more C-terminal to the conventional cleavage site generating a higher-molecular-weight NTF as well as lower-molecular-weight CTFs of PS1 (HARTMANN et al. 1997; CAPELL et al. 1997) and PS2 (CAPELL et al. 1997). Moreover, this apparent alternative cleavage is developmentally regulated. During neuronal differentiation of primary rat hippocampal neurons, an increase of the alternative cleavage was observed (HARTMANN et al. 1997;

Capell et al. 1997). Similar data were also obtained for human brain tissue (Hartmann et al. 1997). These results might, therefore, indicate that proteolytic processing of PS proteins plays a role in neuronal differentiation, a hypothesis that might be supported by the functional analysis of PS proteins described below.

II. Regulation of Fragment Formation

Overexpression in transfected cells (Thinakaran et al. 1996; Podlisny et al. 1997) and transgenic animals (Thinakaran et al. 1996; Lee et al. 1997) led again to a very surprising result. Although the PS holoprotein was observed, the level of unclipped PS protein did not correlate with the amount of the corresponding messenger RNA (Thinakaran et al. 1996; Lee et al. 1997). Moreover, very little increase of fragment formation was observed (Thinakaran et al. 1996). The latter was explained by the finding that proteolytic fragments of overexpressed PS1 can replace the endogenous fragments (Thinakaran et al. 1996). This is most obviously observed in cells and tissue derived from mice. It appears that the human CTF has a slower electrophoretic mobility than the mouse fragment. The opposite is observed for the NTF. When human PS1 is overexpressed in mice, a decrease of the endogenous fragments is observed, which is compensated by a corresponding increase of the human fragments [as judged by the changes in the molecular weight of human vs mouse PS (Thinakaran et al. 1996)]. Therefore, it appears that a highly regulated mechanism controls the amount of PS fragments generated. However, it is still unclear what happens to the endogenous fragments. Are they produced and subsequently degraded or is there a translational control mechanism blocking de novo synthesis? Work on the proteolytic degradation of PS proteins described below might favor the first possibility. In that regard, it is also interesting to note that expression of PS1 Δ9 results in a substantial decrease of the endogenous fragments, indicating, again, a highly regulated mechanism controlling the levels of PS accumulation (Thinakaran et al. 1996, Lee et al. 1997; Walter et al. 1997). One might therefore speculate that a controlled degradation mechanism is involved in the removal of excess PS fragments, allowing only the accumulation of a very limited number of fragments (see below).

III. Effects of PS Mutations on Fragment Formation

After the identification of specific processing products of PS proteins, the question of whether FAD-associated mutations within the PS proteins influence their proteolytic cleavage was raised. A number of recent publications demonstrated conflicting results. The only consistent observation was made for the PS1 Δ-exon-9 mutation. Based on several recent reports (Thinakaran et al. 1996; Citron et al. 1997; Lee et al. 1997; Walter et al. 1997), this mutation inhibits PS1 processing in transgenic animals and all cell lines analyzed so far.

Table 1. The effect of FAD-associated PS-1 mutations on proteolytic processing

Mutation	Tissue/cell line	Effect on processing	Reference
Cells			
Gly384Ala	PC12, C6	No effect	24
Leu 392Val	PC12, C6	No effect	24
Cys410Tyr	PC12, C6	Inhibition	24
Met146Val	PC12, C6	Inhibition	16
Ala246Glu	PC12, C6	Inhibition	16
Δexon10	COS, lymphoblasts	Inhibition	17, 20, 23
Transgenic mice			
Ala246Glu	Transgenic mice (brain)	40–50% Increase	20
Met146Leu	Transgenic mice (brain)	40–50% Increase	20
Δexon10	Transgenic mice (brain)	Inhibition	17, 20
Human brain			
Cys410Tyr	FAD brain	No effect	18
Ile143Thre	FAD brain	No effect	25
Gly384Ala	FAD brain	No effect	25

The analysis of point mutations revealed highly controversial results. Early experiments by MERCKEN et al. (1996) indicated that the PS1 mutations Met146Val and Ala246Glu inhibit proteolytic cleavage. Further work by the same group (MURAYAMA et al. 1997) extended these findings recently by identifying a third mutation (Cys410Tyr), which again inhibited proteolytic processing. However, besides these three mutations, which inhibit PS1 processing, two other mutations were found (Glu348Ala and Leu392Val; Table 1), which had no obvious effect on PS1 processing (MURAYAMA et al. 1997).

LEE and co-workers (1997) found that fragments of mutant PS1 proteins in brains of transgenic animals accumulate to a higher degree (40–50%) than the fragments derived from wild-type PS1 (Table 1). Interestingly, the authors demonstrated that the NTF and the CTF accumulated in parallel, although both mutations analyzed were located within the NTF. This might either indicate that the mutations directly influence the proteolytic cleavage or that both proteolytic fragments accumulate in a stoichiometrically regulated ratio. The latter appears to be more likely, since many point mutations occur far away from the cleavage site (for reviews see VAN BROECKHOVEN 1995; CRUTS et al. 1996; HAASS 1996; TANZI et al. 1996; HAASS 1997; HARDY et al. 1997), therefore probably not influencing the rate of cleavage. Very little is known about PS fragments in the brains of FAD patients. Two recent reports, however, demonstrate no significant difference in fragment accumulation for three different mutations (HENDRIKS et al. 1997; PODLISNY et al. 1997).

Taken together, it appears to be rather difficult to come to a final conclusion. The same mutation (Cys410Tyr) appears not to effect proteolytic processing in a FAD brains (PODLISNY et al. 1997, HENDRIKS et al. 1997); however, upon transfection into PC12 cells, this mutation does inhibit processing

(Murayama et al. 1997). It might, therefore, be possible that tissue-specific and/or cell-type-specific differences occur that could be responsible for the observed controversy.

E. Proteolytic Degradation of PSs

I. PS Holoproteins Are Degraded by the Proteasome

The PS holoproteins are rapidly removed by proteolytic degradation in addition to proteolytic processing. The ubiquitin–proteasome pathway is known to play a major role in the selective degradation of proteins. Proteasomal activity can be blocked with specific inhibitors like lactacystin and others (Ward et al. 1995). By treating cells expressing PS1 or PS2 with proteasome inhibitors, high-molecular-weight, polyubiquitinated PS molecules were identified (Kim et al. 1997a; Marambaud et al. 1998; Steiner et al. 1998). Accumulation of polyubiquitinated PS proteins, therefore, strongly indicates that PS proteins can be degraded in a proteasome-dependent pathway. Proteasomal degradation of membrane-bound proteins is a well-known phenomenon. The proteasome is known to degrade not only cytoplasmic proteins but also proteins located within the membrane of the endoplasmic reticulum (ER; Hochstrasser 1996). The latter is the compartment in which PS proteins were localized previously (Cook et al. 1996; De Strooper et al. 1997; Kovacs et al. 1996; Walter et al. 1996). Therefore, ER-located PS proteins might be another proteasome substrate.

II. PSs Are Death Substrates for Caspases

PS proteins not only undergo the conventional processing pathway described above but are also cleaved in an alternative pathway (Kim et al. 1997b; Loetscher et al. 1997; Grünberg et al. 1998). The identification of a second proteolytic pathway was initiated by the observation that overexpression of PS1 or PS2 can result in the formation of much smaller C-terminal fragments than the previously described CTF_{18-20}. Moreover, two recent reports revealed alternative cleavages in neuronal cells and brain tissue, which might be due to a similar cleavage event (Capell et al. 1997; Hartmann et al. 1997). The proteases involved in this alternative pathway indicate that apoptosis might play a role in alternative cleavage. Activation of apoptosis is associated with neurodegenerative diseases, including AD (Thompson 1995). Moreover, very recently it has been demonstrated that expression of PS proteins might also induce apoptosis (Wolozin et al. 1996; Vito et al. 1996a, b). During apoptosis, cysteine proteases of the caspase family are activated and cleave death substrates such as poly(adenosine diphosphate)ribose polymerase, rho-guanine-nucleotide-dissociation inhibitor, sterol-regulatory-element-binding protein (SREBP), and DNA-dependent protein kinase (Jacobson et al. 1997).

Activation of these proteases is required for the induction of the cell-death program, which finally results in the typical morphological changes associated with apoptosis (JACOBSON et al. 1997) (see also chap. 17).

Based on these observations, PS-expressing cells were treated with selective caspase inhibitors during stimulation of apoptosis. Interestingly, caspase inhibitors completely blocked the generation of the smaller CTF (KIM et al. 1997b; LOETSCHER et al. 1997; GRÜNBERG et al. 1998), thus strongly indicating that these proteases are involved in the additional (alternative) cleavage of PS proteins. Caspases are known to selectively cleave their substrates after aspartates. Mutagenesis of the aspartate at the P1 position of the cleavage site results in the inhibition of proteolytic cleavage (JACOBSON et al. 1997; THORNBERRY et al. 1992; THORNBERRY et al. 1994). When the aspartate at position 345 of PS1 was mutagenized to asparagine, caspase-mediated cleavage was efficiently blocked (GRÜNBERG et al. 1998). Very similar results were obtained for PS2 (LOETSCHER et al. 1997; KIM et al. 1997b), where mutagenesis of aspartates 326 and 329 blocked the caspase cleavage. These results were further confirmed by direct sequencing of the N-termini of the caspase-generated CTFs of PS1 and PS2 (LOETSCHER et al. 1997).

Taken together, these results strongly indicate that stimulation of caspases results in an alternative cleavage of PS1 and PS2, most likely by one of the proteases of the caspase superfamily (Fig. 3). Therefore PS1 and PS2 represent novel death substrates for caspases.

Certainly the question arises whether full-length PS or the conventional CTF_{18-20} is the death substrate, particularly because very little holoprotein but abundant CTF_{18-20} can be detected in vivo. Expression of PS1 Δ-exon-9 still allows the caspase cleavage, suggesting that the holoprotein could be the death substrate (GRÜNBERG et al. 1998). However, recent experiments have demonstrated that the conventional CTF_{18-20} is the major in vivo substrate for caspase-mediated cleavage. In these experiments, untransfected cells were treated with cycloheximide to inhibit de novo protein synthesis. After stimulation of apoptosis, the levels of the CTF_{18-20} were decreased, whereas the levels of the alternative fragment were strongly increased, as in cells not treated with inhibitors of protein synthesis. This precursor–product relationship clearly demonstrates that CTF_{18-20} is the predominant in vivo substrate for caspase-mediated cleavage (GRÜNBERG et al. 1998), which is consistent with the finding that fragments are by far the predominant PS molecules in vivo.

What does that mean for the disease? In the case of PS2, the work by the TANZI and co-workers (KIM et al. 1997b) has demonstrated that a FAD-causing mutation can influence caspase cleavage. Cells transfected with the Asn141Ile mutation of PS2 produced elevated levels of the alternative CTF, thus providing a possible link between caspase-mediated PS cleavage and early-onset AD. Since the alternative cleavage occurs within 3 h after stimulation of apoptosis (KIM et al. 1997b), alternative cleavage precedes the effects of the FAD mutation on Aβ generation. Moreover, the Asn141Ile mutation increases the levels of both the alternative CTF and Aβ42 by a factor of 3–4. Therefore, KIM et al. (1997b) suggested that caspase-mediated cleavage and Aβ42 generation might be intimately related. However, recent experiments demonstrated that caspase-mediated cleavage is not a prerequisite for Aβ production (Brockhaus et al. 1998). In these experiments, the critical aspartates were mutagenized to asparagin. This artificial mutation, which blocks caspase cleavage, was combined with a FAD-associated PS2 mutation. Although caspase-mediated cleavage was completely blocked, the production of Aβ40 and the increased generation of Aβ42 (caused by the FAD-associated PS mutation) were not affected.

Therefore, it appears that the PS proteins are death substrates for proteases of the caspase superfamily, but caspases are not involved in Aβ generation. Interestingly, the caspase-generated PS fragments are rapidly degraded not by the proteasome, but by a cysteine-protease activity (Steiner et al. 1998).

III. A Heterodimeric PS Complex Appears to Be Required for PS Stability and Aβ42 Generation

It was found recently that the CTF and NTF of PS1 and PS2 bind to each other (Capell et al. 1998; Thinakaran et al. 1998; Yu et al. 1998). Fractionating proteins from 3-[(3-cholamidopropyl)dimethylammonio]-1-propane-sulfonate-extracted membrane preparations by velocity sedimentation revealed a high-molecular-weight sodium-dodecyl-sulfate- and Triton X-100-sensitive complex of approximate mass 100–150 kDa (Capell et al. 1998; Yu et al. 1998). To prove whether both proteolytic fragments of PS1 are bound to the same complex, co-immunoprecipitations were performed. These experiments revealed that both fragments of PS occur as a tightly bound non-covalent complex (Fig. 4; Capell et al. 1998; Thinakaran et al. 1998; Yu et al. 1998). Interestingly, PS1 fragments were not found to bind to PS2 and vice versa (Thinakaran et al. 1998; Yu et al. 1998). Formation of a PS complex may explain the recent finding that FAD-associated mutations within the N-terminal portion of PS1 result in the hyperaccumulation not only of the NTF but also of the CTF (Lee et al. 1997). Moreover, these results provide a model that can be used to understand the highly regulated expression and processing of PS proteins (Thinakaran et al. 1996). In both cases, one could argue that a rate-limiting binding protein is required for the stability of the PS complex. Overexpression of one of the PS proteins might then compete for the proteins bound to the corresponding homologous PS protein. PS proteins that lost their binding partner due to competition by an overexpressed PS molecule might then be rapidly degraded (see below).

Since very small amounts of the PS holoprotein are detected in vivo, whereas the fragments accumulate to rather high levels (Thinakaran et al. 1996), it was suggested that the PS complex composed of the two fragments might represent the biologically active unit of PS (Grünberg et al. 1997). In order to answer this question, recombinant fragments (containing a stop codon

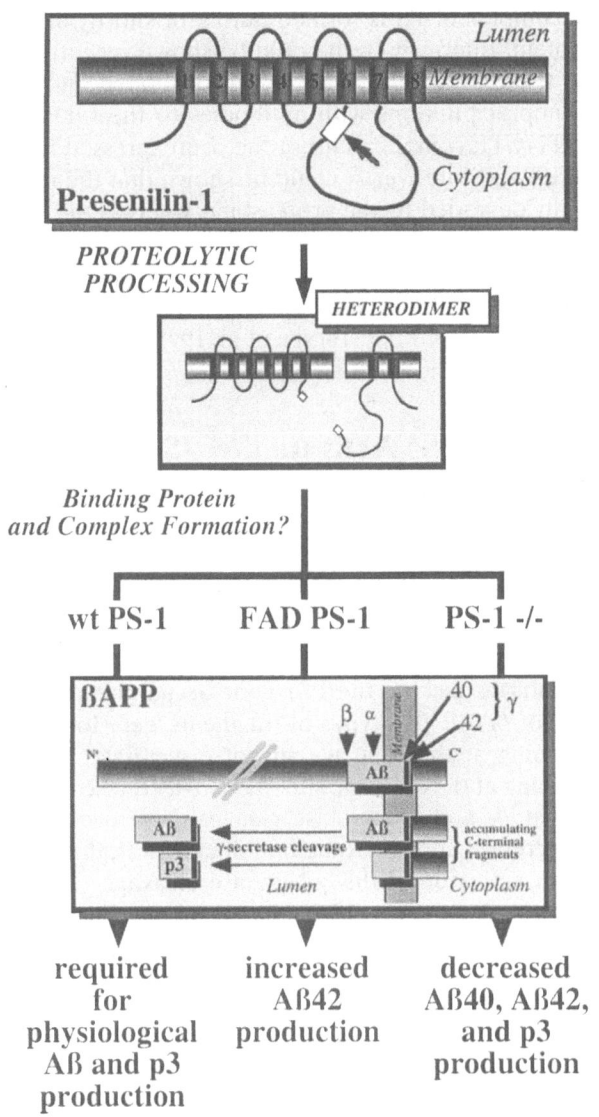

Fig. 4. The role of presenilins (and the heterodimeric complex) in Aβ generation. For details, see text

at the cleavage site), representing the PS1 or PS2 NTF itself, were expressed with and without a FAD-associated mutation. Interestingly, it was found that mutant fragments lost their ability to overproduce Aβ 42 (CITRON et al. 1997; STEINER et al. 1998; TOMITA et al. 1998). Moreover, co-expression of a recombinant mutant NTF and a recombinant CTF also did not result in an enhanced production of Aβ42 (TOMITA et al. 1998). Therefore, it appears that the heterodimeric complex of PS is formed during or shortly after translation of the PS holoprotein. Interestingly, it could be shown recently that the recombinant fragments described above are not incorporated into the PS complex, since co-immunoprecipitations with antibodies to the CTF precipitated the endogenous NTF/CTF complex but not the overexpressed recombinant NTF (STEINER et al. 1998). Moreover, it could be shown that the recombinant fragments are rapidly degraded by the proteasome, whereas the endogenous fragments are stable over very long time periods (up to 24h and longer; STEINER et al. 1998). Therefore, fragments not incorporated into the PS complex are highly unstable and lose their pathological activity even when they are stabilized by proteasome inhibitors (STEINER et al. 1998).

F. Evidence That PSs Activate the γ-Secretase Cleavage

PS1 appears to be directly involved in the γ-secretase cleavage. This became first evident by the analysis of the pathological effect of PS mutations on Aβ42 generation (see above). Moreover, mice which lack the PS1 gene show a significantly reduced production of Aβ40 and Aβ42 (DE STROOPER et al. 1998). This finding suggests that PS1 plays a central role in the general mechanism of Aβ generation. DE STROOPER et al. (1998) also found that cultured neurons derived from embryos lacking the PS1 gene accumulated C-terminal proteolytic fragments of βAPP. Two types of fragments were found: (1) C-terminal fragments beginning at the cleavage site of β-secretase and (2) C-terminal fragments beginning at the cleavage site of the α-secretase (Fig. 4). This finding demonstrates that β- and α-secretase cleavage still occurs very efficiently, whereas the γ-secretase appears to be inhibited. It was, therefore, argued that PS1 might be an activator of the γ-secretase cleavage (DE STROOPER et al. 1998). Interestingly, such a model is supported by the findings of BROWN and GOLDSTEIN (1997), who recently demonstrated a proteolytic mechanism involved in the activation of a transcription factor which is remarkably similar to the potential function of PS1 in the γ-secretase cleavage. SREBP is expressed as an inactive membrane-bound transcription factor, which, dependent on the cholesterol levels, can be activated by proteolytic release from the membrane (Fig. 5). The cleavage occurs in two steps: cleavage 1 takes place in the lumen of the ER and cleavage 2 within the membrane. Cleavage 1 is activated by the SREBP-cleavage-activating protein (SCAP; BROWN and GOLDSTEIN 1997), which has a topology very similar to that of the PS proteins. Moreover, PS proteins are also located within the ER (see above) and the γ-secretase cleavage occurs within the membrane, similar to cleavage 2.

Furthermore, very recently it was demonstrated that the Notch receptor also requires proteolytic release of its cytoplasmic domain from the membrane for its functional activity in cell-fate decisions (SCHROETER et al. 1998; STRUHL and ADACHI 1998). Interestingly, PS1 appears to be functionally involved in Notch signaling. Genetic evidence indicated that the PS gene (sel-12) of C.

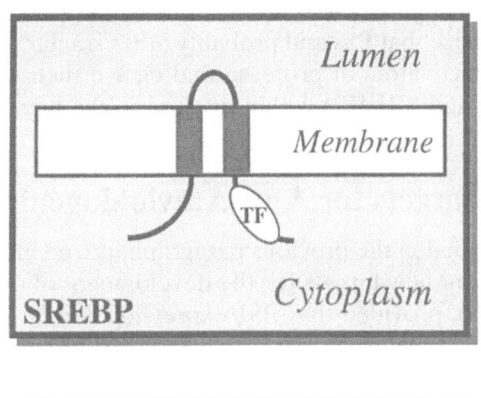

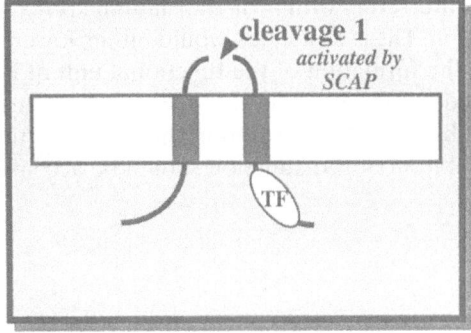

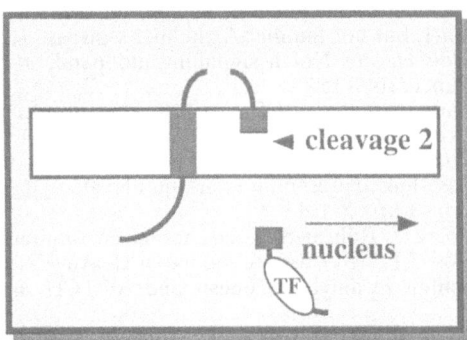

Fig. 5. Sterol-regulatory-element-binding protein (SREBP)-cleavage-activating protein-activated cleavage of SREBP

elegans facilitates Notch signaling (LEVITAN and GREENWALD 1995). Moreover, human PS1 and PS2 can functionally replace a mutant *sel*-12 gene in the worm (LEVITAN et al. 1996; BAUMEISTER et al. 1997), thus demonstrating that human PS genes are probably also involved in cell-fate decisions via the Notch-signaling pathway (for review see HAASS 1997). Furthermore, PS$^{-/-}$ mice display a fatal phenotype that closely resembles the phenotype caused by the deletion of the Notch gene (Shen et al. 1997; Wong et al. 1997). One could, therefore, hypothesize that PSs and probably other similar proteins like SCAP could function as activators of protease that cleave their substrates (such as βAPP, presenilins, and SREBP) within or close to the membrane.

G. PSs: New Targets for Anti-Amyloidogenic Drugs?

The findings described in the previous paragraph indeed indicate that PS proteins might represent new targets for the development of drugs, which inhibit amyloid production, provided that PS1 expression can be specifically reduce (Haass and Selkoe 1998). Expression of PS1 could be repressed at the transcriptional level. The biological function of PS proteins in Aβ generation might also be blocked by interfering with post-translational processing of PSs as well as complex formation. These strategies would either lower PS1 expression or they would inhibit the formation of the functional unit of PS1 (see above). In all cases, this should be accompanied by decreased production of Aβ40 and Aβ42. Since PS1 obviously has an important and essential function during embryogenesis, inhibition of PS1 function should be accomplished late during aging.

References

Arribas J, Massague J (1995) Transforming growth factor-alpha and beta-amyloid precursor protein share a secretory mechanism. J Cell Biol 128:433–441

Baumeister R, Leimer U, Zweckbronner I, Jakubek C, Grünberg J, Haass C (1997) Human presenilin-1, but not familial Alzheimer's disease (FAD) mutants, facilitate *Caenorhabditis elegans* Notch signalling independently of proteolytic processing. Genes Funct 1:149–159

Borchelt DR, Thinakaran G, Eckman CB, Lee MK, Davenport F, Ratovitsky T, Prada C-M, Kim G, Eekins S, Yager D, Slunt HH, Wang R, Seeger M, Levey AI, Gandy SE, Copeland NG, Jenkins NA, Price DL, Younkin SG, Sisodia SS (1996) Familial Alzheimer's disease-linked presenilin 1 variants elevate Aβ1–42/1–40 ratio in vitro and in vivo. Neuron 17:1005–1013

Brockhaus M, Grünberg J, Röhrig S, Loetscher H, Wittenburg N, Baumeister R, Jacobsen H, Haass C (1998) Caspase-mediated cleavage is not required for the activity of presenilins in amyloidogenesis and NOTCH signaling. Neuroreport 9:1481–1486

Brown MS, Goldstein JL (1997) The SREBP pathway: regulation of cholesterolmetabolism by proteolysis of a membrane-bound transcription factor. Cell 89:331–340

Burdick D, Soreghan B, Kwon M, Kosomoski J, Knauer M, Henschen A, Yates J, Cotman C, Glabe C (1992) Assembly and aggregation properties of synthetic Alzheimer's A4/beta amyloid peptide analogs. J Biol Chem 267:546–554

Busciglio J, Gabuzda DH, Matsudaira P, Yankner BA (1993) Generation of β-amyloid in the secretory pathway in neuronal and nonneuronal cells. Proc Natl Acad Sci USA 90:2092–2096

Capell A, Saffrich R, Olivo J-C, Meyn L, Walter J, Grünberg J, Mathews P, Nixon, R, Dotti C, Haass C (1997) Cellular expression and proteolytic processing of Presenilin proteins is developmentally regulated during neuronal differentiation. J Neurochem 69:2432–2440

Capell A, Grünberg J, Pesold B, Diehlmann A, Citron M, Nixon R, Beyreuther B, Selkoe DJ, Haass C (1998) The proteolytic fragments of the Alzheimer's disease-associated presenilin-1 form heterodimers and occur as a 100–150-kDa molecular mass complex. J Biol Chem:3205–3211

Citron M, Westaway D, Xia W, Carlson G, Diehl T, Levesque G, Johnson-Wood K, Lee M, Seubert P, Davis A, Kholodenko D, Motter R, Sherrington R, Perry B, Yao H, Strome R, Lieberburg I, Rommens J, Kim S, Schenk D, Fraser P, St George Hyslop P, Selkoe D (1997) Mutant presenilins of Alzheimer's disease increase production of 42-residue amyloid β-protein in both transfected cells and transgenic mice. Nat Med 3:67–72

Cook D, Sung J, Golde T, Felsenstein K, Wojcyk B, Tanzi R, Trojanowski J, Lee V, Doms R (1996) Expression and analysis of presenilin 1 in a human neuronal system: Localization in cell bodies and dendrites. Proc Natl Acad Sci USA 93:9223–9228

Cruts M, Hendriks L, Van Broeckhoven C (1996) The presenilin genes: a new gene family involved in Alzheimer disease pathology. Hum Mol Genet 5:1449–1455

De Strooper B, Beullens M, Contreras B, Levesque L, Craessaerts K, Cordell B, Moechars, D, Bollen M, Fraser P, St. George-Hyslop P, Van Leuven F (1997) Phosphorylation, subcellular localization and membrane orientation of the Alzheimer's disease-associated presenilins. J Biol Chem 272:3590–3598

De Strooper B, Saftig P, Craessaerts K, Vanderstichele H, Guhde G, Annaert W, Von Figura K, Van Leuven F (1998) Deficiency of presenilin-1 inhibits the normal cleavage of amyloid precursor protein. Nature 391:387–390

Doan A, Thinakaran G, Borchelt DR, Slunt HH, Ratovitsky T, Podlisny M, Selkoe DJ, Seeger M, Gandy SE, Price DL, Sisodia SS (1996) Protein topology of presenilin 1. Neuron 17:1023–1030

Duff K, Eckman C, Zehr C, Yu X, Prada CM, Perez-Tur J, Hutton M, Buee L, Harigaya Y, Yager D, Morgan D, Gordon MN, Holcomb L, Refolo L, Zenk B, Hardy J, Younkin S (1996) Increased amyloid-β42 (43) in brains of mice expressing mutant presenilin 1. Nature 383:710–713

Esch FS, Keim PS, Beattle EC, Blacher RW, Culwell AR, Oltersdorf T, McClure D, Ward PJ (1990) Cleavage of amyloid β peptide during constitutive processing of its precursor. Science 248:1122–1124

Estus S, Golde TE, Kunishita T, Blades D, Lowery D, Eisen M, Usiak M, Qu X, Tabira T, Greenberg BD, Younkin SG (1992) Potentially amyloidogenic, carboxyl-terminal derivatives of the amyloid protein precursor. Science 255:726–730

Golde TE, Estus S, Younkin LH, Selkoe DJ, Younkin SG (1992) Processing of the amyloid protein precursor to potentially amyloidogenic derivatives. Science 255:728–730

Grünberg J, Capell A, Leimer U, Steiner B, Steiner H, Walter J, Haass C (1997) Proteolytic processing of presenilin proteins: degredation or biological activation? Alzheimer's Res 3:253–259

Grünberg J, Walter J, Hendriks L, van Broeckhoven C, Loetscher H, Deuschle U, Jacobsen H, Haass C (1998) Alzheimer's disease associated presenilin-1 holoprotein and its 18–20-kDa C-terminal fragment are death substrates for proteases of the caspase family. Biochemistry 37: 2263–2270

Haass C (1996) The presenilin genes and early dementia. Curr Opin Neurol 9:254–259

Haass C (1997) Presenilins: genes for life and death. Neuron 18:687–690

Haass C, Selkoe DJ (1993) Cellular processing of β-amyloid precursor protein and the genesis of amyloid β-peptide. Cell 75:1039–1042

Haass C, Selkoe DJ (1998) A technical KO of amyloid-β peptide. Nature 391:339–340

Haass C, Koo E, Mellon A, Hung AY, Selkoe DJ (1992a) Targeting of cell-surface β-amyloid precursor protein to lysosomes: alternative processing into amyloid-bearing fragments. Nature 357:500–503

Haass C, Schlossmacher MG, Hung AY, Vigo-Pelfrey C, Mellon A, Ostaszewski BL, Lieberburg I, Koo EH, Schenk D, Teplow DB, Selkoe DJ (1992b) Amyloid β-peptide is produced by cultured cells during normal metabolism. Nature 359: 322–325

Haass C, Hung AY, Schlossmacher MG, Teplow DB, Selkoe DJ (1993) β amyloid peptide and a 3-kDa fragment are derived by distinct cellular mechanisms. J Biol Chem 268:3021–3024

Hardy J (1997) The Alzheimer familiy of diseases: Many etiologies, one pathogenesis? Proc Natl Acad Sci USA 94:2095–2097

Hartmann H, Busciglio J, Baumann K-H, Staufenbiel M, Yankner BA (1997) Developmental regulation of presenilin-1 processing in the brain suggests a role in neuronal differentiation. J Biol Chem 272:14505–14508

Hendriks L, Thinakaran G, Harris CL, De Jonghe C, Martin J-J, Sisodia S, Van Broeckhoven C (1997) Processing of presenilin 1 in brains of Alzheimer's disease patients and controls. Neuroreport 8:1717–1721

Hochstrasser M (1996) Protein degradation or regulation: Ub the judge. Cell 84:813–815

Hung A, Selkoe DJ (1994) Selective extodomain phosphorylation and regulated cleavage of β-amyloid precursor protein. EMBO J 13:534–542

Jacobson MD, Weil M, Raff MC (1997) Programmed cell death in animal development. Cell 88:347–354

Jarrett JT, Berger EP, Lansbury PT Jr (1993) The carboxy terminus of the beta amyloid protein is critical for the seeding of amyloid formation: implications for the pathogenesis of Alzheimer's disease. Biochemistry 32:4693–4697

Kang J, Lemaire H-G, Unterbeck A, Salbaum JM, Masters C, Grezschik G, Multhaup G, Beyreuther K, Müller-Hill B (1987) The precursor of Alzheimer's A4 protein resembles a cell-surface receptor. Nature 325:733–736

Kim T-W, Pettingell WH, Hallmark OG, Moir RD, Wasco W, Tanzi RE (1997a) Endoproteolytic cleavage and proteasomal degradation of presenilin 2 in transfected cells. J Biol Chem 272:11006–11010

Kim T-W, Pettingell H, Jung Y-K, Kovacs D, Tanzi R (1997b) Alternative cleavage of Alzheimer-associated presenilins during apoptosis by a caspase-3 familiy protease. Science 277:373–376

Kovacs DM, Fausett HJ, Page K J, Kim T-W, Moir RD, Merriam DE, Hollister RD, Hallmark OG, Mancini R, Felsenstein KM, Hyman BT, Tanzi RE, Wasco W (1996) Alzheimer-associated presenilins 1 and 2: Neuronal expression in brain and localization to intracellular membranes in mammalian cells. Nat Med 2:224–229

Lammich S, Kojro E, Postina R, Gilbert S, Pfeiffer R, Jasionowski M, Haas C, Fahrenholz F (1999) Constitutive and regulated alpha-secretase cleavage of Alzheimer's amyloid precursor protein by a disintegrin metalloprotease. Proc Acad Sci USA 96:3922–3927

Lee ML, Borchelt DR, Kim G, Thinakaran G, Slunt H, Ratovitski T, Martin LJ, Kittur A, Gandy S, Levey A, Jenkins N, Copeland N, Price DL, Sisodia S (1997) Hyperaccumulation of FAD-linked presenilin 1 variants in vivo. Nat Med 3:756–760

Levitan D, Greenwald I (1995) Facilitation of lin-12-mediated signalling by sel-12, a Caenorhabditis elegans S182 Alzheimer's disease gene. Nature 377:351–354

Levitan D, Doyle T, Brousseau D, Lee M, Thinakaran G, Slunt H, Sisodia S, Greenwald I (1996) Assessment of normal and mutant human presenilin function in Caenorhabditis elegans. Proc Natl Acad Sci USA 93:14940–14944

Levy-Lahad E, Wasco W, Poorkaj P, Romano DM, Oshima J, Pettingell WH, Yu C, Jondro PD, Schmidt SD, Wang K, Crowley AC, Fu Y-H, Guenette SY, Galas D, Nemens E, Wijsman EM, Bird TD, Schellenberg GD, Tanzi RE (1995) Candidate gene for the chromosome 1 familial Alzheimer's disease locus. Science 269:973–977

Li X, Greenwald I (1996) Membrane topology of the C. elegans SEL-12 presenilin. Neuron 17:1015–1021

Loetscher H, Deuschle U, Brockhaus M, Reinhardt D, Nelboeck P, Mous J, Grünberg J, Haass C, Jacobsen H (1997) Presenilins are processed by caspase-type proteases. J Biol Chem 272:20655–20659

Marambaud P, Ancolio K, Lopez-Perez E, Checler F (1998) Proteasome inhibitors prevent the degradation of familial Alzheimer's disease-linked presenilin 1 and potentiate Aβ42 recovery from human cells. Mol Med 4:147–157

Mercken M, Takahashi H, Honda T, Sato K, Murayama M, Nakazato Y, Noguchi K, Imahori K, Takashima A (1996) Characterization of human presenilin 1 using N-terminal specific monoclonal antibodies: evidence that Alzheimer mutations affect proteolytic processing. FEBS Lett 389:297–303

Murayama O, Honda T, Mercken M, Murayama M, Yasutake K, Nihonmatsu N, Nakazato Y, Michel G, Song S, Sato K, Takahashi A (1997) Different effects of Alzheimer-associated mutations of presenilin 1 on its processing. Neurosci Lett 229:61–64

Nitsch RM, Slack BE, Wurtman RJ, Growdon JH (1992) Release of Alzheimer amyloid precursor derivatives stimulated by activation of muscarinic acetylcholine receptors. Science 258:304–307

Nordstedt C, Caporaso GL, Thyberg J, Gandy SE, Greengard P (1993) Identification of the Alzheimer beta/A4 amyloid precursor protein in clathrin-coated vesicles purified from PC12 cells. J Biol Chem 268:608–612

Perez-Tur J, Froehlich S, Prihar G, Crook R, Baker M, Duff K, Wragg M, Busfield F, Lendon C, Clark RF, Roques P, Fuldner RA, Johnston J, Cowburn R, Forsell C, Axelman K, Lilius L, Houlden H, Karran E, Roberts GW, Rossor M, Adams MD, Hardy J, Goate A, Lannfelt L, Hutton M (1995) A mutation in Alzheimer's disease destroying a splice acceptor site in the presenilin-1 gene. Neuroreport 7:297–301

Podlisny M, Citron M, Amarante P, Sherrington R, Xia W, Zhang J, Diehl T, Levesque G, Fraser P, Haass C, Koo E, Seubert P, St. George-Hyslop P, Teplow D, Selkoe DJ (1997) Presenilins proteins undergo heterogeneous endoproteolysis between Thr291 and Ala299 and occur as stable N- and C-terminal fragments in normal and Alzheimer brain tissue. Neurobiol Dis 3:325–337

Rogaev EI, Sherrington R, Rogaeva EA, Levesque G, Ikeda M, Liang Y, Chi H, Lin C, Holamn K, Tsuda T, Mar L, Sorbi S, Nacmias B, Piacentini S, Amaducci L, Chumakkov I, Cohen D, Lannfelt L, Fraser PE, Rommens JM, and St. George-Hyslop PH (1995) Familial Alzheimer's disease in kindreds with missense mutations in a gene on chromosome 1 related to the Alzheimer's disease type 3 gene. Nature 376:775–778

Roses AD (1996) Apolipoprotein E in neurology. Curr Opin Neurol 9:265–270

Scheuner D, Eckman C, Jensen M, Song X, Citron M, Suzuki N, Bird TD, Hardy J, Hutton M, Kukull W, Larson E, Levy-Lahad E, Viitanen M, Peskind E, Poorkaj P, Schellenberg G, Tanzi R, Wasco W, Lannfelt L, Selkoe D, Younkin S (1996) Secreted amyloid β-protein similar to that in the senile plaques of Alzheimer's disease is increased in vivo by the presenilin 1 and 2 and APP mutations linked to familial Alzheimer's disease. Nat Med 2:864–870

Schroeter EH, Kisslinger JA, Kopan R (1998) Notch-1 signalling requires ligand-induced proteolytic release of intracellular domain. Nature 393:382–386

Selkoe DJ (1996) Amyloid β-protein and the genetics of Alzheimer's disease. J Biol Chem 271:18295–18298

Seubert P, Vigo-Pelfrey C, Esch F, Lee M, Dovey H, Davis D, Sinha S, Schlossmacher M, Whaley J, Swindlehurst C, McCormack R, Wolfert R, Selkoe DJ, Lieberburg I, Schenk D (1992) Isolation and quantification of soluble Alzheimer's β-peptide from biological fluids. Nature 359:325–327

Shen J, Bronson RT, Chen DF, Xia W, Selkoe DJ, Tonegawa S (1997) Skeletal and CNS defects in presenilin-1-deficient mice. Cell 89:629–639

Sherrington R, Rogaev EI, Liang Y, Rogaeva EA, Levesque G, Ikeda M, Chi H, Lin C, Li G, Holman K, Tsuda T, Mar L, Foncin J-F, Brni AC, Montesi MP, Sorbi S, Rainero I, Pinessi L, Nee L, Chumakov I, Pollen D, Brookes A, Sanseau P, Polinsky RJ, Wasco W, da Silva HAR, Haines JL, Pericak-Vance MA, Tanzi RE, Roses AD, Fraser PE, Rommens JM, St. George-Hyslop PH (1995) Cloning of a gene bearing missense mutations in early-onset familial Alzheimer's disease. Nature 375:754–760

Shirotani K, Takahashi K, Ozawa K, Kunishita T, Tabira T (1997) Determination of a cleavage site of pesenilin 2 protein in stably transfected SH-SY5Y human neuroblastoma cell lines. Biochem Biophys Res Commun 240:728–731

Shoji M, Golde TE, Ghiso J, Cheung TT, Estus S, Shaffer LM, Cai XD, McKay DM, Tintner, R, Frangione B, Younkin SG (1992) Production of the Alzheimer amyloid β protein by normal proteolytic processing. Science 258:126–129

Steiner H, Capell A, Pesold B, Citron M, Kloetzel PM, Selkoe DJ, Romig H, Mendla K, Haass C (1998) Expression of Alzheimer's disease-associated presenilin-1 is controlled by proteolytic degradation and complex formation. J Biol Chem 273:32322–32331

Struhl G, Adachi A (1998) Nuclear access and action of notch in vivo. Cell 93:649–660

Tanzi R, Kovacs D, Kim T-W, Moir R, Guenette S, Wasco W (1996) The gene defects responsible for Alzheimer's disease. Neurobiol Dis 3:159–168

Thinakaran G, Borchelt DR, Lee MK, Slunt HH, Spitzer L, Kim G, Ratovitsky T, Davenport F, Nordstedt C, Seeger M, Hardy J, Levey AI, Gandy SE, Jenkins NA, Copeland NG, Price DL, Sisodia SS (1996) Endoproteolysis of presenilin 1 and accumulation of processed derivatives in vivo. Neuron 17:181–190

Thinakaran G, Regard JB, Bouton CML, Harris CL, Proce DL, Borchelt DR, Sisodia SS (1998) Stable association of presenilin derivatives and absence of presenilin interactions with APP. Neurobiol Dis 4:438–453

Thompson CB (1995) Apoptosis in the pathogenesis and treatment of disease. Science 267:1456

Thornberry NA, Bull HG, Calaycay JR, Chapman KT, Howard AD, Kostura MJ, Miller DK, Molineaux SM, Weidner JR, Aunins J, Elliston KO, Ayala JM, Casano FJ, Chin J, Ding G J-F, Egger LA, Gaffney EP, Limjuco G, Palyha OC, Raju SM, Rolando AM, Salley JP, Yamin TT, Lee TD, Shively JE, MacCross M, Mumford RA, Schmidt JA, Tocci MJ (1992) A novel heterodimeric cyteine protease is required for interleukin-1 beta processing in monocytes. Nature 356:768–774

Thornberry NA, Peterson EP, Zhao JJ, Howard AD, Griffin PR, Chapman KT (1994) Inactivation of interleukin-1 beta converting enzyme by peptide (acyloxy)methyl ketones. Biochemistry 33:3934–3939

Tomita T, Maruyama K, Saido TC, Kume H, Shinozaki K, Tokuhiro S, Capell A, Walter J, Grünberg J, Haass C, Iwatsubo T, Obata K (1997) The presenilin 2 mutation (N141I) linked to familial Alzheimer disease (Volga German families) increases the secretion of amyloid β protein ending at the 42nd (or 43rd) residue. Proc Natl Acad Sci USA 94:2025–2030

Van Broeckhoven C (1995) Presenilins and Alzheimer disease. Nat Genet 11:230–232

Vito P, Lacana L, D'Adamio L (1996a) Interfering with apoptosis: Ca 2+-binding protein ALG-2 and Alzheimer's disease gene ALG-3. Science 271:521–525

Vito P, Wolozin B, Ganjei JK, Iwasaki K, Lacana E, D'Adamio L (1996b) Requirement of the familial Alzheimer's disease gene PS2 for apoptosis. J Biol Chem 271:31025–31028

Walter J, Capell A, Grünberg J, Pesold B, Schindzielorz A, Prior R, Podlisny MB, Fraser P, St. George Hyslop P, Selkoe DJ, and Haass C (1996) The Alzheimer's disease – associated presenilins are differentially phosphorylated proteins located predominantly within the endoplasmic reticulum. Mol Med 2:673–691

Walter J, Grünberg J, Capell A, Pesold B, Schindzielorz A, Citron M, Mendla K, St George-Hyslop P, Multhaup G, Selkoe DJ, Haass C (1997) Proteolytic processing of the Alzheimer disease-associated presenilin-1 generates an in vivo substrate for protein kinase C. Proc Natl Acad Sci USA 94:5349–5354

Ward CL, Omura S, Kopito RR (1995) Degradation of CFTR by the ubiquitin-proteasome pathway. Cell 83:121–127

Weidemann A, König G, Bunke D, Fischer P, Salbaum JM, Masters CL, Beyreuther K (1989) Identification, biogenesis, and localization of precursors of Alzheimer's disease A4 amyloid protein. Cell 57:115–126

Wolozin B, Iwasaki K, Vito P, Ganjei JK, Lacana E, Sunderland T, Zhao B, Kusiak JW, Wasco W, D'Adamio L (1996) Participation of presenilin 2 in apoptosis: enhanced basal activity conferred by an Alzheimer mutation. Science 274:1710–1713

Wong PC, Zheng H, Chen H, Becher MW, Sirinathsinghji HY, Price DL, Van der Ploeg LHT, Sisodia SS (1997) Presenilin 1 is required for Notch1 and D//1 expression in the paraxial mesoderm. Nature 387:288–292

Xia W, Zhang J, Kholodenko D, Citron M, Teplow D, Haass C, Seubert P, Koo EH, Selkoe DJ (1997) Enhanced production and oligomerization of the 42-residue amyloid β-protein by CHO cells stably expressing mutant presenilins. J Biol Chem 272:7977–7982

Yu G, Chen F, Levesque G, Nishimura M, Zhang D-M, Levesque L, Rogaeva E, Xu D, Liang Y, Duthie M, St. George-Hyslop P, Fraser PE (1998) The presenilin 1 protein is a component of a high molecular weight intracellular complex that contains β-catenin. J Biol Chem 273:16470–16475

Appendix

1 **Fig. 3, Chapter 5:** Solvent-accessible surface of NS3 proteinase with a modeled substrate (AspGluMetGluGluCys-AlaSerHisLeu). Hydrophobic areas on the protein
surface are coloured in *yellow*, basic areas are in *blue* and acidic areas are in *red*.
The solvent-accessible surface area of NS4A is shown in *magenta*

2 **Fig. 4, Chapter 5:** Ribbon representation of the X-ray structure of NS3 proteinase
with the NS4A cofactor in *red*. Also shown, with *green* carbon atoms, is the backbone of a modeled product inhibitor. Hydrogen bonds by P_3 and P_5 are indicated
as *dashed lines*. The carbon atoms of the catalytic triad are shown in *yellow*

3 **Fig. 5, Chapter 8:** Modeling of a substrate in the active site. The Connolly surface
of the adenovirus proteinase–pVIc complex, in *pink*, is shown with the putative
active-site residues, colored *orange*. The substrate, in Corey-Pauling-Koltun-model
form, has the sequence Ala-Ala-Ile-Val-Gly-Leu-Gly-Val; cleavage occurs at the
Leu–Gly bond. The side chains of the P1-P4 amino-acid residues are labeled

4 **Fig. 4, Chapter 8:** Juxtaposition of the active-site residues of papain with equivalent residues of the adenovirus proteinase (AVP)–pVIc complex. The residues
involved in catalysis in papain are shown after alignment of the papain molecule to
fit the equivalent residues in AVP–pVIc. For papain, the active-site residues are
shown with the carbon atoms colored *magenta*; for the AVP–pVIc complex, the
carbon atoms are colored *green*. For both molecules, the nitrogens are colored *blue*
and the oxygens *red*. The important bond distances (in Å) are shown by *dashed
magenta lines* for papain and *dashed green lines* for AVP–pVIc

5 **Fig. 6, Chapter 8:** Possible DNA-binding sites on the adenovirus proteinase
(AVP)–pVIc complex. The charge-potential map of the distribution of positive
(*blue*) and negative (*red*) charges on the surface of the AVP–pVIc complex is shown

6 **Fig. 7, Chapter 8:** Binding interactions between pVIc and adenovirus proteinase
(AVP). The *dashed lines* from atoms of pVIc, which is in ball-and-stick form, indicate the sites of interaction between pVIc and AVP. Hydrogen bonds are represented by *dashed lines*; *dashed yellow lines* represent pVIc side chain interactions
with AVP and *dashed magenta lines* represent the backbone atoms of the β-sheet
of pVIc with AVP. The disulfide bond between Cys10′ of pVIc and Cys104 of AVP
is in *green*. Water molecules are represented by *blue spheres*

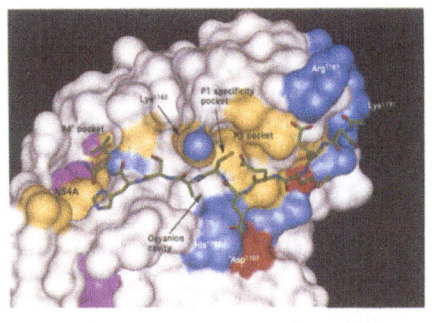

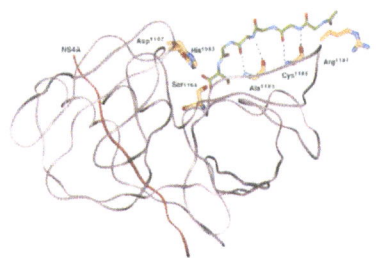

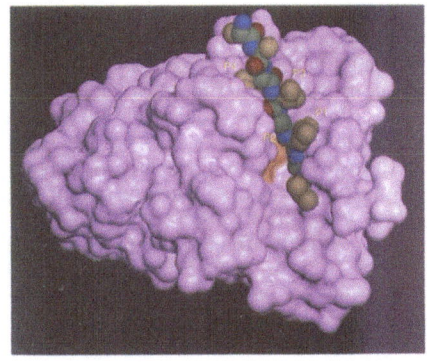

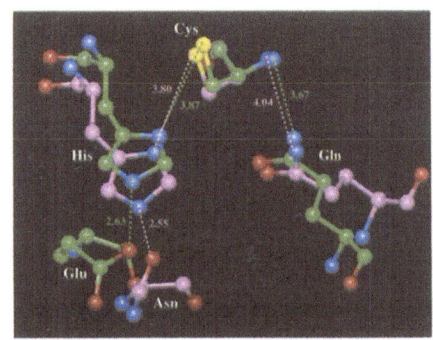

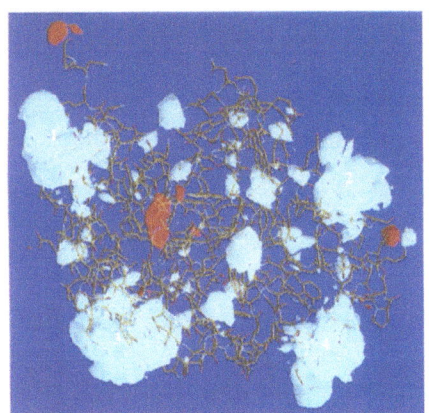

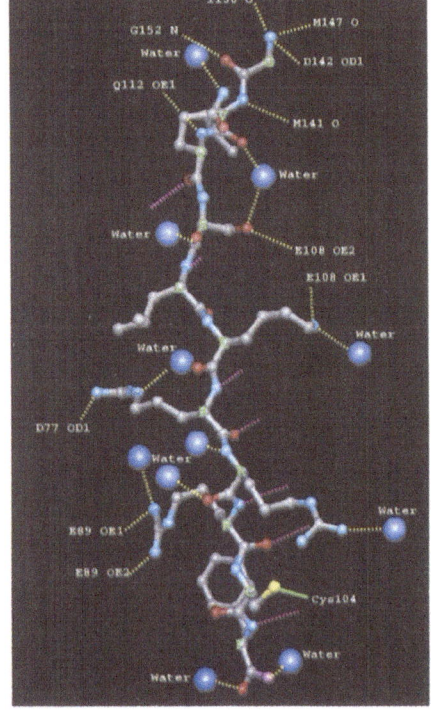

1	2
3	4
5	6

1 **Fig. 2a,b, Chapter 7:** Ribbon secondary-structure representation of the three-dimensional structure of the picornaviral 3C proteinases from (**a**) hepatitis A virus (HAV) and (**b**) poliovirus (PV) in stereo. The views are into the proteolytic active site. Side chains of residues which are involved in the enzymatic mechanism of the 3C proteinase are included and are labeled with the single-letter amino acid code and the residue number. The amino- and carboxy-termini are marked with *N* and *C*. Secondary-structure elements are labeled successively with letters. *Capital letters* designate the helices and *small letters* the β-strands of domains I and II. The amino-terminal β-barrel is colored *cyan* and the carboxy-terminal β-barrel is colored *blue*. Helices are in *green*, reverse turns are in *yellow* and random-coil secondary structures in *red*. The oxyanion hole is colored *black* and the anti-parallel β-ribbon, which extends from the carboxy-terminal β-barrel domain and contributes to the active site, is colored *light gray*. Three water molecules in the structure of HAV 3C are represented by *spheres*

2 **Fig. 3a,b, Chapter 7:** Same as in Fig. 2, but rotated 90° to the left to highlight the RNA-binding site of the 3C proteinase. Hepatitis A virus 3C is shown in (**a**) and poliovirus 3C in (**b**). The main chain of the conserved sequence motif KFRDI is colored *black* and the side chains of theses residues are included. The colors and labels of the secondary structure are similar to those in Fig. 2

3 **Fig. 4a,b, Chapter 14:** Effect of elastase inhibition on coronary-artery transplant vasculopathy. Representative photomicrographs of Movat-pentachrome staining of coronary arteries in the host control, donor control and elafin-treated donor groups. The normal appearing host vessel (*A*) contrasts with the affected donor vessel (*B*) showing a concentric intimal lesion in the control group and (*C*) a more normal appearing artery in the elafin-treated donor group. All vessels are in the medium (100–500 μm diameter)-size range. Original magnification 200×

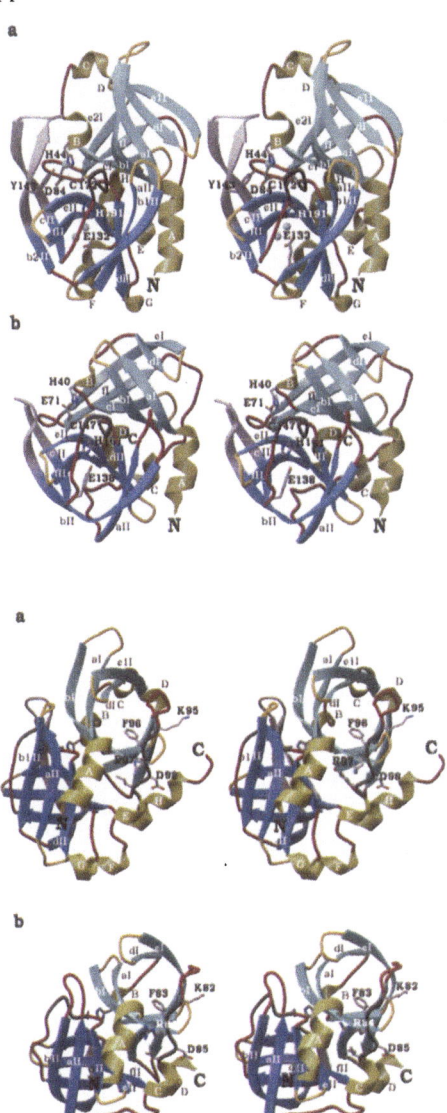

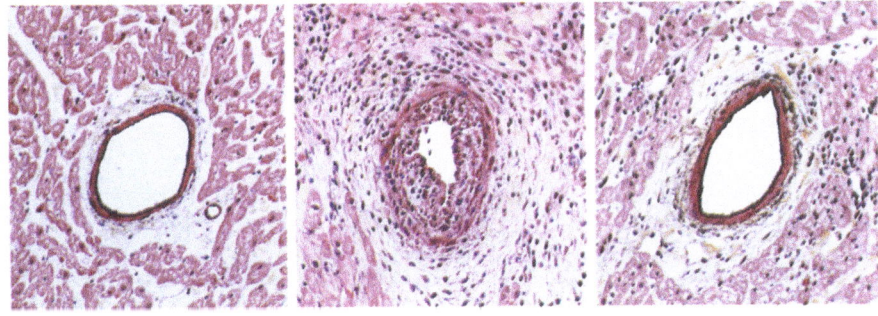

Subject Index